21ST CENTURY INNOVATION IN MUSIC EDUCATION

PROCEEDINGS OF THE 1ST INTERNATIONAL CONFERENCE OF THE MUSIC EDUCATION COMMUNITY (INTERCOME 2018), OCTOBER 25-26, 2018, YOGYAKARTA, INDONESIA

21st Century Innovation in Music Education

Editors

Kun Setyaning Astuti

Universitas Negeri Yogyakarta, Indonesia

Gary E. McPherson

The University of Melbourne, Australia

Bambang Sugeng

Universitas Negeri Yogyakarta, Indonesia

Nila Kurniasari

Universitas Negeri Yogyakarta, Indonesia

Tutut Herawan

Universitas Teknologi Yogyakarta, Indonesia

Christopher Drake

Living Values Education, UK

Ashadi

Universitas Negeri Yogyakarta, Indonesia

Endah Retnowati

Universitas Negeri Yogyakarta, Indonesia

Adi Cilik Pierewan

Universitas Negeri Yogyakarta, Indonesia

Routledge
Taylor & Francis Group
LONDON AND NEW YORK

Routledge is an imprint of the Taylor & Francis Group, an informa business

Typeset by Integra Software Services Pvt. Ltd., Pondicherry, India

Publisher's Note
The publisher has gone to great lengths to ensure the quality of this reprint but points out that some imperfections in the original copies may be apparent.

Published by: CRC Press/Balkema
Schipholweg 107C, 2316XC Leiden, The Netherlands
e-mail: Pub.NL@taylorandfrancis.com
www.crcpress.com – www.taylorandfrancis.com

First issued in paperback 2021

ISBN 13: 978-1-03-223835-7 (pbk)
ISBN 13: 978-0-367-11122-9 (hbk)

DOI: https://doi.org/10.1201/9780429024931

21st Century Innovation in Music Education – Astuti et al. (eds)
© 2020 Taylor & Francis Group, London, ISBN 978-0-367-11122-9

Table of contents

21st Century Innovation in Music Education – Astuti et al. (eds)
© 2020 Taylor & Francis Group, London, ISBN 978-0-367-11122-9

Preface

There are laws of a certain nature in the natural sciences and likewise in music there are structures and procedures, or even rules, that should be followed to produce beautiful music. Thus, music is not only something that has subjective appeal according to the unique taste of each individual's sense of art and beauty but may also be studied in an objective fashion as a form of science. Music is not only something that can be felt but also something that can be counted.

The developmental journey of science that puts music only to the specialization of social sciences and humanities is inhibiting the development of theories that can explain and construct music as part of the discipline, capable of contributing to advancements in other fields such as: natural sciences, psychology, medicine, etc. For most of the music communities, the view that music is only a part of the entertainment media can be disregarded by developing both basic and applied theories that are interconnected with science, including natural science.

Considering the importance of implementation and development of innovation in music education, both basic and applied theories, as well as its connection with other disciplines, we organized an international conference with the theme *21st Century Innovations in Music and Research Education*. The 1st International Conference on Music Education Community (INTERCOME) 2018 was held on 25-26 October 2018 in the Music Department of the Languages and Arts Faculty, Yogyakarta State University, Indonesia. This conference is held once every three years (triennially).

INTERCOME intended to capture the latest musical developments and brought together theoretical and practical views, concepts, theories, and principles that can be used by educators, researchers, observers and the broader community to develop education based on research results.

This volume of proceedings presents 71 selected articles from 127 articles presented at the 2018 INTERCOME seminar attended by participants from 10 countries. The authors who made it through the selection are from six countries, namely Britain, Indonesia, Mali, Nigeria, the Netherlands, and the Philippines. Many papers were written by senior authors such as Prof. Lee Higgins, Ph.D., former President of ISME (International Society for Music Education), Prof. Dr. Suminto A. Sayuti, former Dean of Faculty Languages and Arts of Universitas Negeri Yogyakarta, Prof. Endang Nurhayati. Dean of Languages and Arts of Universitas Negeri Yogyakarta, Prof. Triyono Bramantyo, former Dean of the Faculty of Performing Arts of the Institut Seni Indonesia Yogyakarta Indonesia and Prof. Wiel Veugelers, a moral education expert from the University of Humanistic Studies, Utrecht, the Netherlands who writes about the role of music as a character-forming tool. This proceeding covers seven themes. They are as follow:

1. Picture of 21st century music education in the world
2. The innovation of curriculum, teaching and learning media and technological and environmental evaluation in accordance with the development in the 21st century.
3. The development of music as an educational media in the post globalization era
4. The harmonious existence of eastern, western, traditional, and modern music for all types of music in the digital revolution era
5. The development of critical thinking ability to respond to various phenomena of the latest music

6. Collecting, developing, and constructing the basic theory of the relationship and the effect of music on psychology, physics, social and other disciplines or vise versa and its utilization in modern human life
7. Music research development in the 21st century

We hope that it will encourage theoretical and practical views, concepts, research results and principles to further support the development of music educations and other related fields.

Dr. Kun Setyaning Astuti. M.Pd.

21st Century Innovation in Music Education – Astuti et al. (eds)
© 2020 Taylor & Francis Group, London, ISBN 978-0-367-11122-9

Organizing Committee

Patron
Prof. Dr. Sutrisna Wibawa, M.Pd. (Rector of Universitas Negeri Yogyakarta) UNY

Advisory Committee
Prof. Dr. Margana, M.Hum (Vice Rector of Academic Affairs, Universitas Negeri Yogyakarta)
Prof. Dr. Edi Purwanto, M.Pd. (Vice Rector of Finance, Universitas Negeri Yogyakarta)
Prof. Dr. Sumaryanto, M.Kes (Vice Rector of Student Affair, Universitas Negeri Yogyakarta)
Dr. Rer.Nat. Senam, M.Si. (Vice rector of internationale affairs in partnership, UNY)
Prof. Endang Nurhayati, M.Hum (Dean of Languages and Arts Faculty)
Dr. Maman Suryarman, M.Pd. (Vice Dean of Academic Affairs-Languages and Arts Faculty of Universitas Negeri Yogyakarta)
Rohali, M.Hum.(Vice Dean of financial affairs-Languages and Arts Faculty of UNY)

Head of Organizing Committee
Dr. Kun Setyaning Astuti, M.Pd. (Vice Dean of Students Affairs-Languages and Arts UNY)

Local Organizing Committee:
Dr. Ayu Niza Machfauzia, M.Pd. (Head of Music Education Departement)
Sritanto, M.Pd. (Secretary of Music Education Departement)

Secretary:
Nila Kurniasari, S.Pd. ,M.Pd.
Fransisca Xaveria Diah Kristianingsih, S.Pd., M.A.

Treasure:
Panca Putri Rusdewanti, S.Pd.,M.Pd.
Dra. MG. WidyastutI, M.Sn.

Special Season Chair:
Herwin Yogo W, M.Pd.
Heni Kusumawati, M.Pd.

Sponsorship:
Drijastuti Jogjaningrum. M.A.
Fuadi,M.A.
Dr. Cipto Budi Handoyo, M.Pd.
Agustianto, M.Pd.

Technical support:
Pujiwiyana, M.Pd.
Agus Untung Yulianto, M.Pd.
Mudakir, S.Ip.
Drs. Sugeng Tri Wuryanto.

Program Committee Chair:
Yunike Juniarti Fitria, M.A.

Proceeding chair:
Dr. A.M. Susilo Pradoko, M.Si.
Dr. Hanna Sri Mudjilah, M.Pd.
Suwarta Zebua, M.Pd.
HT. Silaen, M.Hum.

Admin/secretariat:
Virga Renitasari, S.Pd.
Altri Rohmat, A.Md.
Suryati, S.Pd.
Sariyanto, S.Pd.
Sinto Wibowo, S.Pd.
Rini Lestari

Design:
Sugeng Setia Nugroho, A.Md.
Dedy Sartono, S.Pd.

21st Century Innovation in Music Education – Astuti et al. (eds)
© 2020 Taylor & Francis Group, London, ISBN 978-0-367-11122-9

Scientific Committee

21st Century Innovation in Music Education – Astuti et al. (eds)
© 2020 Taylor & Francis Group, London, ISBN 978-0-367-11122-9

Acknowledgement

This page is dedicated to the people who have been a tremendous support in the making of this proceeding. This work would not have been possible without the patronage of the Rector of Yogyakarta State University, Prof. Dr. Sutrisna Wibawa, M.Pd. and his staff, who have been supportive all the way.

I would also like to express our deepest gratitude to the board of editors Prof. Gary E. McPherson, Bambang Sugeng, Ed.D., Tutut Herawan, Ph.D., Endah Retnowati, Ph.D, Christopher Drake, M.A., Ashadi, Ed.D, Nila Kurniasari, M.Pd., Adi Cilik Piereawan and to all reviewers for their suggestions, encouragement, and for ensuring the quality of the papers published.

Beside that, I would also like to express our deepest gratitude to the following people for their contribution for this conference:

Dr. Rernat. Senam (Universitas Negeri Yogyakarta)
Prof. Lee Higgins (York St. John University)
Ass. Prof. Julie Ballantyne (University of Queensland, Australia)
Servaas Schreuders (Hogeschool voor De Kunsten Utrecht, The Netherlands)
Prof. Dr. Suminto A. Sayuti (Universitas Negeri Yogyakarta)
Dr. Clare Suet Ching Chan (Universiti Pendidikan Sultan Idris)
Janjaap Blom (Senior Publisher, CRC Press/Balkema, Taylor & Francis - the Netherlands)
Dr. Cherry Russell (University of Sydney, Australia)
Wayland N Quintero (University of Hawaii)
Supriyanto Wardoyo, SE (Universitas Sanata Dharma)
Prof. Dr. Suwarsih Madya, Ph.D (Universitas Negeri Yogyakarta)
Suhaini M. Saleh (Universitas Negeri Yogyakarta)
Soni Nopembri, Ph.D (Universitas Negeri Yogyakarta)
Mungad (Universitas Negeri Yogyakarta)
Meiyao Wu (National Taiwan Normal University)
Eddy K.M Chong (National Institute of Education, Singapore)
Prof. Wiel Veugelers (University of Humanistic Studies, Utrecht, The Netherlands)
Mei Yee Wong (Education University of Hongkong)
Dr. Marlene Belly (Poitiers University, France)
Edijana Ursula (City University of Hongkong)
Setyo Purwaningsing, M.Sc (University of Bristol)
Rasman (University of Birmingham)

Furthermore, I would also like to acknowledge with much appreciation, the crucial role of the ISME (International Society of Music Education) and APNME (Asia Pacific Network of Moral Education) for their valuable contribution. A very special gratitude goes to the INTERCOME committee and students committee for their dedication and efforts in preparing and hosting the conference.

I am also grateful for the support of English Department alumni of Universitas Negeri Yogyakarta who have been a great help in proofreading the manuscripts. Last but not least, I would like to give my thanks to the publisher for their cooperation and for allowing us the opportunity to publish the conference papers. Appreciation also goes to José van der Veer and Lukas Goosen with their team who worked hard in correcting the draft and thereby making it more perfect.

Sincerely Yours

INTERCOME Organizing Committee Chair

Dr. Kun Setyaning Astuti, M.Pd.

21st Century Innovation in Music Education – Astuti et al. (eds)
© 2020 Taylor & Francis Group, London, ISBN 978-0-367-11122-9

Improvisation as experience in music education: Context & Ideas

L. Higgins
York St John University, York, UK

ABSTRACT: This purpose of this paper is to (1) contextualize improvisation as an experiential activity, and (2) presents four practical examples that can be used in various settings such as music classrooms or community venues. The ideas are drawn from previous work: Free to Be Musical: Group Improvisation in Music (2010) and Improvisation as Ability, Culture, and Experience (2013). These ideas continue to be important in my work as a community musician and my hope is that by sharing them I can reinforce the importance of injecting classrooms and community settings with musical events that embrace artistic-expressive thinking and embodied doing, thus ensuring that those that engage in music education will embark on a musical voyage of life-long musical learning.

1 INTRODUCTION

As music educators, we are aware that creative invention and self-expression is vitally important for our lives. We are also aware of the various philosophical positions that have been offered by musician-educators to politicians, administrators and to the general public, that music-making is a critical means of creative expression. As music educators, we not only want those we work with to embrace their musicking potential but also to connect their encounters with the real world in which they live. In other words, we want people to understand the importance, and the joy, of making and creating music both individually and together.

In this paper, I present some of the ideas from a co-authored book with Patricia Campbell called *Free to Be Musical: Group Improvisation in Music* (Higgins & Campbell, 2010). Some of the ideas align themselves with the spirit of creative composition and the aleatoric movement of the 1960s. John Cage for example created many spontaneous events with his dance. Philip Glass and Steve Reich were the next generation honing influences other cultures and utilizing these within their music. From a music education perspective, we could point to Murray Schafer (Canada) and John Paynter (U.K.), and the ideals of the Contemporary Music Project and Comprehensive Musicianship (U.S.) all resonating with the creative music-making events that are presented here.

The pervasive spirit of the phenomenon known as community music is much of the impetus for this work that forms a key part of my practice (Bartleet & Higgins, 2018; Higgins, 2012). The ideas are enlightened by the perspective of anthropologist John Blacking (1973), that music is a human phenomenon. He reinforced the idea that as a species humans are innately musical and are all capable of making music creatively, especially when the contexts, frameworks and processes are well thought through. The premise that music is improvised in the course of performance (Nettl & Russell, 1998), as well as the belief that musical improvisation can be nurtured in schools and in many societal circumstances (Solis & Nettl, 2009). John Dewey's (1997) belief in "realizing democracy" within the educational process was also paramount for our sense of the importance of making expressive music within the group, with every group member knowing that they are welcome to contribute, such that the group's product results from the contributions of every individual. Paul Friere's *Pedagogy of the Oppressed* (2002) is central to the position that learning is an act of culture and freedom, and that no one person should be privileged in the community effort to create musically.

Seen here as a small cluster of practical activities exampled from the book *Free to be Musical*, the ideas presented aim to provide frameworks through which music educators and the participants they work with can create environments through which journeys of musical discovery can take place. It is understood that these 'events' will both increase musical knowledge but importantly connect the experiences to the participants wider world.

2 LITERATURE REVIEW

2.1 *Be free*

Being free is very important to make the program successful. The short description of the step, group members 1 play a collection of instruments freely, spontaneously, even almost unconsciously (for a while). The loose structure of the event, consisting of a start and a stop, begins with unconscious musical scribbling but then turns toward conscious musical decision-making. One of the important aspects of this event is that students should be encouraged to concentrate on listening to each other. Questions arising from this activity might be: What is music? The activities of the program focus on large group instrumental playing, listening, self-expression, Independency and Group work. The resources of instrument use available instruments, for example drums and other percussions, band and orchestral (wind, brass, and string) instruments, xylophones and marimbas, keyboards, guitars, recorders and penny-whistles.

2.2 *Preparation*

If time permits, lesson precede this event with *Keep Breathing* and arrange chairs in a circle. Try to get all the participants and their instruments into a circle. Wheel the piano into the circle. If electric instruments require amplification, ensure that you use extension cables to enable these performers to be in the space.

2.3 *Event*

The next step, teacher asked the students to arrange themselves into a circle. It is necessary to ensure that they have enough space to sit comfortably while they play their instruments. Furthermore, asked the students to "play" their instrument without thinking too much about what they are doing. In other words, they should initially play unconsciously and might focus on what other members are doing.

Students could focus their attention on listening to others as they play whatever comes to them and should be invited to play anything. Basically, there are no rules, and there is no right or wrong. Group members should begin with silence, and end with a silent signal to stop. It may feel chaotic in this early stage, which is to be expected.

2.4 *Development*

This segment moves away from unconscious musical scribbling towards conscious musical decisions. Begin with silence and end with a silent signal to stop, unless the music-making finds its own resting place. Teacher help students to adjust the focus by asking them to switch on their consciousness. This development requires that students retain the libratory spirit of *the initial segment of the event* but that they concentrate on listening to (a) themselves and (b) each other.

Through this segment, the players will need to make musical choices, posing questions such as: What does this music require me to do? When should I play? How can I make a contribution to the overall sound?

3 DISCUSSION

When starting *Be Free*, the facilitator may find that employing the breath as a structure (as explored in *Keep Breathing* which can be found in the book) will help some students come to terms with the "blank canvas". The initial chaotic character of this type of playing may feel uncomfortable. Yet this uneasiness often leads to a healthy discussion surrounding the question, What is music? Open up a debate with group members without fear of an honest discussion of feelings about the event. In moving from the event to the development of it, there is a dramatic shift of emphasis from, unconsciousness toward consciousness. The demand is very different between the segments, as students begin to make clear and conscious sonic and musical choices.

There may be students who are hesitant to play so freely. They may be worried doing something "wrong". Explain that a blank space can be scary, but that it can be filled in any number of interesting ways. Assume them that there is no right way to "play", and that one can always try again, and again.

3.1 *Ensemble*

Intimate playing experiences demand attentive listening and sonic decision making. The smaller the group, as in the case of any chamber music or 'combo', the more transparent the musical quality. Because it is more exposing than an event like *Be Free,* this event intensifies the communication between and among players and singers. By emphasising critical dialogue after each component experience, group members begin to develop ways of describing what they hear and how they experience music. The focus of this section on small group instrumental playing, listening, and Self-expression. The resources are available instruments, including voices: Drums and other percussions, band and orchestral (wind, brass, and string) instruments, xylophones and marimbas, keyboards, guitars, recorders and penny-whistles.

Precede this event with *Be Free* to liberate players from fixed notions of musical engagement and to stimulate a free flow of musical ideas. Teacher arrange chairs in a circle and try to get all the participants and their instruments into a circle. Wheel the piano into the circle. If electric instruments require amplification, ensure that you use extension cables to enable these performers to be in the space.

3.2 *Event*

The next step, teacher Invite the students to arrange themselves into a circle and ensure that they have enough space to sit comfortably while they play their instruments. Furthermore, teacher select smaller groups from within the larger circle, setting up trios, duos, and quartets. There can be a mixture of these configurations,

Each small group will play in an improvisation for an unspecified period of time. The point is to play—anything. Then, together (but without speaking) they are to find a stopping point. Following each performance, there is a small silence. After the silence, the next group will begin their improvised musical expression. Silence completes the piece. This sequence continues until every small group has played.

Group members should be invited to reflect on their experience, as both players and listeners. The facilitator should ensure that the dialogue has a critical edge, by asking such questions as the following: "What happened in the piece? How did it proceed? What was the experience like? Who were you listening to? What do you think did not work in the piece? What would you do if you had another chance at it?" Invite members of each small group to comment on their musical process. Follow this up by facilitating a larger group discussion. Repeat *Ensemble* again but with different instrumental configurations.

3.3 *Discussion*

In this step, players are held accountable in this event for listening to themselves, to the members of their small group as they play, and to the musical inventions of other groups. Encourage a critical discussion surrounding both the process and resultant product of the musical invention.

3.3.1 *Riff-a-round*

A riff is a small musical phrase, a short series of rhythmic durations or pitches, a musical cell. It is not unrelated to an ostinato, a short phrase or motif that is persistently repeated. A collection of individuals' riffs, contributed by individual players and singers, have the potential for making a cohesive, coherent, and beautiful musical expression. This event creates possibilities for all group member to invent a 'riff' of limited musical material. As a collection of musical ideas, the group can generate an interesting mosaic of moving musicial ideas. The focus of this are 1) Independent rhythmic and melodic awareness; 2) Exploration of a series of notes; 2) Melodic invention; and group sound awareness. The resources in this section are available instruments (including the voice): Drums and other percussion, band and orchestral (wind, brass, and string) instruments, xylophones and marimbas, keyboards, guitars, recorders and penny-whistles.

It is important to make preparation and try to get all the participants and their instruments into a circle. Wheel the piano into the circle. If electric instruments require amplification, ensure that you use extension cables to enable these performers to be in the space. Nex atep is arrange chairs in a circle to invite students to become acquainted with a particular series of pitch, such as those within a pentatonic scale. This invitation could function as a "take-home" or homework activity. If they are instrumentally proficient, suggest that they familiarize themselves with the pitch series by playing in several different octaves. For beginners, playing the five pitches in one octave is sufficient.

Although not essential, 'Drone-on' provides a suitable lead into, or partner to, this event. If the same pitch series is chosen, the two events together provide students with an opportunity to explore musical ideas with different rules, techniques, and challenges..

The next ste*p choos*e an appropriate pentatonic scale. Several examples are provided below, including those referred to as Indian Pentatonic Scale, Minor Pentatonic Scale, and Major Pentatonic Scale. For example Indian pentatonic, minor pentatonic, major pentatonic scale as follows:

Student ask to play (or sing) the selected pitch series slowly in unison. And than perform it twice as fast, and then with repetitions of each pitch. Teacher should inform the group that you

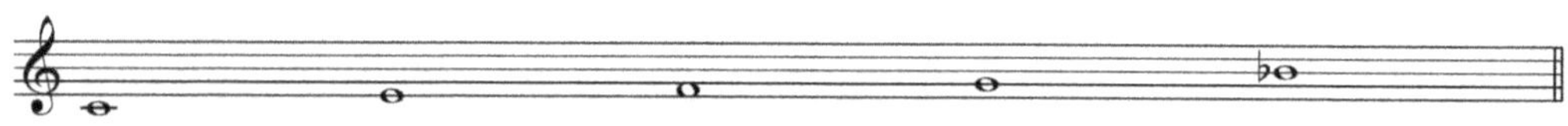

Indian Pentatonic Scale

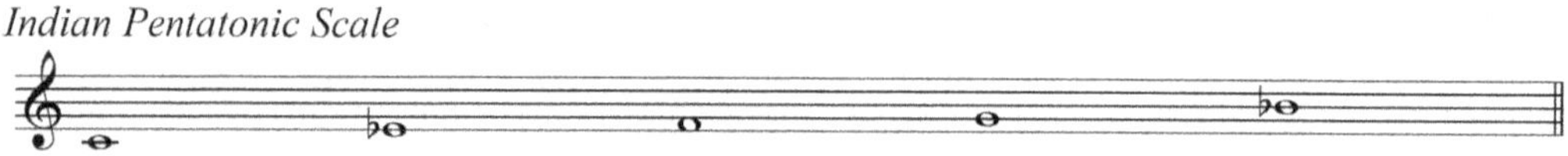

Minor Pentatonic

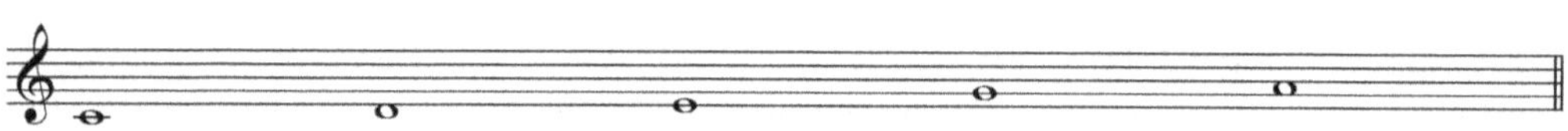

Major Pentatonic

will be working towards creating a musical mosaic. This group invention will be generated through the use of small musical phrases known as riffs. Explain that each player (and singer) will invent a riff that uses a maximum of just three pitches from the pentatonic pitch set. The riffs will fit a selected tempo, and will be repeated.

Teacher must set a clear tempo and initiate the riffing. Moving around the circle each participant joins in one at a time. Importantly, each riff is to be continually repeated as every new player enters. While the first "riffer" is challenged to set the music in motion, every new player should attempt to perform a riff that complements the sounds of the music in motion. Several examples of riffs follow.

Teacher asked student to let the riffs breathe, unfolding in their own time, each new riff adding to the texture until yet another complementary riff is performed. In this way, the overall sound will have a chance to establish its character before the next individual offers his or her contribution.

It is important to remind the group that as the musical mosaic grows, each player (or singer) must think, "What does the music need?" "How can I contribute with my riff?" To finish the event, the first riffer should cease performing. Then, mirroring the order of musical entrances, each group member should exit.

This stage must be repeat to reach establish a tempo and begin the riffing. The facilitator can act as a conductor, making musical decisions such as cueing greater and lesser dynamic intensity, and playing with the orchestration, the nature of the musical texture, by cuing in and out group members in their riffing. The next step is Increase the use of pitches, from 3 to 5, in the invention of new riffs. The further step repeat with the additional pitches. Once everyone is in with their riff, invite performers to alter their riff—just one pitch at a time. An example of the alteration of a riff follows.

From subtle alteration, group members could be invited to leap to extreme alterations,shifting their initial riff entirely to reflect the changing mood of the overall group sound. In the example that follows, the first riff sounds nothing like the second or third riff, and yet a single performer could conceivably alter riffs to this extent in order to find new complements to the collective.

On one hand this event is very simple: Each individual contributes a small musical idea to generate a musical texture of complementary riffs. Group members should reflect on the process of contributing to the musical whole. Recognizing spaces to contribute your musical is a skill to be developed. There are many potential configurations of a few pre-selected pitches, and through the process of doing it, group members should become more equip in answering the questions "What does the music need?" and "How can I contribute?"

Although the overall sound can be likened to "minimalist" compositions such as those from Steve Reich or Philip Glass (and subsequently genres such as Balinese Gamelan, iscathamiya of the South African Zulu, or any number of drumming ensembles out of West Africa), this event can also be easily connected to popular music such as rock, dance, and techno. Encourage the students to find the connections between riff-a-around and the music they know well, ask them to 'bring-in' their favourite riffing music, and lead them into further listening to music that relies on riffs that complement one another.

3.3.2 *Drone-on*

Exploring intervals between key pitches of a given pitch series provides students with a useful process in the creation of generating new musical ideas. Drone-on uses ideas that are reflective of the Hindustani classical tradition. Here, a perpetual drone provides the tonal foundation for individual musical inventions. By exploring the intervallic relationships inherent with a particular pitch series, students can create individual ideas. The ensemble supports this by playing the drone. There is no need to approach the pitch series in the sense of practicing a scale, which in effect is typical a warm-up to the performance of other music. Rather, the drone and the tonal improvisation that ensues is both the means and the end, and the process by which group members can play together and individually, all within the very same event. The activities focus on 1)Individual invention. Approaches to learning and thinking about scales, and group dynamics. The resources are voices and available melodic instrument: Band and orchestral (wind, brass, and string) instruments, xylophones and marimbas, keyboards, guitars, recorders and penny-whistles. If percussion instruments are used, these players should still be considered as a soloists within the event.

The preparations to conduct this activity is ensure that students know the pitch series by engaging the group in singing or playing the sequence. This leads to familiarization. The next step is arrange seating in a circle.'Riff-a-round' provides a suitable stepping stone for this event. If the same pitch series is chosen for the two events, students are then provided with the opportunity to explore musical ideas within different limits.

The guitar is a prime instrument for producing the drone of the sort called for in this event. In an exploration of alternative guitar tunings, try the open C tuning. Beginning on the highest string, play first-string C (2 steps lower than 6th string E), second-string G (1 steps lower than 5th string A), third-string C (1 steps lower than 4th string D), fourth-string G (as is), fifth-string C (1/2 steps higher than 2nd string B), & sixth-string E (as is). The guitarist can explore the pitches on one or two strings even while playing the drone on the remaining strings. The figure below clarifies the full six pitches on the six strings.

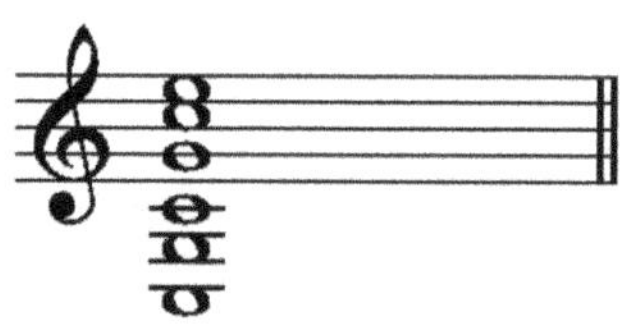

In the event section teacher explain to students that each of them will take a solo improvisation using the pitch series they have previously played. An example is this series, the same series that was used in ‘Riff a-round’, which is sometimes referred to as “the Indian pentatonic”: C, E, F, G, B flat.

To start the activities is begin by establishing the drone. The drone should be the root of the pitch series plus either a fifth or a fourth above the root. Thus, in the key of C, C is the root and either G (the fifth) or F (the fourth) is added. The drone does not set a tempo and may not even be pulsive; the “time” is to be free. As the drone is to support the soloist, it should be played (or sung) in a subtle and light. In Hindustani classical music, the Tampura plays the drone, and while it has a distinctive sound it never overpowers the soloing instruments. Ways to produce the drone include these: Having the keyboard players sound a low root note, adding the guitars on a light drone strum every so often, inviting vocalists to sing root and fifth (or fourth) pitches in a manner that creates a wave-like texture with the rising and falling volumes of crescendos and decrescendos. There are many possibilities for producing the drone. If everyone regardless of instrument or voice, can contribute to the drone, a connectedness is felt to the making of the overall sound.

Once the drone is established, the solo explorations can begin. Move into a circle (if that has not been previously established), with the facilitator modelling what the students will do. Allow the model invention, and others that will follow by the students, to proceed slowly. Encourage students to breathe easily and to establish a relationship between one interval and the next. In a C-drone sound, challenge students to see how many melodic ideas they can invent using just the C and E, sustaining, repeating, moving back and forth in various rhythms. Then, add the F to the C and E, and likewise explore melodic ideas featuring just the three pitches. When all five pitches are in play, consider exploring the pitches in another octave. After the first soloist reaches a resting point, the next soloist can begin. After the final soloist has finished, guide group members into the Riff-a-round event. Without telling them of this development, observe how they respond to a sudden shift from a meditative mood to a moving musical mosaic.

Alongside the students spend some time reflecting on the experience of playing and singing the drones and solos. Encourage them to think about what it was it like to be in the music. Discuss the musical possibilities open to singing and playing a limited set of pitches. Introduce students to the Hindustani tradition through listening experiences, for example recordings by Hariprasad Chaurasia (bansuri), Ram Narayan (sarangi), and Ravi Shankar and Anoushka Shankar (sitar). Discuss the similarities and differences between the music of these great Hindustani musicians and these initial experiences in playing inventions atop a drone.

4 CONCLUSION THOUGHTS

Stressing the importance of establishing an “improvisation culture” Chris Azzara (2002) points towards why improvisation, spontaneity, and interaction should be vital aspects of any music curriculum. From this perspective improvisation becomes not a skill or a cultural practice, but a way of being in and through music, one where teachers and students embark on journeys of musical discovery together. Roger Mantie and I describe this as improvisation as experience, rather than ability or culture (Higgins & Mantie, 2013). In supporting this notion, Vijay Iyer (Miller & Iyer, 2010) have suggested that life itself is improvisatory. They remark that in nature, there is really “no difference between human experience and the act of improvisation.” Similarly, George Lewis (2009) calls improvisation a “practice of thinking” and “a ubiquitous method of meaning exchange in any everyday life interaction." Human existence, he suggests, is "the condition of improvisation." If we employ the term *musicking,* one of Christopher Small's (1998) significant contributions, improvisation can be understood to evokes the anthropological argument that locates the human potential of music making. One might say the capacity to participate in the socially interactive process of making music. From this foundational understanding music-making might be understood as "a trail of no mistakes," becoming instead a celebration of the many and varied musical pathways that both teacher and student can take.

Through the practical activities presented in this paper there is an offering toward an emphasis on self-expression and reflective practice for music educator and the participants they work with. When Patricia Campbell and myself wrote *Free to be Musical* we wanted to encourage teachers (and all music makers) to ask their own questions and through doing so, to connect their musical experiences with the lives they are leading in the surrounding musical world. We wanted to encourage teachers to be musical, and to join *with* their students in crafting collective yet also personally fulfilling music. Even as we ourselves are singers, dancers, and players of music, we uphold the belief that the integration of music and movement are critical in the artistic expressions we facilitate for others. We fully expect that teachers will shape experiences that will suit their students, their spaces, their schedules, and their own priorities, and thus we did not arranged the events nor definitively set them for a particular age or developmental stage. We thought the teacher was best to judge what will work for which setting of students, and do not want to limit the use of our events to specific contexts. My hope is that by injecting classrooms with events that embrace artistic-expressive thinking and embodied doing, those engaged in music education will embark on a musical voyage of life-long musical learning.

REFERENCES

Azzara, C. D. (2002). Improvisation. In R. Colwell & C. Richardson (Eds.), *New Handbook of Research in Music Teaching and Learning* (pp. 171–187): Oxford University Press.

Bartleet, B.-L., & Higgins, L. (Eds.). (2018). *The Oxford Handbook of Community Music.* New York: Oxford University Press.

Blacking, J. (1973). How Musical is Man? London: Faber and Faber.

Dewey, J. (1997). *Experience and Education.* New York: Touchstone.

Freire, P. (2002). *Pedagogy of the Oppressed.* New York: Continuum.

Higgins, L. (2012). *Community Music: In Theory and in Practice.* New York: Oxford University Press.

Higgins, L., & Campbell, P. S. (2010). *Free to be Musical: Group Improvisation in Music.* Lanham, MD: Rowman & Littlefield.

Higgins, L., & Mantie, R. (2013). Improvisation as Ability, Culture, and Experience. *Music Education Journal, 100*(2), 38–44.

Lewis, G. E. (2009). *The Condition of Improvisation.* Paper presented at the International Society for Improvised Music, Santa Cruz, CA.

Miller, D., & Iyer, V. (2010). Improvising Digital Culture: A Conversation. . *Critical Studies in Improvisation/Études critiques en improvisation,, 5*(1), 1–10.

Nettl, B., & Russell, M. (Eds.). (1998). *In the Course of Performance: Studies in the World of Musical Improvisation.* Chicago: University Of Chicago Press.

Small, C. (1998). *Musicking: The meanings of performing and listening.* Hanover: University Press.

Solis, G., & Nettl, B. (2009). *Musical Improvisation: Art, Education, and Society.* Illinois: University of Ilinois Press.

21st Century Innovation in Music Education – Astuti et al. (eds)
© 2020 Taylor & Francis Group, London, ISBN 978-0-367-11122-9

The position of humans in the *Karawitan* lyrics of Ki Narto Sabdo: An ecocritical perspective

S.A. Sayuti
Universitas Negeri Yogyakarta, Yogyakarta, Indonesia

ABSTRACT: This article aims to demonstrate the cenrality of the human subject in the *karawitan* lyrics of Ki Narto Sabdo. It describes how Ki Narto Sabdo presents landscape and considers whether his work contains a socio-political or ethical agenda. Most importantly, it examines the human relationship to space, how humans interact with the environment, and potential sources of conflict in this interaction. The article concludes that the creative texts of Ki Narto Sabdo can be positioned as a defense against the unequal power relations that lead to the loss of self-identity, cultural alienation, and the uprooting of cultural tradition, history, and national character. Ki Narto Sabdo's creative texts are responses to efforts to de-colonize culture by reviving or re-establishing cultural meaning and identity in and through the imaginative world.

1 INTRODUCTION

Historically, when Javanese palaces were the central sites of cultural development, the term *karawitan* was used to refer to various branches of the arts, including dance, *gamelan*, puppet craft, puppetry, music, and carpentry. Etymologically, *karawitan* derives from the word *rawit*, which means small, fine, intricate, or complex (Supanggah, 2002). In the repertoire of world music, the term *gamelan* is used more widely than the term *karawitan* – as, for instance, in the title of the International Gamelan Festival.

Nowadays, the term *karawitan* tends to be used to refer to one of the musical genres that uses gamelan instruments partly or wholly tuned to both *laras pelog* (pentatonic scale) and *laras slendro* (heptatonic scale), and accompanied by singing. The vocal element in a *karawitan* performance is called *cakepan* (lyrics) and comes in many forms, including *bawa*, *celuk*, *cakepan* of related *macapat* songs (such as *Kinanthi, Mijil,* and the like), *rumpakan*, *sindhenan*, *alok*, and *senggakan*.

Ki Narto Sabdo is one of the most prominent Javanese *karawitan* artists. He has composed hundreds of *gendhing* (*gamelan* musical compositions). He is famous not only as a legendary Javanese *karawitan* figure, but also as a master of the art of skin puppetry. This brief article discusses a number of his *karawitan* lyrics from an ecocritical perspective. It aims to remind readers that the arts – in this case, the popular Javanese traditional music genre of *karawitan* – contain educative and cultural values that should not be ignored in the teaching of music within education as a whole.

2 LITERATURE REVIEW

A state, city, village, neighborhood, house, and so forth is a locality that can conceptualized as an environment with mythical, physical, and psychological characteristics. According to Berry (2015), there has been a cultural shift in contemporary society whereby familial relationships have been displaced and people have become strangers to one another.

When we gaze at the 'place' in which we live we create an ontological distance between the place and ourselves. This distance is important for self-understanding and self-realization. Without any distance in between, we tend to be *rumangsa-bisa* (feeling that we have understood it very well), and simultaneously we lose what is referred to as locality sensitivity. A further consequence is that the significance of locality is eradicated from our language because, in annihilating distance, we have removed locality from our thinking. Therefore, the ability to *mulat-salira* (introspect) can be considered as an attempt to achieve self-understanding and, at the same time, as a means of *manjing-kahanan* (immersing oneself in the situation). Such self-understanding makes us *angrasa-wani* (having bravery to act) in realizing ourselves with *greget* (courage) and *ora-mingkuh* (not avoiding responsibility).

We are mutually created by our surroundings. Soil and water enter and leave our body, just as our body enters and leaves the ground and water. People, flora, and fauna are mutual parts of each other: the expression *suket godhong dadi rowang* (the weeds and leaves are acquaintances) means that they cannot exist separately. Therefore, our cultural behavior must be a responsible response to our living environment, since they are *loro-loroning atunggal* (inseparable) – one part cannot become better without the presence of the other. Otherwise, our lives can easily become *kasandhung ing rata, kabentus ing awang-awang* (stumbling on bumpy roads and bumping [our heads] on the ceiling).

This brief discussion has presented an ideal picture of a harmonious relationship between people and the natural environment, a situation that is increasingly difficult to find in industrial, consumerist societies. In this context, understanding of our environment can help us to achieve self-understanding and self-realization which, according to Snyder (1995), is the central aim in life. Or, in Berry's (2010) words: "to know who you are, first of all know where you are."

Whether or not we realize it, our living environment plays an important role in configuring who we are physically, emotionally, and even spiritually. Our awareness develops in, with, and through the places that become our environment. Therefore, one may be able to love a certain area very much or, conversely, feel alienated in others. The particularities of place can have a psychological impact on individuals or groups. In the Javanese cultural context, the mottos *gemblung-gemblung anggere rubung* (despite madness, togetherness is important) and *mangan ora mangan anggere kumpul* (whether or not we have any food to eat is not as important as our togetherness) capture this idea.

All of us may have experienced how the particularity of a certain place makes us *krasan* (feel at home). This may be due to the fact that the place always provides a feeling of 'lightness', which is difficult to describe because 'feeling at home' is a complex phenomenon and has multiple relationships with other aspects of life.

Not everyone, however, is aware of the effect of the environment on them. This lack of 'environmental awareness' may reflect the absence of experiences in other places that can serve as sources of comparison, such as being born and growing up in places other than where we currently live. Some people have never known the experience of life in a house in which they do not feel at home. Therefore, self-understanding will be achieved whenever the understanding of what is not oneself (read: *liyan*, the other) is achieved. The exploration of 'self-space' and 'non-self-space' is important for a more profound self-understanding. When a certain place has affected someone, he or she will respond according to that effect, even if he or she does not realize it.

When the significance of a place is felt and realized as a factor that affects every aspect of our daily lives, all of our senses - seeing, tasting, hearing, and feeling – will be involved in experiencing the various aspects of the place we call 'home'. Therefore, a home has not only a physical meaning but also mental, emotional, spiritual, and even cultural meanings. In this context, however, if a certain locality and its associated culture have been imposed, we do not feel at home and may seek to leave it. In short, a place is inherent in self-configuration, community, and holistic life.

It must be acknowledged, as Hall (2014) writes in *Cultural Identity and Diaspora,* that identities always change in a dynamic process that cannot return to past reality. Every effort to re-establish historical identities is believed to be essentially symbolic and unreal. Revitalizing identities through creative text writing is actually the same as formatting the basis for a new collective identity, and it is quite justifiable.

3 METHOD

The previous discussion has indicated that place and its variants can be considered as a 'matrix' of creation and that they can be converted and expanded in a creative text, such as the *karawitan* lyrics of Ki Narto Sabdo. If this is the case, to what extent does the landscape assume a central position in the text in order to emphasize the position of human beings? To what extent does Ki Narto Sabdo identify landscape or environment as an intrinsic element of a *karawitan* poem for the purpose of self-conceptualization as a creator through mythical, physical, and geographical elements? Or, more specifically: How does Ki Narto Sabdo present landscape? What kind of attitude toward landscape is represented? Is there a socio-political or ethical agenda in this representation? Above all, what is the relationship of human beings to place? How do human beings interact with their environment? Is there any conflict in the interaction?

This article is based on the paradigm of Popper's realism, which identifies truth from its coherence with the universal objective (Muhadjir, 1996). The questions are answered by using linguistics logic syntactically and semantically or hermeneutically (Keeves, 1997).

4 RESULTS

Human beings should learn to live harmoniously with nature by adopting sustainable processes of acculturation and civilization. The following poem *Subakastawa Winangun* by Ki Narto Sabdo (Sugiarto, 1998) is illustrative.

Angripta rengganing gunung
imbange, jurange yen kadulu saking tebih
warna biru maya-maya watune alase
kang jenar sinaput wilis
e e gawe lam-laming paningal
lho... lho... suwe-suwe tandukira angrespati

E endah temen megane
Putih memplak mayungi
Gunung-gunung lha... lha
Pancurane katon banyune bening
Mili mintir ing sawah tan ana cicir
Pari jagung ketela pohung
Kacange canthele grojogane sor sor
Suh sah anggetak nggusah manuk
Per... per... kleper manuke mabur

The poem represents nature by mythically relating human objects to the 'beauty' of nature, which simultaneously affects them. The absence of an explicit reference to a human subject indicates the presence of its influence; the human melts away in the environmental universe, and this influence is not only felt or realized, but also appreciated (*gawe lam laming paningal dan suwe-suwe ... angrespati*). The problem is: Is it true that the beauty of nature in the poem is realistic, or is it only something that is expected to re-exist, a human's feeling of nostalgia for a formative environment that has vanished (it is part of history) for various reasons? If such a beauty really exists, why does the poem begin with the word *angripta,* which can be interpreted as 'creating a textual world'? Furthermore, why do the birds fly away (*per kleper manuke mabur*)? The factual world illustrated in the poem may very likely exist, but in another place, a place 'far away' from the creator's position. Therefore, for him it is only *yen kadulu saking tebih.*

Whatever the possibility is, the creator has responded to the environment. Narto Sabdo has chosen his position in his relationship with the natural environment for the sake of his self-understanding and self-realization as a human being. Even if it is true that such beauty in nature is part of history, he has revived it in the creative universe. In so doing, he awakens a collective memory and shares with readers the importance of environmental awareness in order that *pancurane* remains *katon banyune bening,* far from the crisis of pollution. The voice of nature has been eternalized in the creative universe known as a *karawitan* poem.

The environmental aspects can also be considered spiritually, as the figure that radiates great wisdom and, probably, divinity. Some cultures emphasize tradition and greater concern for certain landscapes through which dialectical and reciprocal relationships are established. This may be manifested in the form of prayer, ritual, and ceremonies intended to maintain balance. Examples include *Merti dhusun, ruwat bumi* and the like in the Javanese tradition. The overall interpretation is that abusing and 'hurting' the earth (our mother, mother nature) will undoubtedly disrupt the balance, thereby causing physical and spiritual illness and misery among the forms of life in it. All living beings are interdependent and interrelated. Human beings achieve their full existence when they can position themselves in their environmental context, as can be seen in the following *Syair* poem, *Singa-singa.*

Manguwuh peksi manyura
sawung kluruk amelungi
wancine wus gagat enjang
ayo rowang amurwani
netepi rukun lima
manembah hyang maha suci
mrih yuwana kang sinandhang
ing donya tumekeng akhir

In the poem cited above, *sawung kluruk amelungi,* as the voice of nature, not only functions as a natural indicator that *wancine wus gagat enjang* but also at the same time wakes up (physically) and reminds (spiritually) those who consider themselves as Muslims to hasten to morning worship, *Subuh* prayer. The benefit of all of this is the achievement of happiness in the world and in the Hereafter (*mrih yuwana. . .ing donya tumekeng akhir*).

Landscape, spatial meaning, and identities are integral aspects of all human development, but they also reflect the particularities of different cultures. The main theme of many *karawitan* poems by Ki Narto Sabdo explores identities and spaces. The identities rooted in Javanese culture, which were previously determined by oral tradition, are recycled and represented as *karawitan* poems. The natural life situation in agrarian villages, for instance, can be read in the trilogy of poems reproduced below.

Syair *Lesung Jumengglung*
Lesung jumengglung sru imbal-imbalan
Lesung jumengglung maneker mangungkung
Ngumandhang ngebeki sajroning padesan
Tok tok tek tok tokgung tok
Tok tek tok tek tok gung
Syair *Lumbung Desa*
Lumbung desa pra tani padha makarya, ayo ca
Nutu pari nata lesung nyandhak alu, ayo yu
Padha maju yen wis rampung nuli adang, ayo kang
Dha tumandang nyosoh beras aneng lumpang

Syair *Aja Ewa*
Esuk-esuk mbakayu yu padha nutu
Pari kang wis tuwa
Lesung alu wis cumawis
Pancen nyata Bathari Sri dadi boga

Kok ewa aku kok ewa aku
Nginteri kok ngono ngel angel dang berase
Mbok ya ngati-ati sinau gemi
Padha ngelingana dhek nyeblokke winih
Tandang cancut taliwanda
Aja keneng ama

Ama kabur tandur subur loh jinawi

The three poems above are interrelated, each representing the harmony between nature's voice and the typical village in a farming community environment. The three poems describe imaginatively the dialectical relationship between man and nature that makes life possible, including memory and acknowledgement of the existence of *Bathari Sri*, as the Goddess of Rice, who is *dadi boga.* Therefore, agrarian-village men need to respect her by way of, among other things, displaying *ngati-ati* and *sinau gemi.* The appearance of the legend of *Bathari Sri* as the Goddess of Rice in this trilogy needs to be interpreted in terms of earthly habituation. The presence of such legends in agrarian oral tradition as a specific landscape, from which people derive a certain meaning, indicates its importance for the sustainability of life. It also becomes a cultural symbol that is closely related to self and spiritual meaning.

With *angripta* (look again at the word initiating the lyric of *Subakastawa Winangun*), the *karawitan* lyrics based on traditional landscape, Ki Narto Sabdo has maintained cultural identities. His creative texts become historical references and documents which, if appropriately elaborated, show that his lyrics are not only a creation but, even more importantly, also hold spiritual and symbolic meaningfulness along with their particularities. In other words, Ki Narto Sabdo wants to share the idea that *angripta*, or the creation of creative texts is truly "a wishful act," that our everyday lives are affected by the landscape, and that we should put (back) the earth as a *shelter*, and not destroy it.

5 CONCLUSION

In the context of the arbitrary exploitation of the earth, *narto-sabdan* lyrics achieve their significance and relevance. Such texts remind us of the importance of interweaving the process of self-understanding and self-realization with the place we inhabit. Therefore, spatial awareness is key to efforts associated with the politics of identity and self-representation.

The problem becomes more important when loss of power is manifested in the distortion of identities, persons, and meaning on a global scale, as particular cultures are marginalized and uprooted and lose their acknowledgement as cultural heritage.

Creative texts such as those of Ki Narto Sabdo can be positioned as a defense against the unequal power relations that lead to loss of self-identity, cultural alienation, and the uprooting of cultural tradition, history, and national character. Ki Narto Sabdo's creative texts are responses to efforts to de-colonize culture by reviving or re-establishing cultural identity and meaning in and through the imaginative world. In other words, Narto Sabdo's *karawitan* lyrics represent efforts to re-territorialize, to revive the missing identities in and through nostalgic reflection and imaginative construction. He also seeks to integrate the past and the present to reintroduce the mythical/historical elements of spatial meaning that are missing in contemporary usage of the term. Overall, the aim is to remind us that "one will never be able to come back through the door one has entered".

REFERENCES

Berry, W. (2010). *The selected poems of Wendell Berry*. ReadHowYouWant. com.

Berry, W. (2015). *The unsettling of America: Culture & agriculture*. Catapult.

Hall, S. (2014). Cultural identity and diaspora. In *Diaspora and visual culture* (pp. 35–47). Routledge.

Keeves, J. P. (Ed.). (1997). Educational research, methodology, and measurement: An international handbook.

Muhadjir, N. (1996). *Metodologi penelitian kualitatif: pendekatan positivistik, rasionalistik, phenomenologik, dan realisme metaphisik telaah studi teks dan penelitian agama* [Qualitative research method: Positivistic, Rasionalistic, phenomenologic, and realisme methaphisic. Study text and religion] Yogyakarta:Rake Sarasin.

Snyder, G. (1995). *A place in space: Ethics aesthetics, and watersheds* (Washington, DC: Counterpoint.

Sugiarto, A. (1998). Kumpulan Gendhing Jawa Karya Ki Narto Sabdo [Gendhing Jawa art work by Ki Narto Sabdo].

Supanggah, R. (2002). *Bothekan karawitan*. Masyarakat Pertunjukan Indonesia.

21st Century Innovation in Music Education – Astuti et al. (eds)

Ethnomusicology of The Dawn: Understanding the intergenerational impact of their song 'Salamat'

E.V. Lianita
La Salle Green Hills, Mandaluyong, Philippines

ABSTRACT: This paper provides a narrative of the impact created by The Dawn in echoing Filipino values. The paper uses ethnography to explain the cultural values behind their musicology. Data were generated from online surveys and social media platforms. Blogs, surveys and anecdotes from crowdsourced elements, in the form of comments and perceptions, played a pivotal role in understanding the compound ideas that established the band's esteemed position in Filipino musical history. This is a perception-based study of the artistry and functionality of music in building communities that weaves among the stale posturing of bands after the ousting of a dictator. The height of their popularity came at the end of the Marcos regime, when most bands were still immersed in romance and escapist art. The collective critical reflections of their followers served as channels of connectivity among this polarised generation.

1 INTRODUCTION

The popularity of Filipino rock band The Dawn's song 'Salamat' (Thank you) has withstood the test of time because of its musicality and winning theme of friendship and gratitude. The high rate of response to this music had been advocated in various social media platforms. The song has created a meaningful discourse between the audience and the artists. The common reaction when it is heard and listened to will always be to sing it with both friends and strangers. It has cut across the emotional stream of both young and old (Henderson & Hodges, 2006). The correlation between the emergence and re-emergence of the band and the historical period when they earned their first commercial success is an adage, but the way that The Dawn sustained their following has not come as a surprise. They are a tribute to the musicology of alternative rock bands in the Philippines (Tietler in Nettl, 1989).

Phenomenology, combined with ethnographic techniques, is used to understand The Dawn's musicology in this first and non-patterned study of the band. Thus, at the end of this paper, a recommendation to look further into the breadth and depth of their music will be included. The paper defines ethnomusicology as the study of music in its primal state or from its cultural aspects, and as the study of all of the world's musical cultures from a broadly comparative perspective.

The artists' characters were studied from the perspective of their levels of interaction, discipline, and musical craft as they performed, which was challenged by the set of audiences they came into contact with. Their inner thoughts were seen from their statements and level of concentration. Beyond anthropology, visual artists were consulted in order to understand the hermeneutic identity of the band. A large proportion of this literature will be continually improved in order to add more information and meaning to the very young legend that The Dawn has created in the Philippine music industry.

The most significant rationale for conducting this research was to establish a concrete and more scientific explanation for the phenomenal success of a rock band that has sustained their musical popularity without the polarising challenges faced by other short-lived rock bands. Besides this primary reason, this paper will also review the importance of popular culture in setting a tone of community among Filipinos who are greatly divided by either physical distance or more abstract political bickering. While ethnography is the main basis for the qualitative platform that will be used in this study, vantage points were accessed using the literature of the band's music. The use of ethnomusicology will help future researchers in establishing the relevance of cultural analysis in musicology. However, this paper is unique because it offers critical reflections about the hidden meanings in The Dawn's music and how they set the standard for rock music bands after the dictator was deposed. The symbolic representation of The Dawn was used as one of the focuses of this descriptive research. It presents a blistered image of Filipino ingenuity in music, which carved an impact in the minds of different generations through their overall experiences. This paper made use of statistics in order to interpret the deeper and more objective insights regarding a contemporary legend in musical art history.

2 METHODOLOGY

> *Musicology is so broad that it's difficult to define without restricting it. Although many musicologists are also trained in performance, it isn't a performance-oriented field. Instead, musicology focuses on the history and cultural contexts of music.* (Peterkin, 2016)

This is mixed qualitative and quantitative research, which defines the character and identity of the paper. The tools of anthropological research were used to analyse the cultural impact of the song(s) that the band popularised. This did not just define the genre of their music, but also its overall ability to entice the audience into following it in all ways (whether they are performing it on stage or playing it in different modalities). The first part of the study deals with an analysis of the musicology and discography of The Dawn's music using interviews and published music record narratives of the band from specific sources, such as news articles, blogs and fan pages.

Vital to the descriptive portion of this paper is the analysis of variance (ANOVA) and *t*-tests derived from IBM's Statistics Package for the Social Sciences (SPSS) software. A total of 100 volunteer participants were used as controlled participants, as they were given the same survey questions and demographic retrieval forms. Their age and sex were recorded, and also their corresponding music listening situations. The data generated from the online and offline surveys were analysed and interpreted using the package in order to validate the null hypothesis constructed during the longitudinal research about the band. This was done using Hierarchical Linear Model (HLM) software. Regarding the impact, respondents were asked three questions: (1) How well do you remember the music of The Dawn, particularly 'Salamat'? (2) How important was the *value of the song* 'Salamat' for your intention to listen to music in any situation? (3) How important was *arousal and mood regulation* for your intention to listen to music in that situation? Regarding the impact, respondents were asked: (1) How strongly did the function of enjoyment actually occur through listening to this music? (2) How strongly did the function of *social relatedness* actually occur through listening to this music? (3) How strongly did the function of *enjoyment and ease of use* actually occur through listening to this music? All of the questions were answered on a scale of 1 (*Not at all*) to 10 (*Very much*).

After generating the quantitative data that validates the intergenerational impact of 'Salamat' among the 100 participants, multiple varieties of interview were conducted with band members and followers in order to shed light on some of the issues and grey areas regarding the 32-year-old band. This was done deliberately so as not to neglect the "blind spot" in this research. Ethnomusicology is about people making music and music making people. Narrative and descriptive insights were collected to support The Dawn's local identity and recognition. Some forms of coding were used to categorise the qualitative responses of the participants of the interviews.

The number of respondents was randomised from the willing sample that participated in the interviews. One hundred ($n1 = 100$) of the $N = 600$ persons interviewed were placed under qualitative interaction, while the remaining $n2 = 500$ respondents were asked questions by designated research participants/assistants. Thirty of hundred participants were subjected to patterned qualitative description/narrative about The Dawn and their music 'Salamat' from the descriptive answers they gave.

Respondents/interviewees were asked to identify their overall experience of The Dawn's music by giving short answers to explain the usefulness, the ease of use and their enjoyment of their favourite song by The Dawn. Responses were charted and categorised according to the parameters set in the conceptual framework.

The other part of the study was conducted using an online survey app, in which an accurate and more easily facilitated online survey was obtained from www.surveymonkey.com. The respondents were asked three questions in order to get a referenced result. The survey covered their perception of The Dawn's most popular music, listing five songs that were randomly selected according to their Billboard record (1987–2017). Similar surveys were conducted through Facebook and Twitter with the same data collection objectives. In these surveys, three varieties of modality of data collection were used, namely Gmail, Facebook and Messenger. After five days of data gathering, 100 respondents had sent their responses and the votes of more than 70% (in SurveyMonkey) and 64% (in a Twitter poll) were achieved by 'Salamat'. Respondents were asked about the reasons why they made these choices and two sets of questions were asked in order to retrieve the qualitative data. The result categorised the responses regarding the specific and generic preference perceptions of the song(s) of The Dawn (Hailstone et.al, 2009). Responses were categorised based on the questions. Question 2 asks why they preferred 'Salamat' and question 3 asks about the generality of choosing The Dawn. Responses to question 2 were labelled as Specific Preference Perception (SPP) and responses to question 3 were presented as Generic Preference Perception (GPP).

This paper also looked into symbolism, phenomena and participant observations regarding how the music 'Salamat' is played by the artists, followers and other artists. This range of observations was intended to widen the interpretation and to construct a fuller meaning of the culture behind the music of The Dawn. A discography was used to chart the band's musical record and commercial details.

Casual and informal types of interviews were conducted among four subjects who represented four decades. Each interviewee engaged in a discussion about their perception of the band with reference to their genre and musical experience. The interviewees, whose identities were withheld in observance of ethical researching, were asked about their take on The Dawn's musicology, and internal confidentiality (Tolich, 2004) was observed during this research attempt. However, I have made their context clues identifiable in this research paper (Sieber, 1992). Given that qualitative studies often contain rich descriptions of the study's participants, confidentiality breaches via deductive disclosure are of particular concern to this qualitative researcher. As such, the researcher faced a conflict between conveying detailed, accurate accounts of the social world and protecting the identities of the individuals who participated in this research, unless they allowed themselves to be identified or named.

The qualitative research interview seeks to describe the meanings of central themes in the lives of the subjects. The main task in interviewing them is to understand the meaning of what the interviewees say (Kvale, 1996). As such, the depth or undertone of their stated expressions was noted as a framework for analysis in the subsequent pages of this paper. It covers both a factual and a meaning level, though it is usually more difficult to interview on a meaning level (Kvale, 1996). The song was played in order to gather reactions and insights about its musicality and its impact on their lives. These interviews are particularly useful for getting the story behind the participant's experiences of The Dawn's music. The interviewer pursued in-depth information around the topic and the interviews were conducted as a follow-up to the questionnaires of certain respondents, for example to further investigate either their responses or those of others (McNamara & Ballard, 1999).

The interviews/conversations were completed according to what the respondent said. This was relied on because interviews are a far more personal form of research than questionnaires. In the personal interviews, I worked directly with the respondents to identify the major components of the research framework. Unlike mail surveys via SurveyMonkey or Twitter, there was the opportunity to probe or ask follow-up questions. From the researcher's experience, this is generally easier for respondents, especially if seeking to gather opinions and

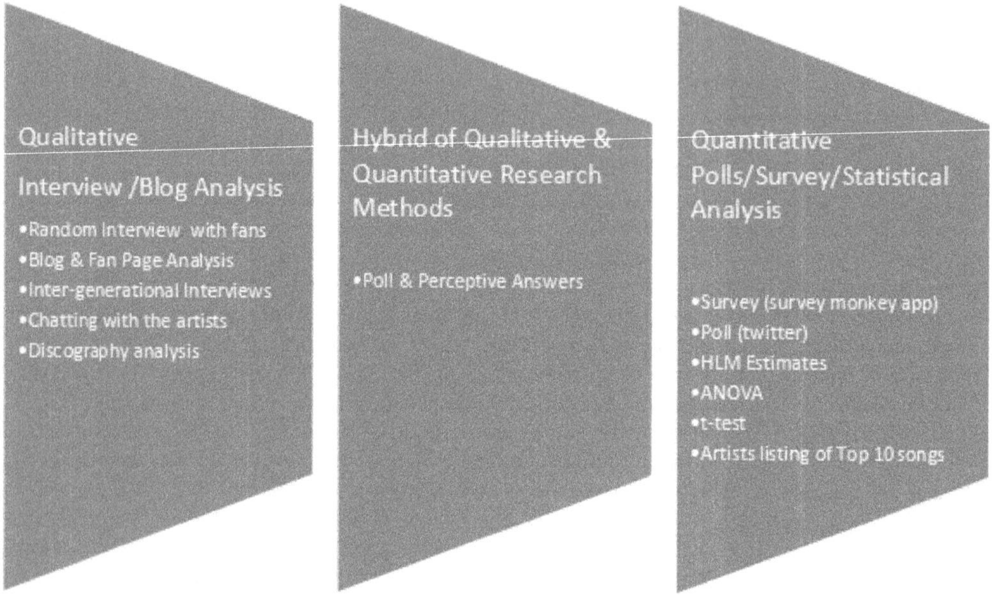

Figure 1. Design of mixed method research (Creswell, 2015).

impressions. While it is time-consuming and resource-intensive, the impact component was examined more closely because I became part of the measurement instrument.

To illustrate the rich data gathering in this paper, Figure 1 shows the different layers of data gathering in this research (Creswell, 2015).

A study of perceptions was also conducted among the artists to assess their wide range of work preference and to find out how 'Salamat' impacts on them as artists. Music scholars were interviewed about 'Salamat' and The Dawn in order to triangulate their perspectives about the musicology of the band.

3 FINDINGS AND CONCLUSION

The Dawn's popularity is an evolving concept that transcends craftsmanship and illustrative virtues. The perception of themselves as an epic *anito* (local term for animist deity) is a self-portrait of grit and excellence. A former member who continues to think of the 3Is (Imitate, Integrate and Innovate) manifested a distinct passion for craft improvement. The concept of character in their music has added more values. It is a clear fact that their music has dominated for around three decades. It is the reason why a 16-year-old school singing group member can relate to their music and also why a 20-year-old university graduate treats their music as classic in the same way as Spandau Ballet and Tears For Fears, combined in the tradition of a U2-like boy band experiment or The Cure of Southeast Asia. The characters in the Dawn are able to penetrate the soul of the listener as an epitome of music deity.

The name of their band has lived as the epitome of new hope for a new day, a characteristic that came into being after the ousting of the dictator who had killed artistic freedom from the outset. The hidden identity of members has contributed to spawned decent posturing contributed to the Dawn. There is no idolatry intended in pursuing this paper, but the tremendous impact it generated among the 40s age group in my list of interviewees showed that the strong impact of the character of the band cut across the fabric of social formation.

Fans and followers created an aura of reverence through blogs and social media statements rather than by concert attendance and album patronage. 'Salamat' has assumed a sense of vitality and is re-lived in many local and diaspora gatherings because of its echoes of Queen's 'We Are The Champions'.

They are seen by Grab (taxi) drivers in their 30s as the most illustrious band to create the alternative rock tradition in the 90s and the following years. People in their 40s and 50s share the same thoughts about the beautiful lyrics and the base experimentation in their music. 'Salamat' is ranked as an example of the modern type of folkloric music that spells out the tradition of reverence for valuable relationships. It has an embedded appeal to mass singers as the ultimate piece that would not allow anyone to be left out. However, the artists themselves are not aware of the intergenerational impact of the song 'Salamat'. To their minds, the preference listing is like a sanctuary that houses their character. The unpopularity of the songs that the artists have chosen reflects their connection to the music that they create (see Table 1). These songs are the ones that strengthen their basis for a private choice that is not shared by others. When asked the questions, "Why didn't you choose 'Salamat' in your list?" and "Are you aware of the intergenerational impact of this song?", one of the artists responded, "No, I'm not aware, though I would rather listen on my own to the songs that I chose on the list and others not mentioned, because I enjoy listening more to these 'unpopular' ones" (from a chat with Carlos Balcells, 1 June 2018).

This is open to wide interpretation and I have relied on the story about the impressionists whose works were rejected in their own time but were appreciated much later (Diaz, 2012). Apparently, this kind of art, which emerged in the early 1900s, was viewed as an insult to the dignity of their contemporaries. The rejection cause of the generation mostly identifies with the Western and American New Wave genre and their music has a certain branding, known as "Pinoy rock". There is a margin of value between the popularity of 'Salamat' and the unpopularity of their other songs, which contain unique themes. One example is 'Hey Isabel', which is about a pretty young prostitute who peddles her flesh among rich patrons. This appears to be in the same context as Ben Cabrera's (one of the country's national visual artists) multiple visual art work for Isabel. This would also be personified during the 90s by Eraserheads' (a popular band) 'Toyang'. Thus, artists are complicit in and are caretakers of their treasured works: those that are patronised take independent flight, but those that are made but not followed by others are gems that are favoured by the artists.

Table 1. Perception-based survey of the artists' personal preferences of their music.

Jett Pangan – Vocals	Axeman Sancho – Guitar	JB Leonor – Drums	Carlos Balcells – Bass
1. 'Susi' (first album)	1 .'Habulan'	1. 'Hatak'	1. 'Let Me Dream'
2. 'I Stand With You' (second album)	2. 'Enveloped Ideas'	2. 'Pedestal', Later Half of Day	2. 'Runaway'
3. 'Habulan' (latest for 2014)	3. 'I Stand With You'	3. 'Little Paradise'	3.'Will We Ever Learn'
4. 'I Saw You Coming' (Prodigal Sun)	4. 'I Saw You Coming In'	4. 'Abot Kamay'	4. 'Take Me Back'
5. 'Standing By Your Heart' (Prodigal Sun)	5. 'Standing By Your Heart'	5. 'Dreams'	5. 'Jolt'
6. 'Love Will Set Us Free' (2nd album)	6. 'Dreams'	6. 'Talaga Naman'	6. 'Holiday'
7. 'Abot Kamay' (re-recorded version, Tulad ng Dati)	7. 'Love Will Set Us Free'	7. 'Runaway'	7. 'Salimpusa'
8. 'Tulad ng Dati' (Harapin album)	8. 'Salamat'	8. 'Laging Narito'	8. 'Can You Tell'
9. 'Harapin ang Liwanag' (Harapin album)	9. 'Lisang Bangka Tayo'	9. 'Traffic'	9. 'It Will Never Be The Same'
10. 'Salimpusa' (Puno't Dulo album)	10. 'Beyond the Bend'	10. 'Change is Breaking Us Apart'	10. 'I Stand With You'

Indeed, The Dawn's 'Salamat' has chemistry, both in maintaining friendships and helping forge new ones. This song emerged as a favourite among Filipinos because everybody can relate to its charm in expressing that bonding with friends is a divine gift. It is not a religious song, but it has become like a cult that makes sense in gatherings. Filipinos, being religious and personable, see this song as a type of magical trance music that is expressed in a techno tempo, which offers a seductive beat to bond around in terms of endearing friendship.

However, this study is still some way from understanding music's power to create social bonds. The intergenerational preference for this song made the point that, to this day, there must have been tremendous evolutionary pressure for music that binds people together (Janata, 2013). I surmised that there is a real connection between the collective brain function that causes people to present themselves as submitting to the lyrics of the song. The mixing of sounds created some neural elements that orient willing participation to the tune and character of the music.

The actual experience of people in relation to the usefulness of the music, enjoyment and ease of use is clearly denoted by the success of this beer-commercial music. The pattern of recall and the flaunting of images of friendship in the advertisement show how relevant friendship is in times of hardship and collective struggle. The liberating component of the music centres around the themes of team spirit and the support of friendship. These themes are captured by the audience and this has created a constant following through the repeated playing/showing of the commercial.

The response generated from the survey shows the meaningful value and the impact of the musical piece on those who are familiar with it or those who have followed the progress of the band. Enjoyment and ease of use are two subsets that completed the experience of The Dawn's 'Salamat', and of their music in general. It gave a voice to the invisible hard-working people and politicians who are joined together in their enjoyment of the music.

The difficult component of this ethnomusicology of a specific song from a particular band was due to the combined musical experiences of all of the respondents, which was evident in relation to the tools of analysis and interpretation. The perceptive mind was tapped in order to express their own take on this particular music and music in general. It was carved out by the sensible characters of the musicians, who are serious yet playful in giving their all when they perform.

The discography study yielded a confusing index because of the strong presence of The Dawn's music on the internet. This new platform, where music can be marketed to a wider audience using multiple apps (i.e. Deezer, iTunes and Spotify), yielded an ease of search and catalogued music according to their chronological sequence of release. The ease of downloading their music from YouTube and other channels must be a growing advantage for the popularity and commercial marketability of the band. Beyond this convenience, however, is a gnawing question of influx on market value.

The following of The Dawn has not dictated the terms of their elite popularity as "lords" of local rock music. They mitigated the cult of reverence by holding free concerts in the state universities and national parks. Their big followers are not disturbed by the bar hops (gigs) that they do to play their music. The Dawn, being a popular people's band of Filipino music, is a young legend that has set the standard for Pinoy rock. Their songs, such as 'Iisang Bangka' (One Boat) and 'Enveloped Ideas', can be the subject of literary stirrings. Beyond poetry and lyrics, they convey the depth of life and the breadth of the creed that belongs to a true nationalist, but they go beyond politics and have transitioned to be the writers of songs for the masses – their music rides the wave of a new beginning.

On the statistical side of the research, data regarding the knowledge and effects caused by listening to The Dawn's music were analysed by computing the averages for the five most popular songs. The fundamental functions across all participants and all listening situations were gathered on the basis of the questionnaires. The mean differences between goals and effects were analysed using *t*-tests. Then the influence of the listener's past functional experiences of the effects of listening to 'Salamat' and the strength it gives them as memorable music was analysed. For these analyses, data regarding the reasons why they remembered listening to the music were averaged. Because the variables involved in the predictions unfold on different levels of measurement (i.e. within and between individuals), hierarchic multilevel modelling was used to analyse the data. Level 1 contains the enjoyment level within the different music listening situations of each participant. Level 2 contains the participant's preference for the music because of its ease of use.

Parameters were estimated using Hierarchical Linear Model (HLM) software. Estimates were calculated for the following coefficients. In Model 1, the dependent variable was the enjoyment of 'Salamat'. The intercept of this variable was predicted by past functional experiences (means-as-outcomes model: see Path A in Figure 2. In Model 2, the dependent variable was the ease of use of the music. The intercept of this variable was predicted by past functional experiences (means-as-outcomes model: see Path B in Figure 2. The variable itself was predicted by the effects of listening to music (simple regression model: see Path C in Figure 2). Finally, the slope of the relationship between the effects of listening to music and the strength of 'Salamat' was predicted by past functional experiences (slopes as outcomes: see Path D in Figure 2. The effects of the Level 2 predictor for past functional experiences were controlled for other Level 2 variables that might affect the strength of 'Salamat' as favourite music: age and musical experience. Musical experience comprised two dummy coded variables: knowledge of the song (0 = no, 1 = yes), and familiarity with the artist(s) of the song (0 = no, 1 = yes). For each predictor variable, the amount of variance this variable is able to explain in the dependent variable ($\Delta\sigma^2$) was estimated by comparing the residual variance of the model without ($\sigma2\varepsilon0$) this predictor and the residual variance of the model with ($\sigma2\varepsilon$) this predictor: $\Delta\sigma2 = 1 - \sigma2\varepsilon\ \sigma2\varepsilon0$. Model fit was estimated by comparing the conditional model (containing all predictors and controls) with the unconditional model, resulting in a difference in deviance, which in HLM is inferentially tested using a χ^2 statistic. While there is a leverage on the preferences of each participant because of their musical inclination or activities in relation to music, this was clearly indicated by the strong correlation between the knowledge of the song and past experiences with the participants enjoyment of the music. Because the functionality of listening to music and its impact are described as a learning process that unfolds over time, it was predicted that higher values in

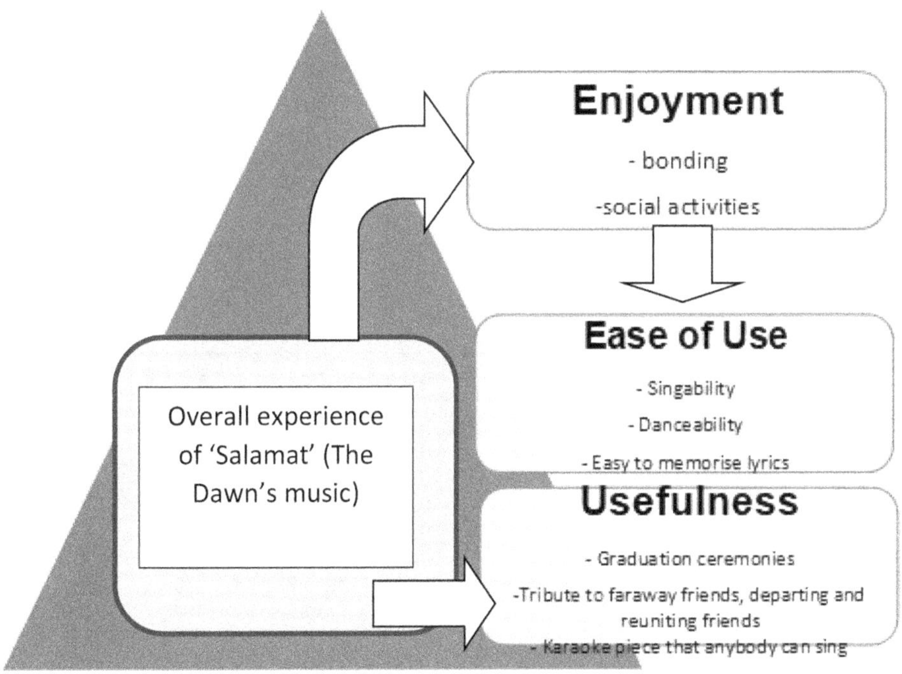

Figure 2. Intergenerational impact of The Dawn's 'Salamat'.

Table 2. HLM estimate – knowledge of the song 'Salamat' (enjoyment).

Level 2 predictors of the intercept ($\chi2$ = 3.8, p = .048).

Predictor	b	SE	t	p	df	Δσ2
Past functional musical experiences	0.23	0.05	4.6	< .001	100	14.8
Control: age	-0.01	0.01	-0.7	.47	100	
Control: sex	0.26	0.19	1.4	.17	100	
Control: singing	0.25	0.21	1.2	.24	100	
Control: instrument	-0.07	0.18	-0.4	.70	100	
Control: music studies	0.44	0.26	1.7	.09	100	
Control: musician	-0.26	0.26	-1.0	.30	100	

Note: the **b statistic** is the unstandardised regression coefficient. this information was considered to be demographic data and incidental to the research. $\Delta\sigma^2$ is the amount of variance in the criterion the predictor is able to explain. the χ^2 statistic is calculated from the difference between the deviance of the unconditional model and the deviance of the conditional model.

past functional experiences would result in higher values in the mean effects of music on the listener.

According to the HLM, there is a strong correlation between the impact of music preference for 'Salamat' and the effects of listening to music. Among the 100 participants that were used for this quantitative interpretation, there was a strong relationship between the effects of listening to music and the strength of the impact (see Table 3, Part B). That is, the strength of Musical Preference (MP) is a function of the music's potential to help the listener attain specific goals – enjoying or using music (Path C in Figure 2. This result ties in with previous

Table 3. HLM estimate – knowledge of the song 'Salamat' (ease of use of the music).

Part A: Level 2 predictors of the intercept ($\chi2$ = 37.1, p < .001).

Predictor	b	SE	t	p	df	Δσ2
Past functional musical experiences	0.42	0.04	9.4	< .001	100	49.1
Control: age	-0.01	0.01	-1.5	.14	100	
Control: sex	0.09	0.18	0.5	.63	100	
Control: singing	0.22	0.22	1.0	.32	100	
Control: instrument	0.01	0.10	0.1	.92	100	
Control: music studies	-0.16	0.23	-0.7	.49	100	
Control: musician	0.37	0.22	1.7	.10	100	

Part B: Level 1 predictor ($\chi2$ = 277.8, p < .001).

Predictor	b	SE	t	p	df	Δσ2
Effects of music listening	0.58	0.05	12.0	< .001	100	20.5
Past functional musical experiences	0.02	0.03	0.7	.51	100	0.0
Control: age	0.00	0.05	0.2	.81	100	
Control: sex	0.01	0.10	0.1	.92	100	
Control: singing	-0.10	0.10	-1.0	.30	100	
Control: instrument	0.23	0.10	2.4	.02	100	
Control: music studies	-0.08	0.11	-0.7	.49	100	
Control: musician	0.02	0.10	0.2	.84	100	

findings about the important role that the function of listening to music plays in the prediction of MP.

4 RECOMMENDATIONS

While the future will reveal the true understanding of the intergenerational impact of music on human society, and an interest in ethnomusicology remains the biggest part of this challenge, it is nonetheless reasonable now to accept this premise: that the intersection of ethnomusicology and language education is a point at which the means for understanding music, language and culture may be found. It is at this juncture that two dynamic fields and their considerable histories merge, and where new knowledge may be developed. From the pure to the practical, this crossroads of specialisations may be critical to future insights in each of these distinctive fields. The focus of this study is an alternative rock band that seemed to either consciously or unconsciously pave the way for a new beginning, starting from the dark days of artistic music history. Added efforts can be cultivated to support future claims about the impact of music in understanding culture and in communicating the challenges of polarity and division.

What lies ahead is a unifying insight into the musicology and discography of other music groups. Further interest in performance analysis and narrative regarding the classical impact of music on the evolving dynamics of ethnomusicology and the study of culture in general could be seen to be imminent. The most important achievement in this study is the provision of a critical reflection on music as an exponential element in the evolution of culture. The intergenerational gaps were bridged by a common preference for the entire musical experience. Future research should include an examination of several musical groups and explore the possibility of building a cohesive identity.

The uniqueness of ethnomusicology in presenting a literary analysis might allow future researches to understand the standard research models of music education, which have been fashioned from those in other fields, such as anthropology, education, psychology and sociology. As such, this might equip future researchers with some useful techniques and processes for the study of musical cultures. In addition, an examination of diverse methods may hold potential for new perspectives on questions of continuing ethnomusicological concern. It may even be that the quantification of data can be useful and relevant to understanding the behaviour of musicians, literary experts and audience members of a given repertoire, and this should not be dismissed in deference to the traditional means of research within the field. Further, while the interpretation of systems of musical culture for the sake of science is a noble goal, possibilities for the practical application of ethnomusicology research through collaboration with educators and literary scholars can now be realised in a more level playing field.

From this study, it was proven that there is great harmony between the generations. This was bridged by the anthem that links them. 'Salamat' may not be a solution, but it can be used as a clue to identify what common threads connect the basis of common identity. The output in this ethnomusicology study can be used to determine the pattern of connection among the different ages. Further studies are required to amplify the worth of a collective mindset about the values that are communicated through music.

The Dawn's music, specifically 'Salamat', is an enduring legacy that needs thorough understanding. The study needs to be followed up in relation to non-structural heritage preservation. The music is a cultural symbol of Filipino values because it provides a social context. Thus, beyond literature and language, it needs a deeper insight in order to be fully appreciated.

REFERENCES

Creswell, J. W. (2015). A concise introduction to mixed methods research. Los Angeles: Sage.

Diaz, S. (2012). Why were the Impressionists initially unpopular? *Owlcation*. Retrieved from https://owlcation.com/humanities/Why-were-the-Impressionists-Initially-Unpopular.

Hailstone, J.C., Omar, R. & Warren, J.D. (2009). Relatively preserved knowledge of music in semantic dementia. *Journal of Neurology, Neurosurgery and Psychiatry*, *80*(7), 808–809. doi:10.1136/jnnp.2008.153130

Henderson, S. & Hodges, S.H. (2006). Music, song, and the creation of community and community spirit by a gay subculture. *Sociological Spectrum*, *27*(1), 57–80.

Janata, P. (2013). Strong musical experiences and the brain. UC Davis, USA.

McNamara, L. & Ballard, M.E. (1999). Resting arousal, sensation seeking, and music preference. Genetic, Social, and General Psychology Monographs. 125: 229-250.

Nettl, B. (1989). *Blackfoot musical thought*. Kent, OH: Kent State University Press.

Nettl, B. (2002). *Encounters in ethnomusicology*. Warren, MI: Harmonie Park Press.

Peterkin, C. (2016). Musicology: A World of Possibilities.Retrieved from https://majoringinmusic.com/musicology on 13/09/2018.

Sieber, J.E. (1992). Planning ethically responsible research: A guide for students and internal review boards. Thousand Oaks: Sage Publications, Inc. Kvale, S. (1996). Interviews: An Introduction to Qualitative Research Interviewing, Thousand Oaks: Sage Publications.

Tolich M. (2004) Internal confidentiality: when confidentiality assurances fail relational informants. Qualitative Sociology 27(1): 101–106.

21st Century Innovation in Music Education – Astuti et al. (eds)

The existence of Hailee Steinfeld ('Haiz') fans in Indonesia: An academic perspective

J. Francoise
Indonesia Defense University, Bogor, Indonesia

ABSTRACT: This study offers a hypothesis about the moral reasoning behind the existence of Hailee Steinfeld ('Haiz') fans in Indonesia ("Haizers"/"Steinfans") and the challenges they may face in the future. This paper references Tourraine's theory of modernism (1992), Malhotra theory of self-conception (1988), and the culture of Indonesian people described by Kuntowijoyo (1985). Tourraine's (1992) concept of modernism is related to the behaviour of Haiz fans with regard to any updates from Hailee. The purpose of this paper is to show that, among several famous American pop stars and Korean stars being idolised by young Indonesian people, Hailee enjoys a special status among Indonesian fans, which began after Hailee performed music and was involved in the Untouchable tour with Meghan Trainor. Despite the culture of politeness among Indonesian people and prominent ideas of nationalism, young Indonesians still follow Western culture. In addition, Hollywood figures are still favoured over high-profile Asian figures, including Koreans.

1 INTRODUCTION

The name Hailee Steinfeld ('Haiz') is now very famous in Indonesia. She is the typical face of beautiful Asian-Americans and her performance at NET TV's Indonesian Choice Awards (ICA) 5.0 NET in April 2018 has successfully marked Haiz out as one of the American idols for young Indonesian millennials. The ICA tickets were sold out and people came from all corners of Indonesia, suggesting that ICA 5's centre of attention was largely Hailee Steinfeld.

From 2018 until 2019, young Indonesian people aged 17–35 have experienced a year dominated by politics, because the provincial elections began in June 2018. In 2019, Indonesians have a presidential election. The seriousness of this year is linked to their behaviour with regard to entertainment, because while young Indonesian people are at war over hoaxes or the "fake news" of politics, they are still searching for updates regarding their idol.

In front of the stage at the ICA 5.0 NET concert, there were more than 200 "Haizers"/"Steinfans", who stood for more than three hours just to get a closer look at their idol, Hailee Steinfeld. If they are asked about Hailee, they know everything, but also, if they are asked who would be the best politician to become their governor or next president, they have their own candidate. Politics and Hailee are equally important for young Haiz fans in Indonesia.

The fact that Hailee is never involved in politics, and that she never tweets or posts anything related to politics, makes young Indonesian people happy to have found a figure who is not political, even though it is known that Hailee was involved in the "I survive" videos with other artists after the Trump election and that Hailee eventually tweeted about the Women's March, an event that is broadly assumed to be a campaign against Trump's attitude to women.

Perhaps this is the reality of the millennial generation. Indonesia will be in a demographic bonus period until 2045; this means that during the next 50 years it will have more young people than old ones. This condition is very different to that of Japan, even though the

Japanese music industry has a greater degree of recognition than that of Indonesia. The effect of this demographic bonus on the music industry is that Indonesia is becoming one of the largest potential markets in Asia and many artists will perform in Indonesia to promote their music.

2 LITERATURE REVIEW

Related works for this research are basically associated with the analysis of Hollywood figures, using the theory of idolisation by Raviv et al. (1996). However, this is the first academic paper that has as its main focus the idolisation of Hailee Steinfeld among young Indonesians.

Zillmann's theory has been used as the basis for the study of how audience members engage and react to a wide range of dramatic and comedy-oriented entertainment content (cited in Holbert, 2005, p. 7). Raney and Bryant's (2009) discussion of the processes leading to the formation of media enjoyment focuses on the intersection between what an audience brings to a mediated communication experience and the mediated messages being supplied to that audience (Holbert, 2005, p. 7).

Self-conception has been defined in a number of ways. One of these definitions states that self-conception is 'the totality of the individual's thoughts and feelings having reference to themselves as subjects as well as objects' (Malhotra, in Abel et al., 2013, p. 2). Another definition relates to a "person's perception of his own abilities, limitations, appearance, and characteristics, including his own personality, and people's impression of the type of person they are" (Graeff, in Abel et al., 2013, p. 2). The image congruence hypothesis states that an individual will prefer products/brands that are perceived as having an image that is similar to the self-concept of the individual.

The idolisation of pop singers is based mostly on two important components: worship and modelling. Worship refers to an unusually intense admiration and reverence of an idol. This can be expressed in such behaviours as actively collecting information and artefacts related to the idol or trying to meet him/her personally. Modelling refers to the desire to be like an idol, which may involve imitation of the idolised figure by, for example, copying their dress, hairstyle, speech, activities and any other social behavioural patterns (Raviv et al., 1996, p. 632). Above is the table indicating how pop stars are idolised.

3 RESULTS AND DISCUSSION

Haiz fans in Indonesia were first recorded in the middle of 2015, with a group of high-school students creating Twitter accounts named @haizersindonesia and @haizindonesia. Then, in 2016, several Haiz social media fan bases began to appear, such as @haizersid or @haizers_id, and everyone said that "this is the official Indonesian fan base of Hailee Steinfeld". In the academic perspective of entertainment theory, there is no such recognition of "being official". What is the standard? Is the idol following that account? Based on the observations, none of the Indonesian fan base accounts are being followed by Hailee, so that recognition of "official" is a one-sided recognition.

The ICA 5.0 NET concert was an important event for Indonesian Haizers and Steinfans because, for the first time, they could gather together and take a closer look at Hailee. Some longstanding Haiz fans in Indonesia have created a Line Square group with the aim of collecting all of the Haiz fans together. To date there are two large Line Square groups of Haiz fans in Indonesia. One is being managed by high-school and university students and the other one is managed by more senior people. The first one is quite popular because they use social media to quickly distribute anything about Hailee.

The second one is not so popular because it looks like a more multiple fandom and in that Line Square it is not only Hailee who is being discussed, but also other artists, such as Lauren Jauregui, Camila Cabello, Taylor Swift, Shawn Mendes, Alessia Cara, Ariana Grande, and so on. Moreover, the seniors who handle that Line Square are working on weekdays, so that the group is only active after work or at weekends.

Haiz fans in Indonesia also have several WhatsApp (WA) groups. The first one began in late 2017, but then disappeared without explanation. A new WA group started at the beginning of 2018 that consisted not only of Indonesian fans, but also fans from Sri Lanka, India, Pakistan, UK, USA and Senegal. They confirmed that they knew about the WA group link from Facebook. There are a great many Facebook pages about Hailee Steinfeld. There are, perhaps, more than 300 accounts that have been created by fans from all around the world.

A Line Square or WA group created by Indonesian fans reflects the self-conceptions of the young Indonesian millennial generation. On the one hand, they are still religious, so they give greetings such as "*Assalamualaikum*" every time that they want to post something about Hailee and say "*Walaikumsalam*" to welcome new members, but at the same time they make comments about Hailee's "sexy" outfits at the Biggest Weekend or on Katy Perry's Witness tour.

Based on Zillmann's theory, it is true that the young Indonesian millennial generation are still religious, but at the same time are going through the normal stages of being young. This means that they have idols and they make comments on what their idols wear, how they behave, what they say, and so on. This is academically normal.

Kuntowijoyo (1985) said that Indonesian people exhibit the behaviour of *aji mumpung*. This special personality cannot be easily translated into English, but *aji mumpung* here means that Indonesians love to have other benefits allied to what they are doing, for example visiting a museum because it is free or because the museum gives souvenirs.

In the context of Hailee, the tickets for ICA 5.0 NET were free of charge, so this type of *aji mumpung* personality was invoked naturally. No wonder that some fans were arguing or even competing to get ICA 5.0 NET tickets via quizzes and games. Everybody wanted to see Hailee for free.

Hailee fans in Indonesia are having trouble getting Haiz merchandise, because it cannot be purchased from www.haileesteinfeldofficial.com and shipped to Indonesia. Therefore, what happened at the ICA 5.0 NET concert was that everybody was sporting unofficial Haiz merchandise, or they had created their own version of Haiz merchandise through their own desire to design.

Based on the idolising table of Raviv (Figure 1), Hailee fans in Indonesia are idolising Hailee in very small ways. They enjoy Hailee's video performances using free WiFi, but they never have original Haiz merch, never go to a Haiz concert (except for the free one at ICA 5.0 NET) and rarely stream or buy Hailee's songs on Spotify, even though Spotify now has some discounts for Indonesian users at only Rupiah 49.000 (USD 3.5) per month. Perhaps Spotify has noticed that Indonesian people love discounted rates.

Based on the self-concept theory, Hailee is more famous as a singer in Indonesia than as a Hollywood actress, although established Haiz fans are already educating other Haiz fans via social media that Hailee first found fame from movies, not from music. However, Indonesian fans are still more familiar with Hailee's songs than her movie title list.

Hailee's lyrics are also related to modernism, for example 'Love Myself' and 'Most Girls'. Girls make up 70 per cent of Indonesian Haiz fans, but do they ever realise what message Hailee is conveying with her lyrics? In Indonesia, girls are still undereducated, they are still forced to marry before they are 20 years old, and there are still some gender-based problems. Indonesia is still struggling to be a modern country and Indonesian girls cannot apply the idea of 'Love Myself' in an easy way.

Based on the concept of modernism of Tourraine (1992), democracy must be in line with justice or human rights. Indonesia cannot be called a modern nation, but young Indonesian people are absorbing more external culture, for example young Indonesians are more familiar with Hollywood or Korean movies than with Indonesian movies. To counteract this, the Indonesian Ministry of Culture has been developing a programme of financing local movies that include the content/story of Indonesian local wisdom and heritage.

With regard to Hailee fans in Indonesia, the overall fan base consists of more than 500.000 young Indonesian people, not including Instagram accounts, and this has the level

Table 1.Items of Expression of Idolization.Causes for idolization, and Reliance on Idols. Arranged by the Factors Obtained for Each Part

Expression of Idolization
I. Singer Imitation
(3) Adoption of singer's style of dressing
(4) Adoption of singer's style hair style
(5) Adoption of singer's opinions
(6) Adoption of singer's mode of speech
(7) Adoption of singer's behavior
(19) Attempts to resemble the singer
II. Worship
(1) Hanging singer's posters
(11) Buying souvenirs related to the singer
(12) Buying printed material related to the singer
(13) Searching for information about the singer n magazines and newspapers
(14) Getting in touch with others fans of the singer
(15) Collecting personal details about the singer
III. Knowledge and Consumerism
(2) A number of singer's records and cassettes purchased
(8) Listening to singer's music
(9) Attending singer's performances
(10) Watching the singer on TV or VCR
(16) Getting to know the melodies of the singer's songs
(17) Getting to know the lyrics of the singer's songs.
Causes for idolization
I. Singer's Features
(4)Singer's voice
(5) Singer's stage presentation
(6) Singer's overall appearance
(7) Singer's mode of dressing
II. Singer' Evaluation
(9) Subject knowledge of the singer's biography
(10) Friends' influence
(11) Singer's personality
(13) Singer's reputation among youth
(14) Singer's uniqueness compared to other singers
(15) Singer's uniqueness compared to other people
III. Singer's Ways to Expression
(1) Singer's lyrics
(2) Singer's melodies
(3) Singer's opinion
(8) Singer's resemblance to the respondent
Reliance on Singer
I. Personal Matters
(3)Pastime
(4)Future planning
(5)Physical appearance
(6)Social relation
(8)Personal Feelings
(10)General reliance
II. Overall Knowledge and Values
(1) School studies
(2) Polities
(7) Values
(8) Sciences

Figure 1. Items of idolisation (Raviv et al., 1996, p. 637).

of potency to develop Indonesia in a very good way into a modern nation. Perhaps America will still be the main superpower over the next 50 years, but Indonesia could be the next emerging rich country during this time because of the demographic bonus that is happening in Indonesia now. Young people have a significant impact on society. As Raviv (1996) stated ever that idolising pop star depends on the young people's behaviour. This means that the future of Western idols also depends on the behaviour of young people in Indonesia, as young Indonesian people are easily connected to what is happening in the Western world.

Hailee's fan base could have expired within the next 10 years, as Hailee gets older, marries, and is replaced by new idols, but the relationship between members of her fan base is long-lived because they have a special connection. For example, Indonesian diaspora could be friends with Mexican diaspora in Toronto, just because they have the same idol: Hailee Steinfeld. So, in this context, Hailee Steinfeld is also a figure of modernism, as she unintentionally directs people from different backgrounds, beliefs, religions, nationalities and ages to unite due to their shared desire to idolise Hailee.

It is optimistically assumed that the positive aspects of modernist behaviour will happen for all Haiz fans in Indonesia, but again there have been several challenges, including 2018 and 2019 being critical political years, and it is becoming natural on Indonesian social media accounts to direct hate speech at other users, just to support one candidate. Haiz fans in Indonesia have no special rule not to discuss politics, because politics is now important for Indonesians, but at the same time Haiz fans in Indonesia are also ready to leave negative comments or fight with other members if they support different political candidates. This is millennial behaviour; if they do not like someone in the social media group, they just leave the group.

While waiting for Hailee's first album, Haiz fans in Indonesia were ready to make several comments on politics, because there was nothing to discuss regarding Hailee. ICA 5.0 NET is no longer a top subject of conversation; Hailee's private life and gossip has already been published by many fans on Instagram and there are no new songs from Hailee. Therefore, Haiz fans in Indonesia will be ready to discuss things other than Hailee and the timing means that many people are just interested in politics until the Indonesian presidential election in 2019.

4 CONCLUSION

The Haiz fan base in Indonesia began with high-school students with good access to the Internet, group know more about Hailee's music than her movies. Even though Hailee is loved by many Indonesians, it is not the culture of Indonesians to buy concert tickets or merchandise related to Hailee. Indonesian fans love to get everything free without any expense, which is why they loved ICA 5.0 NET so much, because they could see Hailee for free.

Indonesians now also provide a market for Korean artists, but Western culture frequently substitutes for Indonesian culture, because Indonesians have recognised Western movies and figures since the colonisation period. Based on the idolising concept of Raviv et al. (1996), Haiz fans in Indonesia are idolising Hailee as a Western figure and appreciated it when Hailee wore a demure uniform during her performance at ICA 5.0 NET, far less provocative than in her supporting role for Katy Perry's Witness tour in the UK and Portugal.

Haiz fans in Indonesia are a peaceful group, although they have to be ready to be divided in the next two years as the political years are coming and there are no new updates from Hailee; there is still no album, and only the *Bumblebee* movie in December. With regard to *Spider Gwen*, perhaps Indonesians are not so interested in accessing it because it is not free on Netflix. Although Emily Dickinson starred by Hailee series broadcasted, it is only for segmented users, in Apple. Not many Indonesians use Apple.

Haiz fans in Indonesia are still supporting Hailee with their unique Indonesian personality and culture, and whenever Hailee returns to Indonesia, Indonesians will still be struggling to be one of the richest countries in the world. Currently, injustice, unemployment, a poor literacy rate and poverty are still critical factors, and loving Hailee is just diverting young

Indonesians' minds from being the next generation of Indonesian leaders, able to lead Indonesia amid an uncertain future.

ACKNOWLEDGEMENT

Thank you to Hailee Steinfeld who inspires me to write this paper.

REFERENCES

Abel, J.I., Buff, C.L. & O'Neill, J.C. (2013). Actual self-concept versus ideal self-concept. *Sports, Business, and Management: An International Journal*, *3*(1), 24-30.

Harvey, C.W. (2005). Epoche, entertainment, and ethics: On the hyperreality of everyday life. *Ethics and Information Technology*. *6*(4), 261-269.

Holbert, R.L. (2005). A typology for the study of entertainment television and politics. *The American Behavioral Scientist*, *49*(3), 125-132.

Kelley, R.P. (2016). *Connecting audience motivations and public relations effectiveness of an organization working with the entertainment industry* (Master's thesis, San Diego State University, CA, USA).

Kuntowijoyo. (1985). *Indonesian human culture*. Jakarta, Indonesia: Pustaka Jaya.

Malhotra, N. K. (1988). Selfconcept and product choice: an integrated perspective. *Journal of Economic Psychology*, 9(1),1-28.

Raney, A.A. & Bryant, J. (2009). *Handbook of sports and media*. New York, NY: Routledge.

Raviv, A., Bar-Tal, D., Raviv, A. & Ben-Horin, A. (1996). Adolescent idolization of pop singers: Causes, expressions, reliance. *Journal of Youth and Adolescence*, *25*(5), 631–650.

Tourraine, A. (1992). *Critique de la modernité* [Criticism of modernity]. Paris: Librarie Arthème Fayard.

21st Century Innovation in Music Education – Astuti et al. (eds)
© 2020 Taylor & Francis Group, London, ISBN 978-0-367-11122-9

The pervasiveness of K-pop in 21st century Indonesia

P.V. Langit, K.S. Astuti, A. Yuliyanti & R.M. Nasution
Universitas Negeri Yogyakarta, Daerah Istimewa Yogyakarta, Indonesia

ABSTRACT: Pop music has hit the top of the global charts nowadays. Korean pop music (K-pop) is widespread around the world, including Indonesia. How much K-pop has pervaded the Indonesian youth of today is the research question addressed in this study. The approach used is a quantitative one using a survey design. Data was collected with questionnaires, interviews and observations. Data analysis used descriptive statistics with mean values, and inferential statistics using chi-squared tests. The results of the study show that: 1) most Indonesian teenagers are familiar with K-pop; 2) the teenagers who like K-pop are predominantly female; 3) there is no difference in knowledge about K-pop between male and female fans; 4) K-pop is not the only music favoured by Indonesian teenagers.

1 INTRODUCTION

1.1 *Background of the study*

In the era of globalisation, with technology that has become increasingly sophisticated, it is very easy to exchange information from one region to another and even to other countries. David et al. (1999) suggested that globalisation may be thought of, initially, as the widening, deepening and speeding up of worldwide interconnectedness in all aspects. Besides information, advances in technology can also spread culture from one country to another. For example, music that is a trend in one country can also be heard in other countries. Astuti (2003, p. 27) revealed that one of the consequences of globalisation is the transformation of culture, which flows more rapidly.

Korean pop music (K-pop) has hit the top of the global chart nowadays. The beginning of the 21st century can be said to be the rising age of K-pop. The truth is that world music was shocked by the boom in K-pop. This phenomenon can be said to be a monumental event, considering that in previous times there had never been Asian music so loved by teenagers around the world; it was Western music that always dominated.

K-pop is not only favoured by Asian teens but also by teenagers in almost all over the world. Tuk (2012) said that K-pop YouTube videos have been watched up to almost two billion times by viewers from more than 21 countries. The number of views of K-pop videos on YouTube as of January 2, 2011 was: Germany 20,114,996; UK 22,705,547; France 26,591,412; Saudi Arabia 42,719,685; Thailand 224,813,564; Malaysia 98,693,969; Singapore 68,546,360; Vietnam 117,358,800; Indonesia 81,128,637; Australia 27,132,121; Canada 46,477,940; US 240,748,112; Japan 423,683,759. The data shows that K-pop is currently the most popular music in the world.

Around the world, pop consumers are increasingly accessing popular products such as music, film, television, and other audiovisual media content through online social media (Jung, 2010). According to Choi (2011), "The recent news of Korean boy and girl bands making headlines in Eastern and Southern Asia, Europe and the US are sending new hopes and energy through the K-pop industry following the surge of pride when the Korean Wave first hit Japan and China in the late 1990s and early 2000s". K-pop's popularity and its participatory fan culture have expanded beyond Asia and become significant in Europe in the past few years (Oh, 2013). Based on Ju and Lee (2015), in the USA, Korean TV dramas have been provided by online streaming services such as Hulu, DramaFever and Netflix. YouTube is another source for Korean TV dramas, films and, in particular, K-pop.

The popularity of K-pop was confirmed by the achievement of awards by South Korean artists. One of them, the South Korean boy band BTS, has won prestigious awards and topped two categories on the Billboard chart, the number one album on the Billboard 200 chart, and number one artist on a 2018 Billboard 100 artist chart. This achievement is a result of the large number of fans spread all over the world. It proves that many teenagers beyond South Korea also like K-pop. The transnational fandom of K-pop indicates the increasing significance of cultural translation in global media consumption. The overseas flow of K-pop has been identified as an intra-Asian phenomenon and more recently as a global process (Yoon, 2017); one country affected is Indonesia.

Many Indonesian teenagers prove that they like K-pop by their attendance at various events related to K-pop. According to Jung (2011), "With youth fan groups as the backbone of the phenomenon, K-pop, predominantly the product of idol girl and boy band music, has also gained recognition in Indonesia. In 2010, over 120 fan-operated K-pop-related events were held, including fan gatherings and Korean pop festivals and concerts". More and more teenagers in Indonesia like K-pop, causing Indonesian promoters to invite boy bands and girl groups from South Korea. Recently, in September 2018, one South Korean boy band, BTOB, gave a concert in Jakarta (Puspa, 2018). Of course, at the concert there were many sellers of K-pop merchandise ('K-stuff'), because there are many fans who will buy it. The interest K-poppers have towards K-stuff causes them to become very acquisitive. Besides this, being a K-pop fan can satisfy the mind, and also encourages K-poppers to start their own business such as opening a store that sells many kinds of K-pop souvenirs/merchandise/albums. One K-pop fan from Indonesia has been successful because her business sells K-pop items. Thus, Orchidia Cindy, known as Cindy Noona, started from a liking for EXO (a K-pop boy band), and now has become the owner of an online shop, attending dozens of EXO concerts.

The popularity of K-pop is increasing with the opening of the Star Museum global auditions and others. The opportunity for teenagers to become Korean artists makes K-pop increasingly a part of their lives because the teens feel "embraced" by K-pop.

Many K-pop fans become fanatical, and many even become hysterical when they see their idol artists in person. Whatever is done and worn by K-pop artists becomes the centre of conversation and establishes a trend.

From the perspective of politics and economics, this phenomenon is very beneficial for South Korea; with the fanaticism of the world's youth towards K-pop it is easier for South Korea to market its products and for South Korean culture to influence the lifestyle of teenagers elsewhere. The spread of K-pop music and the impact of K-pop culture on teenagers in Indonesia will be the focus of this research.

1.2 *Research objectives*

From the outline above, we can see that fanaticism towards K-pop has an impact on the lives of teenagers. Many of them have acquisitive lifestyles in order to buy K-pop merchandise but this can also have a positive side, as seen from their productivity in working independently and even creating new jobs. The aim of the research was to understand the extent of the pervasiveness of K-pop in 21st century Indonesia and the reasons why Indonesian teenagers like K-pop.

2 LITERATURE REVIEW

2.1 *Korean pop*

Korean culture has been spreading since the 1990s and the big boom of the Korean drama brought a cultural and social phenomenon called the Korean Wave, known as *Hallyu*. K-pop is modern music from Korea. Lie (2011) explained that at the beginning of the 20th century,

Korean people were not familiar with Western music. Until the 1970s, Koreans did not have music with diatonic scales but used the pentatonic scale. However, in the 1970s diatonic Western music began to become known to the Korean people. Western musicians influence Korean music, so for the Korean community diatonic scales are familiar. Korean music is even able to meet the world's musical tastes, especially those of the United States, such that Korean music is able to dominate the world. Although there is also J-pop (Japanese pop), K-pop is more able to satisfy the tastes of today's generation so that K-pop is more popular.

Raden (2014) mentioned that Korean society has already been successful in embedding the importance of music in daily life. As a result, music consumption in Korea is very high in terms of traditional, pop, rock, jazz, or classical music. Modern K-pop appeared in March 1992, when Seo Taiji and Boys released their first album; the first song was 'I Know' (Culture, 2013, p. 63), followed by many new producers and groups between 1994 and 1997 that caused K-pop to become more developed.

2.2 *The presence of K-pop music*

The pervasiveness of K-pop music is a reflection of the presence and acceptability of K-pop in a society. Presence can be seen in quantitative and qualitative terms. Quantitatively, presence can be seen in numbers, while qualitatively it can be viewed in terms of depth. The presence of K-pop in society can be interpreted as the number of fans and the number of K-pop events that exist; how far the fans feel the pervasiveness of K-pop can be seen from how deeply K-pop becomes part of life.

2.3 *Interference of the Korean government*

It can be said that the success of the dissemination and development of Korean music cannot be separated from the influence of the state. The Korean government is facilitating the boom in Korean music, even using K-pop as an agent, both in cultural and product terms. One of the leading Korean cosmetics companies achieved a 44% increase in its overseas sales, and Korean games accounted for $3 billion in Korean exports for the year 2014 (Jung-a, 2016).

The Korean government's support for K-pop plays a big role in the spread of Korean culture and products. This is because programmes related to the arts sector become more formal, so transactions become very obvious. This is something that was not done by the governments of other countries. Lee (2011) reported that the Korean government played an active role by aiding the Korean media in publicising Korean popular culture around the world; Medina (2017) found that K-pop is a soft diplomatic tool.

As part of the arts, K-pop is a very effective tool of diplomacy. The beauty displayed in the art of Korea by both artists and their artworks is a very effective tool of ingratiation, because if someone loves something more, then they will ignore other things. That is what happens to Indonesian teenagers.

This intelligence on the part of the Korean government in using art as a means of diplomacy is something that is not done by any other country. In general, the state does not interfere in the affairs of the arts. Indeed, many arts are becoming marginalized because the arts, including music, are not considered important. The marginalization of music compared to other fields is evidenced by the relatively small number of hours of music lessons.

Art teachers in Indonesia also complain that if there are meetings or community service events, they often use time intended for art lessons. Another frequent example is that art subjects are not taught by art teachers but by teachers from other fields who are considered able to play music.

This is not only happening in Indonesia; according to one teacher in The Netherlands, in the beginning the appreciation of parent guardians in their schools for music was also lacking, so the school did not have adequate musical equipment. But after music teachers were able to show their success by students giving musical performances, the views of parents of students changed and parents voluntarily buy musical instruments for the school (Astuti et al., 2012).

The seriousness of the Korean government in supporting K-pop has succeeded in making Korea a country whose culture is known in other countries. In Indonesia, many teenagers feel proud if they use Korean products and adopt Korean styles, in terms of hairstyles, clothes and favourite foods.

2.4 *The inclusion of K-pop culture in Indonesia*

The popularity of Asian art for Indonesian teenagers begins with the success of teenage soap operas from China, Taiwan and Korea. Soap operas such as *Meteor Garden*, *Princess Hours Full House*, *Dae Jang Geum*, and *My Love from The Star* are opening up insights for Indonesian teens in relation to the culture and history of these countries. This is because, for example, Korean culture is close to Indonesian culture, which is from East Asia. This is in accordance with the opinion of Ryoo (Lee, 2011), who said that "South Korean popular culture is much more readily relatable and acceptable to Asian audiences".

For Indonesian teenagers, Korean culture had previously been a "mystery", so that when they could see Korean culture through these dramas, it was felt to be something interesting because it answered their curiosity. The "smoother" politeness, ethics, and practices of Korean culture displayed in these dramas are very interesting to Indonesian teenagers.

If in the 2000s, Indonesian teenagers were excited by soap opera figures from China, Taiwan and especially Korea, in the decade of the 2010s Indonesian teenagers greatly admired K-pop artists such as Super Junior, Big Bang, Girls' Generation, and 2NE1, who were K-pop pioneers. The K-pop boom is still felt today. BTS, EXO, Blackpink, and Twice are K-pop groups that are very popular with Indonesian teenagers in the second decade of the 21st century.

Things that happened in Indonesia apparently also occurred in neighbouring Asian countries closer to Korea, such as Japan, Vietnam and China. This is implied in Lie's (2011) statement that K-pop from South Korea first spread to neighbouring Asian countries such as Japan and Taiwan, and from there spread to America and the Middle East. The same thing was stated by Bok-rae (2015), who described how *Hallyu* (the phenomenon of Korean popular culture) is very popular among young people in China, Taiwan, Hong Kong and Vietnam.

2.5 *Characteristics of adolescents*

Teendom encompasses puberty and transitioning both physically and mentally. Physically, there are very large changes such as height, body shape, voice pitch, and the arrival of menstruation, often causing lack of confidence. Mentally, teenagers need to establish their identity. This causes teenagers to look for idols that can be used as a guide and role model. Most adolescents like to follow the 'peer group culture'. This group adopts a particular way of talking, walking and behaving that can seem very strange to adults. In general, adolescents tend to behave according to their feelings, bypassing analytical thought.

It can be said that adolescence is a critical period because it involves the search for identity. At that age, teenagers tend to judge things on the basis of appearance, and will readily imitate things without appreciating their implicit values.

The K-pop culture that is so widespread among teenagers can have both positive and negative impacts. The positive impact is that K-pop artists can inspire teenagers to achieve, and be creative. But if teenage fanaticism for K-pop goes beyond certain limits, with most time and money spent on K-pop and other more useful things being neglected, K-pop can have a negative impact on teens.

2.6 *Factors that influence adolescent attitudes*

Broadly speaking, there are two types of factor that influence attitudes: intrinsic factors and extrinsic factors. Thus, an intrinsic factor is a factor of the adolescents themselves, such as interest and motivation. Extrinsic factors are factors outside of adolescents, such as the environment, including family, school and society.

K-pop is an extrinsic factor for teens that can affect adolescent attitudes. If an extrinsic factor such as K-pop is stronger than pre-existing intrinsic factors, then K-pop culture will be reflected in the attitudes and behaviours of Indonesian youth.

2.7 *The influence of K-pop on Indonesian youth culture*

Geertz (1973, p. 89) sees culture as "a system of conceptions inherited [from previous generations] and expressed in symbolic form; with the help of human culture communicating, perpetuating". While according to Lévi-Strauss (quoted in Ortner, 1994, p. 387), cultural pluralism is realised not because of the isolation of social groups, but precisely because of the continuous contact between these groups. In addressing this in the era of globalisation, Alam (2014) stated that the process of "globalisation and cultural change" does not need to be faced with extreme self-closing attitudes. On the contrary, by understanding how culture is constructed through discourse and practice, for example, we can also use the process of globalisation as a means to enrich the diversity of our cultures. Habieb (2017) shows that K-pop influences the lives of adolescents between 17 and 21 years of age, and he recommends that teenage Indonesia should be able to filter the culture of good and bad; to develop and promote a culture of Indonesia itself.

This is very reasonable because not all things in K-pop are in accordance with Indonesian culture. For example, the culture of drinking alcoholic beverages together, which is common in K-pop, is something that is prohibited in Indonesia because the majority of Indonesians are Muslims.

2.8 *Framework of thinking*

The age of adolescence is the age at which teenagers are still in the process of establishing their identity and is an unstable period because teenagers do not yet have a strong guide for judging and deciding something. The presence of idol figures in their lives provides an answer for them, being easy to imitate whatever the idol characters do in terms of lifestyle and behaviour, regardless of economic or physical circumstances, or conformity with their personality or the values of their social society.

In general, adolescents will imitate idol figures who are of the same age as themselves. K-pop is often performed by artists of similar age to teenagers, so that teens will be more interested and like K-pop artists. Moreover, the promotion of K-pop is so intense that it "constrains other idol figures including Indonesian artists themselves". It can be said that K-pop has a strong pervasiveness in Indonesia, because the K-pop culture arrived in force, being different from popular music in Indonesia, which is not managed jointly between private and government institutions as is the case in Korea. Thus, it can be said that the close working of Korean arts people has allowed K-pop to successfully pervade the Indonesian music scene. This is in accordance with the theory of Mursell (2007), who said that working together will produce accuracy and speed.

Accuracy will make a work more perfect, while speed will make a work the most timely. Perfection and timing are the keys to the success of selling a product.

The products that incessantly enter Indonesia have caused Indonesian teenagers to love K-pop. Given the age and immaturity of adolescents, their reason to idolise K-pop could be driven by extrinsic factors that only consider appearance, and not for deeper reasons such as ideology.

2.9 *Hypotheses*

Based on the description, the following hypotheses are proposed:

1. K-pop is a genre of music that being appreciated by Indonesian teenagers.
2. The reason Indonesian teenagers like K-pop is based on appearance, not on ideological grounds.

2.10 *Relevant research*

Research relevant to this study includes that of Adnan Habieb on "The influence of K-pop in Indonesia's students behaviour", which found that Korean drama has an influence on the lifestyle of Indonesian youth. Our study, however, focuses on the influence of Korean music.

3 METHODOLOGY

This research is survey-based research with data collection techniques involving a questionnaire, observation, and internet surfing. The questionnaire was used to reveal the pervasiveness of K-pop in Indonesian adolescents, which included their knowledge of K-pop, how long they had known it, their frequency of listening, and how much they spent on it. Observation techniques were used to determine the behaviour of Indonesian adolescents in terms of lifestyle, and interviews were used to explore the rationale of teenagers for liking K-pop. The respondents sampled were 65 adolescents between 17 and 23 years of age. Data analysis techniques involved descriptive and inferential statistics. The descriptive statistics used were the maximum mean values, while inferential statistics used chi-squared hypothesis tests.

4 FINDINGS AND DISCUSSION

The following describes the results of the research and discussion. The data revealed by respondents included gender, age, knowledge of K-pop, how they knew K-pop, indicators of their affection for K-pop, the frequency of listening to K-pop, and whether they idolised artists other than K-pop. Table 1 shows the populations involved in each part of our survey.

Explanations of more detailed analysis of the data can be found in the following subsections.

4.1 *Gender relationship to K-pop*

Based on Table 2, it can be seen that Indonesian K-pop fans are predominantly (92.3%) female.

This is confirmed by the chi-squared test presented in Table 3.

Table 1. K-pop survey population.

Case Processing Summary

		Cases					
		Valid		Missing		Total	
	Gender	N	Percent	N	Percent	N	Percent
Knowing K-Pop	Male	5	100.0%	0	0.0%	5	100.0%
	Female	60	100.0%	0	0.0%	60	100.0%
Source K-Pop	Male	5	100.0%	0	0.0%	5	100.0%
	Female	60	100.0%	0	0.0%	60	100.0%
Frequency of Watching K-Pop	Male	5	100.0%	0	0.0%	5	100.0%
	Female	60	100.0%	0	0.0%	60	100.0%
Love towards K-Pop	Male	5	100.0%	0	0.0%	5	100.0%
	Female	60	100.0%	0	0.0%	60	100.0%
Budget for K-Pop	Male	5	100.0%	0	0.0%	5	100.0%
	Female	60	100.0%	0	0.0%	60	100.0%
When start loving K-Pop	Male	5	100.0%	0	0.0%	5	100.0%
	Female	60	100.0%	0	0.0%	60	100.0%
OTHER FAVORITE	Male	5	100.0%	0	0.0%	5	100.0%
	Female	60	100.0%	0	0.0%	60	100.0%

Table 2. Gender of Indonesian K-pop fans.

gender					
		Frequency	Percent	Valid Percent	Cumulative Precent
Valid	Male	5	7.7	7.7	7.7
	Female	60	92.3	92.3	100.0
	Total	65	100.0	100.0	

Table 3. Gender significance of Indonesian K-pop fans.

gender			
	Observed N	Expected N	Residual
Male	5	21.7	-16.7
Female	60	43.3	16.7
Total	65		

In Table 3, it can be seen that there are five male fans of K-pop, while 60 are female. It can also be seen that the expectation of the number of men is 21.7, but there are just 5, which is lower by 16.7.

Based on the results of statistical tests (see Table 4), the significance is 0.00. This means the number of K-pop fans among males and females is significantly different. Thus, it can be said that K-pop fans in Indonesia are predominantly female.

The next step is to test chi-squared independence. First, we check whether the distribution violates the related chi-squared assumption of "minimum expected cell frequency, which is 5 or greater 80% of cells expected frequencies of more" (Pallant & Manual, 2007, p. 216)

Based on the chi-squared analysis in Table 6, it can be seen that it shows less than 5, so that the distribution does not violate the assumption (more than 0.08). Thus the analysis can be continued by looking at the Pearson chi-squared value. Because we have 2 × 2 tables, we obtain Table 7.

The continuity correction value is 2.560 which associated with 0.110 level of asymptotic significance 2-sided; thus not significant, so there is no relationship between gender and knowledge about K-pop. Even though there are many more female K-pop fans than male ones, among men and women who like K-pop there are not significantly different levels of knowledge about K-pop.

Table 4. Statistical tests results.

Test Statistics		
	Gender	Knowing K-Pop
Chi-square	19.231[a]	124.069[a]
df	1	1
Asymp. Sig.	.000	.000

a. 0 cells (0.0%) have expected frequencies less than 5.
The minimum expected cell fequency is 21.7.

Table 5. Crosstab results.

Crosstab

			Love towards K-Pop				
			Streaming video	Merchandise	Watching concert	Others	Total
Gender	Male	Count	5	0	0	0	5
		% within gender	100.0%	0.0%	0.0%	0.0%	100.0%
		% within love towards k-Pop	13.2%	0.%	0.0%	0.0%	7.7%
		% of Total	7.7%	0.0%	0.0%	0.0%	7.7%
	Female	Count	33	24	2	1	60
		% within gender	55.0%	40.0%	3.3%	1.7%	100.0%
		% within love towards k-Pop	86.8%	100.0%	100.0%	100.0%	92.3%
		% of Total	50.8%	36.9%	3.1%	1.5%	92.3%
Total		Count	38	24	2	1	65
		% within gender	58.5%	36.9%	3.1%	1.5%	100.0%
		% within love towards k-Pop	100.0%	100.0%	100.0%	100.0%	100.0%
		% of Total	58.5%	36.9%	3.1%	1.5%	100.0%

Table 6. Chi-Square tests 1 results.

Chi-Square Tests

	Value	df	Asymp. Sig. (2-sided)
Pearson Chi-Square	3.849[a]	3	.278
Likelihood Ratio	5.662	3	.129
Linear-by-Linear Association	3.008	1	.083
N of Valid Cases	65		

a. 6 cells (75.0%) have expected count less than 5. The minimum expected count is .08.

Table 7. Chi-Square tests 1 results.

Chi-Square Tests

	Value	df	Asymp. Sig. (2-sided)	Exact Sig (2-sided)	Exact Sig. (1-sided)
Pearson Chi-Square	12.188[a]	1	.000		
Continuity Correction[b]	2.560	1	.110		
Likelihood Ratio	5.329	1	.021		
Fisher's Exact Test				.077	.077
Linear-by-Linear Association	12.000	1	.001		
N of valid Cases	65				

a. 3 cells(75.0%) have expected count less than 5. The minimum expected count is .0.8.
b. Computed only for a 2*2 table

4.2 *The pervasiveness of K-pop in Indonesia*

The pervasiveness of K-pop in Indonesia is tested via chi-squared to establish whether Indonesian teenagers only like K-pop or whether they also like other music and/or idol artists. Statistical data is shown in Table 8.

Table 8. Desriptive statistics results.

Descriptive Statistics					
	N	Minimum	Maximum	Mean	Std. Deviation
OTHER FAVORITE	65	1.00	3.00	1.2615	.50858
Valid N (listwise)	65				

Based on these data it can be seen that the mean of 1.26 (minimum value of 1 and maximum of 3) means that it is below the average. Thus it can be said that K-pop artists are not the only idols for Indonesian teenagers.

Based on the open question about the reason why teenagers enjoy K-pop, all respondents answered that the reason for watching K-pop was because it is cool and entertaining.

5 CONCLUSIONS AND RECOMMENDATIONS

Based on the results of the study showing that K-pop succeeded in pervading Indonesia, it was shown that a majority of Indonesian teenagers were familiar with K-pop. In general, K-pop fans are women between 17 and 23 years old. However, Indonesian teenagers not only like K-pop, but also like other types of music and artists. The types of music that Indonesian teenagers also like needs to be explored in future studies. The second conclusion is that Indonesian teenagers like K-pop largely because of its appearance, rather than on ideological grounds.

REFERENCES

Alam, B. (2014). *Globalisasi dan perubahan budaya: perspektif teori kebudayaan* [Globalization and cultural change: Cultural theory perspective]. *Antropologi Indonesia* [Anthropology of Indonesia].54, 1-11.

Astuti, K.S. (2003). *Pembelajaran Ansambel Musik Mempersiapkan Anak Didik Memasuki Masyarakat Multikultural* [Ansambel Music Learning in preparing students to anticipate multicultural community]. *Cakrawala Pendidikan* [The Horizon of Education], *2*, 179-187.

Astuti, K.S., Zuchdi, D., Sudiyono, Wubbles, T. & Hoogeveen, K. (2012). Developing model for teaching and learning music in public schools based on comparative study between Indonesia and The Netherlands. *Laporan Penelitian* [Research Report]. Retrieved from https://core.ac.uk/display/33527042

Bok-rae, K. (2015). Past, present and future of Hallyu (Korean Wave). *American International Journal of Contemporary Research*, *5*(5), 154–160.

Choi, Y.-J. (2011). The globalization of K-pop: Is K-pop losing its Korean-ness? *Situations*, *5*, 69–75.

Culture, K. (2013). Information Service. (2011). *The Korean Wave: A new pop culture phenomenon.*

David, H., McGrew, A., Goldblatt, D. & Perraton, J. (1999). *Global transformation.* Stanford, CA: Stanford Press.

Geertz, C. (1973). *The interpretation of cultures.* New York, NY: Basic Books.

Habieb, A. (2017). The influence of K-pop in Indonesia's students behaviour. *International Journal of Management and Applied Science*, *3*(3), 124-132.

Ju, H. & Lee, S. (2015). The Korean Wave and Asian Americans: The ethnic meanings of transnational Korean pop culture in the USA. *Continuum: Journal of Media & Cultural Studies*, *29*(3), 323–338.

Jung-a, S. (2016, April 12). China awash with Korean wave fever. Financial Times. Retrieved from https://www.ft.com/content/167338ec-fa0b-11e5-8e04-8600cef2ca75

Jung, S. (2010). *Korean masculinities and transcultural consumption: Yonsama, Rain, Oldboy, K-pop idols*. Hong Kong: Hong Kong University Press.

Jung, S. (2011). K-pop, Indonesian fandom, and social media. *Transformative Works and Cultures*, *8*(1), 25-31.

Lee, S.J. (2011). The Korean Wave: The Seoul of Asia. *The Elon Journal of Undergraduate Research in Communications*, *2*(1), 85–93.

Lie, J. (2011). What is the K in K-pop? South Korean popular music, the culture industry, and national identity. *Korea Observer*, *43*(3), 339–363.

Medina, A.A. (2017). The analysis of Hallyu's sustainability as a form of pop culture and soft power diplomacy. In *Ateneo Korean Studies Conference Proceedings* (vol. 4, pp. 124–140). Quezon City, Philippines: Ateneo de Manila University.

Mursell, J.L. (2007). *Successful teaching*. Lanham, MD: Littlefield Press.

Oh, I. (2013). The globalization of K-pop: Korea's place in global music industry. *Korea Observer*, *44*(3), 389–409.

Ortner, S.B. (1994). Theory in anthropology since the sixties. In N.B. Dirks, G. Eley & S.B. Ortner (Eds.), *Culture/Power/History: A reader in contemporary social theory* (pp. 372–411). Princeton, NJ: Princeton University Press.

Pallant, J., & Manual, S. S. (2007). A step by step guide to data analysis using SPSS for windows. In *SPSS Survival manual*. Berkshire UK: Open University Press.

Puspa, A. (2018, May 28). Hore! BTOB Konser di Jakarta 21 September 2018 [Hurray! BTOB Concert in Jakarta September 21 2018]. *Detik*. Retrieved from https://hot.detik.com/kpop/d-4042485/hore-btob-konser-di-jakarta-21-september-2018

Raden, F. (2014). Mission and transmission of Indonesian traditional music in the changing social and global contexts. *Korean Traditional Music Education*, *8*(2), 133–148.

Sung, S.-Y. (2013). K-pop reception and participatory fan culture in Austria. *Cross-Currents: East Asian History and Culture Review E-Journal*, *9(1)*, 90–104.

Tuk, W. (2012). *The Korean Wave: Who are behind the success of Korean popular culture?* (Master's thesis, Faculty of Humanities, Leiden University, The Netherlands). Leiden: Leiden University.

Yoon, K. (2017). Cultural translation of K-pop among Asian Canadian fans. *International Journal of Communication*, *11(1)*, 2350–2366.

21st Century Innovation in Music Education – Astuti et al. (eds)
© 2020 Taylor & Francis Group, London, ISBN 978-0-367-11122-9

Teaching music for communicating and sharing information in the Malian traditional society

Tchello Kasse
University of Bamako, Bamako, Mali

K.S. Astuti
Universitas Negeri Yogyakarta, Indonesia

ABSTRACT: This paper presents a review of the literature concerning music taught in Mali, which serves as a means of transmitting information and communication. In Malian oral tradition, songs were used as forms of messages shared between communities. *Jelis* or griots were particular groups of people who were selected to share knowledge across generations without writing. Although Mali is a land of music and culture, very little research has been conducted on musical knowledge for the current generations. Therefore, this paper intends to expand the current issues on music teaching and learning. The goal is to equip future educators with knowledge and skills that they can use while teaching in a multicultural society. Data in this research are sourced from document study. The results will offer future educators an opportunity to teach students the importance of music and deepen their understanding of Malian traditional music.

1 INTRODUCTION

Music has paramount importance in everyday life situations. Many people in the four corners of the globe enjoy listening to music in different places, whether they are home, at work or at the gym. It is played everywhere, particularly in night clubs, concert places, family events, shopping malls, cafés, restaurants and so on. In addition, people frequently listen to music while travelling by plane, trains, buses and even on foot. For example, studies have revealed that the typical American listens to music for approximately 18 hours on average in a week (Rentfrow, 2012).

Eight years ago, technology made music available everywhere, which pushed people to spend much money on it. In 2010, Rentfrow (2012) argued that people spent almost 4.7 billion USD in the music industry for business reasons. More importantly, Rentfrow (2012) revealed that music enables considerable effects on cognition, emotion and behaviour for students. It develops learners' listening skills, it gives and controls emotions during times of depression, and changes the negative characteristics of people who listen to it into positive ones. It facilitates self-expression for social connection.

Besides music sounding pleasant to our ears, it is also used by people to have fun and it enables people to be very emotive. Moreover, it interprets happiness, sadness, anger or fear. In addition, when music is played, musicians manipulate specific features (loudness, tempo, timbre, etc.) to describe emotions. Therefore, all the above specificities of music show how people feel about it.

Music has developed through the years particularly in Mali where it is considered as a critical part of everyday life. Although many people enjoy listening to music, interpreting songs, spreading it to the largest possible areas in the world, research concerning music remains scarce. This shortcoming resulted in Malian music being unknown to many countries. Lack of research prevents local artists and international researchers from developing new skills. Moreover, teachers of music in Mali will not be able to teach efficiently and effectively, while in other areas of research, things are progressing thanks to the use of technology. Therefore, any research that improves the situation is valuable at this moment. Scholars and researchers need to be present in the field for research purposes.

Combining lots of research will make Malian music stronger in the field of education. In addition, the traditional value interpreted by musicians will be strengthened so that it reaches more people in the world. It is in this perspective that the current research is going to be carried out in order to respond to the current needs of music. Given what has been said, the researcher seeks to respond to the following question: how is music used in the Malian traditional society to communicate and share information?

This paper has a qualitative research form. Data provided here stem from document study. The results of this research will be used by teachers of music, learners and performing artists all around the world to increase knowledge. In addition, the results will facilitate future researchers to conduct research to help develop music in terms of education and culture.

2 LITERATURE REVIEW

2.1 *Definition of music*

Music can be defined as vocal and instrument sounds that are combined with each other to produce beauty of form, harmony and expression of emotion. Another definition refers to music as a pattern of sounds made by musical instruments, voices, computers or a combination of these, intended to give pleasure to those listening to it. Kouakou (2017) argued that, in reality, music is a means of communication in West Africa. People use music to share information, to have dialogues or to spread news in the community.

2.2 *The role of music in the classroom*

One of the best ways to put music into learners' classroom activities is to relate it to other parts of the curriculum. Music can be included in lesson plans by teachers. To do this, teachers can add song titles to the content of the lesson being prepared. According to developmentally convenient practices, song is one of the vital components of any early childhood curriculum (Jones, 2010).

With songs, learners can develop their learning skills, social and cultural understanding, emotional and cognitive skills. Music is very important for learners to show their feelings and later on, acquire basic concepts and skills (Bredekamp & Copple, 1995, as cited in Jones, 2010). In addition, songs can be used in teaching to help students learn new concepts and vocabulary. Through listening activities, students can understand the meaning of terms and their interpretations in specific contexts. Additionally, songs enable students to learn adult roles. They develop learners' awareness of their bodies and what movements they can do. Further, music develops our feelings of self-confidence and fine motor skills. Moreover, listening to music helps to teach students about the foundations of musical properties such as loud, soft, fast tempo, slow tempo, and finally pitch (Jones, 2010). In the classroom, the music we play can help with transitions like clean-up time, circle time or from lunchtime to the end of break time.

Furthermore, North et al. (2000) argue that music can change the manner in which a learner feels, thinks and acts. Music enables students to define themselves in relation to others, their friends, colleagues, social networks and the cultures in which they live. Teaching music strengthens and extends experiences on a daily basis, also giving chance to further opportunities and making strong important links between the home, the school and other parts of the world.

2.3 *The notion of ngaraya in Mali*

Ngaraya is a term used in Mali to describe a specific breed of griots who are skilled in recounting the *Tariku* or oral recitation, which is reserved for male griots or *jelikew*. Some researchers thought that only men can be *ngaraw* (Johnson, 1986, p. 25, as cited in Duran, 2007). *Ngaraw* are ubiquitous in Mande society. They are consulted by many people and have a high status in the society for diverse reasons: singing, telling stories, performing dance, acting as messengers in difficult situations and so on (Duran, 2007).

2.4 *Ngaraya*

Duran (2007) defined *ngaraya* as when a *jeli* becomes a significant practitioner and makes a significant achievement. However, only *jeli* can be *ngaraw*. Anyone who is from a *jeli* family is classified as *jelike* (male) or *jelimuso* (female). A *ngara* has specific qualities regarding information concerning Mande history, genealogy, musical repertoire, skill, confidence and authority in performance lineage, destiny, the power to make things happen without explanation, moral behaviour, fearless character and abrasiveness. However, all *jelis* are attributed the title *ngara* when they are in their late 40s, particularly women. Some prominent Malian *ngaraw* include Bazoumana Cissoko, Kandia Kouyate, Ami Koita, Jalou Damba, Koni Koumare, and Mah Damba. Those *ngaraw* sang for their country, the brave men and women of Mali, and the great figures of the history of Mali who fought for independence.

3 DISCUSSION

3.1 *The function of the drum for delivering messages*

In Malian traditional society, the griot's fundamental function was to inform and spread news that happened in the community. Griots were in charge of playing drums in circumstances when there was some important information to disseminate. For instance, the griot could take his drum and play it around the village with a specific message that the entire community would then embrace and relay. The message could be about a gathering, a delegation visiting, youth club meetings, annual traditional and spiritual celebrations and so on. The drum was used because the sound it produces was capable of reaching large crowds. This era preceded the telephone or any audio-visual communication system in those traditional communities.

3.2 *Teaching oral tradition*

In a small village in Mali – Kela – oral tradition is taught in two forms: the first is named social general training, which covers some aspects of cultural topics dealing with social interaction with elder people. The second is professional training, which covers disciplinary subjects such as music and public oral expression for instance (Toulou, 2008).

While in formal education we can see that lessons are planned and organised, in traditional society teaching of cultural values is informal. Teaching is done orally, and does not require any written documents. Griots receive education outside schools; that is to say, in the families they are from. They are informally taught to memorise words and phrases, including the meaning of some sayings. However, education would refer to any structured lesson that offers opportunities to develop learning through transmission of information, cognition, knowledge and know-how.

3.3 *Traditional role of griots*

Griots played important roles in the traditional society of Mali. For example, during wars, griots were used to motivate warriors by playing music. Griots interpreted songs which recall the courage of the ancestral parents of the people who were engaged in the fighting. When these people heard such interpretations, they go forward to fight using all means they could.

In addition to that, a griot played a key role as the king's spokesperson (in charge of delivering the speech that the kings made for the public). In addition, griots were considered as messengers during official ceremonies. They were appointed by the traditional leaders, kings, village and religious leaders and so on. They were officially recognised by the community. Therefore, they played important roles in the society. In villages, griots were also in charge of reporting traditional celebrations like naming, marriage or mourning events (Sauvé, 2007).

Griots had the mission to tell the stories of things and events for which no written documents were available. The communities relied much on them in terms of information sharing, and also on the identification of their ancestors, their origins and their accomplishments. The roles they played were very well structured. The accuracy of the information to be given was paramount. They were not allowed to offend or demean anyone. As a result, the nobles were very kind to the griots because of the good appreciation that the griots gave them.

Traditionally, griots' voices were used to honour one or many individuals; for example, the king, the chief, or the noble. When the griots made praises, usually they are accompanied by a musical instrument called a *xalam* or the *tama*, which is a small drum placed in the armpit. Moreover, griots' songs express useful expressions that the person being praised would appreciate. During events such as naming ceremonies or other social celebrations, one individual receives praise from someone else who knows them well. That person would praise them by listing the background of the family in front of the people who do not know them very well: such a person ends up being glorified.

A griot is a messenger who organises big meetings in the villages. More importantly, the education system did not divert the griots from their traditional roles and responsibilities. A griot is a social historian, a messenger and a peacemaker. He or she can give an oral account of legendary tales, without any notes, with the goal of educating and conveying key messages to their listeners. They can always explain what happened years before in oral form. All the students who went to universities for education make sure the griot also went there.

Hale (1998) suggested that the role of the griot in West Africa is determined by praise songs, storytelling, instrumental performances and social mediation. In society, these roles are very precious because the griot always tries to create songs that fit with the culture, using traditional musical instruments such as the kora, balafon, djembe and so on, and also tries to bring peace between people who have disagreements. This role was very important for the communities living together because modern justice systems were not in existence at that time. Therefore, the griots also served as justices concerning mediation in the society in terms of traditions.

Research has shown that griots were considered to be the oral tradition specialists in West Africa. Thanks to griots, for example, the oral tradition of the village of Kela has been maintained for generations, since the thirteenth century without any documentation. All the history was memorised by the griots who have taught their children in the same way up until the present.

4 CONCLUSION

Music use in the Malian context is accepted by everyone regardless of their social status, ethnicity, religion or leadership. As a result, the griot in Malian traditional society plays a key role in the development of social bonding. They are storytellers and world tradition keepers, which they inherited from their parents over generations. Consequently, the griot influences society to invoke courage and peace through the songs they perform for the community. They transmit the messages of the rulers. They intervene when there are disagreements and, finally, they are decent.

REFERENCES

Duran, L. (2007). Ngaraya: Women and musical mastery in Mali. *School of Oriental and African Studies, 70*(3), 569–602.

Hale, T.A. (1998). *Griots and griottes: Masters of words and music*. Bloomington, IN: Indiana University Press.

Jones, J. (2010). The Role of Music in Your Classroom. *Exchange: The Early Childhood Leaders' Magazine Since 1978*, *192*, 90-92.

Kouakou, O.K.B. (2017). Revue Scientifique en Sciences de l'Information et de la Communication (*Scientific Review in Science of Information and Communication*). Abidjan, Nigeria: CERCOM.

North, A.C., Hargreaves, D.J. & O'Neil, S.A. (2000). The importance of music to adolescents. *British Journal of Educational Psychology*, *70*, 250–272.

Rentfrow, P.J. (2012). The role of music in everyday life: Current direction in the social psychology of music. *Social and Personality Psychology Compass*, *6*(5), 402–416.

Sauvé, J.P. (2007). *La culture des griots face à la modernisation* (The culture of griots towards modernisation) (Master's dissertation, Université de Québec, Montréal, Canada). Retrieved from https://archipel.uqam.ca/4404/1/M9829.pdf

Toulou, S. (2008). Teaching oral tradition: What type of professional training for the Mande bards? *Revue Suisse des Sciences de l'Éducation*, *30*(2), 325–347.

21st Century Innovation in Music Education – Astuti et al. (eds)
© 2020 Taylor & Francis Group, London, ISBN 978-0-367-11122-9

Music bibliography of South Kalimantan (1978–2017)

S. Hadi
Lambung Mangkurat University, Banjarmasin, South Kalimantan, Indonesia

ABSTRACT: This study aims to obtain music bibliographic data in South Kalimantan. This qualitative-descriptive type study uses a literature study method. Data sources in the form of library materials concerning music in South Kalimantan were obtained from various libraries in Banjarmasin City. In addition, in collecting data, researchers also conducted observations, interviews and document studies related to the subject of this study. The validity of the data was determined through extension of the participation of researchers, persistence in observing, triangulation, peer/expert discussions and referential adequacy. The results of the research obtained are in the form of a bibliographic compilation of musical art in South Kalimantan. It was noted that in almost 40 years (1978–2017), there were 17 publications on the subject of music in South Kalimantan. It is known that most of the musical forms of traditional music in South Kalimantan, such as gamelan, *panting*, and *Kintung,* are still being studied.

1 INTRODUCTION

Music literature in Indonesia, both scientific and popular, is important to measure the quality of musical development. In 2016, a book was published in Indonesia: *Membaca Musik dari Masa ke Masa: Katalog Literatur Musik Berbahasa Indonesia dalam Lima Dekade* (Felicianata & Setiawan, 2016). This book tries to browse Indonesian-language music books and was written by two literary writers. From this book, it can be seen that, concerning music, there was only a small amount of literature in Indonesia in five decades. From the authors' explanation, it is known that the data in the book are still incomplete considering the very broad Indonesian territory. Thus, the search for music literature in Indonesia needs to be deeper by tracking data in various regions. One such area of Indonesia formed the focus of this study: the province of South Kalimantan.

Before further investigation, it was found that, in South Kalimantan, there is still a lack of quantity and quality of data on music literature. This was caused by four factors: first, there were only a few music education institutions (universities) in South Kalimantan. Recently, there is at least one significant institution, namely the Performing Arts Education Study Programme at Lambung Mangkurat University, where the researchers work. Second, this has also resulted in a lack of academic or music researchers in South Kalimantan. Third, because there are only a few universities and a lack of scholars and researchers, the results of music research in South Kalimantan are still very few. Fourth, there is no research output or scientific publication (bibliography) that comprehensively documents music literature on South Kalimantan. Based on the background and problems noted, it is necessary to conduct a study of the bibliographic data in the musical field in South Kalimantan. Therefore, the compilation of South Kalimantan's music bibliography is important.

2 RESEARCH METHOD

The method used in this research was literature research (Sugiyono, 2010). The primary data source used were the literature publications concerning music arts in South Kalimantan which were published extensively and not the results of academic research such as unpublished theses. This research was conducted through the stages of draft preparation, implementation (data collection and analysis) and preparation of reports (Moleong, 2010).

Data collection was conducted through observation, document study and interviews. Observations were conducted by observing documents/data sources located in libraries or significant places in South Kalimantan. Interviews with relevant informants were also conducted to strengthen and lead the researcher in finding and interpreting the data. The analysis and interpretation were carried out through a descriptive analysis model: data reduction, data presentation and conclusion. Then, it was strengthened by having cross checks and discussions with colleagues or experts to draw conclusions. The validity of the data (Patton, 1987) was obtained through an extension of the participation and perseverance of research observers, triangulation, peer/expert discussion and referential adequacy.

3 RESULTS AND DISCUSSION

3.1 *Results*

The music bibliography in South Kalimantan can be recorded as having as many as 17 titles. Of these, 15 can be found physically, while two titles have not yet been located: (1) *Hasil Penggalian Musik Kintung Tahun* (Mahendra, 2018); (2) *Penyelengaraan Eksperimentasi Musik Panting* (Sirajudin, 1996, as cited in Suryana, 2015). From the data obtained, it is also known that literature or publications concerning music in South Kalimantan have been developing since 1978.

3.2 *Discussion*

After describing the main data of the music bibliography in South Kalimantan, it is necessary to review or discuss it. The following discussion is presented by giving annotations or notes that are relevant to the bibliographic data. The following annotations are discussed chronologically based on the order in which they were published.

1978: *Encyclopedia of Regional Music and Dance in South Kalimantan* (Soenarto, 256 pages, Jakarta: Ministry of Education and Culture). This book was the result of research by the Department of Education and Culture in 1977/1978, namely the *Proyek Penelitian dan Pencatatan Kebudayaan Daerah* (P3KD). The specific research on music and dance in the South Kalimantan area was conducted by a team led by Soenarto and his colleagues: Budra Djamal, Amat Asnawi, Lamri Bustami and M. Donie. The main content of the book presents two parts of data about music and dance that exist and have existed in South Kalimantan.

Music is divided into two categories, namely instrumental music and vocal music. There are 13 instrumental musics: *Gambus*, *Gamelan Banjar*, *Kangkanong*, *Kangkurung Batang*, *Katok Bambu*, *Kecapi*, *Kuriding*, *Kurung-kurung Kuda Kepang*, *Solong Bambu*, *Solong Kayu*, *Sarunai*, *Suling* and *Tarbang*. Furthermore, vocal music is noted as having 11 types: *Ahooi*, *Ampar-ampar Pisang*, *Andi-andi*, *Ayun Paran*, *Basyair*, *Kuriding Patah*, *Maayun Anak*, *Madihin*, *Mamuai Wanyi*, *Si Buang Anak* and *Ulak Marabahan*. Completing the data on dance, this music chapter also mentioned the names of the teachers or musical figures. Meanwhile, dance is classified into two types, namely, traditional dance and new creative dance. It is also equipped with teachers' names or dance figures.

Specifically, in terms of the discussion of instrumental music, it is an interesting and comprehensive description, because it presents several very important aspects, such as historical background, area of distribution, form of music, song composition/repertoire, organology (structure/parts, materials and construction, how to play), good musical instruments/composers and skilled players. In relation to vocal music songs, it discusses historical background, area of distribution, form of music and how to perform, composition/repertoire, skilled singers and how to learn it, as well as the notation and text of a song.

1984: *Banjar Puppet* and 1981: *Banjar Gamelan* (Saleh (1984), 45 pages, Banjarbaru: *Museum Negeri Lambung Mangkurat*). This is a series of three written pieces, published by the

Museum Negeri Lambung Mangkurat. They contain a description of one of the traditional music pieces of South Kalimantan, the *Gamelan Banjar*. In this book, the history of development, types of instruments, kinds of songs and classification of uses and objectives are described, then a ceremony or ritual is described before the *Gamelan* is sounded. This book was said by the author to use data sources from previous research conducted by the *Masyarakat Sejarawan Indonesia* (MSI) of Banjarmasin Branch in 1976.

1984: *Banjar Puppet and Gamelan* (Saleh (1984) 97 pages, Banjarbaru: *Museum Negeri Lambung Mangkurat*, South Kalimantan). This book is a special publication of the Lambung Mangkurat State Museum of South Kalimantan, in 1983/84. Although this book deals more with Banjar puppetry, the subtopic of the gamelan is quite important and valuable (Chapter VII). This describes much important ethnographic data about the *Gamelan Banjar*. For example, the data comparison on the number of instruments and songs in *Gamelan Banjar* that existed in the early twentieth century and in the 1950s. In addition, there is data on the names of the gamelan musicians in *Gamelan Banjar* who were still active in the 1970s. For enthusiasts/researchers of *Gamelan Banjar*, this book is quite important for use as material for further study.

1985: *Panting Music from Tapin* (Sjarifuddin, 65 pages, Banjarbaru: *Museum Negeri Lambung Mangkurat*). This book is a special publication of the State Museum of Lambung Mangkurat, South Kalimantan, in 1984/1985. Substantially, this study presents related matters, namely *panting* music from the Tapin Regency, including understanding the terms '*panting*', the development of *panting* music, methods of creation including its types, art creations, venues and performances/performance equipment, technical implementers and procedures for performances, background of the society and supporting artists, as well as people's views of *panting* music. This research on *panting* music is quite interesting, especially its history. It is stated that *panting* music in South Kalimantan comes from the Tapin District, especially Rantau Bujur Village, Sungai Pinang District. The issue of the origin of *panting* music is quite interesting and requires further investigation, given the volume of qualitative data (outside this research) from various groups (artists) stating that the beginning of the development of *panting* music came from Barikin Village, Hulu Sungai Tengah Regency.

1986: *Guriding: Traditional Musical Instruments in South Kalimantan* (Mohd. Saperi Kadir, 67 pages, Banjarbaru: Lambung Mangkurat State Museum). This book is a special publication of the State Museum of Lambung Mangkurat, South Kalimantan, in 1985/1986. It is the result of field research conducted in Ulu Benteng Village, Bakumpai District, Barito Kuala Regency. Regarding the four chapters of this book, the subject of *Guriding* is found in Chapter III. The contents of this chapter include: (1) *Guriding* naming, *Guriding* history, *Guriding* forms and uses, how to make and how to store *Guriding*, classification of *Guriding* types of music instruments, holder names and song names in *Guriding* music performances, the lives of *Guriding* artists, *Guriding* conditions in the present, the cause of *Guriding* scarcity and preservation of *Guriding*.

In the context of scientific literature, this book has important value, because there were still very few publications or scientific books about *Guriding* or *Kuriding*. In addition, the value of this book is based on the sociological data side of the music, especially the traditional music that developed in the Bakumpai community in Barito Kuala Regency.

1986: *Encyclopedia of Regional Music and Dance in South Kalimantan* (Soenarto, 210 pages, Jakarta: Ministry of Education and Culture). This book, published in 1986, was the result of a research project by the Ministry of Education and Culture in 1976/1979 entitled *Proyek Penelitian dan Pencatatan Kebudayaan Daerah* (P3KD). This particular study on music and dance was carried out by two teams: a central team and a regional team. The central team consisted of Suwandono, R. Prasmadji and Binsar Sitompul, while the regional team consisted of Soenarto, M. Donie, Rustam A.A. and M. Saperi K. With sufficient research data sources of actors/artists (ten people), the contents of this book are very good and quite detailed in describing South Kalimantan's music and dance. However, there was also a weakness of this research as stated by the project leader, Yuztan Aziddin, in his introduction. He said that the shortcomings of this study were the lack of music and dance experts that led to unsatisfactory results. This statement was derived from *Titi Mangsa* to the Speech of the Head of the Regional Office of the Ministry of Education and Culture of the South Kalimantan (S.M. Idroes). The book was written in August 1986 which means that it was only published about ten years after the study was conducted in 1976.

The description of music is categorised into two types, namely instrumental music (which contains organological musical instruments such as *Agung*, *Babun*, *Kecapi*, and *Tarbang*) and vocal music that contains songs in ritual/South Kalimantan arts (such as *Badudus*, *Burung*, *Dua Marindu*, *Gandut*, *Lalan*, *Paris Tangkawang*, *Rantau*, *Sumur Kili*, *Timbal Laung*, and *Tirik*). Meanwhile, dance is categorised into two types, namely dance before the independence of the Republic of Indonesia (such as *Gandut*, *Gerbang*, *Japin Sisit*, *Kenanga Dalam*, *Kuda Gepang Siba*, *Simbangan Burung Laut*, and *Tirik Pandahan*) and dance after independence (such as *Bogam*, *Burung Tarabang*, *Maiwak*, *Pengantin Banjar*, and *Tirik Kuala*). The description of music and dance in this book is quite good because it is accompanied by a historical review and presentation of musical notation, motion descriptions and supporting images (song notation, musical instrument images, floor patterns and dance costumes).

1997: *Tarbang Hadrah from South Kalimantan* (Agus Triatno and Agus Wahyudi, 66 pages, Banjarbaru: South Kalimantan State Lambung Mangkurat Museum). This book is the 34th in the series of special publications of the State Museum of Lambung Mangkurat, South Kalimantan, in 1983/1984. It is the result of field research on *Tarbang Hadrah* music in South Kalimantan, especially in Candi Laras Selatan District, Tapin Regency. In addition, to describe the development of *Tarbang Hadrah* music in South Kalimantan at a glance (Chapter III), the important topics in this book are about technology of manufacture (Chapter IV), types and rhythms (Chapter V) and the role of *Tarbang Hadrah's* music in traditional arts and ceremonies (Chapter VI). The forms of arts mentioned are *Sinoman Hadrah* and *Barangkup*, while the traditional ceremonies are the *Balamburan* and *Baayun Mulud* ceremonies. This book is quite important as documentation (scientific-empirical) of one of the traditional arts of South Kalimantan which still continues to develop today, despite its various changes.

1997: *The Collection of Panting Music Groups throughout the City of Banjarmasin* (Cultural Park of South Kalimantan, 11 pages, Banjarmasin: Cultural Park of South Kalimantan). This book is in a stencil-type form that can be categorised as art demographic literature. It contains the results of an inventory of important artists or music players in the Banjarmasin Municipality area. Quantitative data on the distribution number of important artists in this book are presented in the form of tables with four components. These components are names of music players/artists, mastered instruments, group addresses and gender information. In this study, there are ten important music groups with a total of 105 artists. The following are important music group data and artists in this book: (1) *Saraba Sanggam*, 20 artists; (2) *Parimata*, ten artists; (3) *Mining*, 11 artists; (4) *Manikan* (?), 12 artists; (5) *Haur Gading*, ten artists; (6) *Bunga Anggrek*, eight artists; (7) *Rumah Banjar*, ten artists; (8) *Rumah Baanjung*, seven artists; (9) *Junjung Buih*, eight artists; (10) *Kambang Barenteng*, nine artists.

2007: *Banjar Gamelan in South Kalimantan* (Mukhlis Maman, Aman Waluyo and M.N. Hilal, 51 pages, Banjarmasin: UPTD (*Unit Pelaksana Teknis Daerah* - The Regional Technical Implementation Unit) Taman Budaya South Kalimantan). This book is an effort by the author team to describe *Banjar Gamelan* music as one of the traditional arts of the Banjar community in South Kalimantan. The description of this book comes from a series of research activities, discussions and workshops related to *Banjar Gamelan*. This book consists of four main parts. Part one is the introduction. In this introduction, there are some interesting and rare aspects discussed, as in books about other Banjar arts. The book is about the issue of musical (gamelan) aesthetics used as a critique for artists who pay less attention to the historical traditions of *Banjar Gamelan* music. This section also presents Barikin Village in Hulu Sungai Selatan Regency, as an important site that supports *Banjar Gamelan*.

In the second part, it describes clarification of terms related to *Banjar Gamelan* such as 'music' and '*karawitan* art'. This section emphasises that 'Banjar *karawitan* art' is a broad scope of traditional music arts living in the South Kalimantan area such as gamelan, *panting*, *bumbung*, *Kintung*, *kangkanong* and so on. Furthermore, in the third part, it discusses the history of the *Banjar Gamelan* with an explanation of the classification of the *Gamelan Keraton* instruments (16 instruments) and the *Gamelan Rakyatan* (eight instruments), *karawitan* songs and notations. In the last part, the team of writers also give critical notes about the threat of extinction of *Banjar Gamelan*.

2008: *Folk Music of South Kalimantan: Kintung* (Cultural Park of South Kalimantan, 12 pages, Banjarmasin: UPTD of South Kalimantan Art Space). This 12-page book is a pocket-book type. Although a pocket-book is indeed compact and concise, the contents of the book are quite informative. It can be regarded as a scientific book because it is compiled using clear references (presented as a bibliography) and described in continuous writing systematics. The content system in question consists of at least seven parts of discussion about *Kintung* music, those are: (1) introduction; (2) form of *Kintung* music; (3); presentation that contains how to play and the songs played (here are mentioned two songs/rhythms, namely, *Lagu Undangan* and *Lagu Mahu Cepat*); (4) *Kintung* music function; (5) supporting artists; (6) ways and processes of making *Kintung*; (7) conclusions. There are three important conclusions presented in this book, which are: (1) *Kintung* music grows and develops in Banjar's society; (2) *Kintung* music was originally sacred and became a medium of traditional rituals asking for rain; (3) *Kintung* music is threatened with extinction as the young generation does not have any interest in it.

2011: *The Maestro: H. Anang Ardiansyah 'The Paris Barantai'* (Amien Santang and Thamrin Yunus, 183 pages, Banjarmasin: The Government of South Kalimantan). This book is basically a biography presented in autobiographical terms. In this book, the life of H. Anang Ardiansyah is written from the perspective of 'me'. The role of the two authors of this book can be clarified in that Amien Santang is the initiator, while the author is Thamrin Yunus. Although this book focuses on just one figure of Banjar music, H. Anang Ardiansyah, the exposures of the journey of his art are also depicted in characters, groups and musical events that play a significant role in the development of music in South Kalimantan. The important value of this book is in the primary data about the creative journey of the music artist, H. Anang Ardiansyah, coming from the character himself (from an interview). In addition, this book is also complemented by testimony from various Banjar leaders/artists about H. Anang Ardiansyah.

From the primary data in this book, it is worth conducting further research on the role of Banjar music artists in the national arena such as the figures Hamiedan and Zaglulsyah (Sam Sarosa), or a musicological study of the structure of Banjar songs based on H. Anang Ardiansyah's conception that it had never been done.

2013: *Qasidah & Rebana: History of Development and Procedures for Arranging Tambourine Arrangement* (Mudjahidin S., 93 pages, Banjarmasin: UPT South Kalimantan Art Space in collaboration with Banua Library). This book is one of the scarce literature on *qasidah* and *rebana* music in South Kalimantan. The discussion of this book consists of six chapters: (1) various reviews of art which discuss the concept of art from various philosophers; (2) the development of *qasidah* and *rebana* was originally in Indonesia and specifically from South Kalimantan; (3) the element of *rebana* sound discussing various techniques of playing and the character of the sound of a *rebana* instrument; (4) the structure of *Qasidah Rebana,* discussing the form/structure of the performance; (5) the arrangement and appearance techniques and dress; (6) the etiquette to be a judge or judge of a *Qasidah Rebana* competition. At the end of this book, it is equipped with an attachment containing a variety of *rebana* techniques.

The discussion of this book basically focuses on *rebana* arrangement techniques. The background of the author as a practitioner, who has a lot of experience, makes the material of the arrangement techniques authentic. Regarding the writer as a music practitioner of *Qasidah Rebana*, there has been no research results focusing on him so far. This needs to be conducted immediately by interested parties. Regarding the history of *Qasidah Rebana* in this book, it is not yet profound. This means that the matter of historical studies needs to be continued by further historical research with feasible data. One of the weaknesses of this book is not in the substance of the discussion, but the editing of the narrative.

2014: *An Overview of Ethnomusicology of Kuriding Music of Bakumpai Dayak Tribe, Barito Kuala Regency, South Kalimantan* (Maryanto, Dwi Wahyu Candra Dewi and Syahlan Mattiro, 50 pages, Yogyakarta: Aswaja Pressindo). This book is the result of research conducted in Ulu Banteng village (Marabahan Subdistrict, Barito Kuala Regency) by a team of researchers/lecturers from FKIP Lambung Mangkurat University for three months (September to December 2013). The results of this study indicate that *kuriding* music composition is presented in three forms, namely *Ayakan, Tiukang* and *Sabulu Tinjau.* In addition, this study also includes a semiotic analysis of *kuriding* song in its poem verses. However, there has been no clear and deep analysis. In addition, this study also discusses some of the *kuriding*'s musical functions, such as emotional expression, aesthetic pleasure, entertainment facilities, the support of social integration and dance accompaniment.

2014: *H. Achmad Thamrin 'Si Utuh Kincir'* (Thamrin Yunus, 70 pages, Banjarmasin: Government of South Kalimantan). This book is a biographical writing about H. Achmad Thamrin, a South Kalimantan music artist. This book is quite interesting for understanding the sociological aspects of the music artist, H. Achmad Thamrin, especially to understand the connection between the process of art (music) and the cultural environment of the artist. Although the narrative style of this book is unfocused and the narrator's point of view often switches, the data obtained are quite important, especially in terms of the artist's close family. In addition to the important value in the data-narrative based on the perspective of relatives such as the wife, child, brother and friend of the artist, the author of this book also includes his critical views. In particular, regarding the condition of the current 'Banjar Song', which is seen as experiencing an existential decline when compared to conditions in the 1960s and 1990s. Although this book specifically discusses the biography of artists, it also deals with the sociological conditions (fate) of contemporary Banjar songs in South Kalimantan.

2017: *Kuriding Music: A Socio-Cultural Study on the Preservation of Banjar Traditional Music* (Muhammad Najamudin; Sumasno Hadi (Editor), 68 pages, Banjarmasin: Lambung Mangkurat University Press). This book is the result of the academic research (thesis) of the author (Muhammad Najamudin) in the Postgraduate Programme of Semarang State University. The topic of this book concerns the preservation of *kuriding* as one type of traditional music of South Kalimantan. This *Kuriding* music preservation study uses a sociological approach in which aspects of social change have been included. Regarding the forms and processes of preservation, this book examines several topics. These are: (1) the role of artists (Mukhlis Maman) and the Saradifa Community in South Kalimantan; (2) the role of the South Kalimantan Regional Government and Educational Institutions (for example from the Sendratasik Education Study Programme FKIP ULM and STIKIP PGRI Banjarmasin; (3) musical innovations (making Kurbit instruments; (4) performances of *kuriding* musical ensembles with various musical compositions. The conclusion of this book discusses the preservation of *kuriding* music in South Kalimantan, which is very dependent on the roles of the artists themselves (internal factors).

4 CONCLUSION

Regarding the results and discussion of this study, some notes need to be presented. In almost 40 years (1978–2017) in South Kalimantan there were 17 publications concerning the subject of music. Most of the treasures of traditional music in South Kalimantan, such as the music of *Gamelan Banjar*, *panting*, and *Kintung*, are still being studied. Other publications that discuss/study modern popular music have been conducted infrequently. In addition, from the total of 17 publications in the past four decades, it can be concluded that in South Kalimantan, the ideal conditions for the development of its scientific musical study are still not reflected. Therefore, it is necessary to increase the amount of research, number of studies and publications in the field of music.

ACKNOWLEDGEMENT

This research is fully supported by FKIP Lambung Mangkurat University.

REFERENCES

Felicianata, O. & Setiawan, E. (2016). *Membaca Musik dari Masa ke Masa: Katalog Literatur Musik Berbahasa Indonesia dalam Lima Dekade* [Reading music of Indonesian languages in five decades]. Yogyakarta, Indonesia: AMT Publisher.

Kadir, M.S. (1986). *Guriding: Traditional Musical Instruments in South Kalimantan*. Banjabaru: Lambung Mangkurat Museum.

Mahendra, B. (2018). Pembelajaran Musik Kintung Berbasis Kreativitas pada Peserta Didik di Dapur Teater [Kintung Music Learning Based on Creativity in Dapur Theater Students]. *Pelataran Seni* (Art Stage), *3*(1), 53–67.

Maman, M., Waluyo, A. & Hilal, M.N. (2007). *Banjar Gamelan in South Kalimantan*. Banjabaru: Lambung Mangkurat Museum.

Maryanto, M., Dewi, D.W.C., Mattiro, S. (2014) *An Overview of Ethnomusicology of Kuriding Music of Bakumpai Dayak Tribe, Barito Kuala Regency, South Kalimantan*. Yogyakarta: Aswaja Pressindo.

Moleong, L.J. (2010). *Metode Penelitian Kualitatif* [Qualitative research method]. Bandung, Indonesia: PT Remaja Rosdakarya.

Mudjahidin, S. (2013). *Qasidah & Rebana: History of Development and Procedures for Arranging Tambourine Arrangement*. Banjarmasin: South Kalimantan Art Space in collaboration with Banua Library.

Najamudin, M. & Hadi, S. (2017) *Kuriding Music: A Socio-Cultural Study on the Preservation of Banjar Traditional Music*. Banjarmasin: Lambung Mangkurat University Press.

Park, C. (2008). *Folk Music of South Kalimantan: Kintung*. Banjarmasin: South Kalimantan Art Space.

Patton, M.Q. (1987). *How to use qualitative methods in evaluation*. Thousand Oaks, CA: Sage Publications.

Saleh, H. (1984). *Banjar Puppet* [Puppet Series]. Banjabaru: Lambung Mangkurat Museum.

Saleh, H. (1981). *Banjar Gamelan* [Gamelan Series]. Banjarbaru: Lambung Mangkurat Museum.

Saleh, M.I., Kutoyo, S. & Sutjiatiningsih, S. (1984). *Sejarah daerah Kalimantan selatan* [South Kalimantan History]. Jakarta: Department of Education and Culture.

Santang, A. & Yunus, T. (2011). *The Maestro: H. Anang Ardiansyah 'The Paris Barantai'*. Banjarmasin: The Government of South Kalimantan.

Sjarifuddin, H. (1985). *Panting Music from Tapin*. Banjarbaru: Lambung Mangkurat Museum.

Soenarto, M. (1978). *Ensiklopedi Musik dan Tari Daerah Kalimantan Selatan* [Encyclopaedia of Regional Music and Dance in South Kalimantan]. Jakarta: Ministry of Education and Culture.

Soenarto, M. (1986). *Ensiklopedi Musik dan Tari Daerah Kalimantan Selatan* [Encyclopaedia of Regional Music and Dance in South Kalimantan]. Jakarta: Ministry of Education and Culture.

Sugiyono. (2010). *Metode Penelitian Kuantitatif, Kualitatif, dan R&D* [Quantitative, qualitative, and R&D research methods]. Bandung, Indonesia: CV Alfabet.

Suryana, R. (2015). Nilai–nilai sosial dalam penyajian musik Panting di Banjarmasin [Social values in the presentation of *Panting* music in Banjarmasin]. *Socius Journal*, *4*(2), 121-129.

Triatno, A & Wahyudi, A. (1997). *Tarbang Hadrah from South Kalimantan*. Banjabaru: Lambung Mangkurat Museum.

Yunus, T. (2014) *H. Achmad Thamrin 'Si Utuh Kincir'*. Banjarmasin: Government of South Kalimantan.

21st Century Innovation in Music Education – Astuti et al. (eds)
© 2020 Taylor & Francis Group, London, ISBN 978-0-367-11122-9

Differences in developing musical competencies in Indonesia, The Netherlands, and France

K.S. Astuti, A. Armini & H. Sri Mudjilah
Yogyakarta State University, Yogyakarta, Indonesia

E.Th.M. Bijsterveld
Royal Conservatoire, The Hague, The Netherlands

S. Vrede
DaCosta Hoograven, Utrecht, The Netherlands

W. Veugelers
University for Humanistiek, Utrecht, The Netherlands

ABSTRACT: The study was aimed at revealing the differences in the development of students' musical competencies in the Netherlands, Indonesia, and France. The study adopted a mixed methods research design. Data were collected through observation of 30 music classes in the Netherlands, 30 in Indonesia, and seven in France. Observational data were supplemented by interviews with teachers, principals, students, parents, and other stakeholders. This research was conducted using a sequence of qualitative and quantitative methods. The qualitative data were analysed descriptively, while the quantitative data were analysed using descriptive statistic. The results showed that the students in the Netherlands had higher levels of musical competency than Indonesian and French counterparts. Music education in the Netherlands places more emphasis on the affective domain, while in Indonesia and France it focuses more on the cognitive and psychomotor domains.

1 INTRODUCTION

1.1 *Background*

Over recent decades, research in various parts of the world has led to increasing awareness of the importance of the art of music. In the 1980s and 1990s, scientific journals published studies showing that music literally changes the structure of the developing brain in the fetus and which identified a range of benefits for children in learning music. For instance, children who receive music training regularly have increased IQ scores; music therapy for half an hour a day can improve children's immune function; and music can relieve tension, encourage social interaction, stimulate language development, and improve motor skills among children (Campbell, 2001: 4).

Music also has a considerable influence on affective development. Siswa (1977) cites two examples. First, Ki Hadjar Dewantara (cited in Siswa, 1977: 303) argued that music can help to refine hearing to produce feelings of calmness. The exercise he developed involves a combination of language and music training. Second, Rudolf Steiner's theory of *Antroposofisch onderwijs* holds that music (in this case, rhythm) can facilitate physical work, support mental flexibility, develop character, and reinvigorate the human soul. Khan and Timur (2002: 121) have proposed that sound has a certain psychological value, unique to; each voice, and that a sensitive person can grasp another's personality from the sound of their voice alone. Others, like Mudjilah (cited in Imaji, 2003: 147) argued that the poems found in vocal music play a role in influencing a person's psychological condition, even one's morality. Perkins, Yorke, and Fancourt (2018) found that music can be an effective therapy for mothers with symptoms of postnatal depression: the mothers in their arts-in-health program reported that this was the only time they did not feel the need to be getting on with some other activity.

These studies suggest that music, apart from its intrinsic value, can also have an impact in other fields, including intelligence, morals, psychology, and health. School is the most effective learning environment for the transmission of knowledge of science and the arts. Schools and colleges, therefore, are ideal sites in which to investigate the process of music development and learning. Hence this study explored musical development and learning in schools in three different countries, Indonesia, the Netherlands, and France.

Indonesia has a population of more than 250 million dispersed over 18,307 islands. Its 1,128 tribes maintain as much of their traditional arts and languages. For this reason, Indonesia can be used as a model for different countries with varies cultures. Several recent research studies (Astuti et.al, 2010) and (Astuti et.al, 2017) found that learning in the Netherlands pays more attention to the needs of students, especially many psychological needs. Teachers are able to develop and adapt teaching strategies to the needs of individual students, and the curriculum is flexible. These findings indicate that the Netherlands represents best practice in current music education and can be used as a benchmark. France is a country with a long artistic history and reputation and can also be used as a benchmark for achievement in music teaching.

The findings are expected to contribute to music educators' understanding of best practice in the field. Bloom, Simpson, Krathwohl and Anderson classified learning outcomes into three domains: cognitive, psychomotor, and affective. This study compared the development of musical competency in three countries using this framework.

1.2 *Research aim*

The aim of the present study was to describe the development of and differences in musical competencies in Indonesia, the Netherlands, and France in order to identify best practice in music teaching and learning that can be used to guide music educators in developing their students' musical competencies.

2 LITERATURE REVIEW

2.1 *Musical intelligence*

Decades before Gardner (1993) proposed the concept of musical intelligence, Sheasore (1938: 8) identified musical ability as part of intelligence in general:

> Intelligence is musical when its background is a storehouse of musical knowledge, a dynamo of musical interests, an outlet in a musical task, and a worth of musical experiences and responses. Here, as in the case of imagination, the type and degree of intelligence may characterize or set limits to the musical achievement.

Musical intelligence refers to tone sensitivity (Gardner, 1993: 37), that is, one's ability to distinguish pitch (high and low tone) and sing the tone appropriately. The development of musical intelligence, according to Gordon, is the result of innate processing that depends on the interaction of genetic and early environmental factors. Gordon's research results (cited in Pherson, 1997: 69) indicated that the environment has a significant influence on musical talent among nine-year-old children:

> Gordon believes that music aptitude results from both nature and nurture, because of genetic and early environmental influences (including those before birth), interact.... There is also evidence that the age of nine represents an important step in the process of developmental. This statement means that up until age nine the quality of the

environment exerts a "push and pull upon younger children's innate potential at rates that vary among children". If this view is correct, then a baby with a high level of music aptitude at birth whose environment is musically impoverished will lose much of this potential forever. Alternatively, quality musical experiences will enhance children's music aptitude and impact on their future musical development in a variety of ways.

This idea is consistent with Suzuki's philosophy, which states that talent does not arise by itself. Suzuki and Nagata (2014) said that talent is not inborn. Suppose Mozart had been given to me to care for soon after his birth instead of having the influence of his wonderful musician father.

Therefore, musical abilities can develop rapidly if a child is in the right environment. Thus, a child who, from the beginning, experiences a rich musical environment will have more opportunity to reach a high standard of musical ability. Similarly, Djohan (2005: 194) asserted that the failure of students who have the talent to achieve success is not the result of inadequate study and practice but lack of support and encouragement from parents or teachers.

In summary, the environment plays a key role in the development of musical competence. This is in accordance with the behaviourist learning theory, which states that learning outcomes are the result of habituation.

2.2 *Musical competence*

The main aim of the learning process is to acquire knowledge and skills that are useful throughout life. The basic elements of music are rhythm, melody, and harmony. The essence of rhythm is a recurring pattern of tension and release, and time (Kamien 1988, 2). Dewantara (1977) recommends that early learning among young - children should be in the form of a game. In Piaget's (2013) conceptual framework, children aged 8-12 years are in the pre-operational stage, while teenagers (aged 12-18) are in the operational stage. In this framework, the development of the musical competence of children 1st to 3rd-grade should be based on knowing, feeling, and practising the basic elements of music and that of 4th to 6th graders on understanding and practising basic rules of music. As for 7th to 9th grade students should be able to display artistry in accordance with the techniques and procedures, while 10th to 12th graders should be able to implement art as part of social life at local, national, and international Levels.

2.3 *Bloom's taxonomy*

In 1956, Bloom, Simpson, and Gagne proposed a taxonomy of the cognitive domain that had six levels: *knowledge, comprehension, application, analysis, synthesis and evaluation.* In 1964, Krathwohl categorised the affective domain into *characterising, organising, valuing, responding, and receiving.* Between 1964 and 1967, Simpson (1971: 25) and others divided the psychomotor domain into seven levels: *perception, set, guided response, mechanism, complex overt response, adaptation, and origination.* These taxonomies have been used extensively in the field of education. Almost all learning outcomes, from planning to evaluation, use them as reference points in assessing the success of learning.

The learning outcome taxonomy theory was well established for half a century. In 2004, however, it was revised by Anderson, who re-ordered the cognitive domain by replacing 'evaluation' as the highest ability with 'creativity'.

In Indonesia, Anderson's revision of the theory had a considerable impact. Previously, the ability to critique was valued most highly, whereas now the ability to produce a new form or product was ranked higher. This trend seems logical since someone who is good at evaluating is not necessarily able to create something, while someone who is able to make something will find it easier to evaluate because they already have experience in making things.

Greenstein (2012: 13) built on Bloom's taxonomy to identify the skills that must be developed and applied by 21st-century students. For example, Greenstein relates the lowest level of

Bloom's taxonomy, knowing/remembering, to the development of the skill of "embedding learning in memory" and its application the ability to label a diagram, draw a picture of it, search for and bookmark it. The highest level of Bloom's taxonomy, creating, is translated into skills in relation to "generating something new, putting elements into a coherent whole, and reorganising ideas into new patterns". The implementation level of this skill refers to the ability to "create a new ending; design another way to convey the message of the story; come up with a new way to do, prepare, fix something, and create digital products like movies, simulation, games, and podcasts" (Anderson et al., 2001: 32).

2.4 *Learning theory*

In the middle of the 20th century, two different conceptualisations of learning were proposed —the empirical and the rational—each grounded in different learning theory. The former derived from behaviourism, which emphasises learning through habituation, and cognitive learning theory, which prioritises the discovery of meaning. In the behaviourists' empirical approach, learning occurs from a series of events or ideas, while rationalists see common sense as the main source of learning.

Behavioural psychologists focus on observed behaviours and behavioural changes and neglect discussion of unobserved thoughts and emotions. Cognitive psychologists, in contrast, argue that learning involves a change in one's ability to respond to a situation. Visible changes are a reflection of changes that occur within a person. It can state the nature of the environment, such as social attitudes towards music, can support or hinder the development of musicality.

The difference between these two views is bridged by the neo behaviourists, who claim that learning is not just a visible change in behaviour but also a process of change in a person. Considering all three views, in this research, learning is defined as "the formation of internal changes or new responses" (Hoy, 1980: 161).

In the mid-20th century, a new humanist stream of learning psychology emerged, followed more recently by cybernetics. Ausubel, a seminal figure in the humanistic school of cognitive learning psychology, suggested that students will learn well if "advance organisers" are appropriately defined and presented to students (Uno et al., 2004: 18).

The humanistic approach argues that the learning process must reflect the nature of human beings themselves (Uno et al., 2004: 20). The theory is eclectic, meaning that any theory can be used as long as it aims to "humanise humans" (Uno, 2004: 21). Humanistic theory provided the foundation for the work of Ausubel, Bloom and Krathwohl, Kolb, and Habermas. Ausubel (cited in Canas & Novak, 2010) described an approach to "meaningful learning". Bloom and Krathwohl (Anderson, 2001) developed a taxonomy of learning. In his theory of the experiential learning cycle, Kolb (cited in Sugarman, 1985) identified four stages of learning: concrete experience, active and reflective experience, conceptualisation, and active experimentation. Habermas argued that the highest level of learning is achieving understanding and awareness of cultural transformation. Mursell (2007) developed the approaches of Ausubel and Kolb into a more concrete learning theory using the principles of context, focus, socialisation, individualisation, sequence, and evaluation to achieve authentic learning outcomes.

2.5 *Environment*

As previously discussed, the environment has a significant influence on the development of musicality. Broadly speaking, there are three types of educational environments; informal or family education; non-formal education; and formal education. These environments are responsible for developing the students' potential, including musical development. The government, which is responsible for the implementation of formal education, needs to provide an atmosphere that is conducive to the development of all forms of intelligence, including optimal musical intelligence. France and the Netherlands provide education for music-talented students from childhood. It expects music lecturers in its universities to be able to help music-gifted children to become professional artists. In Indonesia, however, guidance for music-gifted children is still managed by private institutions which present courses that are not formally accredited. As a result, the development of musically gifted children is left to these institutions and lacks government monitoring and control.

2.6 *Expected competencies*

Each country has different social values and goals, so it is not surprising that learning objectives vary. Education is the most effective vehicle for instilling these values and developing relevant competencies among its young people, and educational curricula are structured accordingly. The goals in each subject are described in relation to the competency that must be achieved by each student at a certain level.

In Indonesia, the learning objectives in arts subjects focus on the development of aesthetic sensibilities through which traditional arts can be preserved and non-traditional arts can be nurtured. In addition, art is widely used as a medium for moral development. By contrast, in the Netherlands, the purpose of learning the art of music is for learners to be able to express themselves through the creation of a musical composition, and in France, mastery of the art of classical Western music is emphasised.

2.7 *Conceptual framework*

Indonesia, France and the Netherlands have different learning objectives in music education and different approaches to achieving these goals. The Netherlands is concerned with freedom of self-expression and provides educators with flexibility in the learning process. In Indonesia, the curriculum requires teachers to use a scientific approach in all areas, since learning is seen to be related to the acquisition of other skills and characteristics, such as moral development. France continues to teach the traditional arts in the same way it always has.

These different objectives impact on the development of musicality and the competence of learners, as well as on the art forms that result. The Netherlands is a country that is quick to respond to the new approaches, while Indonesia lags somewhat behind changing times, and France maintains its traditions. Hence this study addresses the following research question: What is the effect of these differences on the outcomes of music education in Indonesia, France and the Netherlands?

3 METHOD

3.1 *Research design*

The study adopted a mixed methods approach involving the sequenced collection of qualitative and quantitative data (Day, 2008). In the first stage, qualitative data were collected using the methods of observation and interview. In each country, music lessons were observed and interviews were conducted with teachers, principals, students, parents, and education experts.

In both Indonesia and the Netherlands, 30 lessons were observed. Observations were made in classes from the elementary school to university level. In Indonesia, the research was conducted in the Special Region of Yogyakarta (DIY) and Central Java provinces. In the Netherlands, data were collected in the provinces of Utrecht and Den Haag. In France, however, access was more limited. Observations were conducted on seven music lessons at the conservatory in Poitiers with middle school, high school and University of Poitiers students.

Interviews were conducted with teachers, principals, students, parents, and other stakeholders. The interviews to reveal their perspectives on the implementation of learning as observed in the classrooms.

Quantitative data were collected by identifying the level of student's musical competency in the three countries. The researcher observed the musical competency displayed by students and classified their abilities according to Bloom', Simpson, Krathwohl, Anderson's taxonomy, and Greenstein's development theory. Based on this researcher develop research instrument to collect the level of musical competency. There are six levels cognitive domain, level 1st, C1 (cognitive 1) is knowledge; level 2nd, C2 (Cognitive 2) is comprehension; level 3rd, C3 (Cognitive 3) is apply level 4th, C4 (Cognitive 4) is analyse; level 5th, C5 (Cognitive 5) is evaluation, and level 6th, C6 (cognitive 6) is creativity. Meanwhile psychomotor domain have eight levels. Level 1st, P1 (psychomotor 1) is perception; level 2nd, P2 (Psychomotor 2) is setting; level 3rd, P3 (psychomotor 3) is imitation; level 4th, P4 (psychomotor 4) is guide response; level 5th,P5 (Psychomotor 5) is Mechanism; Level 6th, P6 (psychomotor 6) is complex response; level 7th, P7(psychomotor 7) is adaptation; and level 8th, P8 (psychomotor 8) is originality. There are 4 level in Affective domain. Level 1st, A1(affective 1) is receiving; level 2nd, A2 (affective 2) is responding; level 3rd, A3 (affective 3 is valuing); and level 4th, A4 (affecktive 4 is organization.

3.2 *Research data*

In this analysis, there were three independent variables: country, domain, and school level. The dependent variable was the level of musical competency in relation to the cognitive, psychomotor, and affective domains. The results are presented in the form of distributions of competencies in the cognitive, psychomotor, and affective domains.

3.3 *Data analysis technique*

The qualitative data were analysed descriptively to categorise the observed musical competencies in relation to the cognitive, psychomotor, and affective domains. The qualitative data were used to supplement the quantitative data in terms of categorising each domain of the taxonomy. Quantitative analyse by using descriptive statistic of tendency central.

4 FINDINGS AND DISCUSSION

4.1 *Goal lines of the music lesson*

In Indonesia, the Central Body for the National Curriculum states that the aim of arts education is to develop aesthetic sensibilities in relation to expressing, exploring, creating and appreciating all forms of the arts, including literature and the performing and visual arts. Therefore, the objective of music teaching at schools is not to turn students into artists but to enhance their aesthetic sensibilities. Since the objective of music teaching is to provide an aesthetic experience, the teaching processes are organised in such a way that students enjoy the teaching/learning process.

Interviews with the music educator from Utrecht University, Barensend Mitchel, and the head of the Music Department of Hogeschool Voor de Kunstent Utrecht, Christian Boel, indicated that the Netherlands government gives teachers the freedom to develop their teaching methods as long as the curriculum goal is achieved. In the Netherlands, education has experienced a great deal of change. The overall mission and vision most of the schools in the Netherlands, however, are fully supported by the teaching staff. In Indonesia, by contrast, educators are required to adopt the scientific approach to teaching. The Dutch government's policy means that teachers can be creative in developing their learning strategies, and the country has produced a rich variety of teaching methods. In 'regular' music lessons, children are taught to simultaneously read notes and play them, individually or in groups. They start by playing simple tunes with a straightforward rhythm and gradually progress to more complex pieces. However, many schools have introduced variations. For example, the Royal Conservatoire, The Hague, has developed a program for pre-school children in which they learn music in the same way as they learn a language. In language learning, children's development goes through the stages of listening, babbling, speaking words, speaking short phrases, and, finally, speaking entire sentences. For example, a child only learns to write the word 'tree' at around the age of six, when he or she can speak properly and can recognise different trees, such as a chestnut tree, an oak tree, a Christmas tree, an apple tree. When shown a tree, the child can then make the logical connection: that is a 'tree'.

France, however, still adheres to traditional values to which music learning objectives are directed. The repertoire of the baroque, classical and romantic eras still dominates music education in France.

In Indonesian curriculum, music learning is generally associated with moral development. Similarly, in the Netherlands, the music teacher seeks to generate positive emotions around the relations between self and other. According to the interview with a lecturer in France, music learning is only focused on mastering music, not on morals.

4.2 *Teaching and learning process*

In Indonesia, observations were made of more than 30 music learning classes in Public Elementary School 1 (Kranggan, Temanggung Regency, Central Java), IT Lukmanul Hakim Elementary School (Kota Gede, Yogyakarta), Muhammadiyah Middle School (Wonosari, Gunung Kidul, Yogyakarta), State Middle School 2 (Temanggung), and Middle School (Depok, Sleman, Yogyakarta). In general, the process of music learning begins with theory and moves on to practice. Although the approach recommended by the government is student-centred, its implementation remains teacher-centred.

In the Netherlands, observations were made of 30 classes at 1) Dacosta basis school Hoograven, Utrecht, 2) Cals College Nieuwegein, 3) Junior School, Keesboeken, Bilthoven, 4) Montessori Lyceum Herman Jordan, zeist 5) Hogeschool Voor De Kunstent Utrecht, and 6). The Royal Conservatoire, The Hague. Here, theory and practice are taught in an integrated way.

In France, observations are made on seven music classes in the Conservatoire de Poitiers and the University of Poitiers. Here, the process of music learning was observed to be similar to that in Indonesia, that is, theory and practice were separated, and the instructional approach was teacher-centred. However, the difference between the Indonesian and French systems is that the Indonesian instructional system uses the "fixed-do", while the French instructional system uses the "movable-do".

A comparison of the learning process in the three countries is shown in Table 1.

4.3 *Musical competency development*

As stated earlier, Bloom *et al.* classified learning outcomes into three domains and related levels: cognitive (six levels), psychomotor (seven levels), and affective (four levels). Greenstein developed Bloom's taxonomy to incorporate 21st-century skill needs. Table 2 shows the descriptive data of the type of musical competency development in each results of analysis of the musical competencies observed in the three countries based on this classification.

Based on factorial design in Table 2 shows that both in Indonesia and the Netherlands we have 90 data. This data base on 30 observation in Indonesia, dan 30 observation in the Netherlands. Each observation have cognitif, psichomotor, and affective data, that means for one observation have 3 type of data that is cognitive, psichomotor, and affective of musical competency.Mean while we have 21 data of cognitive, pyschomotor, and affective of musical competency base on 7 observations in France.

Regarding the level of domain competency and duration in each countries depected on Table 3, as follows:

These results show that, on average, the Netherlands has a higher level of competencies compare to both Indonesia and France. However, the proportion of musical development duration of cognitive, psychomotor, and affective domains among the three countries is similar.

Table 1. Comparison of music teaching and learning models in Indonesia, the Netherlands and France.

No	Component	The Netherlands	Indonesia	France
1.	Curriculum	Curricula at different educational levels are closely integrated. A curriculum is established by Parliament based on input from experts and other stake holders i.e. a bottom-up approach.	Curricula at different educational levels are less well integrated. Each educational level has its own authority and the curriculum is set in a top-down approach.	Learning music in primary and secondary education is coordinated by the Director of Provincial Music Education and the curriculum is supervised by the directorate.
2	Learning materials	Learning materials are organised thematically and integrated. More time is devoted to practice than theory.	Learning materials are subject-t specific and separate. More time is devoted to theory than practice.	Theory and practice are studied separately, similar to the approach in Indonesia.
3.	Connecting music material with other fields	Music learning is expected to be associated with other fields, especially moral education.	Learning music is linked to other fields in accordance with the nature of the subject matter.	Learning music is not associated with any other field.
4.	Learning approaches	Learning approach is student-centered and well-implemented.	Learning approach is student-centered in theory but it is not yet implemented optimally.	Teacher-centre at the level of implementation, similar to Indonesia.
5.	Learning method	Many methods are used to encourage students' participation e.g discussion, presentation, demonstration, etc.	Previously the lecturing method dominated, but since 2013, learning is oriented towards the scientific method.	Mainly lecturing, with some demonstration and drills.
6.	Learning procedures	The learning process is similar at all education levels: 1) Teacher sets assignment; 2) Teacher gives students the opportunity to work in groups; 3) Students work in groups to produce a work; 4) Each group presents their work; 5) Teacher and students evaluate the work presented.	Learning process is as follows. 1) Teacher sets assignment; 2) Teacher presents learning materials through lecture or demonstration; 3) Teacher conducts evaluation without involving students.	Similar to the procedure in Indonesia.
7.	Culture	The relation between teachers and students is close, informal, and open.	The relation between teachers and students is very formal, which is reflected in the teaching and learning process.	Like Indonesia, the relationship between teacher and student is formal.

5 CONCLUSION

The results of the study show that there are significant differences in the development of students' musical abilities in the Netherlands, Indonesia, and France. The differences lie in the depth or the competency level of the learning. From the perspective of Bloom's taxonomy, the Netherlands achieves higher levels of development in students' musical potential in the cognitive, psychomotor, and affective domains. Analysis of the qualitative data indicates that music education in the Netherlands places greater emphasis on the affective dimension, while Indonesia and France focus more on the cognitive and psychomotor domains.

Table 2. Descriptive data of the type of musical competency development.

		Factorial	N
Country	1	Indonesia	90
	2	The Netherlands	90
	3	France	21
Domain	1	Cognitive	66
	2	Psychomotoric	68
	3	Affective	67
level of School	1	Elementary	30
	2	Junior school	96
	3	Senior High School	36
	4	University	39

Table 3. Statistical description.

Descriptive Statistics				
	Country	Mean	Std. Deviation	N
Level of domin competency	Indonesia	3.29	1.359	90
	The netherlands	4.26	1.107	90
	France	2.90	1.300	21
	Total	3.68	1.348	201
Durasi (%)	Indonesia	33.56	23.202	90
	The netherlands	33.56	22.389	90
	France	33.33	56.426	21
	Total	33.53	23.075	201

It can be argued that the affective domain plays a fundamental role in human learning by motivating students to engage with an activity (or not). In this way, affective development can impact on other domains. The results of this study, then, indicate that music learning should emphasise affective development, followed by cognitive and psychomotor development.

REFERENCES

Anderson, L. W., Krathwohl, D. R., Airasian, P. W., Cruikshank, K. A., Mayer, R. E., Pintrich, P. R., … & Wittrock, M. C. (2001). A taxonomy for learning, teaching, and assessing: A revision of Bloom's taxonomy of educational objectives, abridged edition. *White Plains, NY: Longman.*

Astuti, K. S. (2010). Pengembangan model evaluasi pembelajaran seni budaya SMP [Developing an arts and cultures evaluation model for Junior School]. *Jurnal Kependidikan: Penelitian Inovasi Pembelajaran, 40*(1).

Astuti, K. S., Zuchdi, D., Sudiyono, T. Wubbels., & Hoogeveen, K. (2010). Developing Model for Teaching and Learning Music in Public School Based on Comparative Study between Indonesia and the Netherlands. *Laporan Penelitian. Yogyakarta: Lemlit UNY.*

Astuti, K. S., Widyantoro, A., Boel, C., & Berendsen, M. (2017). Developing an integrated music teaching model in Indonesia based on the Dutch music teaching model as the implementation of the 2013 curriculum. *Journal of Arts, Science & Commerce, 8*(2), 118.

Campbell, D. (2001). Efek Mozart bagi anak-anak [The effect of Mozart for children]. *Jakarta: PT Gramedia Pustaka Utama*, 331.

Cañas, A. J., & Novak, J. D. (2010). The theory underlying concept maps and how to construct and use them. *Práxis Educativa, 5*(1), 9-29.

Day, C., Sammons, P., & Gu, Q. (2008). Combining qualitative and quantitative methodologies in research on teachers' lives, work, and effectiveness: From integration to synergy. *Educational Researcher*, *37*(6), 330-342.

Djohan (2005). *Psikologi musik* [Psychology of music]. Yogyakarta: Buku Baik.

Gardner, H. E. (2008). *Multiple intelligences: New horizons in theory and practice*. Basic Books.

Greenstein, L. M. (2012). *Assessing 21st century skills: A guide to evaluating mastery and authentic learning*. Corwin Press.

Hoy, A. W., & McCune-Nicolich, L. (1980). *Educational psychology for teachers*. Prentice-Hall.

Khan, H. I., Subagijono, & Timur, F. K. (2002). *Dimensi mistik musik dan bunyi* [Mystic dimension of music and sound]. Penerbit Pustaka Sufi.

McPherson, G. E. (1997). Giftedness and talent in music. *Journal of Aesthetic Education*, *31*(4), 65-77.

Mudjilah, H. S. (2003). Upaya peningkatan prestasi belajar mahasiswa melalui interaksi kelas pada mata kuliah metodologi penelitian [Efforts to enhance student achievements: Teaching students through class interactions in course research methodology]. *Imaji*, *14*(2), 120-128.

Mursell, J. L. (2007). *Successful teaching*. Littlefield Press.

Pallant, J. (2007). SPSS survival manual, 3rd. *Edition. McGrath Hill.*

Perkins, R., Yorke, S., & Fancourt, D. (2018). Learning to facilitate arts-in-health programmes: A case study of musicians facilitating creative interventions for mothers with symptoms of postnatal depression. *International Journal of Music Education*, 0255761418771092.

Seashore, C. E. (1937). The psychology of music. *Music Educators Journal*, *23*(4), 30-33.

Simpson, E. (1971). Educational objectives in the psychomotor domain. *Behavioral Objectives in Curriculum Development: Selected Readings and Bibliography*, *60*, Page 25.

Siswa, M. L. P. T. (1977). Bagian I Karya Ki Hadjar Dewantara: Pendidikan [*The first part of the work of Ki Hadjar Dewantara: Education*]. Yogyakarta: Taman Siswa.

Sugarman, L. (1985). Kolb's model of experiential learning: Touchstone for trainers, students, counselors, and clients. *Journal of Counseling & Development*, *64*(4), 264-268.

Suzuki, S., & Nagata, M. L. (2014). *Ability development from age zero*. Alfred Music.

Sutapa, M., Hantoro, S. S., & Astuti, K. S. (2008). *Analisis komparatif competitive advantageterhadap pendidikan Belanda dan Jepang*[Comparative analysis of advantage competitive toward The Netherlands and Japan's Education]. Yogyakarta: UNY.

Uno, H. B., Sofyan, H., & Atmowidjojo, S. (2004). *Landasan pembelajaran (teori dan praktek)* [The foundation of learning (theory and practice)]. Yogyakarta: Universitas Negeri Yogyakarta press.

21st Century Innovation in Music Education – Astuti et al. (eds)
© 2020 Taylor & Francis Group, London, ISBN 978-0-367-11122-9

The role of music education in promoting moral behaviour among children

M.U. Okeke
Department of Music, Federal College of Education, Eha-Amufu, Enugu State, Nigeria

ABSTRACT: Music is a product of the society which also reflects the needs of the society. The culture of any society can also be observed from their music. Music and culture go hand in hand with each other. Children especially in developing countries are today living in cultural poverty. Many factors, such as industrialization and urbanization, have contributed to the cultural alienation of the children. Music education should be used to inculcate ideas that will help to shape the morals of every child in any given society. Music, as a tool in fostering moral behaviour, can be used as an instrument to give cultural orientation to children by establishing a link between tradition and modern life. Music, if properly planned and taught according to National Policy on Education, would help the development of the moral, physical, mental and social education for the children.

1 INTRODUCTION

Moral has been defined as a measure of good or otherwise of human character or behavior. Osifeso (2008) defines moral as the resistance to temptation or inhabiting behavior that is considered to be wrong. Moral can be described as the distinction between right and wrong. Morality has also been defined by Bull (1983) as "living together in a human society which enables an individual to be concerned with himself and with others." Morality can also be defined as the degree of conformity to moral principles, moral right, conduct, etc. Many of the negative attitudes of Nigerian children could have been developed through indirect consequences of ignorance about traditional values. Through music, children can be socialized into social and moral laws. This paper, therefore, examines the role of music in promoting moral behaviour among Nigerian Children.

2 LITERATURE REVIEW

2.1 *Moral decadence in society*

One of the basic social problems of the Nigerian society today is moral decadence. Moral decadence describes a social situation in which societal values and ethnics are readily violated. Moral decadence is manifested in the following acts or behaviour: cheating (419), telling lies, breaking promises, and disrespecting the elderly or even the law. Others include shameful sexual behaviour, stealing, and so forth. In lamenting the standard of moral conduct of the society, Adeyinka (2003), expressed that "over the years, the actual standards of conduct of children and adults do not meet the desirable standards, also the conducts in general are disturbingly poor."

2.2 *Possible causes of moral decadence in society*

Some of the immediate causes of moral decadence in the society include: materialism, drug abuse, and poor socialization at the home level. Other causes are economic hardships, conflicting social values, and imported cultures through films. Some homes hold liberal values and give children unlimited freedom. Materialism and economic hardship contribute to making some parents abandon their homes from morning till evening. In some of such abandoned homes, even delinquent house-helps are not available, only the children are left at home to labor, feed, and take care of themselves. If the children are too young to stay alone at home, house-helps that are also delinquent adolescents are employed to take care of the young ones. Such house-helps contribute to transferring their delinquent behaviours to young children who get so much attached to them due to the absence of their parents at home. Foreign films and global TV stations through cable satellite play their own role too. When parents are not at home, the youths watch programs of their choice which are not necessarily appropriate for their age. Drugs too are freely used by youths when they are not adequately supervised. There are thousands of delinquent adults who assist the youths in getting drugs and other narcotics. The effects of the drugs are so dangerous because intoxication can lead a person to multiple offences, such as stealing, lying, aggression, and a host of others.

2.3 *Effects of moral decadence on national development*

The effects of moral decadence on the society are detrimental to national development. It can lead to lack of trust even among relatives. Insecurity of lives and property can also be a likely effect. It results in the un productivity. When young, productive adults are engaged in antisocial behaviours, they normally become unemployable and security risk to their society. It is the fear of this effect that makes leaders and responsible members of the society continue calling for moral reformation of the society through any means possible. This could possibly be the basic reason for the need for Sharia implementation on Muslim dominated states in the county.

2.4 *Moral education and morality*

The encyclopedia of education enumerated some aims of moral education as: a) To inculcate moral ways of judging actions; b) To foster a belief for the adoption of some fundamental principles or ideas or values as a final basis; c) To foster a belief in or an adoption of a number of concrete norms, values, and virtues; d) To develop a disposition to do what is morally right or good; e) To promote the achievement of reflective autonomy, self-government, or spiritual freedom even if it leads the individual to criticizing prevailing ideas, principles, or rules; and f) No doubt these objectives are laudable and worth pursuing.

Masqud (2007) specifically mentions some basic characteristics expected from a morally educated person. The characteristics include a) A good deal of factual knowledge; b) A positive commitment towards the values of morality; c) The ability to communicate with others; and d) Ability to understand the feelings of others and those of his own

In spite of the laudable objectives, moral education, the Nigerian society is still bedevilled by moral decadence. This leads one to ask the following questions on moral education:

1) Who becomes the biggest responsibility of moral training? the home, the community, or the school?
2) To what extent has moral instruction given in the schools achieved morality in the society or even within the school environments?
3) Is there a way of reforming moral education programs of the schools to achieve the desired morals?

Research and theories of moral development tend to support the role of the home (family) in moral development either directly or indirectly. Bull (1983) and Osifeso (2008) are in full support of the strong influence of the home on moral education. The Encyclopaedia of Education noted that the school had a role to play but failures in morals would be blamed on the home. In support of the influence of the home, Masqud (2007) cited Freud's psychoanalytical theory which emphasizes the relevance of early childhood attachment with parents in the development of the child's super ego. The Encyclopaedia of Education cited research reports of prominent psychologists, such as Freud, Kohlberg, and Hoffman. The reports too tend to support the role of the home in the moral character formation, pointing out the ineffectiveness of the school. The Encyclopaedia of Education further also noted that reports that moral and religion education classes were found to have no direct influence on the moral character development.

In addition, psychologists believe that a specific act of misconduct such as stealing or lying is largely determined by situational factors. Looking at Skinner's, Bandura's and others' applications of learning theories to habit formation cited by Masqud (1983), the role of the society now dominates. This is because the learning theory emphasized the effect of the reinforcement in habit formation. According to the theory, society is in control because morality is inconformity to societal expectation based in reinforcement.

3 DISCUSSION

3.1 *The importance of music*

Music plays important roles in the lives of people. From the most primitive tribes, with their war chants and tom-toms, to modern day societies, with their huge choirs and orchestra, full of fascinating instruments, music has provided people with opportunities for self-expression and rich, rewarding enjoyment. All races of all kinds, all types of human beings have enjoyed one musical expression or the other. The history of development of music is closely connected with the history of mankind. Each particular group of people in every period in history has used musical sounds to satisfy their needs for human expression, love, loyalty, worship, feelings of joy and sorrow, and the desire to share beauty and they have all been expressed through music for centuries.

Our entire social lives are centered around music. Dancing is undoubtedly one of the most popular social activities. It is a body motion inspired by and added to different types of music. Being able to identify the type of music played and being sensitive to the beat of the music are absolutely necessary if one wants to dance correctly.

Music also plays an important role in many of the major types of religious services. Musical sounds and prayerful words are joined together to give greater meaning to a particular method of worship. Individuals, choirs, instrumentalists, and the whole congregation make sounds resembling the glory of God through inspirations typical of every religious types. The church or synagogue is still another place where one seldom realizes the importance of music but music nevertheless plays a vital role.

3.2 *Music as an instrument for promoting moral behaviors among children*

Music is a product of the society which also reflects the needs of the society. The kinds of music which can be used as an instrument for socio-cultural development are those related to the problem within the environment. Most Nigerian children of today are living in cultural poverty. Many factors, such as industrialization and urbanization, have contributed to the cultural alienation of children. Music is a tool which can be used to give culture orientation to children by establishing a link between tradition and modern life, from which a powerful message for them evolves. One of the ways of using music for the socio-cultural development of children is through music education. The benefits would include, among others, the preservation and transmission of cultural values, the provision of enjoyable recreation, and the promotion of creativity. Since music is one of the school subjects recognized and recommended by the National Policy on Education, the subject, if properly planned and taught, would help in the moral, physical, mental, and social education of the Nigerian children. As Nye and Nye (1985) observe, the aims of music education in elementary schools include:

1) To foster enthusiasm and love for music
2) To lay the foundation for music literacy especially in tonic sol-fa notation
3) To enjoy and develop an interest in Nigerian music
4) To develop a soft and pleasing vocal tone

Bearing all these in mind, various states of the nation have drawn up prescribed schemes of work for instructions in the primary schools. This scheme of work is divided into four sections: local trainings, percussion work, listening activities, and theories of music. Children in primary one are first introduced to vocal and percussion work and as they progress in their education, listening activities and theories are added. By the end of their primary school, they would have acquired knowledge in many areas of music, such as basic musical terms, rhythmic values, and sing by sight. They also would have been able to write in sol-fa notation and play some musical instruments.

Music plays an important role in the education of children particularly because it can be used to impact and reinforce knowledge. Many school activities are performed with music accompaniment and many school subjects are taught with the use of music. Music is used, for instance, during morning assembly, singing lessons, games, and recreation activities. Songs of hygiene, discipline, social conduct, and moral disposition help to mound children into responsible adults. For instance, songs of hygiene learnt at school help to make children observe health rules. For instance:

This is the way we brush our teeth,
Brush our teeth, brush our teeth,
This is the way we brush our teeth,
Early in the morning.

Music as a carrier of cultural and social influences encourages children to strive to exercise some control over their environment and to influence many of the situation in which they find themselves. Music encourages children to manipulate situations and make alteration through symbolic transactions with others. It persuades them to exchange meaningful ideas with people, to expose their views about particular issues, and to seek information to maintain contact with other people in the society. Music, such as the national anthem and school's anthem, help to instill the spirit of patriotism on children and of course, when the interest is ignited and a good foundation is laid, music can provide a fulfilling future career for children. Bank (1976) observes that: 'Music is often described, particularly during periods when its romantic aspects are uppermost, as emotional expression'.

Perhaps this is one of the reasons why children identify themselves with music. It is a vital tool for reaching out to them because it can provide information which cannot be effectively disseminated through other means. Through music, social and moral laws are transmitted to the children and traditional education is achieved and impacted. The role of music in promoting cultural development of the Nigerian child is therefore vital and significant.

4 RECOMMENDATION

Based on the discussion of this paper, it is recommended that music should be used to address issues which directly affect the lives and situations of children. To this end, the arts council should encourage a popular involvement in drama and music activities by funding more community music groups and children's drama. Music should be used to build up confidence in children as a step towards socio-cultural development. Music must become an instrument in the hands of the performers, teachers, community leaders, parents, guardians, and society as a whole for the socio-cultural development of Nigerian children.

5 CONCLUSION

Morality can be defined as the degree of conformity to the moral principles, moral right, and conduct. Many of the negative attitude of the Nigerian children could have been developed through indirect consequences of ignorance about traditional values. Music, as an instrument for promoting moral behavior, has encouraged children to exercise some control over their environment and to influence many of the situations in which they find themselves.

REFERENCES

Adeyinka, L.A (2003). The Role of Science in Moral Education in Ojonti, A.N. *New Perspective in Moral Education* Ibadan: Evans Brothers (Nig) Ltd p. 158–159.

Bull, N.J. (1983) *Moral Education*, London, Routuedge and Kegan Paul. Encyclopedia of Education (1971). Vol. 6 U.S.A. Macmilliam Company and Tree Press.

Masqud, M. (2007). *Research in Moral Education.* A Survey of some Psychological Explanation in Otonit, A.N. New Perspective in Moral Education, Ibadan: Evans Brothers (Nig) Publishing Ltd. P. 115–151.

Nye R. and Nye V. (1985). *Music in the Elementary School.* New Jersey Prentice Hall Inc.

Osifeso, G.A.T. (2008). Enhancing Moral Values in the Society. The Role of Women in Yetune, *Women in Colleges of Education Review.*

21st Century Innovation in Music Education – Astuti et al. (eds)
 ISBN 978-0-367-11122-9

Music learners and the profile of music teachers in the digital world: The Nigerian scenario

P.C. Ozoaghuta
Department of Music, Federal College of Education, Eha-Amufu, Enugu State, Nigeria

ABSTRACT: The teaching and learning of music has significantly changed in this millennium. Surprisingly, some music educators are uninformed about these changes in music teaching and learning. Most of them are intellectually deficient in this regard. The way they taught twenty years ago is still the same as the way they teach today. They all believe that the approach they used in the past can be used in the present and will always be relevant in the future. This can be accounted a main reason as to why increasing number of music students are still struggling to develop basic musical abilities despite the fact that students of today are born into the age of technology. This paper examines the differences between the conventional music learners and teachers and the 21st century music learners and teachers, taking into consideration the historical perspective and philosophy of music education in Nigeria.

1 INTRODUCTION

The world in which we live is dynamic in nature. It is a changing world because so many changes have taken place and more are yet to come. What existed during the time of our forefather is no longer what we have now. The present society has clinched to so many changes and innovations. As it evolves, so do we or otherwise, we are left behind. Jegede (1983) agrees with this assertion explaining that the current development in science and technology has so greatly affected the lives of every human being that to be ignorant of the basic knowledge of these developments is to live an empty, meaningless, and probably unrealistic life. Deducing from the above assertion, we have no choice but to grow with it. These changes and innovations are also noticeable in the area of music education in that the music of our forefathers has been eroded and new genres have emerged. Technology has been advancing and there will not be any dull moment again in music education. Okafor (2005: 210) observes that music education has come a long way from the colonial times when most of the music education in an institution consists of a singing class in primary or secondary schools. Days have gone when learners are viewed as empty vessels whose primary role is to receive passively any information from the teachers whom they see as masters of knowledge and wisdom.

2 LITERATURE REVIEW

2.1 *Historical perspective of music education in Nigeria*

Music education in Nigeria has a long and intricate history. This is because a typical African man is indissoluble to his communal music. In other words, Nigerian societies see music as a veritable way of life and no occasion, whether sad or joyful, is without its peculiar musical expressions (Okafor, 2005). That notwithstanding, the history of music education in Nigeria would not be complete without enumerating and briefly describing various ways in which Nigerian communities have been imparting musical knowledge to their young ones before the colonial era. Therefore, music education in Nigeria exists in the following forms;

1) Informal music education
2) Apprenticeship system
3) Formal music education

2.1.1 *Informal music education*

This is one of the ways music education takes place in Nigerian societies. It is extremely common to African countries especially Nigeria. It is a way of transmitting musical knowledge to the younger generation through informal settings without taking cognizance that music education is taken place. It takes the form of careful listening to and repetition of songs (Mapoma, 2001). Okafor (2005: 189) and Agu in Yekini-Ajenifuja (2011: 69) argue that learning traditional music is a lifelong journey and starts quite early in life. So, under this platform, the young people in Nigerian societies are innocently introduced to musical rhythms through their mother. They learn some musical skills from their mother who is a chief maker of lullabies, game songs, and folk songs. Nigerian mothers are good at backing their babies while engaging in house chores. They often do this with some musical rhythms to ease their tension, not knowing they are, at the same time, passing on the music to the babies. And when such child grows, the level of his or her musical knowledge and skills magnifies. They are also introduced to the do's and don'ts of the societies with the help of their mothers through lullabies.

2.1.2 *Apprenticeship system*

The second way of imparting musical skills to the young Nigerian is through the system of apprenticeship. In Nigerian societies, some people are naturally gifted in music. Others inherit the skills from their father even without going to formal school where music is taught. So, this system is somehow related to informal music education but the only difference is that the students are attached to an expert that are greatly endowed with musical skills for coaching within a specified time agreed by both parties. Learning in this system usually occurs through participant observation and imitation. The trainee is expected to observe and possibly imitate the trainer as he or she performs. Gradually, the trainee will be trained to improve their musical skills that could enable him to fit in with the society in which he lives. Okafor (2005: 302) argues that this system always relies on the expert to teach.

2.1.3 *Formal music education in Nigeria*

The history has it that Nigeria was colonized by the British. Also recorded was that western education started in Nigeria with the arrival of the British traders and colonists. Parts of these British colonists were missionaries that came exclusively for preaching the word of God. When they came, they saw that the best way to achieve their aims and objectives was to incorporate music in their teachings. They discovered that music making was imperative if they were to capture the minds of Nigerians and, at the same time, to achieve their aims and objectives. This led to the introduction of mission schools where music was taught. In line with this observation, Okafor (2005) noted that music education in Nigeria started to be formal with the introduction of mission schools. Based on this, some steps were taken by western-trained Nigerian Musicians, like T.K.E Philip, Fela Sowande, William Wilberforce Echezona, and Akin Euba, to incorporate the teaching and learning of music at primary, secondary, and tertiary levels in Nigerian educational system. This was why formal music education in Nigeria was accorded huge respect. It began to expand. Teacher training colleges and other tertiary institutions where music was taught began to emerge. The primary school was the lowest level while the tertiary institutions were the peak of the formal music education in Nigeria. Robert A. Coker, one of the products of the mission school was said to be the first to study music in Germany as far back as 1871. The colonization nevertheless brought some changes. The system was based on western contents which tremendously did not reflect the African ideology and philosophy of music. Vidal in Aninwene (2011: 82) noted that University of Nigeria, Nsukka produced its first baccalaureate degree graduates in music in June, 1965. He went further to say that its' music program considered the cultural milieu in which it operated and incorporated courses in African music, alongside western music with operatic workshop on its curriculum. It was started with a good balance with the introduction of courses in African music. It was a great step towards the dignity of music and the musicians in Nigeria (Okafor, 2005).

2.2 *Philosophy of music education in Nigeria*

The philosophy of music education in Nigeria is nothing but the study of fundamental questions about the nature and value of music education in Nigeria. It deals with what is basic to learning as well as the nature of what is to be learned. It serves as a guide for daily professional decisions and choices about music education. According to the National Policy on Education.

Education in Nigeria is no longer a private enterprise but huge Government venture that has witnessed a progressive evolution of Government's complete and dynamic intervention and active participation. The Federal Government of Nigeria has adopted education as an important par-excellence for affecting national development. Government has also stated that for the benefit of all citizens, the country's educational goals in terms of its relevance to the needs of the individual, as well as in terms of the kind of society desired in relation to the environment and the realities of the modern world and rapid social changes should be clearly set out.

Therefore, Nigeria's philosophy of music education is based on the integration of the individual into a sound and effective citizen. Nigerian societies believe that formal system of music education should be based on indigenous model in generating the contents and the pedagogy. And when this is done, it assists the young people to understand their immediate setting with a grand view of their socio-cultural base. This will in no doubt help to develop young people to be a complete being. This is to say that a person who has adequately received music instructions in Nigeria both inside and outside the formal school system must be able to function effectively in that regard and in addition, must be able to contribute her quota towards national development through the musical skills he or she has acquired.

3 DISCUSSION

3.1 *Who are the conventional music learners?*

As the name implies, conventional music learners are those who, even with the advent of information technology, still solely depend on their teachers for musical knowledge. In other words, they are completely dependent on their teachers for musical knowledge. They ask little or no questions during the teaching–learning process. They are left to memorize facts, recall details without having a concrete view of what is learned. These types of learners are resistant to innovations and changes in teaching and learning. They are expected to submissively receive and believe all that their master is saying; whether it is the right thing or not is none of their business.

3.2 *Who are the 21st century music learners?*

They are those learners who go extra-mile in sourcing for information for themselves despite the fact they have received instructions from their teachers. They see their teachers as guides or facilitators for their quest for knowledge. Unlike conventional learners, they ask intelligent and timely questions where necessary during the teaching –learning process and always demand for clarifications from their teachers when they do not understand. They are always excited at every innovations and changes made in their learning styles. They are flexible and contributive during teaching-learning process. Because of the zeal they have, always wanting to explore, sometimes they attempt to take over the control of the class from awkward teachers. They have quick access to knowledge for all their challenges through the use of information technology or other related ICT platforms. They are also creative and have higher engagement rate during the teaching–learning process.

3.2.1 *Who are the conventional music teachers?*
They are those teachers who are resistant to changes and innovations in their mode of teaching. The way they taught twenty years ago is still the same as the way they teach today. They are regarded as knowledge dispensers rather than facilitators during teaching-learning process. They have expert mindsets and not learner mindsets. They initiate discussions in the classroom and exclusively focus on textbooks contents and notes. They feel ashamed to ask their colleagues for help when necessary despite the fact that no one knows everything. This is to say that they often shy away from their weakness and sometimes skip the contents they are not familiar with. Many times, they often resort to the use of corporal punishment to maintain classroom discipline.

3.2.2 *Who are the 21st century music teachers?*
In the past, we valued teachers for their "expert mindsets". Now we have to look for ones with "learner mindsets." The 21st century music teachers are those who are delighted at any innovations and changes made in their mode of teaching. They are flexible in their teachings and see themselves as co-learners. They are knowledge facilitators and not teachers per se. They take challenges and run towards their weakness. They feel confident to ask for help from their colleagues when necessary.

3.2.3 *The profile of music teachers in the digital world of the 21st century*
The aspiring music teachers in this digital world of the 21st century should, as a matter of exigency, drop the long-aged garment of teaching and embrace contemporary ways of teaching because technological innovations have extremely changed the learning styles and aspirations of music learners and the best way to connect with them is to see themselves as a co-learners and not teachers. Nye and Nye (1970) observed that 'learners learn best when they see meaning and have a part in establishing their purposes (p.78). This means that if music students are allowed to make their own choices, they own their learning and increase intrinsic motivation. This is an ideal recipe for better learning outcomes. However, there is a need as music teachers to examine what we do and how we go about doing it especially in the classroom. In summary, below are some of the profiles of music teachers in this new age.

- See themselves as co-learners
- Embrace change
- Go globally
- Feel secure asking their colleagues for help in areas they are not familiar with
- Allow themselves to fail often
- Run towards their weakness not away from it
- Allow the students to teach others
- Step outside their comfort zone

3.2.4 *Methodological approaches to be adopted by the 21st century music teachers*
Below are some innovative approaches the 21st century music educators should adopt in order to have an exceptional music teaching in the classroom.

1) Integrating technology
   In order to connect with the students of these days and make teaching and learning of music more inspiring and motivating, music educators must learn how to incorporate technology in their teaching. They must be conversant with technology available in their area of specializations that comes so naturally to the young. They should know that music teaching and learning has become trouble-free unlike before when technology was not too advanced.
2) Differentiating instruction
   There are three main learning styles of the 21st century music learners: visual, auditory, and kinaesthetic. This is to say that some students learn better when they see images or graphs of the learned materials. Others learn through what they hear while the last group learned when they put the learned materials into practice. To connect with them, music educators should first study the background of each student thoroughly to determine the kind of

attention such child needs from them. They should also know that students from rural background require more attention from them than those from the urban and well-educated families.

Having studied the learners, they can now select the learning activities that best suit the learners. They can also differentiate instruction by matching assignments or piano pieces exercises to readiness levels, offering appropriate extension activities as required.

3) Planned repetition

About 25-30 students in some schools make up a class. During teaching-learning process, it is not possible that everybody in the class understands all that the teacher is saying. Some students may be absent-minded. Some would hear but would not understand while others may hear and understood things which are different from what the teacher wishes that they would understand. In this type of situation, music educators are bound to repeat some words in such a way that it will make meaning to the students. The more often a concept is purposefully and meaningfully repeated, the more chances students have in remembering them.

4) Pair-share approach

Pair-share approach is another methodological approach the 21st century music teachers should adopt. It is a learning structure that demands that students learn in pairs or groups. The teachers' job here is that of a facilitator. He/she allows the student to discuss and share ideas about the concept under discussion with partners or in groups having been introduced with the concept to maximize the level of student's participation and consolidate what they have said.

5) Component teaching

This approach demands that music educators teach only those aspects of music they are specialized in. The teaching of keyboard harmony should not be allocated to someone who is not fit in that regard. Someone who has not been a conductor for once should not be made to handle conducting class. Round peg must be put on a round hole and not the other way around. The allocation of courses to music educators should be based on their specializations.

4 CONCLUSION

This paper has carefully examined the differences between the conventional music learners and teachers and the 21st century music learners and teachers, taking into consideration the historical perspective and philosophy of music education in Nigeria. It is obvious that these days, music learners are technologically savy and the best means to connect to them is to integrate technology into music teaching. I am confident that if these approaches of teaching recommended above are properly adopted by music teachers, there will be radical improvements in music education.

REFERENCES

Aninwene, E.O. (2011). "Music is theatre: Meki Nzewi's Approach to operatic workshop and production at the University of Nigeria, Nsukka. *Mekism and Knowledge sharing of the musical Arts of Africa:* A book of general reading in honour of Professor Meki Nzewi. 82–83.

Eze, A.E. (2003). "Assessment of laboratory facilities and equipment for the teaching and learning of secondary school physics". *Journal of School of Education, Federal College of Education, Eha-Amufu.* Vol 3. (1) 116–126.

Jegede, J.O. (1983). "Integrated science in Nigeria. A review of the problem and prospect," *In the Lead Proceeding of 24th Annual Conference of Science Teachers Associations, Press,*. 209–219.

Mapoma, M.I. (2001). "A bird eye view of music education in Zambia". In C. Van Niekerk (Ed.), *PASMEC Conference Proceedings* Press, 4–17.

Okafor, R.C. (2005). *"Music in Nigerian society"*. Enugu: New Generation.

Yekini-Ajenifuja, I.A. (2011). "Music education in Nigeria: Meki's philosophy for modern music.

Education". *Mekism and Knowledge sharing of the musical Arts of Africa:* A book of general reading in honour of Professor Meki Nzewi. 69–72.

21st Century Innovation in Music Education – Astuti et al. (eds)
© 2020 Taylor & Francis Group, London, ISBN 978-0-367-11122-9

The influence of deep learning model on musicality and character through *Dolanan* songs

K.S. Astuti, M.G. Widyastuti & HT. Silaen
Universitas Negeri Yogyakarta, Yogyakarta, Daerah Istimewa Yogyakarta, Indonesia

ABSTRACT: Deep learning is a learning model that provides experiences for students to master the subject matter in depth by examining problems up to the level of value. The researcher developed a deep learning model with the subject matter of *dolanan anak* (children's games) songs. This needs to be done because learning,, has only reached the level of knowing and understanding without reaching the level of living experience. The research method is quasi-experimental . The study was conducted at the Elementary School Kranggan Temanggung, Central Java, Indonesia. Data analysis used MANOVA and the results of the analysis obtained Wilks' Lamda values in sig 0.00 (below 0.05) which showed a positive and significant difference between musicality and character abilities in the pretest and posttest. Thus, it can be said that deep learning with *dolanan anak* material can improve the musicality and character of students.

1 INTRODUCTION

1.1 *Background of the study*

The function of music is not only as a medium of entertainment. Since a long time ago music was used as a medium for religious rituals and customs. In nearly all religious ritual activities and traditional ceremonies music is a medium of accompaniment and also functions as a medium of education. Ki Hadjar Dewantara, the father of Education in Indonesia has used music to refine student character since the beginning of the 20th century. Dewantara (1977) revealed that the work of Rabindrnath Tagore, an Indian education figure, served as a model for what was implemented at the Taman Siswa School, founded by Ki Hadjar Dewantara, after visiting the Rabindranath school in 1936, and established a school called *Kerti Niketan*. Within its curriculum there is material called *gendhing-gendhing Jawa* (Javanese gamelan music) that is used to refine the character of Indian students.

Music as a medium of education not only involves musical elements such as rhythm, tone and harmony that are believed to control emotions but includes lyrics contained in the song. In folk songs and traditional songs in general, song lyrics contain life values upheld by the local community such that instilling noble values is very effective by using regional songsIndonesia is comprised of 18,307 islands and 1,128 tribes with at least as many traditional arts as there are tribes. In Central Java and the DIY (Special Region of Yogyakarta), there are more than 600 types of *dolanan anak* (children's games) songs. Wang (2015) revealed that in Indonesia, especially in the Special Region of Yogyakarta and Central Java, *dolanan anak* songs were found and that these songs have been around since ancient times. In 1930, Hans Overbeck identified at least 690 *dolanan anak* songs.In the Indonesian territories outside Java, there are also many folk music that are used by the community for hereditary character education. In Riau, there is a *Batimang* tradition which is the art of lulling children to sleep with certain tones and free lyrics but the contents give advice and prayers. Adeliani (2014) states that the tradition of lulling children with singing is also found in the Banjar community, Kalimantan. Likewise, Afandi (2015) revealed that the art of lulling children to sleep by singing is also a tradition of the Kaili people in the Donggala district, Central Sulawesi.Indonesia's geographical location in the middle of the traffic of Western and Eastern countries makes Indonesia a multi-cultural country. Hindu, Buddhist, Christian and Islamic religious values greatly influence the life values embraced by the Indonesian people in addition to the cultural values of the local community which are manifested in the arts. Yousof (2010: 84) suggests that the manifestation of religious values in art is spread by imported performance through music.The cultivation of values through songs was also expressed by Desyandri (2015) who revealed that Minangkabau songs (West Sumatra) contained education or moral values. Basandi Syarak custom - Syarak Basandi Kitabullah (ABS-SBK) and Alam Takambang are for those who become Teachers. The basic values found in ABS-SBK show that in Minangkabau tradition ABS-SBK was formed based on Shari'a, and Shari'a based on the Kitabullah (Al-Qur'an). Karomah (2016) also found that the Rejang Lebong folk songs also contain the noble values of the Rejang people.Thus, traditional songs have been used by people in Indonesia as a medium to instill adopted values. Until the end of the 20th century, Indonesian children who are now in their 50s are very familiar with *dolanan anak* songs, because they were always available during every play activity both at school and outside of school.The rapid flow of globalisation and technological progress today has an influence on the lives of people including children. Gadgets and mobile phones are more attractive to children than traditional games including games with *dolanan anak* songs.

Thus, it can be said that the children's song is outdated and it is possible that at this time it is very rare for the younger generation to know songs. In the children's song there are values f character education. Thus, it is necessary to strive to maintain *dolanan anak* songs so that these songs again become a favourite game of children.

Schools are very effective education channels and therefore *dolanan anak* songs can be introduced to students through formal education. Astuti et al. (2018) found that in the Dutch conservatory, music was used as a medium for character building. Many folk songs are used for the learning process by feeling the presence of musical elements in depth. Learning something in depth involving knowledge, skills and feelings will also produce long-lasting learning outcomes and students are able to interpret their learning outcomes. Mursell (2003) mentions that such results are called authentic learning outcomes, to wit, long-lasting learning outcomes that are useful and that students use in life.The development of communication and information technology has rapidly changed the criteria for learning success. Greenstein (2012: 50) suggests that the abilities needed in the 21st century are different from those of the 20th century. These differences among others include: the 20th century ability of "knowledge recall information", while in the 21st century the ability needed is "mastery of core content"; in the 20th century, "understanding making sense of content" while for the 21st century, the ability of communication and collaboration is needed to express ideas and working with others; next, ability related to analysis in the 20th century was about thoughtfully considering the information, "whereas in the 21st century it was meta cognition (learning and self-aware thinking); and if in the 20th century the expected ability is the creation and production of original works through innovation, in the 21st century, the capabilities needed are "creativity and applied purposes at work and in life".Changes in these needs require that the learning process in the classroom must also change in accordance with the demands of the times. In this 21st century, students are required to be better able to recognize their potential, express themselves to understand and accept others. Bengio (2014) proposed a deep learning model to teach students to be able to achieve authentic learning outcomes because students are able to interpret something that is learned and expressed in collaboration with other students to achieve learning goals.Referring to the 21st century learning criteria above, the purpose of learning music at school is not only to develop musicality but also for students to be able to express themselves and to be able to with other students. Considering that music education in schools in Indonesia is a medium for developing personality, teaching music must also be able to simultaneously shape the personality of students.

1.2 *Formulation of the problem*

Based on the background description above, the following problem statement is proposed: How do deep learning models implemented through *dolanan anak* songs, help students to develop musicality and express themselves, and how effective is the deep learning model for increasing the musicality of students and in the formation of student character?

2 LITERATURE REVIEW

2.1 *Dolanan anak (children's games songs)*

Dolanan anak songs is basically one of the Indonesian folk songs. Mireskandari & Sharbatian (2015) argue that folklore or popular culture is one of the components of culture and that folklore is the body of expressive culture itself that is related to literary artistic and cultural issues. Thus, children's songs are also an aesthetic expression of language and culture.*Dolanan anak* songs are used to accompany children's games. However, according to revelation (2013) sometimes this children's song accompanies dance, for example, a new dance creation called *Jarog* which is a combination of *Jaranan* dance and *Reog* dance. Nonetheless, according to children's songwriters, *Dolanan anak* songs are typically used for games.According to the name *dolanan* which comes from the Javanese language, these songs come from the Special Region of Yogyakarta, Central Java and East Java that use Javanese language in their daily lives. Examples of *dolanan anak* songs are, "Jamuran", "Tak Petik-petik Kembang Melati" and "Cublak-cublak Suweng". These songs can be used by children to accompany guessing games and to express aesthetic feelings and togetherness.

2.2 *Moral values*

Moral values are values about goodness and must be followed and owned by each person to achieve the quality of life expected by themselves, society and God. Every human being has a conscience that can distinguish between good and bad. Many experts identify good values. There are many values that must be followed and owned by human beings. identifies 14 values that the Indonesian government through *Menko Kesra* (Coordinating Minister of the people's welfare sector), summarises the 14 values into four basic components which are the basic foundation for other values. The four basic values are honesty, intelligence, toughness and caring.

2.3 *Moral values in dolanan anak songs*

Many moral values are contained in *dolanan anak* songs. The song "Cublak-cublak Suweng" teaches children to understand others. "Tak Petik-petik Kembang Melati" can teach aesthetic taste and love for the natural environment. Wang (2014) revealed that the *dolanan anak* song "Jamuran" can help children practice vocabulary and that if studied, the *dolanan anak* songs contain educational values.

As stated earlier that there are more than 600 *dolanan anak* songs and there may be just as many values contained in *dolanan anak* songs. Among dolanan songs that are popular among children today, examples include "Cublak-cublak Suweng", "Gundul-gundul Pacul", "Jamuran", "Buto Galak" and "Padang Bulan". The value of education is contained in the game accompanied by the song "*Cublak-cublak Suweng*" and is used to develop the intelligence to read the thoughts and hearts of others. Games that demand the ability to find friends who hide things through facial gestures that are displayed are exercises to read the thoughts and feelings of others. The song "Gundul-gundul Pacul" has the same meaning as the song "Buto Galak" which describes the arrogant, haughty and trivial nature of losing what has been obtained. The song "Padang Bulan" can increase students' awareness of the majesty of God who has created such a beautiful universe.

2.4 *Deep learning model*

The Deep Learning model was initially used in the classroom. However, Bengio (2014) also developed the model in the cultural context, while Sigrun Kertesvåg (2017) proved that deep learning can improve the quality of learning in literary learning. Cherkowski (2017) also found that deep learning improves teacher understanding.

The eep learning model provides experiences for students to know, understand and appreciate a material, work or problem so that students enter the world of science, social phenomena and art in depth. With a learning model like this, students not only know a phenomenon, but can also live and feel so that they can judge. Ertesvag (2017) provides an illustration in the literature, that by using a deep learning model, students not only know the names of characters, but are also able to judge whether a character can be imitated or not. To answer that, students must read a book until it's finished and not just reading in passing.

Astuti (2010) found that in general, in learning music in the Netherlands there were student activities made experimentally, produced by working in groups and then the students presented their works in front of the class. These activities show that deep learning has also been carried out in music learning.

2.5 *Development of the deep learning model in learning music*

Music is composed of musical elements including rhythm, tone and harmony. The other elements of music include melodies, scales and musical styles. Generally, in music learning teachers usually only emphasise knowledge that is cognitive or intellectual. However, in deep learning students are given experiences that allow them to feel the rhythm, melody, modulation, dynamics and so forth. With the deep learning model by using *dolanan anak* songs, students are invited to sing and play with *dolanan anak* songs. In the Netherlands, to develop musicality Dutch students are invited to sing children's songs with modulation, changes from major to minor scales or vice versa and feel music as a reflection of natural phenomena, social life and individual feelings. With the implementation of *dolanan anak* songs, the teacher provides experiences for students to understand, appreciate and feel the musical aspects and meanings of *dolanan anak* songs with variations.

2.6 *Learning theory*

David Kolb's Experiential Learning Cycle Theory (Paulina Pannen et al., 2001: 111) states that the learning process starts from a concrete impression experienced by the student. The experience is reflected individually and in the process of reflection the individual will try to understand what happened and/or what was experienced. This reflection is the basis for the process of conceptualising principles and predicting possibilities in new situations and contexts.

Based on this learning theory, it can be said that to achieve progress, learning does not move linearly but can move like a spiral, from a concrete direction to an abstract, from simple to complex and from easy to difficult, however not always hierarchical. To understand the meaning of a song one does not only rely on singing the song repeatedly, but that in each repetition there will be a process of deepening the meaning. The process of deepening the meaning can run slowly or quickly depending on each individual's process of exploring songs and music. The teacher can be a facilitator for students to be able to understand the meaning of a musical work.

Based on this researcher's observations of music learning in several schools and colleges in the Netherlands in 2008 until 2018, methods for providing the ability to understand and appreciate music are carried out by analogizing musical elements with natural phenomena and environment.

For example, to show the dynamic crescendo and decrescendo, the teacher invites students to imagine the mouth of a crocodile opening and closing. Melody stepped and jumped up was described as a mountain climbing trip. Songs with major scales describe excitement, while minor tone songs describe sadness. The interesting thing is that the emphasis on learning is not only on musical aspects, but more on feeling. Students feel their feelings by asking the opinions and feelings of students after listening to or playing music. To reinforce this, sometimes teachers in the Netherlands take advantage of the space atmosphere such as turning off lights to provide a dark and calm atmosphere.

2.7 *Framework of thinking*

Deep learning is a method that provides an atmosphere for students to learn naturally in accordance with the characteristics of the 21st century. In this century, new innovations are discovered every minute such that science does not progress linearly but develops in all directions. Repetitive exercises in playing music not only facilitate the achievement of proficiency levels, but is also a process of appreciation to reach the depth of understanding meaning. Considering that at the core of the process of learning music is the expression of feeling. Learning music with the method of deep learning can provide opportunities for students to be able to understand the beauty of the meaning of life in depth. The ability to enjoy beauty will make students more able to appreciate life and achieve happiness. Happy feelings will foster gratitude for the gifts of nature and life and will have a positive influence on students so that they always think positively and provide positive energy for other students. Such conditions will develop an attitude of empathy and cooperation between students. This attitude will be strengthened by the values contained in the children's song lyrics by: planting attitudes of not behaving arrogantly and not underestimating other people; increasing gratitude by instilling awareness of the beauty of the universe created by God; and improving tolerance and cooperation.

Thus, it can be said that in music learning there is a process of developing taste in this case, affective, which can increase feelings of sensitivity comprehensively. Learning music with the method of deep learning can improve musicality and positive attitudes of students.

2.8 *Hypothesis*

Based on the preceding description, the proposed hypothesis is as follows:

a) The development of deep learning models by introducing elements of music through *dolanan anak* songs can improve the musicality of students.
b) The development of a deep learning model through *dolanan anak* songs can improve student character.

3 METHODOLOGY

3.1 *Research design*

This research uses a quantitative approach. This approach was used to collect, interpret and analyse the data using quantitative analysis. However, good quantitative data must be supported by qualitative data so that researchers also collected relevant qualitative data.

The collected data includes information on the development of musicality and changes in students' behaviours to find out the effect of the deep learning model used as part of the research's experimental method. The data analysis uses MANOVA analysis which was intended to determine the change in character of students after receiving the treatment to be applied. The research stages were: first, compiling a deep learning model based on *dolanan anak* songs; secondly, compose research instruments followed by; the third trial of the fourth research instrument, giving deep learning treatment; and the fifth stage involved determining the effectiveness of the implementation of deep learning models based on changes in musicality and changes in students' behaviours.

3.2 *Site of research*

This research was conducted at the Elementary School, Temanggung Regency, Central Java Province. The sampling was performed using cluster random sampling.

3.3 *Development of research instruments*

The research instrument was developed in the form of questionnaires and observation sheets. Observations were used to collect data related to musicality and the moral values of character education possessed by students before and after treatment. The instrument was then tested to determine its validity and reliability with Pearson correlation and interrater reliability.

The research instrument that was used to develop musicality was adopted from the research of Astuti (2008) for Nasyid learning (Islamic music). Musical aspects consist of indicators of rhythm, melody and harmony. Musical instruments consist of rhythm, melody and harmony components. The research instrument was analyzed by confirmatory analysis with the LISREL 8.51 program. The analysis shows that the validity of the rhythm indicator was 0.78 and its reliability was 0.56. The melodic indicator of validity was 0.95 and its reliability 0.9. The validity of the harmony indicator was 0.85 and the reliability was 0.73.

As for the aspects and indicators of the research instrument used to measure students' moral values or character, the method was developed from the Instrument research instrument based on the archipelago-based character education measurement instrument Astuti (2011). Difference with the instrument, if on the initial instrument, the measured aspect is toughness, in this study the aspect is specified as an aspect of nationalism. The following is the measurement rubric of character.

No.	Aspect	Indicator
	(1)	**(2)**
1	Honesty (in speaking and acting)	According to reality Rational Objective Authentic Transparency
2	Intelligence	Speed of receiving stimulus Accuracy of Capturing Stimulus Organize stimulus
3	Nationalism	Physical endurance Mental Resilience Consistency Symphathy Empathy Assist Other Cooperation

Validity and reliability

were analysed by Pearson correlation analysis techniques and interrater reliability. The results of the instrument trial produced the following data.

Correlations

		character rater 1	character rater 2
character rater 1	Pearson Correlation	1	.553**
	Sig. (2-tailed)		.002
	N	30	30
character rater 2	Pearson Correlation	.553**	1
	Sig. (2-tailed)	.002	
	N	30	30

** Correlation is significant at the 0.01 level (2-tailed).

Based on the analysis, it is known that the items of the instrument have a correlation at the significance level of 1%. Thus, the instrument is by construct, valid.

Reliability Statistics	
Cronbach's Alpha	N of Items
.709	2

Based on Cronbach's Alfa analysis, it is known that the instrument has a reliability of 0.709 which means that the reliability of the instrument is high. Thus, the instrument can be used because it has validity and reliability that meet the requirements, namely the validity in the criteria is sufficient (0.553) at a significance level of 1% and that the reliability is in the high criteria.

3.4 *Treatment*

The study was conducted for two months. The first meeting was used to measure the musicality and character of students using interview and observation methods. Interviews were used to determine the cognitive knowledge of students about music and the values contained in *dolanan anak* songs, while the observation method was used to determine the level of musicality of students by asking students to sing *dolanan anak* songs.

At the next meeting, the treatment was carried out by giving materials in the form of *dolanan anak* songs. At the beginning of the meeting, the subject matter included *dolanan anak* songs known to students and towards the end of the lesson *dolanan anak* songs that are less well known by students were taught. During the last meeting which was the eighth meeting, a posttest was conducted to measure improvement in the musicality and character of the students. For the posttest, the same method was used as in the pretest, interviews and observation.

3.5 *Data analysis*

To determine the effects of deep learning using the material of *dolanan anak* songs, on musicality and character the MANOVA analysis was used. For the significance level which is the criterion rejected or not hypothesized, the significance level is 5%.

4 DISCUSSION AND FINDINGS

In this section, we describe the data from research results, both quantitative and qualitative data with discussion.

4.1 *Use of deep learning development model*

The Deep Learning Model was developed based on the new musical learning model that was created by Astuti (2017). Based on the model, the research team redeveloped the model by emphasising the meaning of learning material by students with question and answer and contemplation methods. The following shows this deep learning model for music learning.

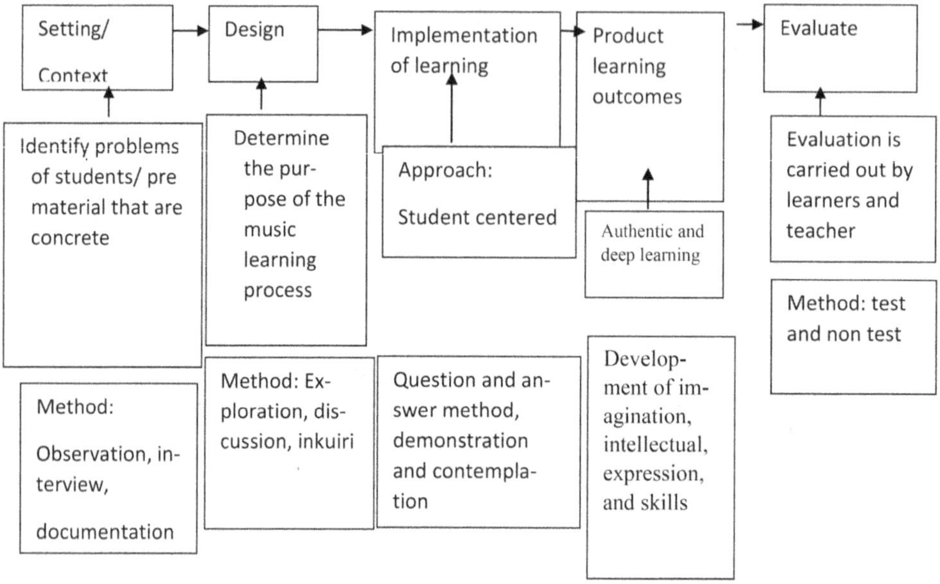

4.1.1 *Qualitative data of musicality*

The study was conducted in SD Negeri 1 Kranggan for two months, to wit, in class V involving 18 students. The research team worked with the students by giving lessons on *dolanan anak* songs, starting with songs that were known to the students. It turns out that the students already knew a number of Javanese songs such as "Buto-buto galak", "Jamuran", "Gundul Pacul", "Cublak-cublak Suweng", "Gambang Suling" (this is not *dolanan anak*), "Pucung" (the children also understood that "Pucung" was a kind of *macapat* that is not *dolanan anak*), in the middle of the month. Following this, songs such as "Jaranan" and "Tak Petik-petik Kembang Melati" were prepared by the research team. The existence of *dolanan anak* songs were known to be used by children to play games, based on interviews with children. However, *dolanan anak* songs are now often used for competitions.

During the research the process of learning music was taught by singing songs without musical accompaniment. In this research activity, researchers used musical instruments to accompany singing activities. It turns out that for elementary students, singing accompanied by music increases enthusiasm and helps them sing with the right intonation. When not accompanied, each student sings in a changing and different basic tone. With the accompaniment, together they can sing with a stable basic tone.

The development of musicality is carried out by introducing students to and training them in musical aspects such as modulation, changes in scales, dynamic changes and changes in tempo. For example, the learning began with singing a song according to the original scales. After that the song is changed to the opposite tone ladder. The "Buto-buto Galak" song is in a minor tone and was changed to a major scale. Likewise, children's songs in a minor tone are changed to major scales.

At the beginning of the learning process, the deep learning method was carried out by giving students the freedom to explore. When students sang *dolanan anak* songs with intonation, rhythm or articulation that is not right, the learning method that was used is imitation. The teacher demonstrated an example that was imitated by students. At the same time, the teacher explored students' understanding of the meaning of each song.

The results of the study show that in general, the fifth-grade students at the elementary school were able to distinguish major and minor scales and distinguish false and not. The problem with 30% of students was that they still had difficulty in singing organs so they could not sing with the right intonation.

4.1.2 *Qualitative data on character*

Based on observations before treatment, the students possessed good character, discipline, courtesy and order. But in general, they were still embarrassed to express their opinions so that these students' understanding of the values of education could not yet be explored optimally.

In subsequent meetings, the students began to express their opinions that revealed their understanding of the values of moral values while the researcher observed the behavior of students during the learning process.

Based on interviews in class, students had understood the importance of mastering national songs, regional songs and *dolanan anak* songs to instill nationalism values. In addition, by singing the songs students could better understand the values that must be embraced and upheld so that in the future they will be successful and useful people for the nation and for their religion.

The results of the study showed that the fifth-grade class students already possessed good character, that they were able to distinguish good and bad characters and that they already held a sense of love for the homeland. This was shown in the students' opinions during discussions about *dolanan anak* songs compared to foreign songs. The following are the kinds of student responses to *dolanan anak* songs: 1) can get to know Indonesia more deeply; 2) preserving Indonesian cultural arts; 2) not forgetting the ancestors; 3) remembering culture; 4) remembering the history of ancestors; 5) get to know culture well and be able to improve culture; 6) can love culture; and 7) love folk songs that have been entrusted to them from ancestors. The students also explained that they must respect Indonesian culture so that the country remains united and not divided, with appreciation for diversity.

4.1.3 *Quantitative data on musicality and students' abilities before and after treatment*

Table 1. Scores of students' musicality and character skills before and after treatment are as follows.

Respondent	Before treatment		After treatment	
	Musicality	Character	Musicality	Character
1	4	36	5	40
2	5	38	5	43
3	5	35	6	41
4	10	40	10	46
5	10	37	10	44
6	6.25	35	10	41
7	2.25	35	3	40
8	1.75	30	2	38
9	2.75	33	3	39
10	6.75	35	7	41
11	7	36	8	43
12	6	35	7	40
13	9	35	7	40
14	9	36	9	46
15	4	33	5	37
16	2	35	3	40
17	7	36	7	43
18	6	35	6.75	41

As for the mean and standard deviation of musicality and character abilities before and after treatment, the results are as follows:

Descriptive Statistics				
	pretest and post test	Mean	Std. Deviation	N
musicality	Pretest	5.5417	2.53831	18
	Posttest	6.0417	2.40748	18
	Total	5.7917	2.45131	36
character	Pretest	35.2778	2.08088	18
	Posttest	41.2778	2.46876	18
	Total	38.2778	3.78426	36

Based on the table above, the mean variable for musicality and character are increasing. This was seen in the increase in pretest on musicality from 5.54 to 6.04. The pretest mean result of 35 for character, rose to 41.28. To find out whether the difference is really an effect of treatment, an analysis test was performed using MANOVA.

The assumption of the MANOVA test is that the distribution must be normal, linear and have equality of variance. Normality was tested with the SPPS program to calculate the Mahalonobis distance (pallant, 278). The following are the results of the normality test based on the Mahalonobis numbers:

Residuals Statistics[a]					
	Minimum	Maximum	Mean	Std. Deviation	N
Predicted Value	.7196	2.2801	1.5000	.45086	36
Std. Predicted Value	-1.731	1.730	.000	1.000	36
Standard Error of Predicted Value	.042	.100	.067	.017	36
Adjusted Predicted Value	.6605	2.3290	1.5015	.45707	36
Residual	-.53453	.51238	.00000	.23209	36
Std. Residual	-2.236	2.144	.000	.971	36
Stud. Residual	-2.272	2.191	-.003	1.009	36
Deleted Residual	-.55171	.53548	-.00151	.25071	36
Stud. Deleted Residual	-2.436	2.335	-.007	1.038	36
Mahal. Distance	.118	5.120	1.944	1.449	36
Cook's Distance	.000	.117	.027	.036	36
Centered Leverage Value	.003	.146	.056	.041	36

[a]Dependent Variable:pretest and posttest

The score of Mahalanobis maximum is 5.120, and for the critical value for evaluating Mahalonobis, the distance value for the two dependent variables is 13.82. (Pallant, 2007: 280). Thus, the distribution does not violate the normality requirements because it is within the range of critical values.

Linearity was tested by a matrix of scatter plots between each pair. The following shows the linearity test.

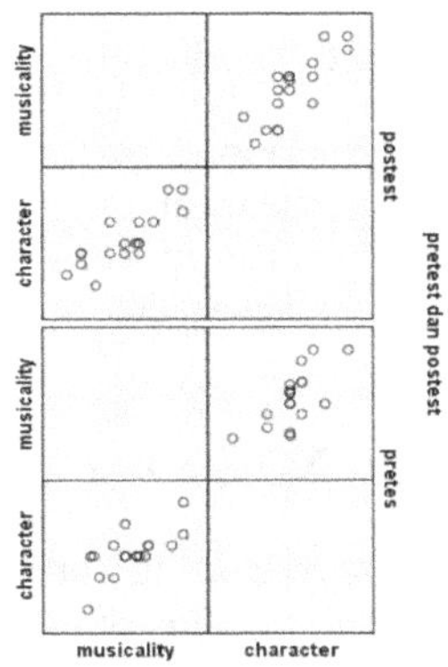

Based on the image above, the distribution is linear because the relationship of each variable forms a straight line.

Equality of variance was tested with Levene's Test of Equality of error variance. Based on calculations with SPSS, the following data was obtained.

Levene's Test of Equality of Error Variances[a]

	F	df1	df2	Sig.
musicality	.031	1	34	.861
character	1.346	1	34	.254

The test shows sig. above 0.05 is good for musicality and character, so the data does not violate the assumption of equality of error variance.

This distribution fulfills the requirements of the MANOVA test assumption. Thus, it was continued with MANOVA analysis. The MANOVA test used the SPSS program and the results were calculated as follows:

Multivariate Tests[a]

Effect		Value	F	Hypothesis df	Error df	Sig.	Partial Eta Squared
Intercept	Pillai's Trace	.998	9213.770[b]	2.000	33.000	.000	.998
	Wilks' Lambda	.002	9213.770[b]	2.000	33.000	.000	.998
	Hotelling's Trace	558.410	9213.770[b]	2.000	33.000	.000	.998
	Roy's Largest Root	558.410	9213.770[b]	2.000	33.000	.000	.998
group	Pillai's Trace	.791	62.267[b]	2.000	33.000	.000	.791
	Wilks' Lambda	.209	62.267[b]	2.000	33.000	.000	.791
	Hotelling's Trace	3.774	62.267[b]	2.000	33.000	.000	.791
	Roy's Largest Root	3.774	62.267[b]	2.000	33.000	.000	.791

[a]Design: Intercept + group
[b]Exact statistic

To find out the effects of deep learning on musicality and character the researchers looked at the significance of the Wilks' Lamda table because the distribution meets the requirements of normality, linearity and equality of variance errors. In the Wilks' Lamda column there is a significance of 0.00 (below 0.05) so that the mean difference between the pretest and posttest is significant. Therefore, it can be said that deep learning by using *dolanan anak* songs has a significant influence on the musicality and character of students.

5 CONCLUSIONS AND RECOMMENDATIONS

The Deep Learning Model provides experiences to students to master the subject matter in depth up to the level of assessment. The Deep Learning Model which was originally used in the field of expression and was developed by researchers in the field of music to develop musicality and the character of students through *dolanan anak* songs.

Based on the results of the MANOVA analysis it is known that there is a significant difference between the ability of musicality and mindfulness of students before and after treatment. Thus, it can be said that the learning of *dolanan anak* songs with deep learning models can improve the musicality and character of students.

With the Deep Learning Model, learning objectives can be achieved effectively. In the future learning objectives need to be developed by applying deep learning models in the music and in various subject matters.

REFERENCES

Adeliani, N. (2015). Lagu Menidurkan Anak pada Masyarakat Banjar: Kajian Bentuk, Makna, dan Fungsi. (lullaby song at Banjar Community: studi of form, meaning, and function) *Al-Banjari: Jurnal Ilmiah Ilmu-Ilmu Keislaman*, *13*(2), 265–284.

Al-Afandi, A. A. FUNGSI DAN NILAI NYANYIAN BUAIAN DALAM SASTRA LISAN KAILI.(a function and values of lulally congs in oral literacy Kaili) *BAHASANTODEA*, *3*(4). 81–92.

Astuti, K. S. (2008). Pengembangan model evaluasi pembelajaran dalam perspektif penciptaan makna baru pada grup nasyid DIY(Developing of teaching evaluation model in new meaning creativity perspective at Nsheed Group).

Astuti, K. S., Zuchdi, D., Sudiyono, T. W., & Hoogeveen, K. (2010). Developing Model for Teaching and Learning Music in Public School Based on Comparative Study between Indonesia and the Netherlands. *Laporan Penelitian. Yogyakarta*: Lemlit UNY.

Astuti, K.S. (2011, November) Developing an evaluation model for character education in Music Learning. In Character Education for Integrity and Justice toward Sustainable Future: Proceedings of the 1st International Conference on Character Education (ICCE 2011) November, 8–9. 2011, Yogyakarta Indonesia. International Office Yogyakarta State University.

Astuti, K. S., & Widyantoro, A. (2018, September). The teaching of morality through musical elements based on learning models of the Netherlands and Indonesia. In *Character Education for 21st Century Global Citizens: Proceedings of the 2nd International Conference on Teacher Education and Professional Development (INCOTEPD 2017), October* 21–22,2017,*Yogyakarta, Indonesia* (p. 389). Routledge.

Cherkowski,Sabre (2017) ICSEI: Flourishing in Adaptive Community: Balancing Structures and Flexibilities for Innovation and Inquiry in Schools. Singapore: NTU.

Dewantara, K.H. (1977). *Pendidikan (Education)*. Yogyakarta, Indonesia: Majelis luhur Persatuan Taman Siswa.

Desyandri, D. (2016). Nilai-nilai edukatif lagu-lagu minang untuk membangun karakter peserta didik (Education Values of Minang's Song to shaping character of students.). *Jurnal Pembangunan Pendidikan: Fondasi dan Aplikasi*, *3*(2), 126–141.

LeCun, Y., Bengio, Y., & Hinton, G. (2015). Deep learning. *nature*, *521*(7553), 436.

Greenstein, L. M. (2012). *Assessing 21st century skills: A guide to evaluating mastery and authentic learning*. Corwin Press.

Ertesvåg,Sigrun K. (2017). ICSEI: Teachers' improvement in instructional support < Singapore: NTU.

Megawangi, R. (2004). Pendidikan Karakter, Solusi yang Tepat untuk Membangun Bangsa(Character Education. A solution to shaping character of nation) . Jakarta: Indonesia Heritage Foundation.

Mireskandari, F., & Sharbatian, Y. (2015). An Anthropological Study of Folklore in Gilan Province with a focus on Lullabies. *IAU International Journal of Social Sciences*, *5*(4), 75–85.
Mursell, J. L. (2007). *Successful teaching*. Littlefield Press.
Pallant, J. (2007). SPSS survival manual, 3rd. *Edition. McGrath Hill.*
Pannen, P. (2001). Penulisan Bahan Ajar. *Jakarta: PAU PPAI. Ditjen Dikti. Depdiknas.*
Wahyu, E. (2013). JAROG DANCE FOR CHILDREN WITH SPECIAL Needs: Case Study in The Celebration of The International Dance Day in Surakarta. *Harmonia: Journal of Arts Research and Education*, *13*(2).
Wang, J. C. (2015). Games unplugged! Dolanan Anak, Traditional Javanese children's singing games in the 21st-Century General Music Classroom. *General Music Today*, *28*(2), 5–12.
Yousof, G. S. (2010). ISLAMIC ELEMENTS IN TRADITIONAL INDONESIAN AND MALAY THEATRE. *Kajian Malaysia: Journal of Malaysian Studies*, *28*(1).97–101.

21st Century Innovation in Music Education – Astuti et al. (eds)
© 2020 Taylor & Francis Group, London, ISBN 978-0-367-11122-9

Developing instructional multimedia for learning traditional music by encyclopedia form

N.A. Anandhita & Ishartiwi
Universitas Negeri Yogyakarta, Daerah Istimewa Yogyakarta, Yogyakarta, Indonesia

ABSTRACT: This research aimed to develop instructional multimedia for learning traditional music and musical instruments in the form of an encyclopedia as a learning supplement for students. This research and development referred to the development model of Alessi and Trollip including: 1) planning, 2) designing, and 3) developing. The result of the research was a product with file format operated by using a computer with any kinds of operating systems. The materials in this product were divided into three categories based on the alphabet, the Indonesian islands, and the instrument classification. There are 413 materials comprising 86 ensembles, 1 genre, 13 kinds of music, and 313 instruments. This product was considered to be appropriate based on the alpha test and the validation of the media experts with the feasibility percentage of 79,65% and from material experts with the feasibility percentage of 88%. Meanwhile, the beta test gave the percentage of 87,05%.

Keywords: encyclopedia, multimedia, traditional music

1 INTRODUCTION

The limited information about music and traditional musical instruments in Indonesia has resulted in a lack of concern from Indonesian people about traditional music. David (2014: 1135) said that this was caused by several factors including the lack of introduction and understanding of traditional musical instruments carried out from an early age and the small number of existing experts. Therefore, we need treatment as stated by Arifitama (2016: 310), *"ideas of preserving cultural heritage become important to keep the identity of a nation which is rich in cultures. The needs of an alternative way of preservation become crucial in order to secure intellectual property that has been handed down by our ancestors which will not fade, disappear and get eroded by time."*

In formal education, the government through the Directorate General of Primary and Secondary Education seeks to further preserve traditional music and musical instruments, as evidenced by the establishment of art vocational education which focuses on the production of graduates who will later work as artists, basic-level experts, or managers in certain vocational fields (Ardipal, 2010: 3).

Based on observations of the curriculum and interviews with the deputy principal in charge of curriculum and teachers of the State Vocational High School 1 (SMK N 1) Kasihan Bantul (the only school in Yogyakarta that has specialization in *karawitan*, the Javanese traditional music) on February 2, 2017, this school prioritizes traditional music. This can be seen from the portion of subjects that specialize in learning Yogyakartan musical instruments and gamelan instruments both in theory and practice. In this school there is no formal learning in the class that introduces music or other traditional musical instruments from the Yogyakarta area, let alone from other provinces.

The data obtained from the analysis of students' needs regarding the initial knowledge related to music and traditional musical instruments showed that students were able to mention the names of Javanese gamelan instruments in detail. There were only three out of thirty-nine students who were able to mention other traditional musical instruments (besides musical instruments forming gamelan) from the Yogyakarta area.

The results of a survey conducted at this school to one teacher of *wawasan nusantara* (archipelago outlook) subject and 11th graders majoring in *karawitan* showed that they wanted to be able to get to know and learn other traditional music from Yogyakarta and other regions. Of the total 39 students, 100% of them want to know music and other traditional musical instruments besides gamelan. The desire to get to know music and other traditional musical instruments is also further indicated by a percentage of 87% of those who even wanted to learn some of the music and musical instruments.

The learning process was hampered by limited support from teaching staff because actually learning music and traditional musical instruments besides Javanese gamelan was not their specialty in school. In addition, original object media (in this case traditional musical instruments) were also very limited. Permana and Sumantri (1999: 202) stated that "original objects are objects that actually help students to get real experiences and attract students' interest and enthusiasm for learning". Using original objects will provide a very important stimulus for students to learn various things, especially regarding the development of certain skills. The observations showed that this school only had traditional Javanese and Balinese gamelan instruments.

In addition, the results of observations in four museums located on the island of Java from March to May 2016, the original media in the form of traditional musical instruments still cannot represent traditional music from 34 provinces in Indonesia. The Indonesian Museum at *Taman Mini Indonesia Indah* (TMII, Beautiful Indonesia in Miniature Park) located in DKI Jakarta only has 30 collections of traditional musical instruments and 2 traditional music ensembles from 12 provinces. The Sabuga Museum of Bandung has a different approach to collection. This museum only displays a collection of traditional musical instruments originating from the area of West Java with 16 collections of traditional musical instruments and 2 sets of traditional musical ensembles. The Indonesian Museum of Music in Malang prioritizes Indonesian pop music collections in the form of digital recordings and articles rather than collections of traditional musical instruments, although there are some traditional musical instruments contributed by observers of traditional music in Indonesia. This museum together with traditional music activists in Malang often holds several traditional music performances at the national level so that there are several collections of videos related to this. Meanwhile, the Sonobudoyo Museum, located in Yogyakarta, once held an exhibition of traditional musical instruments featuring a collection of 209 musical instruments from 28 provinces in Indonesia. This exhibition displayed a collection of original objects but not accompanied by descriptions of the instruments and they were displayed on a short period of time, from April to May 2017.

The process of learning music and/or traditional musical instruments is increasingly difficult given the limited available literature. Based on observations in April 2017, the public library under the Regional Library and Archives Agency (BPAD) in Yogyakarta provides 8 books that mention and discuss some of the well-known traditional musical instruments such as angklung, kolintang, gamelan, and Sasando. In addition, there are 3 books that explain in more details about traditional musical instruments from certain regions but are limited to three provinces outside the province of Yogyakarta. The researcher only found one book by Sri Hastanto published in 2005 entitled "Alat Musik Tradisional Nusantara: Musik-Musik yang Belum Banyak Dikenal" (Traditional Archipelago Musical Instruments: Music that Has Not Been Known Very Much) which was quite extensive in its material studies both in terms of music and traditional musical instruments. The material in this book is quite comprehensive with a total of 143 music and traditional musical instruments and more numbers of musical instruments in the explanation. This book is the result of research under the Deputy for Arts and Film of the Indonesian Ministry of Culture and Tourism. However, researchers can only find this book at the University of Indonesia's central library, which is publicly accessible.

The results of interviews conducted with the deputy principal in charge of curriculum and teachers of the students of Vocational High School 1 Kasihan in February 2017 showed that there was a need for a reference that had a wide range of material. In addition, supporting media such as images, audio and/or video were also really needed to support the materials. For this reason, we need a product development that is able to solve existing problems and provide material-supporting multimedia. Multimedia integrates all elements (audio, video, graphics, text, and animation) that can benefit users compared to the media elements themselves (Reddi & Mishra, 2003: 4). In addition, according to Kemp and Dayton, multimedia is also useful, among others, to shorten the presentation time of learning and can be given anytime and anywhere (Arsyad, 2011: 21). This greatly supports the learning of music and traditional musical instruments in general, whose material coverage is too broad both in terms of space, time, and energy.

With the breadth of material about music and traditional musical instruments, a system is needed that can facilitate users in finding the material they want to learn. As explained by Daryanto (2010: 26), encyclopedia is a source of knowledge in addition to textbooks that will facilitate students because the information in it is arranged alphabetically or according to the field of study. This encyclopedia system will make it easier for developers to group music material either alphabetically or according to the field of study, such as based on the categories and regions the music originated and traditional musical instruments. This will also make it easier for users to find information based on certain categorizations.

Multimedia for learning music and traditional musical instruments combined with the encyclopedia system are expected to be an alternative for users in terms of finding information about music and traditional musical instruments and can be used as additional reference sources and learning supplements for students of SMK N 1 Kasihan (and other vocational students) when looking for information about traditional music and musical instruments especially those outside their realm of learning at school. The purpose of the development in this study is to produce multimedia learning of music and traditional musical instruments in the form of a feasible encyclopedia based on the results of expert judgment and user response.

2 LITERATURE REVIEW

Product development in this study refers to several theories about multimedia, encyclopedias, and the classification of music and musical instruments. Phillips (1997: 8) states that "the multimedia component is characterized by the presence of texts, pictures, sounds, animations, and videos, some of which are organized into some coherent programs." This is in line with the understanding of multimedia proposed by Hackbarth (1996: 228) as follows: "Multimedia suggested as the use of multiple media for memory for the presentation of information, including texts, still or animated graphics, movie segments, video, and audio information. Computer-based interactive multimedia includes hypermedia and hypertext." Furthermore, Newby, et al. (2000: 101) states that "the multimedia term conveys the notion of various system media (e.g., texts, graphics, videos, and audios) that are integrated into a single delivery system under computer control".

According to Hornby (2010: 482), the encyclopedia is a book or set of books that explains all knowledge or about various parts of a particular subject, usually arranged in alphabetical order. Furthermore Hornby (2010: 482) states that in the present time the encyclopedia can take the form of internet sites or applications in compact disks (CDs). The digital encyclopedia system is considered very appropriate when combined with computer-based learning multimedia. The encyclopedia system will greatly help vocational students to search for material especially music and/or traditional musical instruments that are outside the realm of their learning material in school because they are arranged alphabetically, systematically, and based on certain categories. Media elements such as images, audios, and videos can also be integrated so that they can clarify the material. This multimedia-based encyclopedia of learning will later be included in a special encyclopedia type or subject because it is installed in a flashdisk.

The material developed in this product is about traditional music and musical instruments of the archipelago. According to Ali (2006: 2), traditional music is music that is born and developed in a particular area and is inherited from generation to generation. Meanwhile, traditional archipelago musical instruments, when viewed from the source of sound or sound produced, are classified into four, namely: 1) Idiophone, "the substance of the instrument itself, owing to its solidity and elasticity, yields, noise and requiring stretched membranes or strings" (Sachs & Hornbostel, 1961: 14). Suharto (1987: 51) defines idiophone, in accordance with Sach and Hornbostel, as a musical instrument whose source is the body of the instrument itself. In this case, examples include Gong, Angklung, and Kolintang; 2) Aerophone, "the air itself is the vibrator in the primary sense." (Sachs & Hornbostel, 1961: 24). Suharto (1987: 51) defines an aerophone, based on Sach and Hornbostel, as a musical instrument whose source is aero or air. In this case the examples include Kadire, Taleot and Tahuri; 3) Membranophone, "the sound is excited by tightly stretched membranes." (Sachs & Hornbostel, 1961: 17). Suharto (1987: 52) defines membranophone, based on Sach and Hornbostel, as a musical instrument whose source is a membrane or skin membrane. In this case examples include drum, tifa, and tambourine; 4) Chordophone, "one or more strings are strected between fixed points." (Sachs & Hornbostel, 1961: 20). Suharto (1987: 52) defines the chordophone, based on Sach and Hornbostel, as a musical instrument whose source is a chord (metal string) or string. In this case, examples include Rebab, Kecapi, and Sasando.

Learning music and/or traditional musical instruments in SMK N 1 Kasihan in particular requires the support of a computer-based multimedia learning as a learning supplement if students want to know and learn traditional music and musical instruments that are outside the realm of their learning material. Material about music and/or traditional musical instruments is too broad when packaged in the form of a book. Moreover, both media support in the form of images, audios, and videos are also needed to clarify information. All these elements must be integrated in one system so that students can learn them easily and independently outside school hours. For this reason, computer-based multimedia systems are considered more appropriate in product development in this study. In addition to computer-based multimedia systems, we also need a material categorization system that is able to facilitate students in finding the material they want to learn. Materials can be arranged alphabetically or based on certain categorization such as those contained in the encyclopedia system.

3 METHODOLOGY

3.1 *Model*

This research used a research and development model aiming to create a product. Alessi and Trollip's research and development model (2001: 410) has three attributes and three phases. The three attributes include standards, ongoing evaluation, and project management. Meanwhile, the three phases include planning, design, and development. The following is a diagram of the multimedia learning development model based on Alessi and Trollip and the adaptation model used by researchers in this development research:

3.2 *Procedure*

The development procedure includes the steps taken during the research and development. The first phase is planning. This phase consists of: 1) identifying the scope, 2) identifying user characteristics, 3) setting limits, 4) creating time management documents, 5) collecting resources, and 6) brainstorming. The second phase is design. This phase consists of: 1) developing content ideas, 2) conducting competency analysis, 3) describing the implementation of the program, 4) preparing a prototype, and 5) making flowcharts and storyboards. The third phase is development. This phase consists of: 1) preparing the text, 2) writing program code, 3) making

Figure 1. Development model of alessi & trollip.
(Source: Alessi & Trollip, 2001: 410)

graphics, 4) recording audios and making videos, 5) uniting materials, 6) preparing supporting material, 7) conducting alpha test, 8) revising product, and 9) conducting a beta test.

To measure the level of product feasibility, data were obtained from two tests through alpha test and beta test. Alpha test was a series of validation tests of product feasibility both in terms of media and material. Validation was carried out by media experts and material experts by exploring products. Validation by two media experts was conducted to test the feasibility of products that had been produced in terms of media. Assessment aspects of this validation included aspects of display and technical programming. Media experts were competent people in the field of developing learning media both academically and practically. Validation by two material experts was carried out to test the feasibility of products that had been produced in terms of material. Assessment aspects of this validation included aspects of learning and material suitability. Material experts were competent people in the field of developing learning media both academically and practically. Meanwhile, the beta test was conducted on the subject of the study to determine the user's response to the product. This test aimed to test the feasibility of the product obtained from the results of student assessment. This test was carried out after the researchers made improvements based on the results of the validation of media experts and material experts. This test was carried out by 39 students.

The data obtained from research instruments were grouped into two, i.e. qualitative data and quantitative data. Quantitative data in this study were obtained from the results of product assessments contained in the media expert questionnaire, material expert questionnaire, and user response questionnaire. The data were calculated using the following formula (Riduwan, 2013: 41).

$$K = \frac{F}{N \times I \times R} \times 100\%$$

Notes:
K : The Percentage of feasibility
F : The total number of respondents' answers
N : The highest score in the questionnaire
I : The number of questions in the questionnaire
R : The number of respondents

Table 1. Interpretation criteria for assessment of validators (Riduwan, 2013:41).

Percentage	Criteria
0% - 20%	Very Unfeasible
21% - 40%	Less Feasible
41% - 60%	Fairly Feasible
61% - 80%	Feasible
81% - 100%	Very Feasible

The quantitative data that had been calculated using the formula were then analyzed using descriptive techniques. It aimed to convert the data previously in the form of a percentage to more descriptive data. The conversion process was guided by the following Table 1:

4 RESULTS AND DISCUSSION

4.1 *Results of alpha test*

Validation of the aspects of media used for multimedia learning of music and traditional musical instruments in the form of an encyclopedia was conducted by two media experts. They were Prof. Herman Dwi Surjono, Ph.D., and Sigit Pambudi, M. Eng. The first validator, Prof. Herman Dwi Surjono, Ph.D., is Lecturer and Chair of the Learning Technology Study Program of the Graduate School of Yogyakarta State University. He has qualifications in the field of learning media which is proven through scientific fields and research related to what he has done. The second validator is Sigit Pambudi, M. Eng., a lecturer who teaches multimedia courses at the Informatics Education Study Program, Yogyakarta State University. He has expertise in learning multimedia studies as evidenced by his educational and teaching qualifications.

The results of the media expert validation show that broadly the multimedia can enter into the category of "feasible", indicated by an overall percentage of 79%. Several "very feasible" assessment categories were also given on several assessment points such as design attractiveness and clarity of information on packaging covers, suitability of text layout proportions with template size, suitability of text color proportions to background colors, graphic presentation attractiveness, and image compatibility with material. Meanwhile, the results of media expert validation show that broadly the multimedia falls into the category of "feasible", indicated by an overall percentage of 80%. In addition, several "very feasible" assessment categories were also given on several assessment points such as the ease of interacting with the program, the ease of choosing a menu on the program, the ease of choosing material in the program, and the speed of program access.

Validation of aspects of the material used for multimedia learning of music and traditional musical instruments in the form of an encyclopedia was carried out by two material experts. They were Dr. Kun Setyaning Astuti, M.Pd., and Cipto Budy Handoyo, M.Pd. The first validator, Dr. Kun Setyaning Astuti, M.Pd., is Lecturer and vice dean of Languages and Arts Faculty Universitas Negeri Yogyakarta(Yogyakarta State University). She has a qualification in the field of cultural arts education research which is proven through the scientific field and related research that she has done. The second validator is Cipto Budy Handoyo, M.Pd., a lecturer who teaches Karawitan courses in Music Arts Education Study Program and Art Expert Staff who works under the Vice Chancellor III of Universitas Negeri Yogyakarta. He has expertise in traditional music studies as evidenced by his qualifications.

The results of the material expert validation show that broadly the multimedia falls into the category of "very feasible", indicated by the percentage percentage of 86%. Seven of the nine assessment categories in the material aspect fall into the "very feasible" category and the remaining two are in the "feasible" category. The results of the material expert validation show that broadly the multimedia can fall into the category of "very feasible" in the assessment of material aspects, indicated by the percentage percentage of 89%. In addition, eight of the eleven assessment categories in the aspect of material fall into the category of "very feasible" while the remaining three in the category of "feasible."

4.2 *Results of beta test*

This test aims to test the feasibility of the product obtained from student assessment. This test was carried out after the researcher revised the product based on the results of the alpha test. This test consists of two stages, namely small groups and large groups testings. Small group testing was done by randomly selecting six out of thirty-nine students who were the subjects of the study. The results of user responses in small groups in terms of the cumulative assessment of user responses per indicator show that fifteen of the twenty assessment indicators fall in the "very feasible" category while the other five were in the "feasible" category. But if the assessment is seen from the cumulative response of users per individual respondent, it can be seen that there are three assessment categories that fall into the category of "feasible" and the other three falls into the category of "very feasible". But broadly speaking, multimedia products learning music and traditional musical instruments in the form of encyclopedias fall into the category of "very feasible", indicated by an overall percentage of 85%. In this test, there were no suggestions and inputs for product improvement.

Large group testing was carried out for all 39 students. This stage was carried out after it was ascertained that there were no improvements to the product needed based on the results of small group testing. The assessment was carried out using a user response questionnaire. The aspects assessed include ease of use and clarity of material. The results of the response of users in large groups show that overall the assessment of multimedia products for learning music and traditional musical instruments in the form of an encyclopedia falls into the category of "very feasible", indicated by the overall percentage of 87.05%. Cumulation in the assessment of material aspects from 39 respondents showed 88.2% and the ease of use aspects reached 85.90%. Both of these numbers fall into the "very feasible" category. The average rating of each respondent also falls into the "very feasible" category. But when viewed more deeply, there are still four indicators of assessment in aspects of ease of use which fall into the category of "feasible". The assessment indicators include whether the graphics are presented in an interesting way, examples of videos are presented in an interesting manner, animations are presented attractively, and material can be easily selected. However, this did not affect the overall product evaluation results. In this assessment there were no suggestions and inputs for meaningful product improvements. The majority of suggestions and input given contained the accuracy of navigation when selecting material with certain groupings which still had a number of errors. This has been done by researchers with continuous improvement to become the final product.

4.3 *Results of data analysis*

The data from the media expert validation shown in the Table 2 above shows that the average score for the display aspects given by media expert 1 is 4.23 while the assessment score given by media expert 2 is 3.70. The average score for the technical aspects of programming from media expert 1 is 4.42 while the assessment score from media expert 2 is 3.58. In terms of overall overall aspects, media expert 1 and media expert 2 each respectively gave a score of 4.31 and 3.65. Meanwhile, the average overall score given by the media expert is 3.98. The results of the assessment of each media expert based on each aspect can be seen in the following diagram.

The percentage of feasibility of this multimedia for learning music and traditional musical instruments in the form of an encyclopedia according to media experts can be calculated using the following formula.

Table 2. Average assessment of media experts.

Assessment Aspects	Average Scores Media Expert 1	Media Expert 2
Media Disolay	4.23	3.70
Technical Programming	4.42	3.58
Average Score Per Expert	4.31	3.65
Overall Score	3.98	

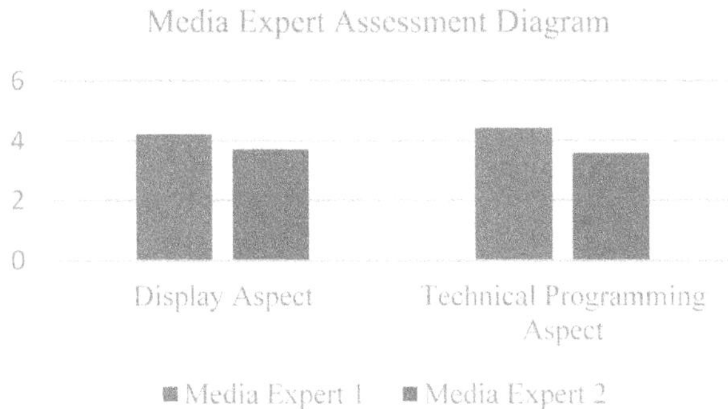

Figure 2. Media expert assessment diagram.

$$K = \frac{F}{N \times I \times R} \times 100\%$$

$$K = \frac{231}{5 \times 29 \times 2} \times 100\%$$

$$K = \frac{231}{290} \times 100\%$$

$$K = 79,65\%$$

Calculation of the feasibility percentage of this multimedia according to media experts shows a percentage of 79.65%. This indicates that the score falls into the "decent" category. Therefore, this product is feasible to use after an improvement is made according to the advice and input of the two media experts.

Based on the data from the material validation experts listed in the Table 3, it can be seen that the average score of the material aspects given by the material expert 1 is 4.44 and the mean score of the material expert 2 is 4.22. The mean score for the assessment of the learning aspects given by the material expert 1 is 4.73 while the average score of the material expert 2 is 4.18. In terms of overall overall aspects, material experts 1 and material experts 2 each gave a score of 4.6 and 4.2. Meanwhile, the average overall score produced by the material expert's

Table 3. Average material assessment.

Assessment Aspects	Average Scores Material Expert 1	Material Expert 2
Material	4.44	4.22
Learning	4.73	4.18
Average Score Per Expert	4.6	4.2
Overall Score	4.4	

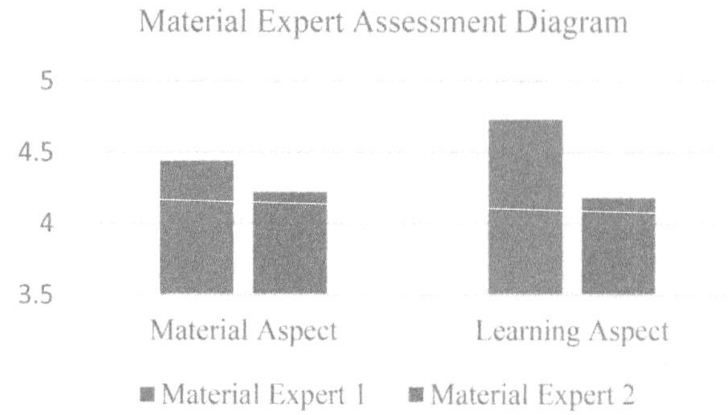

Figure 3. Material expert assessment diagram.

assessment is 4.4. The results of the assessment of each material expert based on each aspect can be seen in the following diagram.

The feasibility percentage of this multimedia according to material experts can be calculated using the following formula.

$$K = \frac{F}{N \times I \times R} \times 100\%$$

$$K = \frac{176}{5 \times 20 \times 2} \times 100\%$$

$$K = \frac{176}{200} \times 100\%$$

$$K = 88\%$$

Calculation of the feasibility percentage of this multimedia according to material experts reached 88%. This score falls into the category of "very feasible", so that this product, after the improvement in accordance with the suggestions and input from the two material experts, is appropriate to use. The feasibility percentage of this multimedia according to user responses can be calculated using the following formula.

$$K = \frac{F}{N \times I \times R} \times 100\%$$

$$K = \frac{3395}{5 \times 20 \times 39} \times 100\%$$

$$K = \frac{3395}{3900} \times 100\%$$

$$K = 87,05\%$$

Calculation of the feasibility percentage of this multimedia according to user responses reached 87.05%. The score falls into the "very feasible" category.

4.4 *Final product review*

The final product is a product that has passed the planning, design, development and testing phases both from experts and users. The products of multimedia learning of music and traditional musical instruments in the form of encyclopedias are installed in an 8GB flashdisk along with the wrapper cover. On the cover there is some information about the product along with the product usage manual that contains: 1) the background of making the product, 2) information about the product, 3) instructions for using the product, and 4) the profile of the developer. The final multimedia product of learning music and traditional musical instruments in the form of an encyclopedia is made in the .swf file format so users do not need to install an application first but they can run it on various operating systems. This product can be run immediately if the computer meets the recommended minimum specifications. The intro page contains the program title (munupedia) and the program title (encyclopedia of traditional archipelago music instruments) which is accompanied by an animation transition effect when the user starts the program by clicking on the file "Munupedia Windows.swf" in the file directory of this product.

4.5 *Research limitations*

The implementation of research and development has several limitations occuring due to several situations and conditions. These limitations include the following: 1) this research is still limited to product feasibility testing and has not yet reached the stage of product effectiveness testing because the main purpose of developing this product is for creating additional supplementation and finding information about traditional music and musical instruments from other regions not taught in at SMK N 1 Kasihan Bantul; 2) the reliability test of the instruments used in this research has not been carried out; 3) the supporting explanation of the material in the form of audio, images and videos on this product is still limited; and 4) the xplanation and grouping of material on this product is based on previous research by Sri Hastanto in 1996/1997 in his book entitled "Traditional Archipelago Musical Instruments: Music that Has Not Been Known Very Much", so there are still many shortcomings especially for grouping music or traditional musical instruments based on the region of origin where at that time was still limited to 26 provinces.

5 CONCLUSIONS AND RECOMMENDATIONS

5.1 *Conclusion*

This study produced a multimedia for learning music and traditional musical instruments in the form of an encyclopedia which is used as an additional learning supplement for students of SMK N 1 Kasihan in particular to help finding information about traditional music and musical instruments in Indonesia. The multimedia product for learning music and traditional musical instruments in the form of an encyclopedia was developed using the Adobe Flash CS6 application. This product was made in the .swf file format. This application was installed in a flashdisk accompanied by cover packaging and product usage manual. This application can be directly run on computers with various operating systems.

In this application there are four main menus: material, quiz, help, and profiles. The material menu is grouped into three grouping categories based on alphabet, island, and instrument classifications. There are 413 materials divided into 83 traditional music ensembles, one traditional music genre, thirteen traditional music, and 316 traditional musical instruments spread in 28 provinces of 34 provinces in Indonesia. Based on the classification of the instruments, the materials are divided into 65 aerophon categories, 141 idiofon categories, 47 kordofon categories, 60 membranophones categories, and 101 special categories. Based on the group of islands, the materials are divided into 136 for Sumatra islands, 100 for Java, 53 for Kalimantan, 56 for Bali and Nusa Tenggara, 37 for Sulawesi, and 31 for Maluku and Papua. In each material there is a descriptive explanation supported by examples of music and traditional musical instruments in the form of audio, images, and/or videos.

This product has been tested in terms of its feasibility through alpha test by two media experts and two material experts, as well as beta testing obtained from the response from 39 users. Based on the results of data analysis, broadly speaking this product is considered to be feasible. This is evidenced by the results of the alpha test from the two media experts which reached the overall feasibility percentage of 79.65%, which falls into the category of "feasible". Furthermore, the alpha test results from the two material experts reached the overall percentage of 88% which falls into the category of "very feasible". Meanwhile, the beta test results from 39 users showed an overall percentage of 87.05% which falls into the "very feasible" category. From the results of alpha and beta tests, it can be concluded that multimedia learning of music and traditional musical instruments in the form of an encyclopedia is feasible to be used as one of the supplementary learning for vocational students in particular and for public in general in terms of finding information about traditional music and musical instruments.

5.2 *Recommendations*

The following is a suggestion on the use of this product based on the results of research and development: 1) This product needs to be implemented further so that its weaknesses and shortcomings are known more thoroughly. Therefore, the effectiveness of this product will be truly tested as one of the supplements in learning traditional music and musical instruments; 2) This product can be run directly on a computer with various operating systems but must be adjusted in advance with computer specifications and usage manuals; and 3) This product can be used as an alternative learning supplement in terms of finding information about traditional music and musical instruments, not only for vocational students, research subjects in this study, but also for the wider community.

This product is far from perfect. There are still many materials that have not been accommodated. Therefore, further product development is needed which aims to improve the product that has been developed before so that the product will get better. The development includes the following: 1) this product needs to be developed in the form of mobile and website so that this product can be accessed in a variety of contexts both offline and online; 2) addition of material needs to be expanded and deepened, along with the grouping system that has been adjusted to the current situation (such as the number of provinces) along with the addition of material samples in the form of audio, images, and videos; 3) the quality of the video supporting media can be increased so that they can represent materials about performances and tutorials on how to play music, ensembles, and traditional musical instruments; 4) products need to be developed by adding combination categorization system so that users will not depend on a single grouping system. For example, grouping can be combined between certain regions (e.g., West Java) with the classification of certain instruments (e.g., membranofon) so that a categorization page will only display membranofone musical instruments from West Java; and 5) this product needs a search function by adding related words and recommendations to make it easier for users to find the desired information.

REFERENCES

Alessi, S.M. & Trollip, S.R. (2001). *Multimedia for learning: methods and development (3rd ed.)*. Massachusetts: Ally & Bacon A Pearson Education Company.

Ali, M. (2006). *Middle school music for class VIII*. Jakarta: Penerbit Erlangga.

Ardipal. (2010). The cultural arts education curriculum that is ideal for students in the future. *Journal of Language and Art*, 11, 1-10.

Arifitama, B. (2016). Preserving traditional instrument angklung using augmented reality technology. *International Multidisciplinary Conference*, 2, 309-315.

Arsyad, A. (2011). *Learning Media*. Jakarta: Rajawali Pers.

Daryanto. (2010). *Learning media has a very important role in achieving educational goals*. Yogyakarta: Gava Media.

David., dkk. (2014). Educational application "traditional arcade music" based on Android. *Jurnal Com-Tech Binus*, 5, 1134–1143.

Hackbarth, S. (1996). *The Educational Technology Handbook*. New Jersey: Educational Technology Publications, Inc.

Hastanto, S. (2005). *Nusantara Traditional Musical Instruments: Music that is not Widely Known*. Jakarta: Deputi Bidang Seni dan Film Kementerian Kebudayaan dan Pariwisata.

Hornby, A.S. (2010). *Oxford Advanced Learner's Dictionary: International Student's Edition*. Oxford: Oxford University Press.

Newby, T.J., et al. (2000). *Instructional Technology for Teaching and Learning: Designing Instruction, Intergrating Computers and Using Media*. New Jersey: Prentice Hall International, Inc.

Permana, J & Sumantri, M. (1999). *Teaching and Learning Strategy*. Jakarta: Direktorat Jenderal Pendidikan Tinggi.

Phillips, R. (1997). *The Developer's Handbook to Interactive Multimedia: A Practical Guide for Educational Applications*. London: Kogan Page.

Reddi, U.V. & Mishra, S. (2003). *Educational Multimedia a Handbook for Teacher-Developers*. New Delhi: Graphic Shield.

Riduwan. (2013). *Measurement Scale of Research Variables*. Bandung: Alfabeta.

Sachs, C., & Horbostel, E.M.von. Classification of musical instruments: translated from the original German by Anthony Baines and Klaus P. Wachsmann. *The Galpin Society Journal*, 14, 3-29.

Soeharto, M., dkk. (1987). *Music Art Lessons for SMTP*. Jakarta: PT. Gramedia.

21st Century Innovation in Music Education – Astuti et al. (eds)
 ISBN 978-0-367-11122-9

Multimedia *Talempong Pacik* learning: Virtual instrument based on Android

B.F.F. Zerky & K.S. Astuti
Universitas Negeri Yogyakarta, Yogyakarta, Indonesia

ABSTRACT: This paper explains the process of developing Talempong Pacik multimedia learning on android-based smart phones, from the production process to product validation. Research and development adapted from Borg & Gall, and Allesi & Trollip include planning, design, development, staging and production. The research and development aim: (1) Produce android-based Talempong Pacik multimedia learning for middle school (SMP) VIII grade students; (2) Test the feasibility of Android-based Talempong Pacik multimedia learning for learning grade VIII SMP students; (3) Find out the attraction product developed; (4) Find out whether students can use Talempong Pacik correcrlt after using the product developed. In the development process using Adobe Flash Professional software and Adobe Animate software, but this software has the disadvantage of being delayed in the Talempong Pacik android game that is ready to be validated by experts. In the end the author used the Android Studio software and the problem found was solved.

1 INTRODUCTION

Talempong Pacik is a traditional musical instrument of the Minangkabau community of West Sumatra. Bahar (2003) Talempong music is the only cultural music that can be found throughout Minangkabau region.

Ikhsan & Astuti (2018) described that Talempong a small gong-shaped instrument made of brass, tin and copper. A song in the talempong pacik game is produced through game interactions of various rhythmic patterns and motives inherent in the three rhythmic units and various rhythmic patterns and motives that are mutually unequal or different from one another with technical games interlocking. Children in middle schools in Padang studied Talempong Pacik, but not all schools have Talempong Pacik. For schools that have Talempong Pacik, they can study and practice while schools that do not only learn theory. Another problem is the absence of traditional music teachers, especially for Talempong Pacik. The author found that there is potential in the fact that most students have an Android smartphone.

The author thinks these problems and potential are important to solve and develop. Talempong Pacik virtual traditional musical instrument is as a simulation tool to play the real Talempong Pacik by utilizing an Android smart phone and opening up the potential of education to utilize smart phones in music learning, especially in the city of Padang. So students can practice learning material in class.

2 LITERATURE REVIEW

There is previous research that is relevant to this research; Mona (2015) entitled *"Multimedia Interaktif Talempong Pacik sebagai Media Pembelajaran Seni di SMP Negeri 3 Kota Solok"(Interactive Multimedia Talempong Pacik as Medium of Learning Arts at Junior School State 3 Kota Solok)* . This study was based on the problems (1) the unavailability of talempong pacik school instruments, (2) teachers did not have an art background so it was impossible to teach pacik talempong to students (3) lack of time in art teaching and culture. In the three problems that exist, teachers can utilize and empower technology to realize effective learning behaviour and increase the motivation to learn talempong, which is using interactive multimedia based talempong. It can be one of the solutions useful to support teachers, and helps students appreciate and express talempong pacik.

3 RESEARCH METHODOLOGY

3.1 *Research methods*

This type of research is development research to obtain a product. The product developed is a Talempong Pacik multimedia learning tool based on android. This research and development adapted from Borg & Gall (1984), Allesi & Trollip (2000) and modifications to R & D artwork by Pradoko, et al (2014).

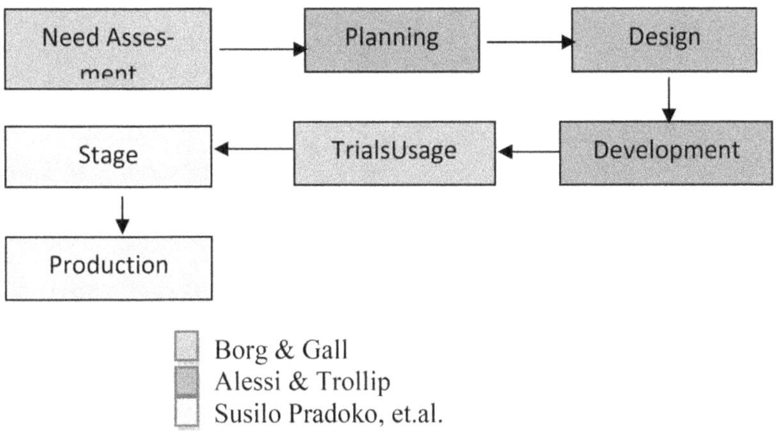

Figure 1. Stages of research and development.

In the needs assessment phase, researchers spent two semesters becoming music arts teachers at Padang State Middle School 5. Problems and potential found show that not all schools have Talempong Pacik and most students have an android smart phone. Another problem is that not all schools have music art teachers. In this step, the researchers also conducted brainstorming with several parties related to product development, so that it was developed in accordance with the results of the needs analysis. At the design stage, the product framework was developed such as flowcharts and storyboards. The framework contains full flow and multimedia content. The development step is the step of assembling all content that has been created and prepared into a multimedia product. After the product is ready, then alpha and beta testing stages are carried out which aim to assess the feasibility of the product, both in terms of appearance and content. The final step, conducting field testing in the school classroom and the Pongcikdroid performance, aimed to find out the effectiveness of multimedia on student learning outcomes.

Pongcikdroid is the concept of combining learning media using technology, which is the basis of the material taught to wit local wisdom in the form of traditional musical instruments. This multimedia was developed based on needs analysis in schools, so that learning materials were adjusted to the level of students' abilities. This multimedia contains material about Talempong Pacik; Pongcikdroid; learning objectives; the songs: Cak Din Din, Tigo Duo, Siamang Tagagau; instructions for playing Pongcikdroid; and Developer.

This multimedia characteristic was adjusted for grade VIII students, by displaying elements of the province of West Sumatra as the place of origin of Talempong Pacik. The opening music used was the song Cak Din Din. This aims to adjust between the background, nuances, and learning material provided. Here are some examples of multimedia displays:

Figure 2. Home screen of pongcikdroid.

Figure 3. Menu preliminary.

3.2 *Data presentation techniques*

To test the feasibility of this learning tool, the validation step was taken by media experts and material experts. Previously, instruments had been prepared and validated which would be used by the two validators. Validation of media instruments and materials was prepared based on variables developed into relevant media and information materials. Assessment is done by setting values in the range 1-5 which are adapted from a scale of five, one of the most diverse can be seen in the following table:

After the two validators provide an assessment of the media and material, the next step is to calculate the average value of the assessment results. The average value is obtained based on the following formula:

Definition 1 (See [1]). *The average number is denoted by X, symbol* $\sum x$ *is defined as the number of scores and symbol "n" is defined for the number of questions.*

Figure 4. About talempong pacik.

Figure 5. Song of talempong pacik on pongcikdroid.

$$X = \frac{\sum X}{n} \tag{1}$$

The results of the average scores that have been obtained are then converted into information in the form of qualitative data using five scale guidelines according to Sukardjo (2009: 84), which is shown in the following table:

Figure 6. Menu preliminary.

Figure 7. Cak din din track practice.

Figure 8. Basic: Cak din din track practice.

Table 1. Guidelines for product assessment.

Scale	Category
5	Very Good
4	Good
3	Sufficient
2	Less
1	Very Less

Table 2. Guidelines to convert quantitative data to qualitative in five scales.

No	Score Range	Calculation Guide	Criteria
1	Mi +1,80Sbi < X	X > 4,21	Very Good
2	Mi + 0,60 Sbi < X ≤ Mi +1,80Sbi	3,40 < X ≤ 4,21	Good
3	Mi - 0,60 Sbi < X ≤ Mi + 0,06Sbi	2,60 < X ≤ 3,40	Sufficient
4	Mi - 1,80 Sbi < X ≤Mi - 0,06Sbi	1,79 < X ≤ 2,60	Less
5	X ≤ Mi -1,80Sbi	X ≤ 1,79	Very Less

Definition 2 (See Table 2). *The scores achieved is denoted by X, symbol Mi is defined as the simulation of "1/2 (ideal maximum score + minimum score ideal)", and symbol Sbi is defined for the simulation of "1/6 (ideal maximum score - ideal minimum score)".*

4 RESULTS AND DISCUSSION

4.1 *Result*

To determine the feasibility of the product in the media and the material aspects, the score for the assessment of media and material experts (validators) was changed to qualitative data using five scale guides. Multimedia will be declared feasible when getting the minimum result, corresponding to the criteria "Good" according to Table 2.

Dr. AM. Susilo Pradoko is a material expert in alpha testing (Lecturer in the Department of Music Education, Faculty of Language and Art, Yogyakarta State University). The assessment process was carried out on May 4, 2018, there were 3 (three) components in the assessment by material experts: the quality of learning, material content, and quality of presentation. These were developed into 20 items of statements, rated on a scale of 1-5.

Based on the assessment of material experts on the products developed, 72 values were obtained with an average score of 3.6 being in the criteria of "good". From the assessment the material expert stated that the material contained in the product was feasible to be developed.

Dr. Kusnadi, M.Pd is a media expert in the alpha test (Lecturer in the Department of Dance Education, Faculty of Language and Art, Yogyakarta State University). The assessment process was carried out on May 9, 2018, there were 3 (three) components in the assessment by material experts: the quality of learning, the content of the material and the quality of presentation. These were developed into 20 items of statements, scoring 1-5.

Based on the assessment of material experts on the products developed, the value of 95 was obtained with an average score of 4.75 being in the criteria of "very good". From the assessment material experts stated that the multimedia contained in the product was feasible to be developed.

4.2 *Discussion*

Material experts provide advice outside the application that has been developed in order to add the Learning Process Plan (*Rencana Proses Pembelajaran/RPP*) in learning to use the product[Nisak, 2013]. One piece of advice from media experts is that song scores are made bigger because they are considered too small. The developer has tried to revise this in accordance with the advice of media experts, but it cannot be done because the screen resolution on Android is not sufficient to be enlarged. If this change is forced it will make the layout on the screen display asymmetrical. Also students are still able to see the score despite its small resolution size.

Suggestions from media validators and material validators are very helpful in developing this product. Everything that has not been done on this final product will be tried to be better developed in the future.

5 CONCLUSION

In today's modern era, technology is becoming more influential and can be developed into learning media, especially based on Android technology. Pongcikdroid is one way to simulate the real Talempong Pacik performance. There is a need to learn more about how to create multimedia that is good and effective, as well as the need to add more content to existing multimedia products. Multimedia *Talempong Pacik* Learning is a Virtual Instrument Based on Android can be use. However This media has many limitations, one of which is that it can only be used on Android. The advantages of this product must be mobile and can be played anytime and anywhere, and ensemble games can be played between offline virtual users. Researchers must be actively involved in the development of these so that multimedia learning in the art of music will be richer.

REFERENCES

Alessi, S. M., & Trollip, S. R. (2000). *Multimedia for learning: Methods and development*. Allyn & Bacon, Inc.

Bahar, M. (2017). *Cultural Uniformity and Diversity of Talempong Music*. The Social Sciences, 12 (10),1897-1910.

Borg, W. R., & Gall, M. D. (1984). *Educational research: An introduction*.

Ikhsan, Y., & Astuti, K. S. (2018). *Efektivitas tradisional Talempong Pacik terhadap peningkatan disiplin, toleransi, dan kerjasama siswa*. [Effectiveness of Talempong Pacik tradisional music instrument teaching to instil discipline, tolerance,and cooperation of students]. *Harmoni Sosial: Jurnal Pendidikan IPS*, *5*(1), 57-66.

Mona, D. (2016). *Multimedia Interaktif Talempong Pacik Sebagai Media Pembelajaran Seni Di SMP N 3 Kota Solok* [Talempong Pacik Interactive Multimedia as an Art Learning Media in Public Middle School 3 at Solok]. (Doctoral dissertation, Universitas Pendidikan Indonesia).

Nisak, K. (2013). *Pengembangan perangkat pembelajaran IPA terpadu tipe connected pada materi pokok sistem ekskresi untuk kelas IX SMP*[Developing of teaching and learning plan of sains for Junior School]. *PENDIDIKAN SAINS*, *1*(01).

Pradoko, S., & Priyanto, W. P. (2014). *Rancang Bangun Musik Angklung Model SATB, Dasar Aransemen Model Orchestra* [Design of SATB Angklung Music Model, Basic Arrangement of Orchestra Model]. Imaji, 14(1),90-97.

21st Century Innovation in Music Education – Astuti et al. (eds)
© 2020 Taylor & Francis Group, London, ISBN 978-0-367-11122-9

Beatbox for teaching in 21st century music education

E.P. Grandena & A.N. Machfauzia
Yogyakarta State University, Yogyakarta, Indonesia

ABSTRACT: This paper discusses the potential for beatbox to be used for teaching music education in the 21st Century. Contemporary teaching and learning are needed in order to fulfill manifestations and adaptations to environments that are much more complex than those of the 19th century. This discussion was made possible by collecting previous studies and literatures and making some observations about musical landscape changes that happened before the 20th century and the present, so that we can predict the next century as well as set future program plans for music education. The conclusion shows that beatbox can be used as an innovative way for teaching music in the 21st century music education classroom. This strategy can also fulfill cultural empowerment in the field of music education. Beatbox in music education can be use as method, strategy, or even an instrument for teaching music in very interesting ways.

1 INTRODUCTION

In "Why We Need Arts Education" Mill states that what constitutes human happiness is "the quality of life we experience and in association the quality of art we embrace knowing what the beneficence of an art educational experience means as a component of welfare maximization overall" (Cannatella, 2015). Anthony Storr adds that the arts enhance our feelings of personal worth, that they enable us to comprehend ourselves and our external world that touches the core of our physical, emotional and mental being (Cannatella, 2015).

From Mill's and Storr's statements, it can be concluded that art has a very important role in life. Based on the above, arts education programs need to reflect on themselves and express the continuation of common experiences in life that allow people to feel honesty and truthfulness in the achievements of art in maximizing the welfare of human life.

Music has a direct route to the movement and expression of one's mood, and is a very effective mood induction procedure. As stated in "The Evolutionary Nature of Musical Meaning" Cross (2009) explains that the idea that music has a direct route to the emotions. This appears to be confirmed in the use of music as a highly effective mood induction procedure (MIP) in experiments that have explored the behavioural, cognitive and neural underpinnings of affective states, as well as in its centrality of music in manipulating the emotional responses of audiences in cinematic context (Cross, 2009). From this statement, it can be concluded that music education is needed and should be taught in schools.

While it is important to learn music in schools, some studies found that learning music in schools is still too formal and less attractive to students. Sometimes students seem busy playing alone or not paying attention to the teacher at all. The methods which are generally used by most teachers are also old methods that seem too monotonous and boring for students. Music teachers in schools generally do not keep up with the times and globalization. Thus, teaching methods have become less innovative, creative and only rely on common teaching strategies. Knowing these problems, we need a learning model that attracts and motivates students.

In teaching and learning strategy, a music therapist must also master the art of music and music engagement (Goodman, 2013). It is important for a music therapist or teacher to master skills to motivate students to be involved in the music itself and the process of learning music.

Another problem in school environments is that there are limitations in musical instrument facilities. Schools usually do not adequately facilitate provisions of musical instruments for teachers of music arts and students. Even if there are music instruments and facilities, the number is very limited and doesn't provide enough for all students.

Considering the various problems above, beatbox can be one solution. Beatbox can be a solution and a unique approach to teaching that will attract students and stimulate them to learn. Beatbox can also be an innovative teaching method that follows the times in the era of globalisation, and at can be a medium that facilitates the teaching of music to be applied to students in music education programs. Beatbox can be an instrument that can be used anytime, anywhere and even saves costs.

Beatbox is one form of art that focuses on producing rhythmic sounds, drum beats, musical instrument sounds and imitations of other sounds through utterances by human vocal organs such as the mouth, tongue and lips for example (Stowell & Plumbley, 2008; Patil, Blaylock, & Narayanan, 2017).

With such a unique sound production, the application of beatbox will be able to stimulate students' interest in music education. This is because the sounds produced from the mouth source will stimulate children to be curious and even try to imitate mouth sounds. In Sam Woolstencroft's (2012) research it was found that beatboxing could increase the skills and enjoyment of phonics learning for students/young people. From the results of the abovementioned studies it can be concluded that an increase in skills and learning pleasure is a result of students' interest in learning models using beatbox. Interest in music learning is very important and this is the key to teaching music for children at a young age. Thus, school music teachers must know how to package the song materials to be interesting. In this case, the teacher's skills in presenting song material attractively is needed for music teaching because intelligent reception results from interesting presentations (Sunderman, 1941).

In "Facing the Music" (Schippers, 2010), the author states that a number of musical and extra-musical factors have influenced what is now widely referred to as World Music and its role in music learning and teaching. These factors extend over a number of interrelated fields: sociology, anthropology, ethnomusicology and also music education. Thus, the development of World Music will tend to influence and play a role in music education. Thus, it is important to adjust music learning and teaching in schools by following musical landscape developments.

Devos (2007) mentions that today rap music is considered to be one of the most popular genres in the music industry. As an extension of hip-hop popularity, rap can seen as background pedagogy by adapting to the musical landscape changes. Petchauer added that hip-hop pedagogy has grown and that scholars and educators have researched and experimented with the use of hip-hop music and culture to improve students' empowerment, cultural responsiveness and skills of literary analysis and critical literacy (Kelly, 2013).

From the statements above, beatbox, which is a branch of hip-hop culture, can be borrowed as one form of pedagogy into music learning. As music teachers, we cannot close our eyes to reality. Borrowing hip-hop cultural pedagogy will be an alternative way for increasing students' interest in up-to-date learning of art music in the classroom.

Related to the problems challenging music educators, in existing school music programs we have no realistic way to fulfill the mission of cultural empowerment. We must break out of the box into which we have confined ourselves (Reimer, 2009).

In "Music Education in the 21st Century" Reimer (1997) mentions that it is necessary for music education to become more relevant to the actualities of musical life in the present and future and thus another step is perhaps even more essential. Finally, we must eliminate the embarrassing and destructive gap between the major way people partake of music in their lives, listening and our neglect of, and disparagement of the teaching of listening in our music programs (Reimer, 1997). In this way the beatbox can be one method that is quite relevant to be applied in music learning for the 21st century, because it applies to everyday life.

The technique of beatboxing consists of elements from words which are in our daily life used to hear and apply (Woolstencroft, 2012). By looking at these perspectives, beatbox can be a potential option for learning music education in the 21st century.

2 LITERATURE REVIEW

2.1 *Music*

Music is very important for human life because music can affect human feelings and can also be a medium for expressing one's feelings. This is in accordance with the views of Kartono, Utomo & Priyono (2007) that music is one medium used to express feelings through sound (human voice or vocals and instrumental sounds or musical instruments).

Music according to Suwarto (2007), consists of several elements: patterns of rhythm, bars, melodies, tempo, dynamics and expressions, and songs and poetry. This paper discusses the beatbox, music that is produced from human mouth organs. Music produced as beatbox mainly creates a rhythmic sound that generally mimics the sound of a percussive musical instrument. Beatbox can also create melodies by imitating sounds such as wind instruments, other melody instruments or by imitating digital melodic sounds such as sound effects.

2.2 *The function of music*

Music function by types (Suwarto, 2007) is categorised as pure music, entertainment music, spiritual or religious music, music as the spirit of people's lives and natural or life music. Another statement from Setyobudi (2007), states that the function of music in general is as a medium of entertainment, treatment, increased intelligence, creating atmosphere for religious ceremonies and accompaniment of dance (Setyobudi, 2009). Based on these two statements, it can be concluded that the role of music cannot be separated from human life and is used to fulfill and complement human needs. Beatbox music functions as entertainment that can be developed as part of a music education learning programs, as learning models, strategies, media and instruments for teaching music in schools of the 21st century.

2.3 *Hip-hop music*

The genre of hip-hop music is the expansion of rap music, which makes this kind of music, is well known as rap music (Devos, 2007). In "Social Significance of Rap & Hip-Hop Culture", the author mentions that rap shares its roots with other forms of traditional African-American music such as jazz, blues and soul (Blanchard, 1999). In addition, Black musical genres including jazz, the blues, funk, soul and rhythm and blues all had unique, pronounced, and indelible impacts on the culture that would eventually come to be known as hip-hop (Rose, in Washington, 2018).

A brief development of hip-hop music culture is also described in "The Role of Rap Hip-Hop Music in the Meaning and Maintenance of Identity in South African Youth" (Cohen, 2009), mentions that hip-hop is the most influential form of music of our time. The reason for this statement is that in the culture of hip-hop music there were remixing techniques. Remixing is a technique that can be used to protect the environment (Cohen, 2009). Therefore, the ability to take any music sample and summarise it into a more accessible format provides opportunities for rap music by connecting and commenting on many different genres. Thus, hip-hop can be one of the most influential music genres across the landscape of music development and opens the boundaries of rap music open to unlimited influences. This idea of remixing is the core dynamic aesthetic of rap music and is also reflected in other aspects of rap culture (Cohen, 2009). As one of the most influential music genres in the musical landscape that develop overtime, the genre of hip-hop can be borrowed and incorporated into a model or program of teaching music education in schools especially in the 21st century.

2.4 *Beatbox*

Stowell & Plumbley (2008) mentioned that beatboxing is a tradition of vocal percussion which originated in 1980s hip-hop, and is closely connected with hip-hop culture. It involves the vocal imitation of drum machines as well as drums and other percussion, and typically uses the simultaneous imitation of bass lines, melodies and vocals to create an illusion of polyphonic music. The author of "*Comparison of Basic Beatboxing Articulations between Expert and Novice Artists using Real-Time Magnetic Resonance Imaging*" mentions that beatboxing is a musical art form in which artists aim to emulate physical instruments using vocal percussion (Patil, Blaylock, & Narayanan, 2017). This physical instrument vocal imitation is also well-known as Human Beatbox (Picart, Brognaux, & Dupont, 2015). The genesis of early mainstream beatboxing or"old school"beatboxing, originated in New York, where the rising hip-hop culture of the 1970s and 1980s gave a platform for artists to emulate drum machines, or "beat boxes,"as accompaniment to singers and rappers (Patil, Blaylock, & Narayanan, 2017). In "Analysis and Automatic Recognition of Human Beatbox Sounds", Picart (2015), Sandrine Brognaux, & Stephane Dupont mention that"Human BeatBox"(HBB) is a newly expanding contemporary singing style where the vocalist imitates drum beats, percussive sounds as well as pitched musical instrument sounds. Drum sounds typically use a notation based on plosives and fricatives, and instrument sounds cover vocalisations that go beyond spoken language vowels (Picart, Brognaux, & Dupont, 2015). This statement is also supported by Michael I. Proctor, Krishna Nayak, & Shrikanth Narayanan in "Paralinguistic mechanisms of production in human "beatboxing": A real-time magnetic resonance imaging study" (2013) that human beatboxing is a performance art in which the vocal organs are used to produce a range of percussive sounds, usually an accompaniment to lyrics spoken, rapped or sung at the same time, sometimes by the same artist (Proctor, Nayak, & Narayanan, 2013). It may be performed a capella or with amplification, with some characteristics of the beatboxing vocal performance style, as relevant for music signal processing and related fields (Stowell & Plumbley, 2008).

Thus, it can be concluded that beatbox is a musical art with techniques for producing musical sounds originating from human vocal organs by imitating sounds or instruments, especially rhythm or percussion. In this paper, beatbox is one of the musical arts and involves music techniques derived from vocal organs that can be used as instruments that contribute to learning program models and strategies for teaching music in the 21st century.

2.5 *The needs of teaching and learning music education in the modern era*

Music provides a window into important cultural and community changes. Technological developments in computing, sound recording and synthesis have changed the face of musical composition, performance and listening (Jorgensen, 2002). The need for music teaching cannot be separated from the effects of changes in culture and society and the development of music along with the development of technology. The tasks faced in learning in the century before 2000 are quite different from those in the present. How do we retain our long-standing, traditional, well-established strengths in teaching performance and at the same time expand our contributions to finally bring it into harmony with the realities of the culture we exist to serve, which is no longer dominantly or even marginally a performing culture (Reimer, 1997). In the 20th century, we still preserve the traditional way (Jorgensen, 2002). In the present, we can continue to offer good performance opportunities and at the same time we must be able to meet the musical needs of cultural performance. A balanced, comprehensive general education in music is obligated to involve students in experiencing and learning more about all the possible ways music can be encountered (Reimer, 1997). We have to be clever enough to preserve these traditional groups while expanding far beyond them in the kinds of performing opportunities we make available that will appeal to far more students than these groups ever have or ever will, especially as our musical culture continues to change in directions having little to do with bands, orchestras, and choruses (Reimer, 1997). However, adapting to 21st century needs means revisiting each dimension and their interplay such as knowledge, skill, character and metacognition (Fadel, 2015). There are the skills that students need to succeed in work, school and life. They include: 1) Core subjects (as defined by NCLB); 2) 21st century content (global awareness, financial, economic, business and entrepreneurial literacy, civic literacy and health and wellness awareness); 3) Learning and thinking skills (critical thinking and problem solving skills, communications skills, creativity, and innovation skills, collaboration skills, contextual learning skills and information and media literacy skills); 4) Information and communications technology literacy; 5) Life skills: leadership, ethics, accountability, adaptability, personal productivity, personal responsibility; and 6) People skills, self-direction and social responsibility (Partnership For 21st Century Skill, 2009).

Fadel (2015) states that there are few explanations about four dimensions that we need to adapt based on 21st century needs such as the terms, knowledge, skill, character, and meta cognition. Revisiting knowledge as a dimension that 21st century needs, is essential, but more relevance is required. Students'lack of motivation, and often disengagement, reflects the inability of education systems to connect the content to real world relevance (Fadel, 2015). The meaning of relevance is critically important to economic and societal needs; it's not only to satisfy students'wishes. In terms of skills, we need higher order skills in the 21st century, such as the "4 C' s" of Creativity, Critical thinking, Communication, Collaboration, also known as "21st Century Skills essential for deeply learning knowledge as well as for demonstrating understanding through performance" (Fadel, 2015). Furthermore, another dimension called "character" is about how we engage in the world: In order to face an increasingly challenging world and to benefit civil and civic society (Fadel, 2015). In addition,"meta cognition is the awareness of one's own learning and cognitive ability essential for activating transference, building expertise, and establishing lifelong learning habits" (Fadel, 2015).

Arostegui(2011) states that in educating music teachers for the 21st century it would take effort to build teacher pedagogical knowledge by identifying factors and conditions that influence the quality of music teacher education programs from the point of view of interests and goals. That goal is to understand global conditions that affect music, community education and ongoing teachers in these local settings and vice versa, and to improve the context of our knowledge of current issues that affect music programs of teacher education throughout the world (Arostegui, 2011). Thus, 21st century teachers are required to be able to integrate their pedagogical skills and adapt to developments that occur in the global world that affect music. As discussed earlier, hip-hop is one of the music genres that is quite influential in the world of music. This is because the hip-hop genre has a distinctive remixing technique that can hybridise various genres and adapt to the existing culture. Therefore, hip-hop pedagogy can be borrowed as a strategy, a music teaching method that can fulfill the preparation of music education learning programs in the 21st century.

3 DISCUSSION

3.1 *Beatbox potential for 21st century music education*

Today, the musical landscape is very different from landscape of the previous century around the world. We have new styles of music and new genres that are abundantly developing over time. We have many new musical instruments capable of producing new sounds and there are also so many musical genres in the world. Sometimes evolution in the music area is in line with developments in society, technology, and trends in globalisation.

The 21st century musical landscape will be very different from the previous century. Thus, it is important to adjust the way teachers teach music in the classroom based on changes in the musical landscape of the 21st century. We cannot ignore the reality of music genres that develop over time. There are certain musical genres that have become popular lately, such as hip-hop music, R n B, EDM. HBB (Human Beatbox) is an interesting and challenging case study related to spoken language technologies (Picart, Brognaux, & Dupont, 2015). In relation to beatbox, hip-hop music is the root and emergence of beatbox, a branch of hip-hop music which is very closely related to music such as R n B and EDM. Thus, we need to start considering a pedagogical approach that might establish more relevance for students. It is important to focus on a pedagogical approach based on what students prefer, what genre of music has become popular in their era, and what is actually developing in their lives.

Nowadays, music is everywhere all the time. Therefore, it would be great if we could teach music for instant access, a way to create music anywhere, anytime, for students. It's important to think about how to do this. This paper suggests that we can use beatbox as a tool to teach music in schools, especially in music education classes. Educators must play an important role in integrating their pedagogical skills to adapt to the music landscape that has developed and changed over time. Music teachers are on the critical threshold of failure to keep pace with the changing needs and interests of students in the 21st century. Contemporary teaching and learning is needed to fulfill the manifestations of adaptation to the environment that are far more complex than in the 19th century. Music teacher education can be a contributor to exploring the balance between curriculum and pedagogy, the power structure that influences music education at all levels, and the role of contemporary music genres that develop over time. These are challenges that surround institutional change especially in music or school education institutions. In this case we need a program model that can bring beatbox into music education. Program models featuring beatbox involvement in contemporary participatory culture can all be offered as a springboard for innovative teaching tools in music education.

3.2 *Music education as cultural empowerment*

For Reimer (2009), music education is for empowerment and gives real authority aimed at the cultural realm. Reimer states that within existing school music programs we have no realistic way to fulfill the mission of cultural empowerment. Reimer argues that the notion of cultural empowerment as capable of encompassing inclusive vision and comprehensive, balanced, and sequential program of music education, is ideal that we have not yet seriously thought through, let alone acted upon (Reimer, 2009). We must break out of the box into which we have confined ourselves, so that general music is reconceived to achieve genuine balance among all the roles, and performance continuing to be an essential component of a reconstituted elective program (Reimer, 2009). Cultural empowerment focuses on enabling each individual to achieve things related to a particular culture where they are members.

According to Reimer's statement, a number of important points can be summarised as follows: first, to be able to share widely with all valuable efforts provided by culture. Second, cultural empowerment enables students to contribute to cultural efforts by applying the breadth of their acquaintances, with opportunities for certain aspects and roles that fulfill their respective interests and capacities, so this elective music program is needed. Third, cultural empowerment recognises the reality of twins over taking part and contributing as a basis for full participation in the possibility of cultural empowerment.

An innovative teaching model or program is needed to guide individuals into cultural empowerment according to their needs and in order to meet the challenges of teaching and learning music in the 21st century. To build character in the 21st century music education we need to teach "how we involve the world" through indicators of cultural needs, and the impact of the media, social and personal challenges, and global challenges.

4 CONCLUSIONS

In music teaching in the schools, music education is an idea for empowerment and giving authority aimed at the cultural realm. Thus, teachers can be required to teach with their pedagogical skills referring to the cultural domain. Thus, in teaching music in the 21st century it is important for teachers to understand global conditions that affect music, as well as the music education program itself.

Music teachers of the 21st century need to improve the performance of their skills not only their abilities in playing music, but also to pay attention to students' engagement in learning music. This can be conducted by considering and predicting the development of the musical landscape era and the phenomena that are relevant in students' daily lives. Innovative teaching strategies are needed and which can include the application of beatboxing in educational music programs in the 21st century. Beatbox can trigger student interest in the process of encourage student engagement in learning music. In this case, beatbox can also be an alternative to innovative teaching methods that follow developments in the era of globalisation (21st century). In fact, this can be applied by incorporating the realm of the elements of local culture into music education programs. This is in line with the aim to continue the achievements of learning via music education programs in the previous century. Beatbox as an instrument can be used anytime, anywhere, and even saves costs.

ACKNOWLEDGMENT

The author would like to thank the Yogyakarta State University for supporting this international seminar. It is an honor for the author to be able to take part in the INTERCOME seminar. In addition, the author also wishes to express his warmest thanks to Ms. Dr. Kun Setyaning Astuti, M.Pd. who have provided guidance and advice in writing this journal article.

REFERENCES

Arostegui, J.L. (2011). Educating music teachers for the 21st century. AW Rotterdam: Sense Publisher.

Blanchard, B. (1999). The social significance of rap & hip-hop culture, Journal of poverty & prejudice, Ethics of Development in a Global Environment (EDGE), 1-10.

Cannatella, H. (2015). Why we need arts education. AW Rotterdam -Sense Publisher.

Cohen, D. (2009).The role of Rap/Hip Hop music in the meaning and maintenance of identity in South African youth. Johannesburg: University of the Witswatersrand.

Cross, I. (2009). The evolutionary nature of musical meaning. Musicae Scientiae, 13(2),179-200.

Devos, J. (2007). The evolution of Hip-Hop culture. Antwerp: Torhout KATHO-Departement Lerarenopleiding-RENO.

Fadel, C. (2015). Redesigning the curricullum for a 21st century education. Center For Curricullum Redesign. Retrieved from http://curriculumredesign.org/wp-content/uploads/CCR-FoundationalPaper_FINAL.pdf.

Goodman, K.D. (2011). Music therapy education and training. Springfield: Charles C. Thomas Publisher, LTD.

Jorgensen, E.R. (2002). Transforming music education. Bloomington: Indiana University Press.

Kartono, A., Utomo, N., & Priyono, A. (2007). Kreasi seni budaya untuk SMA(Art and Cultural Creation for High School). Jakarta: Ganeca Exact.

Kelly, L.L. (2013). Hip-Hop literature: The politics, poetics, and power of Hip-Hop in the english classroom. English Journal, 51-56.

Patil, N., Greer, T., Blaylock, R., & Narayanan, S. (2017). Comparison of basic beatboxing articulations between expert and novice artists using real-time magnetic resonance imaging. Proc. Interspeech 2017, 2277-2281.

Picart, B., Brognaux, S., & Dupont, S. (2015). Analysis and automatic recognition of human beatbox sounds: A comparative study. *InAcoustics, Speech and Signal Processing (ICASSP), In the Proceedings of IEEE International Conference on 2015 Apr 19,* IEEE Press, 4255-4259 (2015).

Proctor, M., Bresch, E., Byrd, D., Nayak, K., & Narayanan, S. (2013). Paralinguistic mechanisms of production in human beatboxing: A real-time magnetic resonance imaging study. The Journal of the Acoustical Society of America, 133(2), 1043-1054.

Reimer, B. (1997). Music education in the twenty-first century: Outgrowing its historically limited purposes, music education needs to be poised for an expanded mission in the twenty-first century. Music Educators Journal, 84(3), 33-38.

Reimer, B. (2009). Seeking the significance of music education. United States of America: MENC: The National Association for Music Education.
Schippers, H. (2010). Facing the music. New York: Oxford University Press.
Setyobudi. (2009). Seni budaya SMP, Jilid 3 (Middle School Arts, Volume 3). Demak: Erlangga.
Stowell, D., & Plumbley, M.D. (2008). Characteristics of the beatboxing vocal style. Dept. of Electronic Engineering, Queen Mary, University of London, Technical Report, Centre for Digital Music C4DMTR-08-01.
Suwarto, H. (2007). Seni budaya musik 2 (Art of Music Culture 2.). Jakarata: PT Galaxy Puspa Mega.
Sunderman, L.F. (1941). A music program for Elementary-School Teachers in training. The Elementary School Journal, 42(3), 207-211.
Washington, A.R. (2018). Integrating Hip-Hop culture and Rap music into social justice counseling with black males. Journal of Counseling & Development, 96(1), 97-105.
Woolstencroft, S. (2012). An independent research study into the effect of beatboxing on the skill and enjoyment of young people in phonics. Retrieved from https://www.scribd.com/document/379854528/An-Independent-Research-Study-Into-the-Effect-of-Beatboxing-on-the-Skill-and-Enjoyment-of-Young-People-in-Phonics. Retrieved 6 July, 2018.

21st Century Innovation in Music Education – Astuti et al. (eds)
© 2020 Taylor & Francis Group, London, ISBN 978-0-367-11122-9

The effects of applying games in music class on character and learning achievements

H. Praptomo & T.H. Retnowati
Universitas Negeri Yogyakarta, Yogyakarta, Daerah Istimewa Yogyakarta, Indonesia

ABSTRACT: This research aimed to discover: (1) the effects of applying games in music class on the learning achievements of grade five students at the SD Joannes Bosco Yogyakarta; (2) the effects of applying games in music class on the character of 42 grade five students at the SD Joannes Bosco Yogyakarta. This research used a quasi-experimental research design. The data that was collected was analysed using Multivariate Analysis of Covariance (MANCOVA). The results of the study are: (1) the application of a gaming method in the learning of music influenced the character and learning outcomes of the participants; (2) the application of the gaming method in the learning of music influenced the character of the participants; (3) the application of the gaming method in the learning of music influenced the learning outcomes of the participants.

1 INTRODUCTION

There are many common problems in the world of education that have yet to be solved. Some of these are related to students, such as cheating, skipping class and not completing school assignments. These problems reflect dishonesty, a lack of discipline and irresponsibility. The students do not admit to cheating (dishonest), are not disciplined to use their time to study at school and do not submit their work before deadlines. Moreover, they are not responsible in doing their jobs and they break school rules.

Character building takes place in the family, society and at school. According to Amri and Poerwati (2013, p. 157), education is an activity that changes the students' behaviours to be positive. Education is not only focused on the development of the way of thinking of the students but also focuses on the development of all potentials possessed by the students. Through education it is expected that the character of each individual can be built. So, school education has an important role for students to help them optimally develop.

Education in Indonesia focuses on character building in every subject learned at any institution, including formal education from the primary levels up to the high school level. At school, character building must be at the core of all school activities, particularly in class during the learning process. This makes character building crucial for the progress of education, especially in primary school.

One of the ways to implement character values in education is by integrating values into each lesson. It is stated in the Constitution No. 20, Year 2003 that the purpose of the national education system is to improve national competency, and build national character and civilisation with a sense of pride and dignity, in an effort to make each individual become part of a smart generation (Indonesia, 2003). It is also intended to develop the students' potentials to be religious and faithful persons who believe in only one God, possess pleasant personalities, have good knowledge, health, competency, creativity and independence, so that they become responsible and democratic people.

Music learning using gaming method that contains character values was one of alternatives to develop the students' character in primary school. The gaming method was one way to deliver a lesson by implementing a game related to the lesson, creating fun learning, motivating students and making them feel excited.

Through the gaming method in learning music, the students' character could be developed to improve their appreciation and knowledge of music and build their skills and creativity in music. Thus, students would be encouraged to grow their musical talents through exploring creative learning activities. Thereby, student character could be built. The students who built their character would have self-confidence and study hard to obtain their best achievements, including in the music class.

Based on the problem background, pre-observation at the SD Joannes Bosco Yogyakarta, and the results of the research, it was found that in order to increase student character values what was needed were song lyrics that contained good character values, and also that the learning had to be exciting. To make the learning music more fun, the researcher implemented the gaming method and used suitable musical instruments for the students.

Considering the problem above, it was necessary to conduct research on the influence of applying gaming method in learning music in developing the students' character and the students' progress reports. The experimental research was conducted at the SD Joannes Bosco Yogyakarta.

2 RESEARCH METHODOLOGY

2.1 *Type of research method*

The type of research used was quasi-experimental. The model experiment that was used involved a pretest and post-test of the design group control with one action. In this model, before doing an action, two groups were given a pretest to measure their starting condition (O1). Then, for the experiment group, i, a sign was given (X); no sign was given for the control group. After the action was completed, the two groups were given one more test as a post-test (O2) (Arikunto, 2010, p. 210).

The research was conducted at the SD Joannes Bosco Yogyakarta between September and November 2017. The research population was the grade five (Grade V) students at the SD Joannes Bosco Yogyakarta. The research population consisted of two classes, the Synergy Class of Grade V and the Intrapersonal Class of Grade V. The research sample consisted of two groups, the experimental group and the control group. The experimental group comprised 21 students of the Intrapersonal Class of Grade V, while the control group was comprised of 21 students from the Synergy Class of Grade V. The data gaining technique consisted of questionnaires and the students' progress reports. The data analysis technique used Multivariate Analysis of Covariance (MANCOVA).

3 RESEARCH RESULTS AND DISCUSSION

3.1 *Research results*

The descriptive analysis result of the pretest with the character variable and the students' progress result variable for the experimental and control groups are presented in Table 1 below.

Table 1. The results of pretest descriptive analysis.

		Pretest		
Variable	Class	Minimum	Maximum	Average
Character	Experimental	2.15	2.73	2.38
	Control	2.15	2.58	2.44
Progress report	Experimental	65.00	73.75	68.10
	Control	62.50	75.00	66.55

Table 1 shows the average pretest result of the character variable of (2.38) for the experimental class, which is not completely different from the average result of the pretest for the control class with a character variable of (2.44). The results shown were collected before implementation of the gaming method in the music class. The students in the experimental class had the same character result as the students in the control class. The pretest average result variable for the experimental class was (68.10), which is close to the pretest average of the control class (66.55). The results that are shown were collected before implementation of a gaming method in the music class. The students in the experimental class had the same test results as the students in the control class.

A post-test was given after the experimental class had implemented the gaming method, while the control class did not implement a gaming method. The analysis results from the descriptive post-test of character variable and the students' progress reports from the experimental class and control class are shown in Table 2.

Table 2 shows the post-test average of the character variable as (3.34) for the experimental class, which is more than the post-test average of the character variable as (2.56) for the control class. The results show that after the implementation of a gaming method in music learning in the experimental class had been conducted, the students showed better character results than the students in the control class. The post-test average of the students' progress results in the control class was (69.38). The results show that after the implementation of a gaming method in music learning in the experimental class, the students in that class had better character results than the students in the control class.

The research demonstrates that the implementation of a gaming method in music learning affected the character and the students' progress reports of the grade five students at the SD Joannes Bosco Yogyakarta. Whether the influence was significant or not, was tested using MANCOVA.

A prerequisite test was administered to test whether or not the research data fulfilled the preconditions to apply MANCOVA. The preconditions that were needed are the normally distributed data and its homogeneous variants.

3.1.1 *Normality pretest*

Table 3 shows that all scores under Asymp. sig. (2-tailed) is more than 0.05. The result shows pretest data for the character of the experimental class, pretest data for the character of the control class and the pretest of the students' progress reports in the experimental class. The pretest results of the students' progress reports of the control class were all normally distributed. So, the normality condition for pretest data had been fulfilled.

Table 2. Analysis results of descriptive post-test.

		Post-test		
Variable	Class	Minimum	Maximum	Average
Character	Experimental	2.85	3.69	3.34
	Control	2.27	2.96	2.56
Progress report	Experimental	90.00	96.50	93.68
	Control	66.25	75.50	69.38

Table 3. The test results of the normality pretest.

	Test statistic	Asymp. sig. (2-tailed)	Note
The pretest character of the experimental class	0.123	0.200	Normal
The pretest character of the control class	0.121	0.200	Normal
The pretest progress results for the experimental class	0.168	0.123	Normal
The pretest progress results for the control class	0.155	0.200	Normal

Table 4. The normality test of post-test.

	Test statistic	Asymp. sig. (2-tailed)	Note
The post-test character for the experimental class	0.164	0.143	Normal
The post-test character for the control class	0.182	0.067	Normal
The post-test progress result for the experimental class	0.181	0.070	Normal
The post-test progress result for the control class	0.178	0.080	Normal

Table 4 shows all of the scores under Asymp. sig. (2-tailed) as being more than 0.05. The results show the post-test data for the experimental class character, the post-test data for the control class character, the post-test of the students' progress report for the experimental class and the post-test of the students' progress report for the control class are all normally distributed. So, the normality precondition for all post data had been fulfilled.

3.1.2 *Homogeneity post-test*

Table 5 shows all scores under sig. are more than 0.05. The test result of Box's M is sig > α (0.219 > 0.05), meaning that the post-test for the character variable and the post-test for the progress report variable altogether have the homogeneous variant and co-variant matrices in the experimental and control classes. The test result of Levene for the post-test of character variable is sig > α (0.060 > 0.05) meaning that the post-test for character variable has homogeneous variant and co-variant matrices in the experimental and control classes. The test result of Levene for the post-test of the progress report variable is sig > α (0.0390 > 0.05) meaning that the post-test of the progress report variable has homogeneous variant and co-variant matrices in the experimental and control classes. So, homogeneity of the precondition has been fulfilled.

3.1.3 *Multivariate analysis of covariance (MANCOVA)*

Refers of Pallant (2007, p.275 and p.291) this research involve more than one dependent variabel and two condition, so that MANCOVA analyse could be performed using IBM SPSS software. The criteria for making a conclusion is sig < α, meaning that there is a significant difference, while if sig > α then there is no significant difference. The research used a level of significance of (α) = 0.05. The computation of MANCOVA was generated with SPSS software, and the results are displayed in the following table.

Table 6 shows the post-test of the character and progress report for the experimental class as being significantly different from the control class (F counted = 671.402 and sig 0.000 < 0.05) and was not significantly different nor affected by the pretest of the character (F counted

Table 5. Test results of the homogeneity post-test.

Variable	F counted	Sig.	Note
Character and progress result	1.475	0.219	Homogeneous
Character	3.738	0.060	Homogeneous
Progress result	0.756	0.390	Homogeneous

Table 6. The results of MANCOVA.

	F counted	Sig.	Note
Pretest character (covariate)	2.802	0.074	Not significant
Pretest progress result (covariate)	14.025	0.000	Significant
Class (factor)	671.402	0.000	Significant

Table 7. The separated test results.

	Dependent variable	F counted	Sig.	Note
Pretest character (covariate)	Post-test character	4.385	0.043	Significant
	Post-test progress result	1.110	0.299	Not significant
Pretest progress result (covariate)	Post-test character	3.377	0.074	Not significant
	Post-test progress result	24.340	0.000	Significant
Class (factor)	Post-test character	135.705	0.000	Significant
	Post-test progress result	1,285.403	0.000	Significant

= 2.802 and sig 0.074 > 0.05). The results show that the implementation of a gaming method in music learning in the experimental class successfully improved the students' character, compared to the control class students where gaming method was not implemented, and thus there was no change in character from the beginning of the research.

The post-test character and the post-test progress results of the experimental class are significantly different from the control class (F counted = 671.402 and sig 0.000 < 0.05) that were influenced (significantly affected) by the pretest students' progress result (F counted = 1.025 and sig 0.000 < 0.05). The results show that the implementation of a gaming method in the music learning for the experimental class successfully increased the students' better character, compared to the control class students where gaming method was not implemented, and thus showed no change in character from the beginning of the research.

Table 7 shows that the post-test result for character in the experimental class was significantly different from the control class (F counted = 135.705 and sig 0.000 < 0.05) and was affected (the influence is significant) by the pretest of character class (F counted = 4.385 and sig 0.043 < 0.05) but it is not affected (the influence is not significant) by the pretest of the progress result (F counted 3.377 and sig 0.074 > 0.05). The results show that the implementation of a gaming method in the music learning for the experimental class increased better student character, compared to the control class where gaming method was not implemented and did not influence character from the beginning of the research, but was instead influenced by the beginning progress result.

The post-test result for the students' progress report in the experimental class was significantly different from the control class (F counted = 1,285.403 and sig 0.000 < 0.05), which was not affected (not significantly influenced) by the pretest for character (F counted = 1.110 and sig 0.299 > 0.05), but was affected (significantly influenced) by the pretest of the progress result (F counted 24.340 and sig 0.000 < 0.05). The results show that the implementation of a gaming method in the experimental class successfully increased the students' better character, compared to the control class students where gaming method was not implemented and without any influence on their character from the beginning of the research, but was influenced by the beginning students' progress results.

3.2 *Discussion*

At the start of the research, student character in the experimental class was considered to be the same as in the control class. The pretest average score of the experimental class (2.38) was not significantly different from the pretest average score of the control class (2.44). After the implementation of a gaming method in the music learning had been accomplished in the experimental class versus no implementation in the control class, the average score of the post-test in the experimental class (3.34) became higher than the average score of the post-test in the control class (2.56). The results show that the implementation of a gaming method in music learning had successfully increased student character if it is compared to the class where there was no implementation of a gaming method. Thus, it can be concluded that the implementation of a gaming method in the music learning classroom influences the character of grade five students at the SD Joannes Bosco Yogyakarta.

At the beginning of the research, student character in the experimental class was considered to be the same as in the control class. It can be analysed that the pretest average score of the experimental class (68.10) was not significantly different from the pretest average score of the control class (66.55). After the implementation of a gaming method had been accomplished in the experimental class, versus the control class where there was no implementation, the average score of the post-test in the experimental class (93.68) became higher than the average score of the post-test in the control class (69.38). The results show that the implementation of a gaming method had successfully increased student character, compared to the class where gaming method was not implemented. So, it can be concluded that the implementation of a gaming method in the music learning influenced the character of the grade five students at the SD Joannes Bosco Yogyakarta.

The MANCOVA results show that the implementation of a gaming method in music learning influenced student character and the progress of the grade five students at the SD Joannes Bosco Yogyakarta (F counted 671.402 and sig 0.000 < 0.05), in which the result was influenced by the progress report before the action was taken (F counted 14.025 and sig 0.000 < 0.05), but it was not influenced by the students' character before implementation (F counted 2.802 and sig 0.074 > 0.05). The implementation of a gaming method in music learning influenced student character of the grade five students at the SD Joannes Bosco Yogyakarta (F counted 135.705 and sig 0.000 < 0.05), in which the result was influenced by student character (F counted 4.385 and sig 0.043 < 0.05) but was not influenced by the progress report before the action was taken (F counted 3.377 and sig 0.074 > 0.05). The implementation of a gaming method influenced the progress report of the grade five students at the SD Joannes Bosco Yogyakarta (F counted 1,285.403 and sig 0.000 < 0.05), in which the result was influenced by the progress report before the implementation action was taken (F counted 24.340 and sig 0.000 < 0.05), but it was not influenced by student character before the implementation action was taken (F counted 1.110 and sig 0.299 > 0.05).

The results show that the increases in character values and progress report of the grade five students were not only influenced by the implementation of a gaming method in the music learning but also influenced by the progress report before the implementation action was taken. The increase of character values of the grade five students at the SD Joannes Bosco Yogyakarta was not only influenced by the implementation of a gaming method in the music learning before the action was conducted but was also influenced by student character before the action was taken. The increased score of the progress report of the grade five students was not only influenced by the implementation of a gaming method in music learning but was also influenced by the students' character before the action was taken.

The research results support Angela Lee's (2014) research results that the use of musical activities enables increases in children's educational character aspects in preschool students in Taiwan. The present research results also support Haeni Nur's research results (2013) that traditional children's games help to build children's character.

A group task for making song lyrics with the themes of discipline and responsibility lead students to learn more discipline values, honesty and responsibility. Students who are able to write song lyrics know good character values.

The research results show that there was an increase of discipline average scores in the experimental class from 2.34 (pretest) to 3.33 (post-test), an increase of 0.99. Discipline average scores in the control class increased from 2.33 (pretest) to 2.49 (post-test), an increase of 0.16. The increase in discipline character scores of the experimental class is more than that for the control class, suggesting that the games enable increases in the students' discipline values. Discipline values were measured in the games in terms of obeying the rules of the games, creating song lyrics that are relevant to the determined theme, to use time allocated for each game step and to queue in a line for each student to wait for his or her turn while playing continuous songs.

The average scores for honest behaviour in the experimental class increased from 2.46 (pretest) to 3.34 (post-test), an increase of 0.88, while the average score for honest behaviour of the students in the control class increased from 2.51 (pretest) to 2.61 (post-test), an increase of

0.10. The increase of honest behaviour in the experimental class is greater than the increase of honest behaviour in the control class, suggesting that the games enable increases in the students' honest behaviour. The honest behaviour was scored by students answering the questions, without cheating, fair game playing and admitting mistakes.

The average score for responsible behaviour in the experimental class increased from 2.35 (pretest) to 3.34 (post-test), an increase of 0.99. The average score for responsible behaviour of the students in the control class increased from 2.48 (pretest) to 2.58 (post-test), an increase of 0.10. The score increase in responsible behaviour in the experimental class is greater than the increase in honest behaviour in the control class, suggesting that the games enable increases in the value of responsibility by the students. Responsible behaviour was scored in terms of not complaining when students were having a trouble, writing song lyrics carefully and correctly, taking part in the games excitedly and actively playing the games.

The games enable increases in the students' character. Each group of students was able to write song lyrics that suited the theme and they did it before the time was up. Group I wrote song lyrics with the theme of responsibility, asked others to take responsibility seriously, did not complain while doing their task and tidied up all instruments. Group II wrote song lyrics with the theme of honesty, asking others to be brave, honest and to admit making mistakes and not tell lies so they could be trusted. Group III wrote song lyrics with the theme of discipline and asked others to be disciplined with time, not to be late to school and submit assignments before deadlines. Group IV wrote song lyrics with the theme of discipline and asked the others to obey the rules and not to forget to do their job. The song lyrics that were written show that the students who used gaming method were able to feel, explore, and apply the character values of discipline, honesty and responsibility.

The progress report assessment showed that gaming method in the music learning helps to increase student competency in making exact notations, rhythms, tempos and staying in harmony. The skills increase because the students have learned better discipline, honesty, and responsible behaviours.

The research results show that gaming method in the music learning is relevant to what Gafur (2012) stated, that gaming method is a method or learning activity which requires that students have competency in both physical and mental activities by following rules that guide the game.

4 CONCLUSION

The research results show that: 1) The implementation of a gaming method in music learning in the experimental class increased the students' character and the students' progress report, compared to the control class in which gaming method was not implemented. It is proven from the results of the MANCOVA that F counted = 671.402 and sig 0.000 < 0.05. The increase is also affected by the students' progress result before the implementation action was taken (F counted = 14.025 and sig 0.000 < 0.05). Based on the results, it is concluded that the implementation of a gaming method in music learning influences the students' character and the students' progress report, and is also influenced by the students' progress results before the implementation action was taken; 2) The implementation of a gaming method in music learning enabled increases in students' character in the experimental class, compared to the control class in which gaming method was not implemented. The result was proven with the results of the MANCOVA that F counted = 135.705 and sig 0.000 < 0.05. The increase is also affected by the students' progress result before the action was taken (F counted = 4.385 and sig 0.043 < 0.05). Based on the results, it is concluded that the implementation of a gaming method in music learning influences the character of the grade five students at the SD Joannes Bosco Yogyakarta and is also influenced by the students' progress result before the implementation action was taken; and 3) The implementation of a gaming method in music learning in the experimental class enabled increases in more of the experimental students' progress results, compared to the control class in which gaming method was not implemented. It is proven with the results of the MANCOVA that F counted = 1,285.403 and sig 0.000 < 0.05. However, the increase is also influenced by the progress report before the implementation action was taken (F counted = 24.340 and sig 0.000 < 0.05). Based on the results it is concluded that the implementation of a gaming method in music learning influenced the grade five students at the SD Joannes Bosco Yogyakarta and was also influenced by the students' progress result before the implementation action was taken.

REFERENCES

Amri, S. & Poerwati, E.L. (2013). *Panduan Memahami Kurikulum 2013 Sebuah Inovasi Struktur Kurikulum Penunjang Pendidikan Masa depan* [Guideline to understanding 2013 curriculum: An innovation in supporting education curriculum structures of the future]. Jakarta, Indonesia: PT Prestasi Pustaka.

Arikunto, S. (2010). *Manajemen Penelitian* [Research management]. Jakarta, Indonesia: Rineka Cipta.

Gafur, A. (2012). *Desain Pembelajaran Konsep, Model, dan Aplikasinya dalam Perencanaan Pelaksanaan Pembelajaran* [Learning concepts, models, and applications design in implementing and planning the learning]. Yogyakarta, Indonesia: Ombak.

Hurlock, B.E. (1978). *Child development* (volume 1). Jakarta, Indonesia: Erlangga.

Indonesia, R. (2003). *Undang-undang Republik Indonesia Nomor 20 Tahun 2003 tentang Sistem Pendidikan Nasional* [Republic of Indonesia Constitution Number 20 of 2003 Concerning the National Education System]. Jakarta: Government of the Republic of Indonesia.

Lee, A. (2014). Implementing character education program through music and integrated activities in early childhood settings in Taiwan. *International Journal of Music Education, 34*(3), 340–351.

Nur, H. (2013). *Membangun Karakter Anak Melalui Permainan Anak Tradisional. Jurnal Pendidikan Karakter*[Building characters of children through traditional children games]. *Jurnal Pendidikan Karakter, 3*(1), 25-32.

Pallant, J., & Manual, S.S. (2007). A step by step guide to data analysis using SPSS for windows. In *SPSS Survival manual*. Berkshire UK: Open University Press.

Pranoto, Y.K.S., Sugiyo, S. & Hong, J.J. (2014). Developing early childhood's character through Javanese traditional game. *Indonesian Journal of Early Childhood Education Studies, 3*(1), 54–58.

21st Century Innovation in Music Education – Astuti et al. (eds)

Learning through an interactive multimedia CD to strengthen musicality and character values

T. Dwipa & K.S. Astuti
Universitas Negeri Yogyakarta, Yogyakarta, Daerah Istimewa Yogyakarta, Indonesia

ABSTRACT: This study aimed to produce a multimedia interactive learning product for senior high-school students, a CD titled Technology Arts Research Arrangement (TARA). The product included music education materials based on 'Mensahang Lah Mirah', a traditional song from Indonesia's Bangka Belitung province. The research used the research and development method of Gall et al. (1996). Data collection took place at the high school, SMAN 2 Pangkalpinang; both the control and experimental classes consisted of 35 students. Data were collected by observation, questionnaire, interview and documentation and analysed using descriptive qualitative techniques and quantitative statistical tests including multivariate analysis of variance. The result of this research is a multimedia interactive learning product. The data show that the developed media is effective in increasing the students' musicality and tolerance, but it is not effective in increasing the students' mutual cooperation.

1 INTRODUCTION

The twenty-first century is an era that differs significantly from previous ones; technological diffusion is rapid and students increasingly have easy access to Internet-based resources via their personal electronic devices such as smartphones and notebooks. This is as true of music as it is of any other domain. Indonesian high-school students are now more familiar with modern songs than with traditional songs. Zulhendri's (Pramudita, 2016) study of children's knowledge of different musical genres showed that students' recognition of the modern pop song, Pop Dewasa (adult), exceeded 60% whereas local or traditional songs had dramatically lower recognition rates, less than 10%.

Traditional songs reinforce important pro-social values, such as kindness. Furthermore, learning traditional songs should be implemented in educational settings, particularly music arts learning.

Lack of familiarity with traditional songs contributes to lower musicality. The authors have observed that many students in the tenth-grade class at SMAN 2 Pangkalpinang school in Indonesia's Bangka Belitung province sang traditional songs incorrectly. Further, students were unable to sing in groups of two to four parts, in other words, Soprano Alto, Tenor and Bass (SATB).

The lack of knowledge of traditional songs affects not only students' musicality, but also their values; with intolerance and individualism a growing problem. Therefore, learning (including music arts learning), which strengthens students' character values, such as tolerance and mutual cooperation, is necessary.

In teaching and learning processes in schools, teachers may attempt to teach students traditional songs; however, teaching methods are frequently limited to speech and texts. Students may become bored and rapidly lose interest. Similarly, teachers may not use classroom facilities to their full advantage, including projectors and speakers. These problems could potentially be overcome by using a multimedia interactive learning programme that encourages students' interest in their local songs.

For these reasons, this study aimed to develop an interactive multimedia learning product, the Technology Arts Research Arrangement (TARA) CD, which contained the song 'Mensahang Lah Mirah', a traditional song from Indonesia's Bangka Belitung province. The purpose of developing the product was to strengthen students' musicality as well as tolerance and mutual cooperation.

2 LITERATURE REVIEW

2.1 *Interactive multimedia learning*

Interactive multimedia systems are computer-based digital environments that enable the user to choose where they want to go next within the environment. According to Bedi et al. (2011), learning via multimedia is more complex than traditional learning in which the medium is speech and/or text. As indicated by the term's morphology, 'multimedia' is the combination of various media: text, sound, graphics, video, animation and so on.

2.2 *Arrangement of the traditional song 'Mensahang Lah Mirah'*

The TARA CD is an interactive multimedia learning environment that contains arrangements of traditional Indonesian songs, of which 'Mensahang Lah Mirah' is one. The arrangement is for four voices and is suitable for a chorus or vocal group. The arrangement is tailored for male and female voices of senior high-school-level students using the ranges and tessituras described by Steinman (2010).

2.3 *Musicality and character values*

2.3.1 *Musicality*

Sture Brandstrom (as cited in Jaffurs, 2004) found two distinct views on musicality: an 'absolute' view and a 'relativistic' view. The absolute view sees musicality as inherited, and able to be measured by achievements such as performing, composing and improvising. The relativistic view sees musicality as socially constructed and something of which all people are capable.

In this study, to avoid subjectivity in the assessment of musicality, an absolute definition is adopted: musicality is measured by musical achievements, including performance. The performance is based on the syllabus and lesson plan, 'Indonesian Music Performance (local song of Bangka Belitung)' which is arranged for a SATB vocal group. The song is performed inside or outside the class, and it is the performance that is evaluated to measure musicality.

Key musical elements are rhythm, melody, harmony and dynamics (Kamien, 1980). 'Rhythm' is the well-ordered flow of the music – time in music can be compared to time in all aspects in life, including speed and intensity. 'Melody' is a series of single tones that have pitch. Melody begins, continues and finishes; melody has direction, form and continuity. Tension and release may be achieved through melodic movement, including through contrasts in pitch. Such tension and release may be experienced as hope and coming. 'Harmony' is the building of chords and how the chords are ordered. A chord is the combination of at least three notes heard together. A chord is a group of simultaneous tones, whereas melody is a series of individual tones that are heard one after the other. 'Dynamics' refers the degree of loudness or tenderness in music. Loudness is the amplitude of the vibration that produces sound. A player may stress a tone by playing it louder than other tones in the phrase – a dynamic accent. The change in the music's dynamics as loudness gradually increases from soft to loud engenders enthusiasm and emotion in the performance. Dynamics are not always written in the music but can be understood and produced by the players.

In skill practice of musicality at school, students may focus on the musical score or music notation at the expense of musical expression or feeling. In evaluating musicality, musical expression needs to be included so students can 'feel' the song and so that it can also be felt by the audience.

Rismawan (2014) stated that 'expression' is the way that performers convey their feelings or thoughts. The way people express their feelings depends on the situation and conditions in which they find themselves. Musical expression reflects the meanings experienced in the heart and mind, which are expressed through inspiration and appreciation of the song.

2.3.2 *Character values*

Lickona (2013) stated that character values consist of operative values – values in action (p. 51). He asserts that the development of good character is made up of three attributes: knowing the good (judging what is right), desiring the good (caring deeply about what is right) and doing the good (acting out what is right). Character consists of 'operative' values and 'action' values.

As someone develops 'good character', they express values such as kindness. As a result, they are able to respond to situations in accordance with those moral values. 'Character' has an 'affective' component. Based on Andersen (as cited in Mardapi, 2011), there are two methods that can be used to measure affect: observation and self-report. The observational method is based on the assumption that affective character can be seen in behaviour and attitudes and/or psychological reactions. The self-report method assumes that the individual is best placed to know their own affect.

Nevertheless, honesty is needed if an individual is to accurately express their affective characteristics. In this research, the self-report method is used. Students complete a questionnaire that contains 30 questions. The questions are arranged based on Mardapi's (2011) indicators:

(1) Tolerance

Tolerance is the attitude and action that respects diversity of backgrounds, opinions and beliefs. Indicators of tolerance values are:

a. does not disturb someone else who has opinions that differ from their own;
b. accepts an opinion even if it is different from their own;
c. can apologise for their mistakes;
d. able and willing to work with those who have different backgrounds, opinions or beliefs;
e. does not compel others to share their opinions or beliefs;
f. willing to learn from other beliefs and opinions, so that they can understand the other person (open-minded); and
g. open toward, or willing to accept, new experiences.

(2) Mutual cooperation

Mutual cooperation is working together to achieve a shared purpose, by sharing tasks and helping each other unconditionally. Indicators of mutual cooperation are:

a. being actively involved in cleaning the classroom with classmates;
b. being willing to fulfil a commitment to do a task;
c. being willing to help others without expecting a reward;
d. being active in group work;
e. focusing on the group's work;
f. not wishing to prioritise personal interest over others' interest;
g. looking for ways to reconcile differences of opinion or thought between themselves and other people; and
h. encouraging others to cooperate to achieve shared goals.

3 METHOD

3.1 *The research and development method of Gall et al.*

The product development in this research used the model of Gall et al. (1996), which consists of ten steps:

a. Research and information collecting: This was the first step in analysing needs and was conducted through observing the teaching and learning process during a music arts lesson.
b. Planning: In this step, the researcher created a plan for developing a multimedia product to enhance music arts learning.
c. Developing a preliminary form of the product: In this step, the researchers arranged 'Mensahang Lah Mirah' in four parts for a SATB vocal group. Finale score-writing software was used for this purpose.
d. Preliminary field testing: The 'Mensahang Lah Mirah' score was trialled with nine students of Candra Gemilang vocal group from SMAN 5 Karawang school. The purpose of the trial was to assess the suitability of the score's vocal range for children at senior high-school level.
e. Main product revision: After the first trial, it was necessary to modify the score by altering the bass part, which was too difficult and too low.
f. Main field testing: After revising the score, the score was then trialled with the larger ensemble of 20 students from Candra Gemilang vocal group from SMAN 5 Karawang school. At this time, the tutorial video was recorded, which consisted of five video segments: four segments contained a student demonstrating each of the four voices (soprano, alto, tenor and bass) and the fifth video segment contained 20 students singing in four parts (SATB) with one teacher as conductor.
g. Operational product revision: In this step, the score and tutorial video from the previous step were integrated into the interactive multimedia learning tool.
h. Operational field testing: Nine students from SMAN 2 Pangkalpinang school's vocal group participated in the operational field testing. In this step, the multimedia tool was used for the first time as a learning tool. This testing aimed to ensure that the product was free of bugs and functioned as expected, so that it could be used successfully in the experimental class in Step j. The researchers also, in this step, sought evaluations from the media and content experts.
i. Final product revision: Based on field observations and expert evaluation obtained during the previous step, the product underwent final modifications.
j. Implementation: The product was tested with tenth-grade students of SMAN 2 Pangkalpinang school. The implementation step was taken with the control class, X IPS 2 class. The step also involved X IPS 5 class through the socialisation of the multimedia learning product development.

3.2 *Data analysis technique*

Assessing the feasibility of the multimedia programme was done using a Likert scale. The scale integers used were: 1 (Very inappropriate), 2 (Inappropriate), 3 (Appropriate enough) and 4 (Very appropriate) (Sugiyono, 2010). The percentage calculation process was done by comparing the observed frequency with the expected frequency. The percentage is calculated using the following formula:

$$\text{Percentage} = \frac{\text{observed frequency}}{\text{expected frequency}} \times 100$$

The resulting percentage was compared with the feasibility scale in Table 1.

Table 1. Feasibility scale (Arikunto, 2012).

Achievement percentage	Scale number	Interpretation
76–100	4	Very appropriate
56–75	3	Appropriate
40–55	2	Less appropriate
0–39	1	Not appropriate

The effectiveness of the multimedia product was determined using IBM SPSS software to test the data for normality and homogeneity and to generate descriptive and multivariate statistics (Multivariate Analysis Of Variance, MANOVA).

4 RESULT

4.1 *Product presentation*

The media contained within the TARA CD are text, sound and video (for a tutorial). The product developed in this study is an interactive multimedia product which enhances the teaching and learning process. The traditional song, 'Mensahang Lah Mirah', comes from Indonesia's Bangka Belitung province, the location of this study. The lyrics of this song are:

Mensahang-sahang lah mulai mirah (The field is already reddish,)

Musim petik ge mulailah (it is time to begin the harvest.)

Banyak keluar bujang kek miak (The man and woman go out,)

Memetik sambil berdendang (harvesting while singing together.)

Tangan memetik mate berpandang (Hands pick, eyes see each other,)

Ati berdebar dak kruan (hearts beat quickly.)

Tingok betingok janji bejanji (They see each other and make a promise)

Kelak pulang bedue (to come back together.)

Surelah dateng, ari nek malem (The afternoon will come and the day will become night,)

Sedenget agik nek pulang (we will be back soon.)

Bulan tersenyum, bujang ge girang (The moon smiles, and the man will be happy;)

Miak seneng ati e (the woman will be happy too.)

La la la la la la la la la la (La la la la la la la la la la)

Bujang kek miak betinten-tinten (Men and women are side by side.)

Pulang ke rumah bedue (They come home together.)

Malem ni cuma beseneng-seneng (Tonight we will be happy together,)

Isok lah begawe ulik (tomorrow we will work again.)

According to Sutrisna Ekowibowo, expert on Bangka Belitung's traditional music, the values of tolerance and mutual cooperation contained in the song make 'Mensahang Lah Mirah' an appropriate choice for music arts learning.

The TARA CD contains the song 'Mensahang Lah Mirah' for four voices, represented in two types of notation, that is, a Western score and a numerical score.

4.2 *The appropriateness evaluation*

The evaluation of appropriateness was made by the media expert and the content expert. The media expert assigned a total score of 75 to the product (35 + 20 + 20). Using the formula presented earlier, the percentage can be calculated:

$$\frac{75}{100} \times 100 = 75\%$$

The appropriateness evaluation made by the content expert was 75 (8 + 39 + 28). The percentage of 75% was calculated by the same means as above. It can be seen that each expert's evaluation led to an appropriateness score of 75%, which equates to level 3 in Table 1. Hence, the multimedia product has been assessed as 'Appropriate'.

4.3 *The effectiveness evaluation*

The normality test results (Table 2) show that the data do not significantly deviate from a normal distribution ($p < .05$) (Pallant, 2013); hence, we concluded that the data are normally distributed.

Homogeneity testing was carried out using the Levene test; as the Levene statistic did not reach statistical significance, we accepted that the variances are equal across the samples. The MANOVA shows that the data have an F score greater than the F table, with $p > .05$; the data from both groups are homogeneous.

As indicated in Table 4, based on Box's test, Box's M was 5.416, which was not statistically significant ($p > .05$), so that the data in both groups are homogeneous.

Table 2. Normality test.

NPar Tests

One-Sample Kolmogorov-Smirnov Test

		musik 2	toleransi 2	gotong 2	musik 5	toleransi 5	gotong 5
N		35	35	35	35	35	35
Normal Parameters[a,b]	Mean	77.8571	3.3743	3.2562	83.4286	3.5352	3.3419
	Std. Deviation	5.97614	.27725	.26269	6.61676	.31736	.29220
Most Extreme Differences	Absolute	.217	.154	.223	.212	.140	.123
	Positive	.217	.154	.223	.212	.112	.123
	Negative	-.154	-.089	-.165	-.188	-.140	-.121
Kolmogorov-Smirnov Z		1.284	.909	1.318	1.255	.831	.727
Asymp. Sig. (2-tailed)		.074	.380	.062	.086	.494	.667

a. Test distribution is Normal.
b. Calculated from data.

Table 3. Homogeneity test.

Test of Homogeneity of Variances

	Levene Statistic	df1	df2	Sig.
Musikalitas	.273	1	68	.603
Toleransi	2.877	1	68	.094
Gotong	.159	1	68	.691

Table 4. MANOVA test.

Box's Test of Equality of Covariance Matrices[a]

Box's M	5.416
F	.859
df1	6
df2	33502.189
Sig.	.524

Tests the null hypothesis that the observed covariance matrices of the dependent variables are equal across groups.
a. Design: Intercept + Kelompok

The effectiveness result showed a statistically significant difference between the experimental and control classes in 'musicality', while there was no statistically significant difference between the classes in the 'mutual cooperation' score.

4 CONCLUSION

The results of the research are:

a. The contents of the TARA CD product, which include:
   1) Learning objectives, with achievement targets for the learning process.
   2) Materials that contain the SATB arrangement for a vocal group of the traditional song 'Mensahang Lah Mirah' from the Indonesian province of Bangka Belitung, including: (a) bar notation of the musical score; (b) numerical notation of the musical score; (c) a tutorial video, which are the tutorials for each of the SATB voices and the combination of all four voices; (d) audio of the original song in one voice in MP3 format.
   3) Song profile that contains the song's lyrics and their meaning.

b. The feasibility results obtained from the experts' evaluations show that the feasibility of the product is 75%, and it is at level 3 with an interpretation of 'Appropriate', meaning that the TARA CD multimedia product is appropriate for implementation in a music arts learning programme in senior high schools.

REFERENCES

Arikunto. (2012). *Dasar-dasar evaluasi pendidikan edisi 2* [The Basic Foundation Of Education Evaluation Edition 2]. Jakarta: PT. Bumi Aksara.

Bedi, K., Coric, A. & Samardzija, D. (2011). Project based learning: Students' design of interactive multimedia CD/DVD with educative content in secondary school. In *Central European Conference on Information and Intelligent Systems* (p. 97). Varazdin, Croatia: Faculty of Organization and Informatics.

Gall, M.D., Borg, W.R. & Gall, J.P. (1996). *Educational research: An introduction* (6th ed.). White Plains, NY: Longman Publishing.

Jaffurs, S.E. (2004). Developing musicality: Formal and informal practices. *Action, Criticism, and Theory for Music Education, 3*(3), 18–27.

Kamien, R. (1980). *Music: An appreciation*. New York, NY: McGraw-Hill.

Lickona, T. (2013). *Educating for character*. Jakarta, Indonesia: Bumi Aksara.

Mardapi, D. (2011). *Penilaian Pendidikan Karakter* [Character education assessment]. Yogyakarta, Indonesia: UNY Press.

Pallant, J. (2013). *SPSS survival manual* (5th ed.). Maidenhead, UK: McGraw-Hill Education.

Pramudita, M. (2016). Pembelajaran Lagu Daerah dalam Menanamkan Apresiasi Siswa Kelas V di SD 3 Blimbing Kidul Kabupaten Kudus *(Learning Local Songs to Instilling Appreciation of Class V Students in Blimbing Kidul Elementary* School, *Kudus Regency)*, Semarang; Semarang States University.

Rismawan, S.A. (2014). *Ekspresi Musikal dan Fungsi Musik Saestu Band Reggae bagi Masyarakat Kota Semarang* [Musical Expression and Music Function of Saestu Reggae Band for the Community in Semarang]. *Jurnal Seni Musik, 3*(1), 20–28.

Steinman, P.M.W. (2010). *Becoming a choral music teacher*. New York, NY: Routledge.

Sugiyono. (2010). *Metode Penelitian Kuantitatif, Kualitatif, dan R&D* [Educational research methods: Quantitative, qualitative, and R&D approaches]. Bandung, Indonesia: Alfabeta.

21st Century Innovation in Music Education – Astuti et al. (eds)
© 2020 Taylor & Francis Group, London, ISBN 978-0-367-11122-9

Gamolan multimedia: An innovative media for traditional musical instrument learning

I.K.S.S. Putra & A.N. Machfauzia
Yogyakarta State University, Yogyakarta, Indonesia

ABSTRACT: This study aimed to explain the procedure for developing and testing the feasibility of a multimedia learning tool. The multimedia was developed using a modified model development procedure derived from Gall et al. (1996) and Alessi and Trollip (2000). The data in this study were quantitative in nature in the form of product assessment scores, which were converted into qualitative data using a five-point scale conversion table. The results of the feasibility test conducted by one teacher showed a value of 4.06 with the category of "Good". The results of the feasibility test conducted by 25 students demonstrated an average value of 4.27, giving a category of "Very Good". These results show that multimedia for learning *Gamolan* falls into a category of "Feasible" for use in teaching music.

1 INTRODUCTION

The *Gamolan* is one of the traditional musical instruments from the province of Lampung, Indonesia, which has official recognition from the Ministry of Education and Culture as part of Indonesia's intangible cultural heritage. This musical instrument is usually played in ritual ceremonies, by playing the *Tabuh* (song) typical of the *Gamolan* which has its own meaning for the people of Lampung. All *Gamolan* raw materials come from *petung* bamboo (*Dendrocalamus asper*). Historically, *Gamolan* originated from the word "begumul" in the Lampung language which means "gathered". This seems appropriate because, at first, a *Gamolan* was a communication tool for the Lampung community called "kelukup" which was nothing but a *kentongan* (slit-drum). The *Gamolan* is a percussion instrument because it is played by hitting its seven tone slats, namely 1, 2, 3, 5, 6, 7, 1. The *Gamolan* is played individually or in a group (ensemble) accompanied by other musical instruments.

Today, the *Gamolan*'s existence in Lampung is increasingly eroded by the development of modern music. Technological developments make it easier for people to get information about music from around the world. It is unfortunate that the young generation, who should have a role in preserving the *Gamolan*, is affected by the globalisation of music. The easiest way to introduce the *Gamolan* to the younger generation is through the learning process at school. Several foundations related to traditional music learning both in the curriculum and regulations have been provided by the government. One of them is contained in the 2013 Curriculum, on basic competencies 3.3 and 3.4 on Arts and Culture subjects for eighth-grade students. In addition, the Regional Regulation of Lampung Province No. 2 of 2008 concerning Maintenance of Lampung Culture requires the use of traditional arts as learning materials in schools. Based on these two references, the *Gamolan* is one of the traditional musical instruments that must be used in music lessons, especially for eighth-grade students.

In Lampung, many schools have implemented these references in teaching music. But some prefer to use non-traditional musical instruments such as pianicas, recorders, and band instruments. This certainly threatens the existence of traditional musical instruments, especially the *Gamolan*, in musical subjects in this area. The reason teachers keep using these non-traditional musical instruments is their lack of understanding of the *Gamolan* playing technique. In addition, schools that have used the *Gamolan* as learning material still face the same obstacles. *Gamolan* learning depends exclusively on the creativity of teachers and students, without any understanding of the tradition of the *Gamolan* itself. Based on these constraints, solutions related to the provision of teaching materials capable of providing theoretical and practical material are urgently needed. One of the solutions is to provide technology-assisted multimedia, so that learning material can be presented in a practical, interactive and interesting way for the students.

In recent years, several applications have been developed that use the *Gamolan* as their object. The first is called *Gamolandroid*, an Android-based application, the second is named *Gamolan*, also an Android-based application, and the third is called Virtual *Gamolan Pekhing 2*, a computer-based application (Windows). The three applications attempt to move the *Gamolan* into an electronic (virtual) form, so users can play the *Gamolan* anywhere easily. However, what needs to be realised is that these applications are not intended for lay users who have limited capabilities. The unavailability of information related to the *Gamolan* and how to play the *Gamolan* and *Tabuh* properly and correctly are important elements that are ignored. Therefore, it is necessary to develop multimedia for *Gamolan* learning that provides information related to the history, function, and form of the *Gamolan*. In addition, this multimedia also needs to provide attitudes and techniques for playing the *Gamolan* and tutorials on playing *Tabuh*. A cognitive evaluation feature is also needed as feedback on learning that has been undertaken.

The research question in this study was "how to develop multimedia for learning that provides *Gamolan* material that is complete, easy and feasible?" Based on this question, the objectives of this study were: (1) introducing and preserving traditional *Gamolan* musical instruments that are closely related to the tradition of Lampung people; (2) developing *Gamolan* learning multimedia that is complete, interactive, interesting, and easy to use for music teaching in schools.

2 LITERATURE REVIEW

2.1 *Gamolan*

According to Margaret J. Kartomi in Hasyim (2011), the *Gamolan* is estimated to have existed since the Hindu era, as she revealed as follows: "thus, this Lampung instrument may be a survivor of a widespread keyed instrument of the Hindu period called gamelan". The *Gamolan* is also named as a musical instrument in the Hindu period because such instruments were not used at the time of the influx of Islam in the area. Music with Islamic influences generally did not use *Gamolan* instruments, instead using instruments such as *gambus*, tambourines, accordions, *marwas*, and violins.

Based on an interview with the Head of Music Education, Lampung University, K.S.S. Hasyim, which was conducted on 20 August 2017, the *Gamolan* is a musical instrument made of bamboo. This instrument consists of several parts, including *bilah*, *baluk*, *ganjal*, and tools made out of bamboo that are used to beat the *Gamolan* when played.

The song for a *Gamolan* is often called a *Tabuh*. A *Tabuh* for a *Gamolan* can be divided into two:

1) *Tabuh Adati*: *Tabuh* that is related to traditional events. Examples of this *Tabuh* include *Tabuh Jakhang, Tabuh Sekeli, Tabuh Sambai Agung, Tabuh Labung Angin,* and *Tabuh Khapot.*

Figure 1. A *Gamolan*.

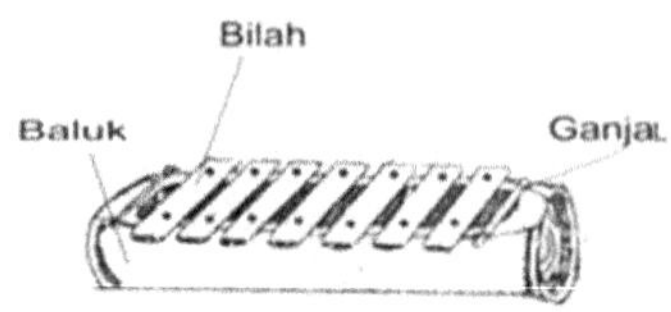

Figure 2. Parts of a *Gamolan*.

2) *Tabuh non Adati*: *Tabuh* that is played at non-traditional events. This *Tabuh* may represent values. Examples of *Tabuh non Adati* which includes *Tabuh Sermendung Sirliah, Takhi, Gedang Gung,* and *Suko Atei.*

2.2 *Application-based traditional musical instruments*

Several applications have been developed in recent years in relation to the *Gamolan*. In 2012, Faktur, a student from the University of Lampung, developed an application called *Gamolan-droid* that can run on Android-based smartphones. This application provides a *Gamolan* in a virtual form, so users can play the *Gamolan* anywhere via their smartphones.

Furthermore, in 2017, a similar application called *Gamolan* was developed by developers under the name of *Labuhan Digital*. This application is Android-based, incorporating the content of a virtual *Gamolan* and a tutorial on how to play a *Tabuh*.

Another application developed to preserve traditional musical instruments is *E-Learning Kolintang,* which was developed by Wikarsa et al. (2017). This application provides content in the form of a virtual Gamolan including its history, types, lyrics, and tutorials in playing kulintang.

The multimedia for *Gamolan* learning developed in this study has some similarities with these applications. Both Android-based *Gamolan* applications and the multimedia share similarities in terms of the background and development goal, which is to preserve traditional musical instruments. However, the multimedia for *Gamolan* learning is a computer-based application, so this application is not intended for Android users. In addition, the virtual *Gamolan* is only intended for playing a *Gamolan* electronically, without any principle of learning in it. The absence of any content that provides information about *Gamolan* is used as a reference in developing this multimedia. Multimedia for *Gamolan* learning contains material about the history, functions, and parts of a *Gamolan*. The material is inspired by *E-Learning Kolintang* by Wikarza et al (2017). The final product has many similarities, the only difference is the base programming and musical instruments used.

2.3 *Principle of material design*

According to Susilana et al. (2008, pp. 33–34), the starting point of material formulation is the formulation of objectives. Material is closely related to the substance of the lesson that must be given. Material needs to be designed by considering certain criteria, including validity, level of importance/usefulness, learnability, and attracting interest. The design of the material in the development of multimedia is based on these principles, so that the material is designed in accordance with the systematic design of the material.

3 MATERIAL & METHODOLOGY

3.1 *Data*

The data in this study were collected from the feasibility test conducted by one teacher and 25 students in SMP N 10 Bandar Lampung. The data were in the form of a score from student assessments of multimedia products carried out by filling out the product assessment instrument. The product assessment instruments for the students contain aspects of assessment related to product feasibility and attractiveness. Each aspect was given an assessment with a range of 1–5. The data obtained were quantitative in nature which were then converted into qualitative data according to Sukardjo's (2008, p. 84) guidelines. After the conversion, the feasibility and attractiveness of the multimedia based on students' opinions were inferred.

3.2 *Method*

This study used a type of development research known as Research and Development (R&D), which aimed to produce a product. The product developed in this study was multimedia for *Gamolan* learning for eighth-grade students. The procedure used in this development adopted two R&D models, namely those of Gall et al. (1996) and Alessi and Trollip (2000). These models were combined to adjust the phases of research and development carried out by the developer. This research combined these two models because both were procedures for developing products for education. The model of Gall et al. (1996) was adopted in the needs analysis and field testing sections. Meanwhile, the model of Alessi and Trollip (2000) was adopted in the entire procedure because this was a special model for the development of multimedia for learning. The method in this development can be seen in Figure 3.

The major steps in this combined R&D cycle used to develop minicourses were as follows:

1. Needs assessment
   In this step, researchers conducted the preliminary study and literature review. This step broadly included observations and interviews conducted directly by researchers. Observations aimed to directly find out the status of the students, the condition of the school, the learning equipment, and the learning process. Interviews were conducted with the teachers and students to get information related to the needs of the product to be developed.
2. Planning
   This step was conducted in four phases: 1) defining the scope; 2) identifying characteristics; 3) producing a planning document; 4) conducting brainstorming. This stage broadly covered the objectives of learning that would be achieved by students and the behaviours of the students when following the learning process. After obtaining that information, a planning document was prepared which aimed to create the content to be used in multimedia learning. Preparing the content was carried out by brainstorming so that criticisms and suggestions could be obtained regarding the product being developed.
3. Design
   This step was conducted in three phases: 1) task and concept analyses; 2) flowcharts and storyboards; and 3) script writing. This stage broadly covered the initial product development and the framework of products such as flowcharts and storyboards. This step also prepared a narrative script that would be used in the video tutorial on how to play the *Gamolan*.

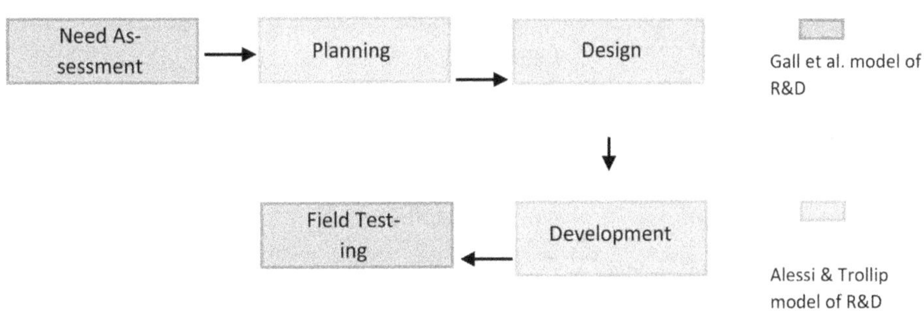

Figure 3. Combined R&D model derived from the models of Gall et al. (1996) and Alessi and Trollip (2000).

4. Development
   There were several phases at this step, which were generally a series of processes to assemble and evaluate multimedia products. In the assembly phase, content that had been prepared such as texts, images, photos, videos, and sounds were arranged according to their respective parts. After all the contents were compiled and packaged in multimedia form, it proceeded to the evaluation phase. The first evaluation was carried out by two experts, namely a media expert and a material expert. The aim was to find out the multimedia feasibility in terms of design, ease of use, and content of the material. After the multimedia had been considered to be feasible, the evaluation was continued in school by involving the teacher and students. The purpose of the evaluation was to find out the attractiveness and ease of the multimedia that had been developed.
5. Field testing
   This step was conducted in the form of field testing in the classroom to find out the effectiveness of the multimedia to improve student learning outcomes.

3.3 *Research instruments*

The research instruments used in this study were the product assessment instruments for teachers and students. The instruments were designed by adapting the principle of designing the material. The instruments were designed in the form of statements related to the content of the material and the attractiveness of the product. Based on the principle of preparing the material, it was then developed into statements given by the teachers and students. Both of these instruments were first validated by an instrument validator, so that they were considered to be suitable for use in product assessment.

3.4 *Data analysis*

The data analysis involved two stages: (1) calculating the product assessment scores and (2) converting the quantitative data into qualitative data. The first phase of the analysis activity calculated the product assessment scores conducted by one teacher and 25 students. This phase was carried out to get the final number of scores and the final average. The assessment was done by assigning values in the range 1–5 adapted from the five-point scale. It can be seen in Table 1:

After all students gave an assessment of the product, the next step was to calculate the average value of the assessment results. The average score was obtained according to the following formula:

$$X = \frac{\sum \mathrm{X}}{\mathrm{n}} \tag{1}$$

where the average number is denoted by X, symbol Σx is defined as the number of scores and symbol n is defined as the number of questions.

Table 1. Guidelines for product assessment.

Scale	Category
5	Very Good
4	Good
3	Sufficient
2	Bad
1	Very Bad

Table 2. Guidelines to convert quantitative data to qualitative using a five-point scale.

No.	Score range	Calculation guide	Category
1	Mi + 1.80Sbi < X	X > 4.21	Very Good
2	Mi + 0.60Sbi < X ≤ Mi +1.80Sbi	3.40 < X ≤ 4.21	Good
3	Mi – 0.60Sbi < X ≤ Mi + 0.06Sbi	2.60 < X ≤ 3.40	Sufficient
4	Mi – 1.80Sbi < X ≤ Mi – 0.06Sbi	1.79 < X ≤ 2.60	Bad
5	X ≤ Mi – 1.80Sbi	X ≤ 1.79	Very Bad

Definition: The score achieved is denoted by X, symbol Mi is defined as the simulation of "1/2 (ideal maximum score + minimum score ideal)", and symbol Sbi is defined as the simulation of "1/6 (ideal maximum score - ideal minimum score)".

The second phase was to convert the quantitative data into qualitative data. The result of the average score that had been obtained was then converted into information in the form of qualitative data using the five-point scale guideline according to Sukardjo (2008, p. 84), which is shown in Table 2.

The five-point scale was used in this study to summarised the results of the eligibility criteria of the teacher and student assessments. By using several statements, the scores obtained needed to be changed into an average value which would later determine the product feasibility category. The scale provided guidance regarding the category from an average of the research data. In this study, data from the teacher and student assessments were analysed to obtain the final result in the form of an average score.

4 RESULTS AND DISCUSSION

4.1 *Interfaces of the multimedia for* Gamolan *learning*

The multimedia for *Gamolan* learning combined learning media with technology, with the local wisdom in the form of traditional musical instruments as the focus of the material to be taught. This multimedia was developed based on the needs analysis carried out in schools, so that learning materials were tailored to the level of student ability. This multimedia contains material about the *Gamolan* including its history, function, and shape. In addition, there are tutorials on how to play the *Gamolan* correctly, as well as tutorials that train students how to play traditional *Tabuh* (song) using a *Gamolan.*

The characteristics of this multimedia were tailored to the needs of eighth-grade students by displaying elements of Lampung province where the *Gamolan* originated. The opening music used is a combination of the sound of the *Gamolan* and digital music, giving a new nuance of traditional music. It aims to adjust between the background, nuance, and learning materials provided. Figures 4 to 9 display some examples of the multimedia.

Figure 4. Home screen.

Figure 5. Main menu.

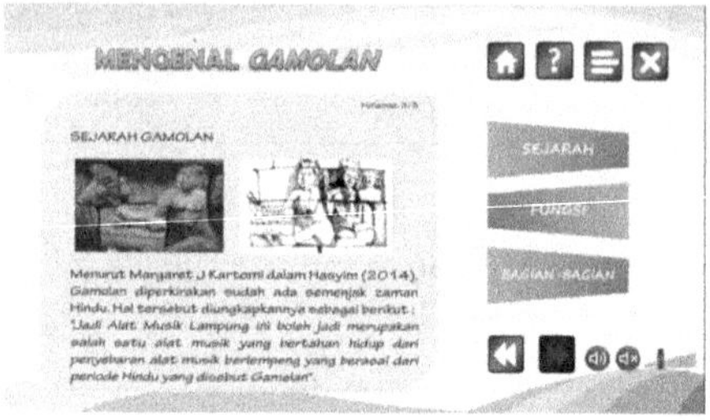

Figure 6. History of the *Gamolan*.

Figure 7. *Tabuh* tutorial.

Figure 8. Playing manner.

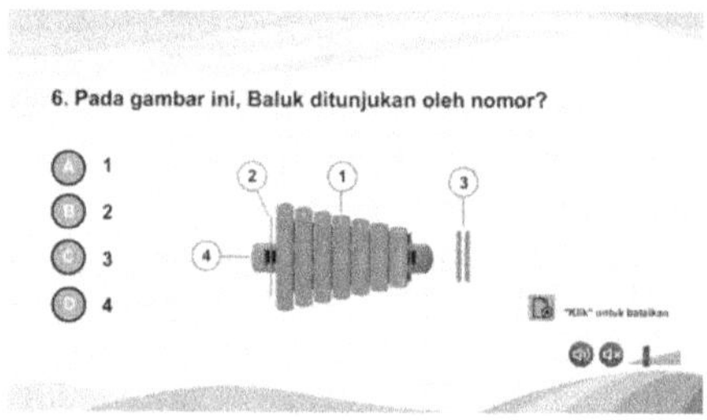

Figure 9. Quiz.

The initial display that appears in the multimedia is the home screen display that contains the background display and the multimedia icon. To start using the multimedia, choose the "Start" button and then press click on the button. Once selected, the user will be directed to the "Main Menu" display which contains menus namely: Introduction, Objectives, Material, Evaluation, and Profile. Each menu contains content including introductory words, learning objectives, learning material, quizzes, and developer profiles. To maximise the use of this multimedia, users are advised to open the menus in sequence starting from Introduction to Profile.

This multimedia was computer-based, thus, the whole process of assembling the contents was done with a computer. The initial step was to make a design display, multimedia icons, and navigation buttons. The process was carried out using Adobe Photoshop CS4 software. The sound editing process was carried out with 2018 Adobe Audition CC software, while video editing was done with 2015 Adobe Premiere Pro CC software. The design and content were then assembled into one file using Macromedia Professional Flash 8 software. The final result of the assembled contents is a multimedia file with a *.swf* (shockwave file) extension that can be run on the computer.

4.2 *Results of the student assessment*

To determine the feasibility of the product based on student opinions, the scores of the assessment from the students were converted into qualitative data using the five-point scale guideline. The multimedia was considered to be feasible when obtaining a minimum result that fell into the "Good" category according to Table 2. The score for the feasibility test conducted by the teacher was 4.06, which fell into the "Good" category. The score for the feasibility test conducted by the students was 10.67, with a total average for all assessment aspects of 42.68.

Furthermore, the researchers calculated the average value of the 25 students to find out the product category. The average product score from all students was 4.27. Based on the table for the conversion of qualitative data using a five-point scale, that average score was in the "Very Good" category. Based on these results, it can be said that the multimedia for *Gamolan* learning is feasible for use in junior high schools.

4.3 *Discussion*

The contents designed in this multimedia were tailored to the abilities of eighth-grade students related to their theoretical and practical understanding. The ease of use, the clarity, and the learning principles of the contents presented in this multimedia can provide a new nuance for learning music. This multimedia provides materials related to the history, functions, and parts of the *Gamolan* that can improve students' insight. Tutorials on the attitude and techniques in playing *Tabuh* were arranged systematically, capable of providing training in playing the *Gamolan* and improving students' skills. Students with little understanding, skills, and experience of playing the *Gamolan* can learn and practice by following the instructions in the multimedia, wherever there is computer access. In other cases, the teacher can use this multimedia as a learning media if the understanding, skills and experiences that they have are still lacking. The multimedia contents compiled from text, images, sounds, videos, and animations are used to provide a more interesting and complete presentation that can improve learning success. The multimedia provides the material needed for teachers and students to learn and practice the *Gamolan*. Instructions related to the technique of playing the *Gamolan* and *Tabuh* address the difficulty of playing *Gamolan* instruments.

4.4 *Final product study*

Based on the results of the discussion on the development, assessment, and product testing, this multimedia has more advantages than those of previous learning media (books). This multimedia is compiled based on preliminary research and the learning syllabus that are devoted to music subjects in the eighth grade. The materials prepared have been adjusted to the learning objectives so that students do not experience difficulties in understanding it. For understanding of the *Gamolan*, the materials are equipped with pictures, photos, and videos that make it easier for students to understand it. This multimedia was developed as one of the solutions to the difficulties found when learning the *Gamolan* and also as an effort to preserve Lampung's traditional musical instruments. The multimedia display presents several icons from the Lampung region such as the Siger Tower, elephants, and typical Lampung traditional clothes. The nuances given are natural so that this product is able to create a refreshing and enjoyable learning atmosphere by always reminding and being proud of the culture of Lampung. Learning activities using this multimedia learning can be done individually and in groups. This requires students to shape characters such as independence, responsibility, cooperation, respect, confidence, and pride about local wisdom.

5 CONCLUSION

Prioritising local wisdom as a learning material in school is important. In today's modern era, the advancement of technology can be a medium of learning. The development of computer-based multimedia for *Gamolan* learning is one way to raise awareness of local wisdom, particularly the traditional musical instruments. We need to learn more about how to create good and effective multimedia and add more contents such as the number of *Tabuh* for the *Gamolan*. One of the limitations of this multimedia is that it can only be used on the computer. This multimedia has to be developed further both in terms of material and mobility of the product. The benefits of this multimedia are its ability to provide a medium for learning traditional musical instrument in an interesting way, and provide materials about the *Gamolan*, both in theory and in practice, in a way that is practical, effective and easy to use. Researchers developed this kind of product so that people are provided with media about their own local wisdom.

REFERENCES

Alessi, S.M. & Trollip, S.R. (2000). *Multimedia for learning: Methods and development* (3rd ed.) Boston, MA: Allyn & Bacon.

Gall, M.D., Borg, W.R. & Gall, J.P. (1996). *Educational research: An introduction* (6th ed.). White Plains, NY: Longman Publishing.

Hasyim, K. S.S. (2011). *Gamolan: instrumen musik tradisional lampung; bentuk, fungsi, dan perkembangannya* [Gamolan: Lampung traditional music instruments; the form, function, and development]. (Doctoral dissertation). Yogyakarta: Gadjah Mada University.

Sukardjo, D. (2008). *Evaluasi pembelajaran* [Learning evaluation]. Yogyakarta: Yogyakarta States University.

Susilana, R., Si, M. & Riyana, C. (2008). *Media pembelajaran: hakikat, pengembangan, pemanfaatan, dan penilaian* [Learning media: Nature, development, utilization]. Bandung, Indonesia: CV. Wacana Prima.

Wikarsa, L., Paseru, D. & Pangemanan, V.T. (2017). Applied e-learning of Kolintang musical instruments case study: University of De La Salle Manado. *Widya Teknik*, *15*(1), 16–23.

21st Century Innovation in Music Education – Astuti et al. (eds)
 ISBN 978-0-367-11122-9

MIDI as a medium for improving student's vocal abilities in the choir learning

F. Pagga & K.S. Astuti
State University of Yogyakarta, Yogyakarta, Daerah Istimewa Yogyakarta, Indonesia

ABSTRACT: This study aims to investigate the effect of MIDI (*Musical Instrument Digital Interface*) learning media in choir teaching for improving the vocal abilities of students in SMK (Vocational High School) Kartika Wirabuan Makasar, who lack a proficient standard of music knowledge. In practice, students are given sheet music, instruments are played using MIDI Software, and then they practice together. The study concluded MIDI has a very important role in increasing the vocal ability of students in choir learning. Playing instruments produced from MIDI software is very effective at improving students' vocal abilities compared to reading sheet music alone. MIDI accelerates students' understanding of choir material that is taught so students can learn more easily and pleasantly.

1 INTRODUCTION

Education is a conscious and systematic effort, carried out by people given the responsibility to influence students so that they align their moral character and behaviour with the ideals of education (Munib, 2004). Therefore, education is expected to help students achieve maturity and independence. Various activities are carried out to support education in Indonesia, such as the recent replacement of the curriculum and developing facilities and infrastructure in the world of education itself.

Learning should be fun for students because in these activities, there will be interaction between students, educators and learning resources in a learning environment. Music learning in schools is always praised for its capacity to provide aesthetic experiences in the form of creative and appreciative activities. Students can use practical activities to achieve aesthetic experience. Students can undergo a creative process by responding to stimuli provided by the teacher, while appreciation of music is obtained by observing and analysing learning materials. However, learning can be a very intimidating prospect for students, due to limitations in students' abilities and the absence of sufficient learning support.

The implementation of learning is the result of integration of several components which have their own functions; this is intended so that the achievement of learning objectives can be fulfilled. The main characteristic of learning activities is interaction, which occurs between students, their learning environment, teachers, peers, tools, learning media and/or other learning resources. Other characteristics of learning are related to the components of learning itself such as goals, materials, strategies, media and evaluation.

Almost all human work requires technological assistance. Technology was developed for education in the form of educational media. Such media functions as an intermediary to channel both audio and visual information to the recipient. The media used in teaching is called learning media. Rapid technological developments bring changes to the role of the teacher as a messenger of the subject matter. The teacher no longer acts as the only authority of knowledge in learning activities. Learners can obtain information from various media and learning resources such as magazines, modules, radio broadcasts, television, and digital media.

According to Suherman (2009) media are a channel used for communication. According to Latuheru (1993) media used in learning in Purbasari can improve the quality of education efficiently and effectively. Media is also one of the factors that supports success in learning at school because it can help the delivery process from teacher to student or vice versa (Arda et.al, 2015). This is where the role of art is tested so that it can be a fun learning forum for students. Learning media is considered able to maximize the efficacy of a teacher to deliver learning material compared to in-person or writing. There are various media that can be used according to the user's needs and capacity. For choir learning, MIDI software was chosen for students at SMK Kartika Wirabuana Makasar due to their musical limitations. MIDI is software that plays various musical instruments, one of which is the piano. This MIDI software media will function as an intermediary that transmits audio information to the recipient. MIDI is expected to facilitate choral learning to be more meaningful and enjoyable.

2 METHODOLOGY

Qualitative research was carried out and described. Research procedures involved collecting data on choir learning using MIDI as the learning medium. The students under study are from SMK Kartika Wirabuana Makassar, who participated in choir learning as an extracurricular program at school. The study investigates how effective the use of MIDI is in these learning activities. In this study, researchers who were also extracurricular teachers at SMK Kartika Wirabuana were directly involved in conducting the research. According to Margono (2000), the task of the researcher gives questions about research and at the same time as observers in the process. Other supporting instruments are observation.

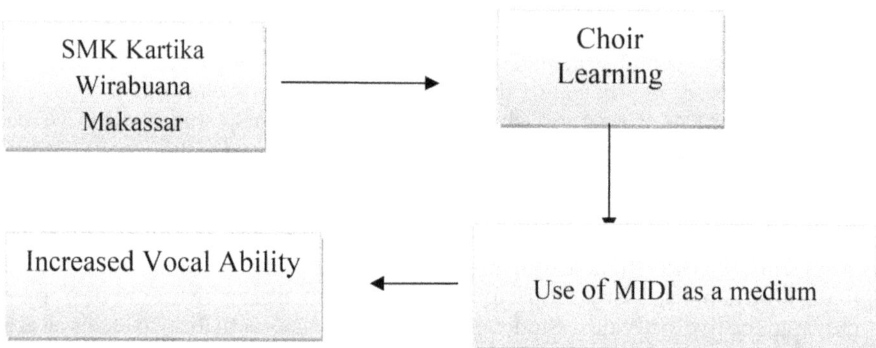

Figure 1. Research process.

3 RESULT & DISCUSSION

3.1 *Result*

3.1.1 *Learning*

In article 1 point 20 of Law Number 20 Year 2003 concerning the National Education System, "Learning is the process of interaction between students and educators and learning resources in a learning environment." Muhaimin (in Riyanto, 2010: 131) says that, "Learning is an effort to teach students to learn. Learning activities will involve students learning something in an effective and efficient way." Gagne, Briggs, and Wager (in Winataputra, 2014: 19) state, "Instruction is a set of events that affect learners in a way that learning is facilitated." Learning is a series of activities designed to enable the learning process to occur in children.

While Morrison (2009: 111) says, "Learning refers to the cognitive and behavioural changes that result from experiences. The experiences that make up the curriculum are the core of the learning process and experiences you provide for children should be based on a theory or theories of how children learn."

Learning refers to cognitive and behavioural changes that result from experience. The experiences that shape the curriculum are at the core of the learning process. The experiences you provide children are based on theories about how they learn.

Kimble (in Hergenhahn & Olson translated by Wibowo, 2009: 8) also says, "Learning is a change in behavior or potential behaviour that is relatively permanent that comes from experience and cannot be attributed to temporary body states such as conditions caused by illness, fatigue or medication."

Learning is a process of interaction between educators and students in a learning environment and using previously prepared learning resources to enable the learning process in children and at the end of learning activities cognitive and behavioural changes occur in children cognitive and behavioral changes come from things experienced by children and are relatively permanent.

3.1.2 *Media*

As for some media meanings by Ahmad (1997), including:

a. Media is all forms of intermediaries used to convey ideas to the recipient.
b. Media is an intermediary used to carry or convey a message, running between communicators and communication.
c. Media are all physical tools that can run messages that are stimulating and appropriate for learning, for example: print media, electronic media.

From some of the interpretation above, it can be concluded that educational media in general are educational facilities or infrastructure that can be used as learning resources, and to support students to best achieve learning goals.

3.1.3 *Choirs and vocals*

According to Sitompul (1986: p.1), "Choir is a collection of a number of singers grouped according to the type of voice". This grouping is generally based on two sound criteria; the pitch and the tone of the voice or the timbre of the singer. Choir is a term that refers to a music ensemble comprising singers who perform music. A choir can also be defined as a collection of singers grouped by type of voice. For children and women, the types of voice include: soprano, mezzo soprano, alto. For men, the types of voice are: tenor, baritone and bass (Sitompul, 1986: 1).

The technique of singing for both solo singers and choir singers is the same. The effectiveness a soloist is determined by the individual, while the effectiveness of the choir is determined by technical mastery, cohesiveness and cooperation built in the choir itself (Listya, 2012: 27).

3.1.4 *MIDI (Musical Instrument Digital Interface) software*

Software is a special term for data formatted and stored digitally including computer programs; documentations and information that can be read and written by a computer. In other words, intangible computer parts are referred to as software. MIDI is a language used by electric music instruments, controllers, computers, and similar devices to communicate. MIDI captures event notations and changes in attributes and tone accents, encodes them into digital messages, and sends the code as a message to other devices to regulate the resulting sound and its parameters. So, MIDI software is interpreted as the embodiment of MIDI which originally existed in the form of an object or physical entity and has manifested into a digital program or application to facilitate music exploration.

In the world of music (including vocals), some software applications that transfer visual files to audio are used to facilitate music activities. Applications that are often used in the realm of MIDI, especially music and also vocals, include Sibelius, Encore, Groovy music, Auralia, Musition and others.

3.2 *learning process using MIDI software*

Choir learning, which usually involves students reading sheet music, is now carried out in combination with learning media in the form of music applications or software, in which there are components of a musical instrument in digital form called MIDI. One of the music application programs used is Sibelius. Sibelius is a special software program for typing musical notations in the form of musical notes.

Sibelius can edit the sheet music and play the results of the song before printing. In addition, musical notes can also be written without typing but by playing the piano or guitar contained in the program. The type of music Sibelius's compositions generate include rub instruments, woodwind instruments, brass instruments, percussion instruments, guitar, piano, vocals, etc.

The display of Sibelius is as follows:

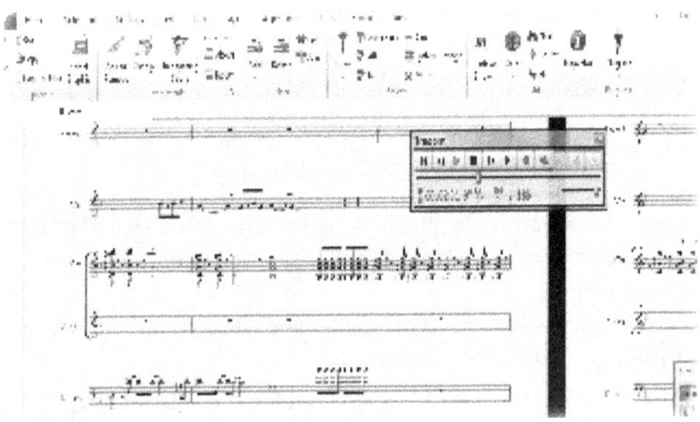

Figure 2. Example of MIDI software (Sibeliu.

In teaching choir using MIDI on Sibelius as a learning medium, students are played MIDI music as a guide for singing are observed very effective to support the implementation of success in the choir learning process. The learning process of using chorus MIDI is described as follows:

1. First, students are given instructions and the choir learning process that uses MIDI as a learning medium is explained.
2. Then students are given sheet music that is taught as an initial guide to singing. In this process, the teacher is expected to provide as simple an explanation as possible so students can understand the activities.
3. The next step is to play the MIDI file, which is played according to the sheet music provided by the teacher. The MIDI played in this activity used a piano instrument that had previously been programmed using the Sibelius software.
4. After playing the MIDI, students are instructed to pay attention to the sheet music so that they can understand the song being played.
5. After listening, the students are then asked to sing without the supporting MIDI.
6. Finally, the MIDI was played back and the students sang together accompanied by the MIDI music.

3.3 *The role of MIDI media software in improving students' vocal ability*

Choir learning using MIDI software is aimed at helping teachers and students during the learning process. With Sibelius, the teacher only needs to listen to the students the instruments that have been made while paying attention to the existing sheet music.

Based on the description above, the role of learning media are:

1. As a tool to clarify learning material.
2. As a source of questions and stimulation of student learning.
3. Learning resources for students. (Sudjana & Rivai, 2010)

Benefits of Learning Media:

1. Teaching lessons becomes more standard.
2. Making the learning process more interesting so that students can remain attentive to the class. Message clarity and coherence, the appeal of changing images, the use of special effects that can lead to curiosity cause students to think, all of which indicate that this method of teaching is motivational and increases the interest of students.
3. Making the learning process more interactive.
4. Improve the quality of student learning. (Arsyad: 2003)

Based on several opinions above, it can be concluded that the learning media has several practical benefits, including:

1. Concretizing abstract concepts, to reduce verbalism. For example, by using images, schemes, graphics, models, and so on.
2. Arouse motivation, so that it can increase the attention of individual students because the course of the lesson is not boring and monotonous.
3. Functioning of all students' senses, so that weaknesses in one of the senses (e.g. the eyes or ears) can be balanced with other senses.
4. Clarifying a theory or concept that is difficult to explain using other methods. For example, to provide knowledge about the pattern of the earth, children are not likely to experience directly. Then the globe was made as a model of the globe.
5. Increasing the likelihood of direct interaction between students and their environment. For example, by using recordings, experiments, field trips, and so on.
6. Providing uniformity in observation, because the understanding of each student will vary depending on the experience and intelligence of each student.
7. Presenting learning information consistently, which can then be repeated or stored according to needs. For example, in the form of recordings, films, slides, pictures, photos, modules, and so on. (Muhson, 2010).

Learning media can be a vehicle for channelling messages and learning information. A well-designed and correct learning media will greatly support students in their efforts to digest and understand the subject matter. Media functions in learning activities are not just teaching for the teacher but as carriers of information/messages of learning. Each type of learning media has its own characteristics, advantages and disadvantages. That is why it is necessary to have systematic planning on the use of each learning media.

Reflecting on the process of learning choirs with MIDI on Sibelius, it is clear that a teaching process that was originally quite difficult, involving a teacher teaching sheet music, is greatly simplified through the use of MIDI. The presence of MIDI media in the choir learning process makes the learning process more interactive, interesting, and improves the quality of learning. The role of MIDI learning media in choir is an innovation that is expected to continue to train sensitivity, stimulate musicality and improve students' vocal abilities.

4 CONCLUSIONS

Hère, learning media are used as support tools that function to assist the teacher in delivering materials to students. Realizing the importance of the role of learning media in the education process, SMK Kartika Wirabuana has also used MIDI learning media in teaching and learning activities, especially choral learning which aims to improve students' vocal abilities. The learning process is carried out as well as learning in general but is use by MIDI technology. Students are given choir learning material with piano MIDI accompaniment in the Sibelius software. After its use in learning, a very significant effect can be observed. Demonstrated by the fact that the enthusiasm of the students in the learning process increases. From the beginning to the end of the process, students seem to focus more on receiving choir material. MIDI that is played makes students concentrate more on choir singing, significantly reducing teaching time. Based on these results, we can observe the important role of a learning media in the success of the choir learning process, so that the learning objectives of creating interesting and meaningful lesson can be achieved.

REFERENCES

Ahmad, R. (1997). *Media instruksional edukatif.* Pt Rineka Cipta.

Arda, S. S. (2015. *Pengembangan Media Interaktif Berbasis Komputer untuk Siswa SMP Kelas VII.* [Development of Computer-Based Interactive Learning Media for Class VIII Middle School Students]. e-Jurnal Mitra Sains, 69–77.

Arsyad, A. (2011). *Media Pembelajaran.* [Instructional Media]. Jakarta: PT Raja Grafindo Persada.

Hergenhahn, B.R. & Matthew H. Olson. (2009). Teori pembelajaran. [*Learning Theory*]. (Penterjemah: Tri Wibowo B.S). Jakarta: Kencana Prenada Media Group.

Listya, Rama Agastya. 2007. *AZ Direksi Paduan Suara.* [A-Z Choir Directions]. Jakarta: Yayasan Musik Gereja di Indonesia.

Latuheru, J. D. (1993). Media pembelajaran dalam proses belajar-mengajar kini[Teaching and Learning Medium at Present]. *Ujung Pandang: Penerbit IKIP Ujung Pandang.*

Margono (2000) *Metode Penelitian Pendidikan.* [Educational Research Methods]. Jakarta: Rineka Cipta.

Morrison, G. S., & Brunson, M. N. (1988). *Early childhood education today.* Columbus, Ohio: Merrill Publishing Company.

Muhson Ali, 2010. *Pengembangan media pembelajaran berbasis teknologi informasi.* [Development of Information Technology-Based Learning Media]. *Jurnal Pendidikan Akuntansi Indonesia*, Vol. VIII. No. 2.

Munib, A. (2004). *Pengantar ilmu pendidikan.* [Introduction to Education] Semarang: UPT MKK UNNES.

Riyanto, Y. (2010). Paradigma Baru Pembelajaran: Sebagai Referensi Bagi Guru *Pendidik dalam Implementasi Pembelajaran yang Efektif dan Berkualitas*[New Learning Paradigm as references for teacher inquality Teaching], *Cetakan ke-2, Jakarta: Kencana.*

Sitompul, B. (1986). Paduan Suara dan Pemimpinnnya. [*Choir and its Conductor*]. Jakarta: PT. BPK Gunung Mulia.

Sudjana, N & Rivai, A. (1990). *Media Pengajaran.* [Learning Media]. Bandung: CV. Sinar Baru.

Suherman, Y. (2009). Pengembangan media pembelajaran bagi ABK[Developing of Learning Media for Special need]. Paper presented in *Diklat Profesi Guru PLB Wilayah X Jawa Barat Bumi Makmur, Lembang Bandung.*

Winataputra, Udin S. Delfi, R., Pannen, P., & Mustafa, D. (2014). *Teori Belajar dan Pembelajaran.* [Teaching and Learning Theory]. Jakarta: Universitas Terbuka.

21st Century Innovation in Music Education – Astuti et al. (eds)
© 2020 Taylor & Francis Group, London, ISBN 978-0-367-11122-9

Experimenting Sing, Sign and Play method in an elementary school to improve musical learning

R.L. Ulfa & A.N. Machfauzia
Universitas Negeri Yogyakarta, Yogyakarta, Daerah Istimewa Yogyakarta, Indonesia

ABSTRACT: This study experimented with classroom techniques by using the Sing, Sign and Play method. This method is commonly used to improve student learning outcomes of preschool children and to improve music learning outcomes in elementary school settings. In this study, we conducted pretest and post test evaluations to determine the significance and correlation of Sing, Sign and Play learning method to student outcomes through music learning. The results show an increase in learning outcomes of elementary school students who use the Sing, Sign and Play Method compared to students who did not use the method. The average post test value of the control class is 63.06 while the average post test value of the experimental class is 73.78, for a difference in value of 10.72. This means that the Sing, Sign and Play Method applied to the experimental class turned out to be more effective in improving student learning achievement.

1 INTRODUCTION

Teaching method is very important for effective learning. According to one analysis (Kristin, 2016), discovery learning methods can increase the learning progress starting from the lowest 9% to the highest 27% with an average of 17.8%.

This suggests that teaching methods play an important role in improving student learning progress. To develop a teaching method, a teacher must think creatively and innovatively in order to be able to compile a lesson plan, carry out learning activities, evaluate, analyze evaluation results, and follow up on learning progress.

Going to school affects a child's core experience. By this period of child development, students are considered to begin to be responsible for their own behavior. School-age is the period during which children obtain knowledge foundations in order to be successful in adapting to adult life and in order to acquire certain skills (Wong et.al: 2009). In general, many children have less than pleasant experiences in a subject matter that is taught in elementary school with inappropriate learning methods that are monotonous. Additionally, learning music and the English language is challenging for elementary school children. Thus, there are impacts to the future of elementary school children. Children may develop a less than pleasant view on learning music and the material about musical notes such that children are reluctant to understand the other elements of music. In this study, we identified correlation methods of learning and learning progress at an elementary school. The remainder of the paper is organized as follows: section 1 describes the introduction; section 2 explains the definition of learning method, the definition of the Sing, Sign and Play Method and learning progress; section 3 describes the research method; section 4 explains and discusses the results; and the conclusions are offered in section 5.

2 RELATED WORKS/LITERATURE REVIEW

2.1 *Learning method*

Learning method is a presentation technique mastered by a teacher to present subject matter to students in the class either individually or in groups so that the subject matter can be absorbed, understood and utilized by students (Prasetyo &Ahmadi,2005). The results that manifest in the near future have a direct impact (Instructional effect) due to the application of methods, while learning results that accrue over time from a long time ago are called the nurturing effect, usually associated with the development of attitudes and values. (Syaiful Bahri Djamarah., 2000).

2.2 *Sing, Sign, and Play Method*

The Sing, Sing, and Play Method is a way of singing and playing while having fun, and is useful in stimulating the child's brain to more quickly understand the material. Music affects teachers and students, can set the mood, change the mental state of students and support the learning environment. Music can also help students work better, remember more and strengthen learning both consciously and unconsciously.

The method of singing uses dubbed verses. Usually, poems are adapted to the material to be taught (Fadlillah, 2012). Play method is a method that applies certain games or toys as a vehicle for student learning (Fadlillah, 2012: 168).

2.2.1 *Preschool prodigies*

The Sing, Sign and Play Method was developed as an education-based platform with an online curriculum on the website, PreschoolProdigies.com. There are many details at PreschoolProdigies.com, a website and a name for an educational institution. More importantly, when a teacher begins to develop music lessons, music can become a fun subject for children. The core goal of the Sing, Sign and Play Method is to create an exciting and fun learning atmosphere so that it is easier for students to use and comprehend all the material, and learned through singing, symbols and colors that can be accessed by anyone with online access.

2.2.2 *Kodaly method*

Zoltan Kodaly was a Hungarian composer, born in Keczkemet, Hungary 16 December 1882. After World War II, Kodaly focused his attention back to music education, a field of learning that began in and continues after 1929 (Jamalus, 1988: 134). Kodaly's method uses practical steps such as: the use of solfa tonic, Rhythm syllables and hand signing which is a combination of practical techniques that have been carried out separately by other music education figures.

The music teaching method applied by Kodaly is very concerned about "deep hearing", a term that means students must be able to "imagine the tone" in their mind or imagination (Jamalus, 1988: 135). According to Jamalus (1988: 134), the Kodaly Method uses practical stages in learning: solfa tonic, rhythm syllables and hand signing.

Base on elementary observation students used a creative program of community service, they were able to understand the material that was well-taught using the sing to play method. The students could also learn independently because the sing to play method can be done anywhere and is not limited to constraints of classroom space and time.

1) Solfa tonic

Solfa tonic is a tonal arrangement in the form of notes do, re, mi, fa, sol, la, ti, do. Actually, solfa tonic is a tonal system introduced and developed by Dalcroze and further developed by Kodaly who added hand signing technique.

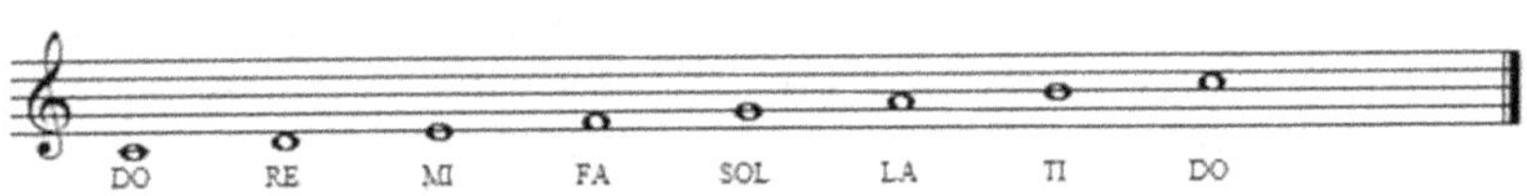

Figure 1. Solfa tonic.

2) Rhythm syllables

Rhythm syllables is a way of learning rhythm using certain syllables. The following is a picture of Rhythm Syllables. According to Taher, Kodaly used the term "rhythm syllables" to facilitate rhythm learning. Each note is given a syllable. As an example, a quarter note is labeled "ta" and a half note is labeled "ta-a".

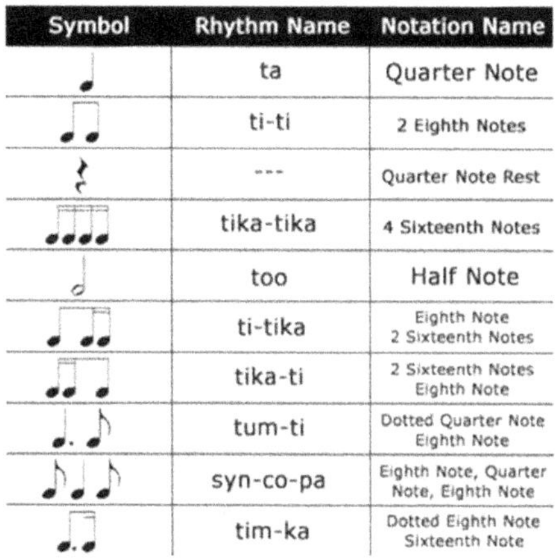

Symbol	Rhythm Name	Notation Name
	ta	Quarter Note
	ti-ti	2 Eighth Notes
	---	Quarter Note Rest
	tika-tika	4 Sixteenth Notes
	too	Half Note
	ti-tika	Eighth Note 2 Sixteenth Notes
	tika-ti	2 Sixteenth Notes Eighth Note
	tum-ti	Dotted Quarter Note Eighth Note
	syn-co-pa	Eighth Note, Quarter Note, Eighth Note
	tim-ka	Dotted Eighth Note Sixteenth Note

Figure 2. Rhythm syllables.
Source: http://www.soundpiper.com/elements/kodalyrhythm.JPG

3) Hand signing
Hand signing is the marking of tone sounds using a pattern of hand movements.

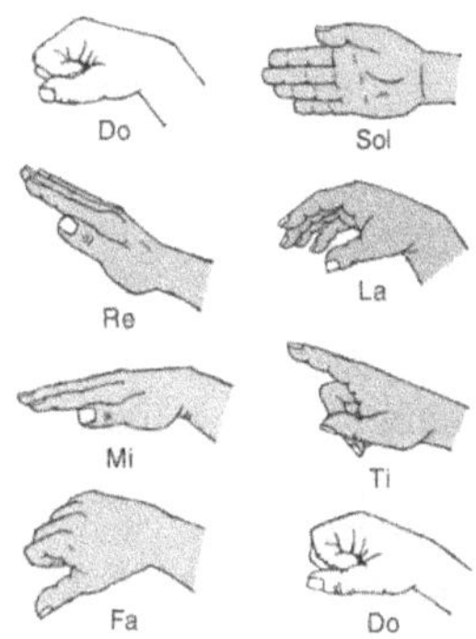

Figure 3. Hand signing.
Source:*http://profileengine.com/groups/profile/421306601/its-not-gang-signs-its-kodaly*.

2.3 *Learning material(call and respond)*

2.3.1 *Rhytmic introduction*

Instruction Tips and Materials: Students will likely request this song again and again. Take advantage of their enthusiasm by letting them clap and sing along as many times as you can allow. Materials include crayons, book sheets and the Beet & Cherry Video. Access the Table of Contents for the Beet & Cherry Song, the sheets on Tas & Ti-Tis Eighth Notes How Many Beats and Write Rhythm.

2.3.2 *Notation introduction with symbols*

Before and after children watch the video, allow children to practice with the sheet music exploring all methods: using hand-signs; singing the words, note names or Solfege names; and playing the bells. Materials include Bells in C, Crayon, Hello C Video Access, and Workbook.

2.4 *Learning progress*

Learning progress is the most important part of learning according to Sudjana & Rivai (2009) and defines student learning progress. In essence changes in behavior are a result of learning in a broader sense covering the fields of cognitive, affective, and psychomotor capabilities. Benjamin S. Bloom (Dimyati, 2006: 26–27) mentions six types of cognitive domain behaviors:

a. Knowledge involves the ability of memory about information that has been learned and stored in memory. Knowledge is concerned with facts, events, understanding rules, theories, principles, or methods.
b. Understanding includes the ability to capture meaning and the meaning of concepts and ideas learned.
c. Implementation is the ability to apply methods and rules to deal with real and new problems. For example, using principles.
d. Analysis includes the ability to specify a unit into parts so that the overall structure can be well understood. For example, breaking up the problem into small parts.
e. Synthesis includes the ability to form a new pattern. For example, the ability to compile a program.
f. Evaluation is the ability to form opinions about several things based on certain criteria. For example, the ability to assess test results.

Based on understanding the learning progress above, it was concluded that the results of learning are based on abilities possessed by students after receiving experience.

3 RESEARCH METHOD

3.1 *Participants*

The population for this study was a group of third grade students at the Al-Falah elementary school in Jambi City, Sumatra, Indonesia, a total of 140 people. The researchers chose third grade students because the 2013 syllabi for 3rd grade elementary schools included music and English learning materials. The sample groups in this study were divided into students of class 3 A with 36 people as the experimental group and class 3 B with 34 people as a control group (comparator). The researcher used purposive sampling, a technique that determines samples with certain considerations (Sugiyono, 2010: 122).

3.2 *Measurement*

In this study, the measurement instrument was a written test. The form of the test that was used was a closed model- multiple choice items. Table 1 shows a grid of instruments that was used in data acquisition.

Base on instruments try outonly 20 questions can be used (valid) while the questions that cannot be used (not valid) are question numbers 16, 12, 13, 18, and 25.

Reliability refers to the understanding of whether an instrument can measure the truth of something that is measured consistently over time (Nurgiyantoro & Gunanwan, 2009: 341). After being tested for reliability with the SPSS 16.0 for Windows program, Cronbach Alpha values were obtained > 0.6 which was 0.706. Thus, it can be concluded that the research instruments that have been validated in this study are reliable and worthy of being used as research data collection tools.

Normality test aims to determine whether or not the research data collected is normally distributed (Nurgiyantoro, et al., 2009: 110). Calculations are carried out with the SPSS 16.0 for Windows. In this case, the data is said to be normally distributed if the significance value (p) > 0.05 and not normal if the significance value (p) < 0.05. The result of the normality test that has been obtained through calculations with the SPSS 16.0 for Windows program can be seen in Table 2.

Based on the data in Table 2, it is known that the experimental class obtained a Z index of 0.809 with a significance value (p) > 0.05 which is 0.530, while the control class obtained a Z

Table 1. Instrument references.

No	Main Material	Sub Main Material	Indicator	Question Number	Cognitive Level
1	Music Theory	a. Notation	• Knowing notation	1, 2, 3	C_2, C_3, C_3
			• Knowing Notation values	4, 5, 6	C_3, C_3, C_3,
			• Knowing key symbol	7, 8	C_3, C_3,
			• Reading notation	9, 10	C_1, C_1
			• Writing and changing notation	11, 12, 13	C_3, C_3, C_3,
		b. scale	• Knowing major scale	14, 15	C_3, C_3
			• Knowing minor scale	16	C_3
		c. Rhytm	• Knowing rhytmic symbol	17	C_2
			• Knowing tempo	18, 19	C_3, C_3
		d. Expression sign	• Introducing tempo symbol and dynamic symbol	20,21	C_2, C_2
			• Knowing tempo symbol and dynamic symbol	22, 23, 24	C_3, C_3, C_3
			• Dynamic symbol application in a song	25	C_3

Table 2. Normality test result.

Pretest Data	Z Index (Kolmogorov-Smirnov)	Significance (p)	Information
Experiment Class	0,809	0,530	Normal Distribution
Control Class	0,901	0,392	Normal Distribution

index of 0.901 with a significance value (p) also > 0.05 that is 0.392. Both are greater than 0.05 and thus the results show that the distribution of obtained data has a normal distribution.

The homogeneity test aims to determine whether or not the study sample comes from a population that has a homogeneous variance. In this study, the homogeneity test was carried out with the SPSS 16.0 for Windows. The sample variant is said to be homogeneous if the significance value (p) > 0.05(Pallant, 2013). Conversely, if the significance value (p) < 0.05, this means that the sample variance is not homogeneous. The results of homogeneity tests that have been obtained through calculations with the SPSS 16.0 for Windows program can be seen in Table 3.

Table 3. Homogeneity test result.

Levene Statistic	df1	df2	Significance	Information
1,298	1	71	0,258	Homogen

Based on Table 3, the homogeneity test obtained a significance value of 0.258 which is greater than 0.05. This means that the sample comes from a population that has a homogeneous variance.

3.3 *Procedure*

The test aimed to find the relationship between the application of the Sing, Sign, and Play Method and the achievements through learning progress of Al-Falah elementary school students in Jambi City. At the immediate outset, the participants were given a pretest. The pretest was used to evaluate the students' level of knowledge about music. Students who were pretested were introduced to the Sing, Sign, and Play Method for 4 learning cycles. After 4 learning cycles, the students were tested again using a post test related to the elements of music with multiple choice questions and essays.

3.4 *Hypothesis test*

Here, a comparative analysis of two correlated samples is used. The collected data from the final test results for the control group and for the experimental group tested for the average difference. To test the difference in averages, t tests were carried out using the SPSS 16.0 for Windows program. The t test is used to test the significance of the mean difference (mean value) between the experimental group and the control group. The results of the t test were obtained via the amount of t count and significance level and will be used to test the hypothesis that has been formulated. The hypothesis that has been tested can be used to draw conclusions in this study. The hypothesis is there are significant differences between the classes taught using the Sing, Sign, and Play Method with classes that were not taught using the method.

4 RESULTS

The results of the study show that there are significant differences between the classes taught using the Sing, Sign, and Play Method and classes that were not taught using the Sing, Sign, and Play Method. This shows that the use of Sing, Sign, and Play Method can improve students' learning achievements in music theory.

Based on observations that researchers have documented at the Al-Falah elementary school, most of the grade 3 students (the sample research subjects) have a sense of pleasure when teaching and learning activities are accompanied by singing, signing and playing. If learning is only centered on the teacher and if a teacher rarely involves the students, students will tend to be bored and passive in their learning. The Sing, Sign and Play Method is a learning model that involves activities such as call and response by students as stated in the description of the Sing, Sign, and Play Method.

In the experimental class, the researchers conveyed music theory using the Sing, Sign, and Play Method. As the learning process takes place, all students participate with each other and have fun while learning the elements of music. This was seen in the concentration of students who really followed what was being explained. There were only a handful of students who still talked to their friends. The questions given were completed with good cooperation. This can be seen in the increase in student learning achievement in the experimental class where there were originally only 6 students who fulfilled the KKM score (Minimum Completion Criteria). However after the Sing, Sign, and Play Method was applied to music theory, there were 24 students who met the KKM value. Thus, initially only 16% of the students met the KKM value which increased by 65% after using the Sing, Sign, and Play Method. In the control class, the researcher delivered the material as is usually prepared by the teacher, using the handbook without using the learning model. During the learning process, students in the control class also paid attention to the teacher's explanations, but they did not really understand the material that was presented. Researchers assumed that most students in the control class are not very interested when the teacher only teaches material in the usual way. This certainly can affect student learning achievements as seen in the differences in achievements between the control class and the experimental class. After receiving the Sing, Sign, and Play treatment, 24 students in the experimental class achieved the KKM score (Minimum Completion Criteria) while only 15 students in the control class achieved the KKM score, an overall increase of 42% compared to only 19% of students meeting the KKM score. Based on the results of the study, both classes experienced an increase in value. In general, most of the post test values were better than the pretest scores. In this study, hypothesis testing was not performed by comparing the average value of the pretest with the post test mean value, but by comparing the average post test scores of the two classes (groups).

The average post test value of the control class is 63.06, and the average post test value of the experimental class is 73.78. There is a difference in value of 10.72. This suggests that applying the Sing, Sign, and Play Method to the experimental class was effective in improving student learning achievement. After carring out a different test with the SPSS 16.0 for Windows program, the value of t count = 2.937 and significance (p) = 0.004. From these results it can be concluded that the proposed hypothesis is accepted, because the value of t count (2.937) > than the value of t table (1.993), and significance (p) is < 0.05.

5 CONCLUSION

Based on the results of the research, the following conclusions are offered: There were differences in class learning achievements using the Sing, Sign, and Play Method (experimental class) compared to learning achievements not using the Sing, Sign and Play Method (control class). This difference is indicated by the average test value in the experimental group of 73.78 while in the control group the average test value is 63.06, a difference in value of 10.72. Furthermore, the results of the t-test type independent sample t-test show the t score count> t table (2,937> 1,993) with significance level <0.05 (0.004 <0.05) which means that the hypothesis which states, "there is a significant difference between classes taught using the Sing, Sign, and Play Method versus classes that are not taught using the Sing, Sign, and Play Method" has been proven.

REFERENCES

Dimyati, M. (2006). Belajar dan pembelajaran. *Jakarta: Rineka Cipta.*

Fadlillah, M. (2012). *Desain Pembelajaran Pendidikan Anak Usia Dini* [Learning design for early childhood education]. Jogjakarta: Ar Ruzz Media.

Jamalus, D. (1988). *Pengajaran Musik Melalui Pengalaman Musik* [Teaching Music Through Music Experience]. Jakarta: Direktoral Jendral.

Kristin, F. (2016). Efektivitas Model Pembelajaran Kooperatif Tipe STAD Ditinjau Dari Hasil Belajar IPS Siswa Kelas 4 SD. *Scholaria: Jurnal Pendidikan dan Kebudayaan, 6*(2), 74–79.

Nurgiyantoro, B., & Gunawan, M. (2004). Statistik terapan untuk penelitian ilmu-ilmu sosial. *Cetakan Ketiga (Revisi), PT Gramedia Pustaka Utama, Jakarta.*

Pallant, J. (2013). *SPSS survival manual.* McGraw-Hill Education (UK).

Prasetya, J. T., & Ahmadi, A. (2005). Strategi belajar mengajar. *Bandung: CV.*

Sudjana, N., & Rivai, A. (2010). Media pengajaran. *Bandung: sinar baru Algensindo.*

Sugiyono, D. (2008). *Metode penelitian bisnis* [Business Research Methods. Fifteenth Edition]. Bandung: Pusat Bahasa Depdiknas.

Wong, D.L, Hockenberry, M, et al. (2009). *Buku Pelajaran Perawatan Anak edisi terjemahan* [Pediatric Nursing Textbook. Translation, Monica Ester; (6th.ed). volumen 2]. Jakarta: EGC.

21st Century Innovation in Music Education – Astuti et al. (eds)

The implementation of the Suzuki method in *Tingkilan* music extracurricular learning

S.N. Hikmah & S. Pradoko
Yogyakarta State University, Yogyakarta, Indonesia

ABSTRACT: The Suzuki method put forward by Dr. Shinichi Suzuki tends to be implemented in violin and orchestra learning. However, in Indonesia this method is seldom implemented in the country particularly in traditional music activities in schools. The SMAN 2 (State Senior High School 2) is the only school in Samarinda, East Kalimantan to have the Kutai people's traditional music and extracurricular activity called Tingkilan, taught with the Suzuki method. The purpose of the research was to discover and then describe how implementation of the Suzuki method within the SMAN 2 Samarinda Tingkilan musical extracurricular practice affected the students. In order to acquire accurate data, the researcher utilized a descriptive, qualitative approach working with research subjects such as the music coach, the counselor and the students within the Tingkilan music extracurricular activity. The data collection techniques implemented were observation, interview, documentary and data validation through triangulation.

1 INTRODUCTION

Music is an emotional language that is constantly connected to the human mind and continues to connect with people's actions, behaviours and morals. Music is a medium to improve and assist in personal skills development. This definition fits with Plato's opinion in Desyandri (2014) stating that:

> Music is the language of emotion. Emotion is connected with thought. Thought is connected with action, action deals with conduct, and the sphere of conduct is connected with morals. Therefore, ladies and gentlemen, if music is connected with emotion, and emotion is connected with thought and thought is connected with action and action is connected with the sphere of conduct or with morals, things which are connected by the same must be connected with one another, and therefore music must be connected with morals.

Therefore, it is suggested that music has the potential to form one's character into a better personality. Thus, if we consider the educational purpose of 2013 Curriculum in Indonesia, today's music learning is necessary. Music education not only transfers and enhances music skills, but is a field that also pays attention to students' behaviors through emotional language that will later affect on their actions, behaviors and morals.

Musical art education could be integrated within the standard curriculum and regular school hours or as an extracurricular activity. The SMAN 2 (State Senior High School 2) is the only school in Samarinda to integrate the Kutai people's traditional music called Tingkilan into extracurricular activities.

Tingkilan is one kind of musical art in Kutai society. This music originated alongside the spreading of Islam in Kutai and has a rough similarity with Malay art. Tingkilan spread through the process of acculturation with the local culture and developed into three kinds of Tingkilan- Hulu Mahakam, Tengah and Pantai each of which have its own characteristics.

The word Tingkilan itself means to satirize through poetry and music. The meaning remains the same despite each kind having its own musical structure that has gone through many changes (Hakim, 2011).

Tingkilan as part of the extracurricular activities within the SMAN 2 Samarinda, was established in 2008 and continues to score achievements. Thus, this extracurricular activity is the most popular within the school and draws a high level of interest from students. This success is possible through the Tingkilan coach's and counselor's determination in implementing Tingkilan method as part of the learning process. Both the coach and counselor believe that in order to make students comfortable, confident, collaborative and constantly in love with Tingkilan, there is a need to form cooperation with the students' parents. This was, the coach, counselor and parents create a natural learning environment for students to acquire knowledge and absorb lessons that will always be modernized but without leaving the traditional practice behind. This coach adopted the learning strategy from the Suzuki method(Thibeault, 2018). The method's goal is to guide every student to grow the sense of love towards music and in this case, toward Tingkilan while developing good character and not merely mastering musical instruments. This idea came to be known as "Talent Education".

Following the description above, the purpose of this research is to reveal, describe and analyse the implementation of Tingkilan using the Suzuki method within the SMAN 2 Samarinda extracurricular musical activities.

2 METHOD

This study was conducted using a qualitative research effort with a descriptive method. Qualitative research explains every element of data with detailed non-numeric form(Sugiyarto, 2017). This opinion is further asserted by Moleong (2007, p. 6) who states that qualitative research is meant to understand the phenomena of what is experienced by research subject, such as behaviour, perception, motivation, action and so forth, holistically and by description in the form of natural words while utilizing various scientific methods. According to Mahmud (2011), the descriptive method is one method used to picture or interpret the existing conditions or relationships between objects or data previously acquired in the field. In this study, a descriptive method is used to acquire data matching the data in the field.

The research was conducted at the SMAN 2 Samarinda in the province of East Kalimantan, Indonesia. The specific location for data collection was the Tingkilan music, extracurricular learning class in the school's trophy room. The researcher followed eight students who participate in Tingkilan. There were two groups, a male Tingkilan musical group and a female Tingkilan musical group. The research subjects were divided into three types: 1) coach; 2) counselor; and 3) students. The data acquisition techniques included observation, interview and documentation.

3 RESULTS AND DISCUSSION

3.1 *General profile of the SMAN 2 samarinda*

The SMAN 2 Samarinda was established on 14 July 1981 by the Minister of Education and Culture, via Republic of Indonesia Decree Number 022/0/198, at Jalan Kemakmuran No. 27 Kelurahan Sungai Pinang Dalam, Kecamatan Sungai Pinang Northern Samarinda, Samarinda city, province of East Kalimantan. The SMAN 2 Samarinda is better known as Smada. The school often receives pride generating awards such as first place in the National Nasyid Putri in Pentas Agama Seni Jakarta in 2008. The school has won many more awards that can be seen in the school trophy room. The school is also famously known because it is the only public school in Samarinda that is strictly Islamic, requiring female Muslim students to wear the veil and carry the Qur'an to be collectively recited for 15 minutes before the first hours of school.

3.2 *Profile of tingkilan music as an extracurricular activity in the SMAN 2 samarinda*

In 2007, one of the art teachers in the SMAN 2 Samarinda named Laila Rachmawati initiated a proposal request to the Headmaster to establish an East Kalimantan traditional music extracurricular activity as a place for naturally gifted students to learn in and as an introduction to traditional music for other students for culture preservation. The Headmaster approved the request and then made a proposal to the Head of the Education agency about the learning goals along with an explanation about necessary equipment for the Tingkilan extracurricular activity. In consideration of the requests from several schools, the Head of the Education agency approved the SMAN 2 Samarinda Headmaster's proposal. Subsequently in 2008, the education agency provided the necessary music instruments. Those instruments included one gambus, one ukulele, one cello, one cable microphone and one sound controller. After receiving all of the instruments, the counselor gathered several students who were interested in learning Tingkilan. The counselor then visited Mr. Triono, his sibling, to become the Tingkilan coach. By the early part of the 2008 semester a group of four female students has been established. The end of 2008 saw the first achievement by the female Tingkilan group. The group won the general championship in culture festival contest over several adult participants. Since the Tingkilan group became more serious and the coach made updates to the Suzuki method. In 2016, the counselor at SMAN 2 Samarinda was replaced by Hadjar Barunawan until today.

The history of establishing the Tingkilan extracurricular activtiy included several supporting actors which is as important to the development of Tingkilan music at the SMAN 2 Samarinda.

1. Coach
   The Tingkilan music coach at SMAN 2 Samarinda is Triono Budiwiryawan, a civil employee in the Education Agency of East Kalimantan and he is also a humanist in Samarinda. His birthplace was Yogyakarta and he was a graduate of the Indonesian Music Academy in Yogyakarta. Mr. Triono began training the students in the Tingkilan extracurricular activity starting in 2007 until today. The coach is responsible in creating arrangements, song partitures, to teach instrument playing and the vocals of Tingkilan. Beside being a coach at SMAN 2, Mr. Triono also continues to coach the East Kalimantan teen choir called Gita Ruhui Rahayu, since 2011 until today under the supervisorship of the Governor of East Kalimantan. Every year he is invited to be the judging coordinator for the provincial level, senior and vocational school FLS2N acoustic guitar, and judging coordinator for the provincial level, junior high school FLS2N of East Kalimantan.
2. Counselor
   Mr. Hadjar Barunawan is the counselor for the Tingkilan music extracurricular activity at SMAN 2 Samarinda and he is also an arts and craft teacher. Hadjar Barunawan is a graduate of the Institute of Education and Teacher Training Yogyakarta in S1 Music Art. He is experienced in music and is annually invited to be a judge for the FLS2N solo vocal and traditional music at the provincial level in Samarinda Bahari.
   3. Students
   Students that participate in the Tingkilan music extracurricular activity of SMAN 2 Samarinda, consist of eight participants divided into four female groups from the students in the X and XI grade levels and four male group students in the XII grade level.
3. Goal
   In an interview on 13 January 2018, Mr. Triono, the coach, shared that, the goal of learning Tingkilan music as an extracurricular activity was to shape the character of the students especially to be disciplined, to decide their own practice schedule so that it does not interrupt other activities, to be punctual at meetings and to concentrate during practice sessions. Other goals include building the students' self esteem by training their minds and building courage to perform in shows/musical competitions, sharpening awareness in playing music and being responsive during practice and understanding new ideas and concepts. It is also important to be sensitive towards this musical art form inherited from their cultural ancestors, by loving

and preserving Tingkilan tradition that is part of East Kalimantan Kutai culture and finally, to implant virtue towards parents, teachers, friends and other people.

4. Material

   In the interview on 13 January 2018 with Mr. Triono, various material was taught for the learning of Tingkilan music. Some example are: traditional songs of East Kalimantan such as Samarinda Tepian Mahakam, Tajong Samarinda, Buah Bolok and Balarut; Indonesia's national songs such as Indonesia Pusaka; Indonesian pop songs such as "Kolam Susu, Telaga Sunyi"; and western pop songs such as "It's Now or Never" and "Boulevard". These songs were arranged using the Tingkilan pentatonic format and *keroncong* and jazz as distinct features of the SMAN 2 Samarinda Tingkilan group. This material was taught for one semester repeatedly.

3.3 *Method in SMAN 2 Samarinda tingkilan learning*

The method that was and continues to be used in the SMAN 2 Samarinda Tingkilan extracurricular learning is the Suzuki method. Suzuki is taught along with several combined learning methods of lecture, demonstration, question and answer, and practice (drill).

3.3.1 *The suzuki method*

The Suzuki method is an approach to teaching and learning put forward by Dr. Shinichi Suzuki, a violin player from Japan. Suzuki's main belief, according to his language acquisition theory, is that everyone is capable of learning from their surroundings. Learning music is like learning a native language so that students learn by hearing the structure of an arrangement before learning to read music notation. Therefore, for the process of learning Tingkilan, the coach implemented this basic Suzuki principle.

3.3.2 *Lecturing method*

The lecturing method takes place during meetings early on to explain information linked to the musical instrument that will be used. For example, lectures introduce organology or parts of the instrument, maintenance, techniques to play each instrument and vocal technique in order to sing traditional songs alongside the playing of Tingkilan. The lecturing method is also used by the coach to implant the sensitivity of feeling, ethics, and morality to every student to become a better person.

Figure 1. Early learning lecture of Tingkilan music: the coach giving explanations concerning Tingkilan, its history, names of instruments and song themes to be played.

Figure 2. Interview session conducted by one of the students during a practice session about how to set up music instruments in this case, the cello.

3.3.3 *Interview method*

The interview method was conducted after the coach explained the notation and song structure to be learned. Students were given the opportunity to ask about notation, chords, vocal and playing techniques. This is done for the purpose of creating a communicative and productive environment in the learning process. The research results suggest that in the process of learning Tingkilan music as an extracurricular activity, students actively ask questions and also respond to questions from the coach.

3.3.4 *Demonstration method*

The demonstration method was used by the coach in the process of teaching Tingkilan: (a) the coach demonstrates the positioning, holding and setting of the instruments properly and correctly; (b) the coach demonstrates the placing of fingers on every

Figure 3. The coach demonstrates how to play instruments according to the arrangement.

instruments in Tingkilan; (c) the coach demonstrates picking techniques for gambus, cello, ukulele and also demonstrates vocal technique in a Tingkilan arrangement; and (d) the coach demonstrates how to play the song to be learned in tempo, patterns and with correct vocal dynamics.

3.4 *The learning process*

In the process of learning Tingkilan, certain stages are necessary. These stages of learning are:

3.4.1 *Introduction of the instrument's parts and maintenance*

During the early stages of learning Tingkilan music the coach explains what Tingkilan is, the history of the art, and the achievements of previous Tingkilan extracurricular students. The coach then introduces the names of the instruments, each of the instrument's parts and how to maintain each instrument in order to minimize deterioration and preserve the good condition of each instrument.

3.4.2 *Tuning*

The tuning of Tingkilan music instruments is carried out by grinding/picking 2/D string on the gambus and setting the strings following the tuning device, then setting the other three strings-1/G, 3/A, 4/D. Next is the cello, setting strings- 1, 2, 3 and 4 on the gambus and finally setting the ukulele strings- 1E, 2B, 3G. The tuning of each Tingkilan instrument is repeated by the coach due to the constant change of weather and the changing conditions of the students' instruments that cause the violin's strings to change its tone. Continuous settings are necessary in every meeting.

3.4.3 *Technique practice*

Notation and technique practice is conducted before students begin to read the song notation. Notation practice is meant for practicing intonation and correct playing technique on every instrument.

3.4.4 *Learning towards the song's material*

The process of teaching the material is involves several stages. These stages are: reading the whole song, learning one piece at a time, song perfection and job splitting. In teaching the material, the coach provides traditional songs such as "Tajong Samarinda" so that the introduction is to be learned first, then followed through stanza one, then through the refrain and coda of the song. After memorizing the song arrangement format, vocals are then allowed to be practiced with the song and together repeated until sufficiently achieved. The goal is to set habits, or automating the students' performance of each song.

3.4.5 *Song perfection*

Song perfection sessions are conducted by the coach and students at different times because each student has her or his own level of competence and determination. There are students who can master the song in two meetings while there are other students that need three to five meetings. After the song perfection is satisfactory, the coach advances the students to new material. Each song is given step by step, beginning from simple song arrangement to more challenging arrangements to increase proficiency.

3.4.6 *Evalution*

An evaluation is conducted by the coach at the conclusion of every meeting. Evaluation is given not by asking questions or giving a test but by observing each student's ability from the beginning of the learning process especially taking note of each student's attitude and practice in playing instruments and practicing the vocal technique. The coach made dedicated notes to assess each student's progress with the purpose of recalling the character and ability of each student in order to improve future meetings in accordance with each student's mindset.

3.5 *Applying the suzuki method*

The research results from January 2018 in the SMAN 2 Samarinda via the data from the coach using the Suzuki method and its implementation are as follows:

3.5.1 *Parental role and involvement*

Parental role and involvement is just as what Suzuki prescribed for the process of learning. During early meeting with students, the coach and the counselor of Tingkilan mandated that students invite their parents to school to discuss the schedule and benefits of the activity. The coach and the counselor asked the students' parents to support their children in arranging the scheduling of other activities so that Tingkilan sessions and other activities would not interrupt each other. It was mandatory that students not dismiss or abandon the extracurricular activity for one school year (two semesters) and that they be ready to follow the rules set for the Tingkilan extracurricular activity at SMAN 2 Samarinda.

3.5.2 *Adaptation towards their music instrument before learning to read music*

The Suzuki method instructs that at the beginning, the material be given by listening rather than reading direct notation(Rupiyono, 2011). During the beginning of the Tingkilan music sessions, examples to listen to and play were given first before learning to read notation. With this approach, students automatically remembered the notes and then learned the notation symbols by guessing. Besides, it is Indonesian tradition that music is learned and spread through the ears rather than by reading with the eyes.

3.5.3 *Playing by memory*

The process of learning that was encouraged by the coach, meant that he guided students to continue playing what the students remember. When students continued to play the playable part, students were then pushed to listen to the tone quality of what they produced.

3.5.4 *Constructive repetition*

Repetition is important in learning music. Students are only allowed to listen and play the song that is given to them to be practiced repeatedly. From the perspective of the coach, the ability to repeat a song will improve playing, the technique becomes more expert and there is an overall improvement in music learning and playing.

3.5.5 *Combination between individual lessons & group lessons*

Every student that participates in Tingkilan is pushed to take on a leadership role, to arrange schedules and to remind and control their peers. The schedule is made merging the male and female students to create a happy and joyful atmosphere. The coach's goal is to form good character in students, to be more disciplined and independent. Also, when students witness others who perform well, the motivation arises to play well together. This is how natural learning happens.

3.5.6 *Public performances*

The coach said that learning occurs not only in school or private studios but also by taking students to attend traditional art competitions usually initiated by the local government to perform Tingkilan. Excursions to public performances or competitons are scheduled with the purpose to motivate students to perform in front of a large crowd, to be more determined to practice, and to strengthen the will to learn.

3.5.7 *Listening*

The arrangement that is learned will be recorded by the coach using the Yamaha PSR 950 keyboard with midi/mp3 format that is then copied for students to learn the parts that are assigned. In the process of learning the coach demonstrate the song. This begin by listening to the song first followed by mimicing what is played the coach. The result of this demonstration is is that playing the arrangement is easier as the students' ears have been trained to listen throughout the learning process for continued progress.

3.5.8 *Encouragement*

Student success cannot be separated from the encouragement, motivation and praise of coaches, counselors, parents and other students.

4 CONCLUSION

The Tingkilan music activity at the SMAN 2 Samarinda takes place during two meetings, every week with a compulsory session every Thursday at 15.00-17.00 and another session according to a schedule agreed upon by all of the students, the coach and the counselor. There are eight students divided into male and female Tingkilan groups. The activity takes place in the SMAN 2 Samarinda trophy room, Jalan Kemakmuran No. 27 Sungai Pinang, Samarinda, East Kalimantan and also in the coach's private studio located in Jalan Proklamasi 1B No. 100 Sungai Pinang, Samarinda, East Kalimantan but only during holidays. The instruments, gambus, ukulele, cello, microphone and sound system are provided by the school to support the facilitation of the SMAN 2 Samarinda extracurricular activity. Teaching materials and sheet music that will be used are provided by the coach before learning takes place.

The learning process involves several stages beginning with the introduction of the instruments' parts, maintenance and tuning, followed by technique practice, learning through the material, song perfection, and evaluation. The research analysis has been made possible by observing the learning process of Tingkilan music extracurricular activity using the Suzuki method. The implementation of the Suzuki method for Tingkilan music learning involved: (1) Parental role and involvement with the coach and the counselor building partnerships with parents to aid students in practicing Tingkilan; (2) Adaptation by students towards each of the music instruments before learning to read music, guiding students to understand notation symbols by memorizing sound; (3) Playing by memory; (4) Constructive repetition; (5) Combination of individual lessons and group lessons; (6) Attending public performances in order to grow strong motivation within students to practice with the determination and courage to perform in public; (7) At first, listening repeatedly to the song that is being learned to increase musicalisation abilities of students; and (8) Encouragement, support and praise for every student's efforts.

REFERENCES

Banoe, P. (2003). *Kamus Musik Kanisius.* [Music Dictionary]. Yogyakarta: Kanisius.

Desyandri (2014)Pendidikan Seni Musik Humanis (Suatu Tinjauan Konseptual)[Humanism Music Education: a conseotual study]. UNP; PGSD.

Hakim, A. Q. (2011). *Tingkilan (Alunan yang Mengarungi Abad).* [Tingkilan (Alunan over Decades)]. Samarinda: Nuansa Harmoni.

Rupiyono, L. A. (2011). *Aplikasi Metode Suzuki Dengan Media Lagu Anak Nusantara.* [Application of the Suzuki Method to the Nusantara Children's Song Media].

Thibeault, M. D. (2018). Learning with sound recordings: A history of Suzuki's mediated pedagogy. *Journal of Research in Music Education, 66*(1), 6-30.

Mahmud, H. (2011). Metode Penelitian Pendidikan. [Educational Research Methods]. *Bandung: CV Pustaka Setia.*

Moleong, L. J. (2007). Metodologi penelitian kualitatif edisi revisi. *Bandung: PT Remaja Rosdakarya*, 103.

Sugiarto, E. (2017). *Menyusun Proposal Penelitian Kualitatif: Skripsi dan Tesis. [Prepare Qualitative Research Proposal for Thesis]. Yogyakarta: Suaka Media.* Diandra Kreatif.

21st Century Innovation in Music Education – Astuti et al. (eds)

 ISBN 978-0-367-11122-9

An effort to increase learning outcomes in musical scale teaching using the 'PILKABE' strategy

Boiman
SMP 2 (Junior High School) Tunjungsari Wonosari, Guning Kidul, Daerah Istimewa Yogyakarta, Indonesia

ABSTRACT: This study aims to describe the steps of teaching musical scales and identify how far the 'PILKABE' strategy increases learning outcomes of Grade 7A students of Junior High School 2 Tanjungsari in Semester 2 in the 2017/18 academic year. This study was conducted by classroom action research. The results showed that the 'PILKABE' strategy could increase students' learning outcomes. In both Cycle I and Cycle II, the highest score was 100 and the lowest score was 50. The average score in Cycle I was 75 and in Cycle II was 85.94, so there was an increase in average score of 10.94. The classical mastery in Cycle I was 80% while that in Cycle II was 93.75%; thus, there was an increase in classical mastery of 13.75%.

1 INTRODUCTION

1.1 *Background of the problem*

Learning the basic concept of musical scales is important in playing music because the sound vocals and instruments will be heard harmoniously and smoothly if played with the right tone. Therefore, it is important to teach students about the basic concept of musical scales correctly because understanding musical scales is the foundation to playing musical instruments. It is also important for learning basic melodies.

In Junior High School 2 Tanjungsari, students face some obstacles in learning musical scales. They find it difficult to memorise subject materials in the form of a list of letters arranged randomly. The alphabet list in musical scales has the function of determining the basic tone, knowing the number of changed tones, and identifying the changed tones in every musical scale. Thus, difficulty in learning musical scales makes it difficult to determine the basic tone, the number of changed tones, and the tones to be changed. Besides, the teaching strategy used is not interesting, making the students reluctant to learn music theories. Because of these challenges, learning outcomes do not meet the target (i.e. a minimum grade of 60 for passing).

To solve this problem, it is important to apply 'PILKABE', a creative and fun strategy. PILKABE is one of the strategies used in learning a musical scale. PILKABE stands for three vowels, namely, PIL, KA and BE. PIL is short for PILIHAN, which means choice; KA is short for KATA, which means word; and BE is short for BEBAS, which means free. Thus, the PILKABE strategy refers to PILIHAN KATA BEBAS translated as 'free word choices'. This strategy can teach the students how to determine the basic tone, the number of tones changed, which tones are changed, how to compose a row of letters/tones and how to form a musical scale.

In the PILKABE strategy, students are given the freedom to choose original words to be arranged as a creative sentence by using the previously determined first letter. Then, is the words are arranged based on the existing sequence of acronyms. The end of this process produces a sentence that may be strange or funny. This stage is interesting and fun for students. This is in accordance with the learning model of *Quantum*

Learning by Bobby De Porter and Mike Hernacki (1999, pp. 236–237). It is about associating learning with comfort and fun.

In this study, the researcher limited the problem to the effort to increase learning outcomes of musical scale teaching using the PILKABE strategy for Grade 7A students of Junior High School 2 Tanjungsari in Semester 2 in the 2017/18 academic year.

1.2 *Objective and significance of the research*

This research aims to describe the steps of applying the PILKABE strategy and identify how far the PILKABE strategy improves students' learning outcomes in musical scales. The application of this strategy was predicted to be one of the ways to reach the aims.

The significance of this research is (1) to increase students' motivation in learning music with correct musical scales and to increase students' participation in the learning process; (2) to inform teachers about various learning strategies, to improve and enhance the learning of arts and cultures and to apply an appropriate learning strategy to a certain material; (3) to establish an active and enjoyable learning atmosphere in schools and achieve students' mastery in learning.

2 CONCEPTUAL FRAMEWORK

The obstacle in learning musical scales is that the students find it difficult to memorise learning materials in the form of a list of letters arranged in random order. This difficulty, in turn, leads to difficulty in determining the musical scale, how many tones are changed and what tones are changed. Besides, the obstacle exists also because the learning strategy used in teaching musical scales is less interesting.

The PILKABE strategy can be a solution to solve the problem. Using this strategy can make it easier for students to memorise without any letters being tucked away. So, the researcher predicted that this strategy could improve the students' learning outcomes in musical scales.

3 RESEARCH METHOD

This classroom action research used Kemmis and McTaggart's model (Kemmis et al., 2013), further developed by Arikunto (2007, p. 16). The research was done in two cycles, with four phases in each cycle: planning, acting, observing and reflecting.

Suryabrata (1983, p. 35) and Wagiran (2005) state that action research aims to develop new skills or a new approach to solve the problem using their direct application in a work field or real-world scenario. Based on this explanation, action research is the research done by an individual or a team which aims to develop new skills or a new approach to solve a problem. The participants of this research were Grade 7A students in Junior High School 2 Tanjungsari. There were 32 students participating, 17 boys and 15 girls. The setting of this research was the Grade 7A classroom of Junior High School 2 Tanjungsari. This research used two cycles conducted on Friday, 2 February 2018 (Cycle I) and on Friday, 9 February 2018 (Cycle II).

3.1 *Subject matter*

In this research, the subject matter presented by the researcher was the basic competence of 3.3: to understand the basic concept of playing a simple musical instrument individually or as a team. The learning material discussed was limited as the students were able to arrange musical scale letters and make a musical sound by using the correct tone (Alfianto et al., 2014)

3.2 *Research design*

This study was carried out by the subject teacher as the researcher, helped by a colleague. The colleague took part as a collaborator and supervisor. This research was planned to be done in two cycles. This research used the spiral model of Kemmis and McTaggart (Kemmis et al., 2013), which consisted of planning, acting, observing and reflecting in each cycle.

3.3 *Research procedure*

This action research followed the procedure of Kemmis and McTaggart's spiral model consisting of two cycles. Each cycle comprised four steps.

1) Planning: The researcher planned actions aimed to improve or increase changes in attitude and affective domains. This planning included subject materials, learning tools, lesson plans, learning methods, learning strategies and teaching approaches to be implemented.
2) Acting: The teacher (researcher) entered the classroom and greeted the students, checked the students' attendance list and delivered the subject material. During the learning process, the teacher observed students' activities and documented these activities in the form of notes and photographs. The learning activities ended with an assessment.
3) Observing: This phase focused on observing the action process and the result or effect of the action performed by the students. The teacher and the collaborator observed if there were any changes in the learning outcomes or the students' attitude related to the learning activities done in this research.
4) Reflecting: The researcher and collaborator discussed and considered the result or the effect of the action from the viewpoint of various criteria. Based on this reflection, they reviewed and evaluated the initial planning (in Phase 1) to make improvements for the next planning phase.

3.4 *Data sources*

Data were collected from the students by using questionnaires and references. The former was administered to students directly related to the learning material in the research. The questionnaire was in the form of written questions asking about the students' response to the research material (Moleong, 1995). The references discussion was with some experts to seek their opinions about the learning material used in this research.

3.5 *Data collection*

Data from this research were collected directly from students and from observation.

1) Observation was conducted by noting students' actions in the learning process presented. Therefore, the learning outcome is as before.
2) Direct data collection was done to learn the students' response to the PILKABE strategy in learning activities.

3.6 *Research instruments*

The research instruments used in this study were as follows.

1) Observation sheets: The sheets measured aspects of students' learning outcomes in this research, including the activities involved in learning, students' skills in arranging creative sentences from the original words that had been chosen and the skill to play a musical scale by using correct tones.

2) Documentation of photographs: Photographs were used to find stronger evidence of the learning process, and included those of the learning process and of students' worksheets and questionnaire sheets.
3) Questionnaire: This was important to get information about students' response to musical scale learning through the PILKABE strategy. The written questionnaire contained questions or problems relating to the students' privations and the learning materials.
4) Testing of learning outcomes: This testing was done at the end of Cycle I and Cycle II.

3.7 *Data analysis techniques*

The data analysis techniques were as follows.

1) The observation data were counted and then presented in percentage. The data were used to study the improvement of learning outcomes, which were presented descriptively.
2) The questionnaire data contained statement points in the group questionnaire, which then were counted to get the total score in every point in percentage. The data were used to study the improvement of learning motivation which was presented descriptively.
3) The learning outcomes were noted before and after each learning activity. They were then described.

3.8 *The success indicator*

The students were considered to be successful if there was a change in their attitude and thinking pattern during the study. The success of the students' learning outcomes was indicated by the students' mastery: a minimum grade of 60 individually and a minimum grade of 85 for classical mastery.

4 RESULT

4.1 *Research finding in Cycle I*

This research was done on Monday, 2 February 2018. The subjects of the research in Cycle I were 32 Grade 7A students; two students were absent at the time of assessment, so research findings in Cycle I are based on 3000 subjects.

The four steps of the learning done in Cycle I are described in the following.

1) Planning: In this phase the teacher wrote a lesson plan, designed observation sheets for students and the teacher, prepared learning materials about the diatonic musical scale 1#–7#, prepared learning instruments (keyboard and pianica), and prepared assessment sheets.
2) Acting: This phase included the following stages.
   a) In the introduction stage, the teacher motivated the students by presenting the PILKABE strategy and the objectives of learning. The students chose original words using the previously determined first letter and arranged these to form a creative sentence.
   b) The students were able to make a diatonic musical scale 1#–7# by using the PILKABE strategy.
   c) The students were able to play a diatonic musical scale 1#–7# by using the correct tones.
   d) The five main activities of the students are described in the following.
      (i) Observing: The students paid attention to the teacher's explanation about the PILKABE strategy; about the letter used as a symbol to determine the basic tone in a musical scale (in the first music sentence); and about the letters used as a guideline to determine the sharp tone in a musical scale (in the second music sentence).

Musical scale 1#–7#
G=1=1# ➔F
D=1=2# ➔F, C
A=1=3# ➔F, C, G
E=1=4# ➔F, C, G, D
B=1=5# ➔F, C, G, D, A
FiS=1=6# ➔F, C, G, D, A, E
Cis = 1=7# ➔F, C, G, D, A, E, B

(ii) Questioning: The students were given a chance to ask questions about the materials explained by the teacher.

(iii) Trying:

- The students chose words using a letter in the basic tone and arranged these words to form a sentence, where the first letter of every word was arranged as determined (the first sentence in Table 1).
- The students chose words from letters of the musical scale which were sharp and arranged these words to form a sentence, where the first letter of every word was arranged as determined (the second sentence in Table 1).

Table 1. Words chosen and sentences created by the students in Cycle I.

First sentence	**G**ajah (elephant)	**D**uduk (seat)	**A**mat (very)	**E**lok (comfortable)	**B**ila (if)	**F**oto (take picture)	**C**is
Second sentence	**F**eri (feri)	**C**ari (looking for)	**G**adis (girl)	**D**apat (met)	**A**nak (girl)	**E**lok (beauty)	**B**anget (very)

- The students made a musical scale using the PILKABE strategy.
Example 1: Gajah=1=1# ➔ Feri
↘
G, A, B, C, D, E, F#, G'
Example 2: Duduk=1=2# ➔ Feri, Cari
↙ ↓
D, E, F#, G, A, B, C#, D'
- The students played the musical scale 1#–7# with the correct tones.
- Reasoning: The students, with the guidance of the teacher, drew conclusions based on the discussion results and the question-and-answer session between the teacher and the students. There were two sentences: in the first sentence, the first letter of every word was arranged as the basic tone; in the second sentence, the first letter of every word showed the tones that were sharp or got changed. On the basis of the two sentences, we can determine the basic tone, the number of tones that were sharp and which tones were sharp in every musical scale.
- Presenting: Students played the musical scale 1#–7# with the correct tones. In Cycle I, there were 32 Grade 7A students but, at the time of assessment, two students were absent. Thus, there were only 30 students taking the assessment in Cycle I. There were nine students who got a score of 100 (30%), eight students who got a score of 90 (26.67%), three students who got a score of 80 (10%), five students who got a score of 70 (16.67%), three students who got a score of 60 (10%) and two students who got a score of 50 (6.67%). The highest score was 100, the lowest score was 50 and the mean score was 75; the classical mastery was 80%. The minimum mastery criterion (KKM) was 60. The assessment results reveal that six students did

not master the materials because they got a score below 60 (20%) and 24 students (80%) mastered the materials because their score passed the minimum mastery criterion (KKM). The students were considered not to have achieved classical mastery in the materials because less than 85% had obtained mastery.

4) Observing: The teacher as the researcher with the collaborator observed both the students' and the teacher's activities during the learning process. In Cycle I, there were some students who initially did not understand the teacher's explanation. Thus, they misunderstood how to choose the words and found it difficult to make sentences. For example, they made a sentence with two or more nouns in order, but it seemed to be overlapping in the order. The students did the activity individually, with no cooperation or teamwork. Some students found it difficult to play the instrument with their fingers correctly. This was because they could not do it yet.
5) Reflecting: Based on the observation results, there were still some students who did not understand well the teacher's explanation earlier, so the teacher had to re-explain until the students understood the learning materials on arranging sentences and on fingering. In Cycle II, it was important to make a team, so they could help each other to understand the learning materials. The teacher had to explain the materials more slowly so that the students understood the learning materials. The teacher reflected on the activities and observations by asking the students about the difficulties regarding the materials learnt. The teacher drew a conclusion about the new materials they had just learnt and gave the students the task to learn the materials again at home. The activity was closed with a greeting.

4.2 *Description of Cycle II*

In Cycle II, there were 32 Grade 7A students consisting of 17 boys and 15 girls. This research was done on Friday, 9 February 2018. The steps of learning were developed based on Moore (2005) and are described in the following.

1) Planning: Before introducing the subject matter, the teacher greeted the students, checked the attendance list and guided the students to make groups of eight (there were four groups). The teacher motivated the students by asking about the previous materials at a glance and conveyed the purpose of learning. The purpose of learning in this cycle was to improve the ability to make a correct sentence, the ability to make a musical scale, the ability to play the musical scale 1b–7b using the correct tone and the ability to finger (play) the pianica and the keyboard.
2) Acting: The teacher followed the steps of learning using the PILKABE strategy. The strategy included the following five steps.
   (i) Observing: The students paid attention to the explanation about how to make the right sentence. They observed the name and the position of chromatics tones in a pianica/keyboard shown by the teacher and observed examples of a musical scale played using the correct tone and fingering technique.

First sentence

F=1=1b ➔ B
Bes =2b ➔ B, E
Es=1=3b ➔ B, E, A
AS=1=4b ➔ B, E, A, D
Des= 1= 5b ➔ B, E, A, D, G
Ges=1-6b ➔ B, E, A, D, G, C
CeS=1=7b ➔ B, E, A, D, G, C, F

Second sentence

(ii) Questioning: The students were given an opportunity to ask questions to clarify the learning materials that the teacher explained.

(iii) Trying:

- The students chose words using a letter in the basic tone and arranged these words to form a sentence, where the first letter of every word was arranged as determined (the first sentence in Table 2).
- The students chose words from letters of the musical scale which were sharp and arranged these words to form a sentence, where the first letter of every word was arranged as determined (the second sentence in Table 2).

Table 2. Words chosen and sentences created by the students in Cycle II.

First sentence	**F**atimah	**B**esuk	**E**suk	**A**kan	**D**apat	**G**aun	**C**antik
Second sentence	**B**eli	**E**ndok	**A**ku	**D**apat	**G**adis	**C**antik	**F**ull

- The students made a musical scale 1b–7b using the PILKABE strategy.
  Example 1: Fatimah=1=1b ➔ Beli

  F, G, A, B^b, C, D, E, F'

  Example 2: Bes = 1= 2b ➔ B, E

  B^b, C, D, E^b, F, G, A, B'
- The students played the musical scale using the correct tone and fingering technique.

(iv) Reasoning: Students with the guidance from the teacher made a conclusion based on the result of the question-and-answer session between the teacher and students.

(v) Presenting: The students, with guidance from the teacher, played the musical scale 1b–7b using the correct tone and fingering technique. Based on the research results of Cycle II, of the 32 students, 12 students got a score of 100 (37.5 %), nine students got a score of 90 (28.13%), three students got a score of 80 (9.38%), four students got a score of 70 (12.5%), two students got a score of 60 (6.25%) and two students got a score of 50 (6.25%). The mean score was 85.94, the highest score was 100 and the lowest score was 50. The students were considered to have achieved classical mastery because the percentage of students who mastered the materials (93.75%) was more than 85%.

3) Observation: The teacher as the researcher with the collaborator observed both the students' and the teacher's activities while the learning activities were in progress. In Cycle II, the students focused more on following learning activities and making sentences correctly. The students did this work in groups, helping each other to understand the lesson, even though there were some students who were passive in the group, making it hard to achieve the optimum result of cooperation.

4) Reflection: Based on the observation in Cycle II, the students' learning outcomes were better than those in Cycle I. This was because the students worked in groups and cooperated with each other in doing the task. However, intensive supervision from the teacher was required to enable active participation in the groups as there were many students who did the task passively.

The teacher reflected on the activities and observations by discussing with the students the materials that were difficult to learn. The teacher reiterated the materials that had been learnt and asked the students to study the materials again at home to help remind them what had been learnt in the school. This activity was closed with a greeting.

4.3 *Students' learning mastery criteria*

The researcher applied learning mastery criteria to assessing students both individually and for classical mastery. The limitation of the students' learning mastery is measured using the minimum mastery criterion (KKM). The students were considered to have mastered the materials individually if they got a minimum score of 60 as stated in KKM; they were considered to have achieved classical mastery of the materials if 85% of the students who were present mastered the materials.

The assessment results of Cycle I and Cycle II can be summarised as follows. Thirty-two students were the subjects of this study, but in Cycle I there were only 30 students because two students were absent during assessment. In Cycle II, all 32 students were present. In both Cycle I and Cycle II, the highest score was 100 and the lowest score was 50. Therefore, the highest and lowest scores in both cycles were the same. The mean score in Cycle I was 75 while that in Cycle II was 85.94. Thus, there was an increase of 10.94. Classical mastery in Cycle I was 80% while that in Cycle II was 93.75%. Thus, there was an increase in the students' classical mastery of 13.75%.

4.4 *Steps in learning musical scales using the PILKABE strategy*

1) Observing: The teacher explains how to choose creative words and make sentences correctly, shows the names and positions of chromatics tones in the pianica and keyboard, and gives examples for how to play a musical scale using the correct tone and fingering technique.
2) Questioning: The teacher gives the students an opportunity to ask questions to clarify the learning materials explained by the teacher.
3) Trying: The students choose original words, make creative sentences, arrange the musical scale and play the musical scale using the correct tone and fingering technique.
4) Reasoning: Conclusions are drawn based on the discussion results and question-and-answer session between the teacher and the students.
5) Presenting: Students play a musical scale using the correct tone and fingering technique.

5 CONCLUSION

5.1 *Conclusion*

The results of the research show that the PILKABE strategy can increase the learning outcome of musical scale teaching for Grade 7A students of Junior High School 2 Tanjungsari, as shown by the comparison of results and the increasing scores gained in Cycle I and Cycle II.

In both Cycle I and Cycle II, the highest score was 100 and the lowest score was 50. Classical mastery in Cycle I was 80% while that in Cycle II was 93.75%. Thus, there was an increase in classical mastery by 13.75%. The mean in Cycle I was 75 while that in Cycle II was 85.94. Thus, there was an increase in mean score of 10.94. The classical mastery level in Cycle I was 93.3% while in Cycle II the classical mastery level was 98.3%. Thus, it can be concluded that there is an increase of 5% in the classical mastery level. The data and results prove that the PILKABE strategy can improve students' activeness (i.e. learning outcomes) in the learning process.

5.2 *Suggestions*

Based on the conclusion in this research, fellow cultural art teachers are invited to do the following things:

1) Do something creative and innovative to help solve the problems in learning.
2) Apply the PILKABE strategy as one of the methods of solving the problems of musical scale learning.

3) Do further research in this area creatively and innovatively, as many issues have not yet been written about by the researcher.
4) Use the results of this research as an inspiration in the learning of art and culture in general.

REFERENCES

Alfianto, F., Florentinus, T.S. & Utomo, U. (2014). *Pengembangan Instrumen Penilaian Apresiasi Seni Musik Materi Seni Budaya Sekolah Menengah Pertama* [Developing of music appreciation assessment instrument for Junior High School degree]. *Innovative Journal of Curriculum and Educational Technology*, *3*(2), 15-22.

Arikunto, S. (2007). *Penelitian tindakan kelas* [Classroom action research]. Jakarta: Bumi Aksara.

DePorter, B. & Hernacki, M. (1999). *Quantum Learning: Make Learning Comfortable and Fun*. Bandung: Kaifa.

Kemmis, S., McTaggart, R. & Nixon, R. (2013). *The action research planner: Doing critical participatory action research*. Singapore: Springer Science & Business Media.

Moleong, L.J. (1995). *Metodologi Penelitian Kualitatif* [Method of qualitative research]. Bandung: PT. Remaja Rosdakarya.

Moore, K.D. (2005). *Effective instructional strategies from theory to practice*. Thousand Oaks, CA: SAGE.

Suryabrata, S. (1983). *Metode Penelitian* [Research methods]. Jakarta: Tarsito.

Wagiran, B.K. (2005). *Pengertian dan prinsip-prinsip Penelitian Tindakan Kelas* [Definition and principles of classroom action research]. Yogyakarta, Indonesia: Satker Pembinaan PLP.

21st Century Innovation in Music Education – Astuti et al. (eds)
© 2020 Taylor & Francis Group, London, ISBN 978-0-367-11122-9

Keroncong multimedia: An alternative music learning media

D.A.P. Anggara & A.N. Machfauzia
Universitas Negeri Yogyakarta, Yogyakarta, Indonesia

ABSTRACT: This research and development aims to produce a multimedia learning product designed around keroncong material in learning arts and culture for class IX students at SMP Negeri (Public Middle School) 2 Wangon, and to establish the feasibility of using this keroncong learning multimedia product for learning. Model and design development references Trollip and Alessi (2001). The subjects of this research and development were 34 grade XI students at SMP N2 Wangon. The data was collected through a questionnaire from the validation of media and material experts, beta tests and student response questionnaires in learning. The data analysis uses a descriptive technique. This development produced a multimedia learning product that is operated on computers as a medium for teachers and students. The results of the research show that the keroncong multimedia produced is worthy of being used as learning media.

1 INTRODUCTION

As one of the formal education institutions in the Banyumas district, SMP N2 Wangon has carried out art and culture learning. Teaching is carried out according to the 2013 curriculum, and the main material implemented in class IX in the 2017/2018 school year at SMP N2 Wangon was taken from the Basic Competencies 3.3 understanding of the concepts, forms and characteristics of popular music. Popular music has a variety of types including popular foreign music such as jazz, blues, reggae and rock, and popular Indonesian music such as pop music, dangdut and keroncong. Keroncong music is a popular type of traditional music because it has grown and developed in Indonesian society since the sixteenth century.

In learning music, students not only learn to play musical instruments or practise singing, but also must understand the music itself through theoretical knowledge. This was also taught by art and culture teachers at SMP N2 Wangon. But in its implementation, it is known that students are more familiar with the types of foreign popular music compared with popular Indonesian music based on the results of a preliminary study conducted on 3 and 5 August 2017, which were interviews with music teachers and observations in SMP N2 Wangon. Knowledge about keroncong music is very important in order to increase students' knowledge and appreciation of traditional music. But in its implementation there are still several problems such as the unavailability of teaching materials or material on keroncong music. Unavailability of musical instruments that support learning to introduce keroncong music, the limited ability of music teachers to play musical instruments, and the lack of familiarity and limitations of teachers in creating learning media for the learning process, making it difficult for teachers to convey the material.

The learning process is essentially a communication process that delivers messages from a source through certain channels/media to a recipient (Sadiman, 2009). Without media, communication will not occur and as a communication process, learning will also not be able to take place optimally (Daryanto, 2016). It can be said that the learning carried out has not been optimal because of the unavailability of supporting learning media. Students tend to lack motivation and appreciation in learning. In line with the advancement of information and communication technology, art and culture learning in the 2013 curriculum can utilise technology as a media and learning resource for students (Kemedikbud, 2017). The function of technology in the world of education is to improve teaching and learning. However, because of the limited ability of teachers to create new learning media, the use of technology that supports learning is still lacking (Tanbeer et al., 2008). With technology it is hoped that learning media can be created that can be used to support the learning process of art and culture so that learning objectives can be achieved optimally.

Keroncong learning multimedia is made more interesting and easier through the use of Adobe Flash CS6 software. Through the use of multimedia keroncong learning, students can learn and deepen their knowledge of the material within a limited time, so students are more interested in learning. The purpose of music education with music technology is to develop a music education system that is easy, interesting and appropriate for developing musical skills (Dittmar et al., 2012). Developing keroncong learning multimedia is expected to increase knowledge and foster student appreciation and motivation toward learning about keroncong music.

2 LITERATURE REVIEW

Learning occurs because of the process of interaction carried out between teachers and students. The presence of technology can be a tool that supports the interaction process. Current technology can be used in learning as a learning media and student learning resources to obtain information. Without communication, media will not occur and the learning process as a communication process will not be able to take place optimally (Daryanto, 2016). The function of technology in the world of education is to improve teaching and learning (Tanbeer et al., 2008). Technology can be considered integrated if used in groups as part of the daily learning process that occurs in the classroom (Tanbeer et al., 2008).

Teaching using technology is more effective than not using technology because new technology techniques do not involve substantial changes in the teaching and learning process. Technology supports certain practical or administrative problems but does not take an important role in pedagogies (Tanbeer et al., 2008).

The use of learning media in the learning process can generate new desires and interests, generate motivation and stimulation of learning activities, and even psychologically influence students (Tanbeer et al., 2008). To overcome the problems of learning art and culture encountered, technology can be the right solution. The development of music technology is part of the information and multimedia technology revolution produced by computer technology. Music software, electronic keyboards and computers are effective tools in facilitating student presentations, compositions and musical notations (Teen et al., 2014). Multimedia can be applied in learning arts and culture in schools as learning media that support teachers and students in learning keroncong music. The multimedia contains keroncong materials that support students in improving their knowledge about keroncong music so that learning objectives can be achieved.

3 METHOD

3.1 *Data collection*

The data collection techniques and instruments used in this research and development are in the form of questionnaires, observation sheets and interviews. Questionnaires were given to media experts, material experts and students both in beta testing and to students when testing products. The data obtained through questionnaires produced responses about the product in terms of material and media so that the feasibility will be used to revise the product.

The observation sheet was used during the preliminary study to observe the condition of the school and the learning that took place, and used as reference material in making the keroncong learning multimedia product. Interviews were conducted during the pre-survey and preliminary study as a reference for the needs analysis. Interviews are conducted as a data collection tool to find out the initial conditions in learning, constraints faced by teachers, the curriculum used, strategies and methods used, teacher knowledge about learning materials, student characteristics and the facilities and infrastructure available at the school.

The data used in this study is in the form of qualitative data and quantitative data. Data obtained based on interviews with music art subject teachers in SMP N2 Wangon were analysed using a descriptively qualitative technique. This analysis is intended to describe the characteristics of student data and needs analysis. These data is used as a basis for the development of keroncong learning multimedia. In addition, qualitative data from media experts, material experts and students is used as a basis for reviewing the development of keroncong learning multimedia products in junior high schools that have already been developed.

Quantitative data in the form of statements falls into the categories excellent, very good, good, satisfactory and unsatisfactory. The data is then converted to qualitative data using a scale of 5 using the reference conversion from Sukardjo (2008), as shown in Table 1.

3.2 *The stages of the research development*

Research and development procedures use a simplified Trollip and Alessi (2001) development research design according to development needs. Development procedures in this study pass through three stages, namely planning, design and development, adapted from the procedure developed by Trollip and Alessi (2001) and simplified in accordance with the needs of the research.

The planning stage is carried out by pre-survey and preliminary study as analysis material for the development of multimedia needs for keroncong learning so that the results can be obtained regarding the condition of the school and the learning of art and culture. The activities carried out at the design stage relate to the selection of forms and multimedia content of the learning to be made. The collected material is selected and developed in accordance with the main ideas of the product. Students perform task and concept analysis with supervisors and subject teachers, and create scripts for making videos. The activities of students carried out at the development stage are making learning multimedia into learning media applications using the Adobe Flash CS6 software.

The design of the trial in research and development includes the formative evaluation phase and summative evaluation. Formative evaluation consists of an alpha test and a beta test. The alpha test is carried out by one media expert and one material expert who are competent in their field, while the beta test is carried out by seven students selected with high, medium and low abilities. Formative evaluation aims to obtain multimedia-related assessments produced before being used in trials.

After going through the formative evaluation stage and product revision, a summative evaluation is then carried out. This summative evaluation was conducted by 34 students aimed at finding out the feasibility and response of students to keroncong learning multimedia

Table 1. Quantitative data conversion to qualitative data scale 5.

Quantitative data	Range	Qualitative data
5	$X > Xi + 1.8\ Sb$	Excellent
4	$Xi + 0.6\ Sb < X \le Xi + 1.8\ Sb$	Very good
3	$Xi - 0.6\ Sb < X \le Xi + 0.6\ Sb$	Good
2	$Xi + 1.8\ Sb < X \le Xi - 0.6\ Sb$	Satisfactory
1	$X \le Xi - 1.8\ Sb$	Unsatisfactory

Conditions:
(Xi) = 1/2 (ideal maximum score + ideal minimum score)
(Sb) = 1/6 (ideal maximum score – ideal minimum score)
where Xi = ideal score average: Sb = ideal standard deviation; X = empirical score.

products. Summative evaluation is carried out by applying keroncong learning multimedia products to classroom learning.

4 RESULT

The multimedia product produced in this research and development is keroncong learning multimedia for class IX students in junior high schools. Broadly speaking, keroncong learning multimedia products are compiled in the form of theoretical and practical learning material about keroncong music in .exe form and packaged into DVDs. The product is made using Adobe Flash CS6 software. The main components of the product produced are proposal, material, evaluation and developer profile. Creating learning multimedia products involves a combination of media components, such as graphics, video and audio, into one. The process of making audio and video is done in stages. Audio creation is done using Cubase software, while video creation and editing is done using the Pinnacle Studio software.

Media expert validation was carried out by one media expert by completing a questionnaire using a scale of 5. Aspects assessed consist of appearance, animation and graphics and programming. The assessment carried out by the media expert produced an average total score across all aspects of appearance, animation and graphics and programming of 3.96 (very good), meaning that the learning multimedia product was very feasible for use. In addition to providing assessment, the media expert also provided some suggestions for improving the keroncong learning multimedia product.

Material expert validation was carried out by one material expert by completing a questionnaire using a scale of 5. The aspects assessed were learning and material aspects. The results of the assessment carried out by the material expert produced an average total score across both aspects of learning and material of 4.39 (very good), which means that it is very feasible to use the learning multimedia product. In addition to providing assessment, the material expert also provided some suggestions for improving the keroncong learning multimedia product.

The beta test was conducted by selecting seven students from class IX of SMP N2 Wangon taken randomly with the student ability level of low, medium and high. This trial was conducted to find out the feasibility of the product in terms of appearance, programming, learning, material and attractiveness. The results of the assessment conducted by students produced an average score across all aspects of 4.46 (very good), which means that the learning multimedia product is very suitable for use. In this beta test there was no input or suggestions from the students, so that no revisions were made from the perspective of product users.

The keroncong multimedia product produced have been through a process of assessment and revision based on suggestions from media and material experts and students. The main page display of the multimedia product can be seen in Figure 1.

The introduction is a page that contains competencies and learning instructions. This includes a competency explanation about core competencies, basic competencies and learning

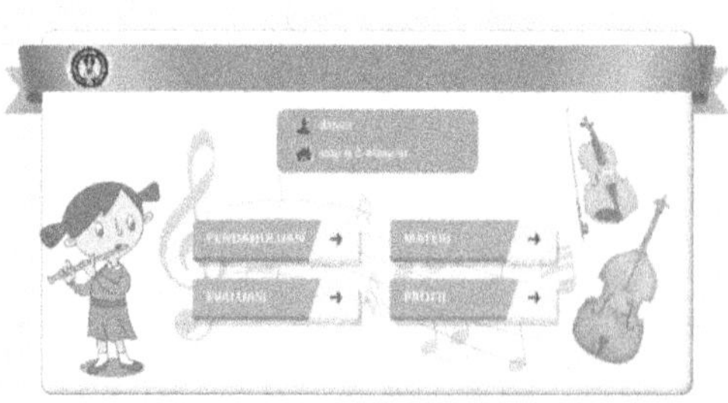

Figure 1. The main page for multimedia product.

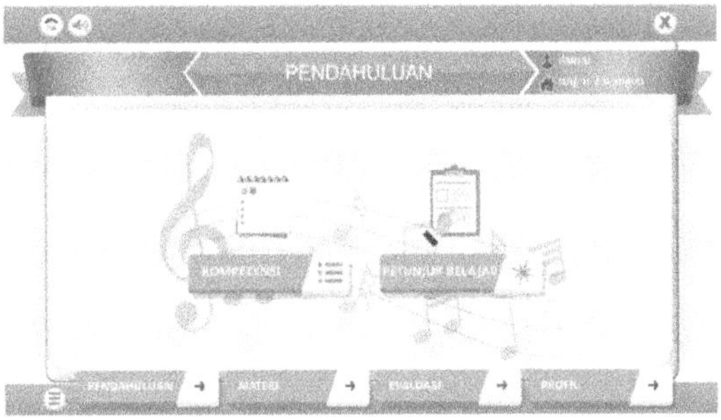

Figure 2. The introductory page of the multimedia product.

objectives in the use of keroncong learning multimedia used by students and learning instructions containing instructions for using keroncong learning multimedia.

The material pages contain two material buttons that act as keys to theoretical and practical material. Theoretical material relates to the theories that exist in keroncong music such as the history of keroncong music, keroncong types, keroncong musical instruments and keroncong figures. Practise material is related to keroncong learning in practice, which can aid students when they practise playing music. In practise material, all material is presented in the form of audio and video tutorials from each musical instrument in keroncong music, such as cuk, cak, cello, contrabass, violin, flute and vocals, using the style of the keroncong song the title is Bengawan Solo.

The evaluation page contains evaluation questions to be answered after using keroncong learning multimedia. The evaluation consists of three types, which are quizzes, guessing lyrics and guessing instruments. One quiz is a collection of multiple-choice questions related to the theory of keroncong music, intended to measure the extent to which students understand the

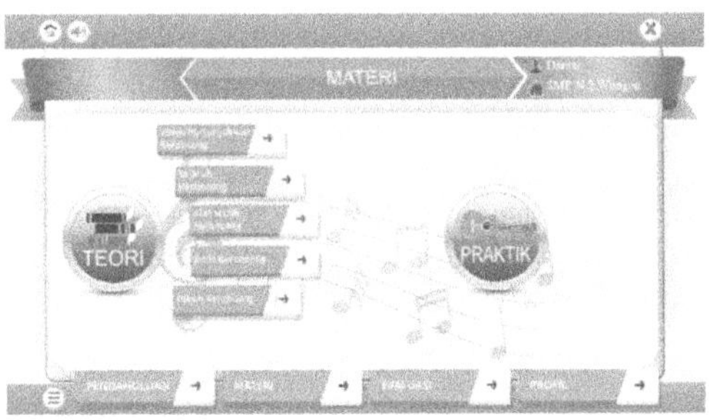

Figure 3. Display of multimedia product material pages.

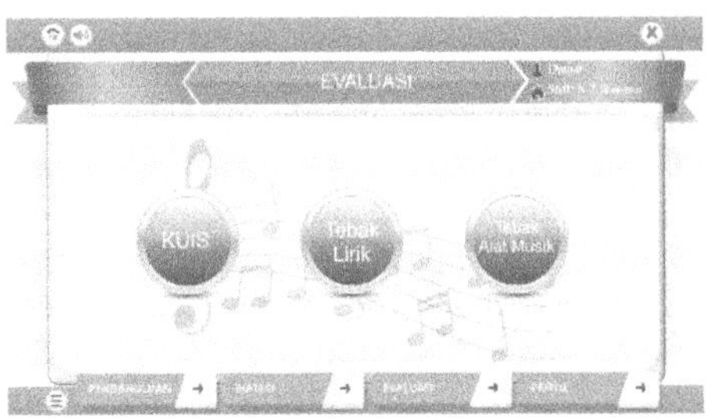

Figure 4. Display evaluation page of multimedia product.

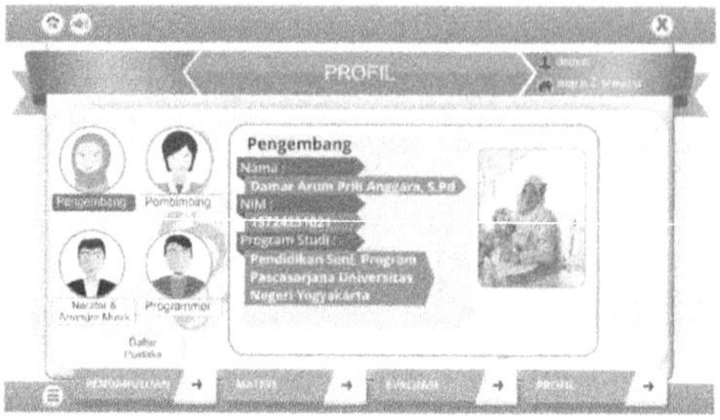

Figure 5. Display of profile page for multimedia product.

learning material related to theoretical material. Guessing the lyrics and guessing melodies are questions in the form of audio pieces of lyrics and melodies taken from Bengawan Solo keroncong song, and are used to measure the extent to which students understand the learning material related to practical material.

The developer profile page contains information about the developer data, thesis supervisor, music arranger and programmer.

After going through the alpha test phase, the beta test and product revision, a summative evaluation was carried out. This summative evaluation was conducted by 34 students and aimed to establish the feasibility of and response of students to the keroncong learning multimedia product in classroom learning. Product assessment was done in terms of appearance, programming, learning, material and attractiveness. The score of all aspects in the summative evaluation produced an average of 4.45 (very good), which means that the keroncong learning multimedia products created are very feasible to use as learning media, and students' responses to the learning multimedia are very good.

In this research and development, the keroncong learning multimedia product that has been developed is considered very feasible to use as learning media, as can be seen from the results of the formative evaluation and summative evaluation. In other words, keroncong learning multimedia can be used by teachers and students as a medium to introduce the theory and practice of keroncong music. Theoretical material in multimedia aids students in learning the history, characters and types of keroncong in an integrated manner. Practise material also aids students in the practice of playing music because it is presented in the form of audio and video tutorials using keroncong musical instruments such as cuk, cak, cello, contrabass, violin, flute and vocals. However, teachers still have a very important role in the learning process. For example, when practising singing keroncong songs, the teacher keeps guiding students in interpreting the songs' notation, melody, lyrics and rhythm correctly.

5 CONCLUSION

Multimedia learning is believed to be a solution for overcoming problems in learning music at school. This research and development generally consists of three stages, which are planning, design and development. The planning stage is carried out by pre-survey and preliminary studies as material for analysing the development needs of multimedia in keroncong learning, so that results can be obtained regarding the condition of the school and the learning of art and culture.

The design stage relates to the selection of the form and content of the learning multimedia. The development stage is the creation of learning multimedia in the form of learning media applications using Adobe Flash CS6 software so as to produce the ultimate multimedia keroncong learning product that includes theoretical and practical material, which is then tested for feasibility. Trials that were conducted to determine the feasibility of the product and the students' response to the keroncong learning multimedia product that has been developed show that it is very feasible to use as learning media. This can be seen from the results of formative evaluations and summative evaluations, and the very good responses from students. The resulting multimedia product is useful for students because it is attractive and easy to use according to the needs of students.

However, this research and development still has several shortcomings, including the use of video and audio in practise material, and not including all related theoretical material. At the trial stage, it was limited to the school where the research was conducted. In practise material, this research has just arrived at understanding students' knowledge of keroncong music, both in singing and playing musical instruments. This research is only limited to testing the feasibility stage of the product produced and further testing has not been carried out to test the effectiveness so that it can be used for further research with a wider trial.

REFERENCES

Daryanto. (2016). *Media Pembelajaran (Edisi ke-2 Revisi)* [Learning media (Revised 2nd ed.)]. Yogyakarta, Indonesia: Penerbit Gava Media.

Dittmar, C., Cano, E., Abeßer, J. & Grollmisch, S. (2012). Music information retrieval meets music education. In *Dagstuhl Follow-Ups* (vol. 3, pp. 95–120). Dagstuhl, Germany: Schloss Dagstuhl-Leibniz-Zentrum fuer Informatik.

Kemedikbud. (2017). *Kurikulum 2013* [2013 Curriculum]. Jakarta: Government of the Republic of Indonesia.

Ruismäki, H., Juvonen, A., & Lehtonen, K. (2013). The iPad and music in the new learning environment. *European Journal of Social & Behavioural Sciences*, *6*, 1084–1096.

Sadiman, A.S. (2009). *Media Pendidikan: Pengertian, Pengembangan, dan Pemanfaatannya* [Educational media: Understanding, developing and utilising it]. Jakarta, Indonesia: Pustekkom Dikbud & PT. Rajagrafindo Persada.

Sukardjo Sari, L.P. (2008). *Penilaian Hasil Belajar Kimia* [Assessment of chemistry learning outcomes]. Yogyakarta, Indonesia: FMIPA UNY.

Tanbeer, S.K., Chowdhury, F.A., Jeong, B.S. & Lee, Y.-K. (2008). CP-Tree: A tree structure for single-pass frequent pattern mining, In T. Washio et al. (Eds.), *Lecture notes in artificial intelligence* (pp. 1022–1027). Berlin, Germany: Springer.

Teen, L.B., bin Haji Ramli, A. & Ying, L.F. (2014). Raising interest with software integration in music notation learning. *Procedia - Social and Behavioral Sciences*, *143*, 379–383.

Trollip, S.R., & Alessi, S.M. (2001). *Multimedia for learning: Methods and development*. Boston, MA: Allyn & Bacon.

21st Century Innovation in Music Education – Astuti et al. (eds)
 ISBN 978-0-367-11122-9

Bringing 21st century music skills to the curriculum through interdisciplinary study

A. Mustikawati & K.S. Astuti
Universitas Negeri Yogyakarta, Yogyakarta, Indonesia

ABSTRACT: The demand to master 21st century skills prompts schools to develop students' skills and competence. The curriculum should be adjusted to accommodate and accomplish this goal. As part of the school curriculum, art is the one of the potential subjects to develop 21st century skills. This paper aims to explain how and why interdisciplinary studies in arts can promote 21st century skills. Several skills that students can develop include critical thinking, technology and information management, collaboration, cultural awareness and broader artistic communication. To accomplish effective interdisciplinary study-based learning, planning is the crucial step. Several aspects to be considered in the planning stage are selecting the appropriate materials, theme, subjects, and activities that accommodate and promote 21st century skills.

1 INTRODUCTION

The rapid development of Technology and Information requires people to employ a strategy in order to survive and compete globally. To fulfill these 'basic' life-skill requirements, the role of school is important to help students acquire the skills needed in the 21st century such as (1) critical thinking and problem solving, (2) technology acquisition, information management and communication skills, (3) collaboration skills and (4) cultural awareness (Saaverda& Opver, 2012; Beers, 2018). Music as part of art education in the school curriculum is not merely about how to sing, how to play an instrument, and other technical matters.

There are many ways to develop music lessons to encourage skill development. However, in this paper, we will focus on interdisciplinary study as an approach that could help high school students develop and improve their skills in relation to an Indonesian art curriculum context. This paper contains three main issues. The first part of the paper includes the concept of art education and interdisciplinary study. The next part deals with how to incorporate interdisciplinary study into the Indonesian arts curriculum which will be preceded by a discussion of the curriculum and music education in detail. Then, the third part comprises the positive impact of interdisciplinary study in developing 21st century skills. This paper is based on different resources such as government policy document, student textbooks, and other relevant literature.

2 ART EDUCATION AND INTERDISCIPLINARY STUDY

There are various definitions of music education and arts education in general. Many objectives and studies conducted by experts and scholars have proved the benefit of art education. From various definitions, objectives, and studies, there is an interesting theory proposed by Swanwick in her work "A Basis for Music Education". He proposed the idea of a comprehensive art experience that consists of *composition*, *audition*, *literature*, *skill*, and *performance* or *CALSP* (Swanwick, 2002). *Audition* concerns about direct interaction with art works in a broader way. *Composing* encourages students to expand their creativity. *Literature study* reinforces students' knowledge about art. *Skill acquisition* is related to the technical aspect of art and *performance* provides a space for students to critically appreciate works of art.

This concept is relevant to UNESCO which proposes three dimensions of art education, namely interaction and knowledge of art and artists, art practice, and art study (UNESCO, 2006). All of these concepts illustrate that the ideal art education should be comprehensive. It spans exploring studying art and appreciation to skill acquisition and composition. In fact, every aspect is actually connected and supports one another. The aspects of audition and appreciation support art study, while composition requires sufficient knowledge of music and vocabulary. Therefore, all the aspects mentioned needed to be developed in order to maximize learning achievement. Furthermore, this concept has great potential in developing 21st century skills through art education.

Using appropriate methods, the concept can be developed as a comprehensive way to learn by improving cognitive, social, and psychomotor faculties. This can be achieved through interdisciplinary study. The term *interdisciplinary study* might sound too theoretic, yet it is a comprehensive study that can be used as the basis of art practice in art education. In addition, interdisciplinary study also benefits other subjects that are involved especially in an academic setting such as a school.

Theoretically, interdisciplinary study can be defined as a way to look at a problem from many points of view and from different subjects (Lattuca, 2001). Integration between subjects is an essential point that differentiates it from similar methods such as multidiscipline (Spelt, Luning, Boekel,& Mulder, 2017). Integration is an ability to connect the knowledge from each subject involved into a whole corpus of meaningful knowledge (Mansilla, 2016). This process involves many skills such as critical thinking, collaboration, and communication (Spelt, et. al,2017; Styron, 2013).

Based on the *CALSP* concept, interdisciplinary study can start from audition or interaction with a work of art. Audition becomes a fertile ground for students to observe a work of art using sensory perception. This process becomes the basis and source of data for interdisciplinary analysis. Next, the observed work of art is studied using interdisciplinary study by involving other relevant subjects and data. After that, the result of the study is employed in practice by using it as a basis for art making. Using these steps, art learning becomes a comprehensive learning technique that accommodates other aspects of art.

3 INTERDISCIPLINARY STUDY TOWARD INDONESIAN ARTS CURRICULUM

According to the Indonesian Senior High School curriculum, art education consists of four branches of art: music, fine art, dance, and theatre. Based on *peraturan menteri pendidikan dan kebudayaan* no.21 tahun 2016 (regulation by ministry of education), there are several competencies that have to be achieved in art education (*Seni dan Budaya* subject) as summarized in

Table 1. Art & culture competence for Senior High School according to regulation (Kemendikbud, 2016).

No	Competency
1	Show curiosity, care for the environment, cooperation, honesty, confidence, and autonomy in relation to art creation
2	Appreciate diversity and the artistic value of arts and cultures
3	Ability to investigate each artistic works and cultural values to identify Their uniqueness, beauty, and aesthetic value
4	Ability to adopt and develop concept, technique, procedure, and media in the process of creating art works
5	Ability to explore concept, technique, procedure, and media in the process of creating art works
6	Examine the diversity and uniqueness of artworks
7	Present the result of the analysis in the form of artworks and cultural studies
8	Create original art works
9	Appraise the diversity and uniqueness of art works
10	Present the result of evaluation in the form of artworks and cultural studies

Table 2. Music Lesson Material based on textbook for X grade students (Kemendikbud, 2017).

Semester	Material	Sub-material
I	Traditional Music	Definition of music Musical symbol and symbolism of music Investigate aesthetic value of music based on its social context Traditional music genre Function of music Music practice
	Music Performance	Introduction & basic concept Music exploration Body movement collaboration (dance in music performance) Compare music
II	Music Performance	Concept of music performance Technical aspect of music performance Performance management Stage performance
	Music Performance	Introduction to music critique criticism Composing music performance critique Present the result

Table 1. These competencies are then elaborated into learning materials such as textbooks listed in Table 2. Table 2 contains music materials for grade X students taken from the 2017 edition of *Seni Budaya* textbook published by the Indonesian government (Kemendikbud - Ministry of Education).

From both tables, it can be seen that the curriculum for music and art in general does not only focus on art practice but also includes cognitive and social process. Connected to the previous *CALSP* concept, this curriculum implements that concept. The interaction with work of art (audition), appreciation, study, and practice are already written in both competencies and textbook. For example, in the chapter of Traditional Music, which consists of several sub-topics such as the definition of music, symbol, and aesthetic value of music that relates to social context, type of traditional music, function of music, and practice of music. From these sub-topics, there is audition or direct interaction with art work, art study, and art practice, which will be continued by *performance* in the next semester. It is interesting to note that theoretic art study does not only discuss the theory of music but also critical discussion regarding art and its social environment. This concept becomes a prototype of comprehensive art learning if conducted appropriately.

Regarding interdisciplinary study, choosing material is essential. The multi-perspective nature of interdisciplinary study requires a special strategy in its realization especially in a school context. Mansilla proposed four epistemological principles in interdisciplinary study that can be used as a basis in creating interdisciplinary learning (Mansilla, 2016). Firstly, the study's framework must be plural in terms of how it can accommodate the different form and nature of the subjects involved. Secondly, the framework of the study must be relevant to the interdisciplinary phenomenon that includes integration. Thirdly, the framework must provide a clear description related to the progress of knowledge in the learning process. Finally, the framework must guarantee quality learning by enclosing a solid basis and relevant standard for each subject involved (Mansilla, 2016).

There are many models that could be used as a tool to develop interdisciplinary learning such as ADDIE, ASSURE and many more. However, based on the principle proposed by Mansilla, we must pay attention to the planning phase because is an important step in interdisciplinary-based learning in school. The topic or materials selected must be relevant to the subjects involved. This point is also related to students' background knowledge in which they need to have sufficient knowledge for the learning to take place. Additionally, the choice of learning activities must enhance the development of 21st century skills such as group activities and discussion. Through group activities, students are expected to be cooperative in collective work becomes a comprehensive learning that accommodate other aspects of art.

4 WHY WE HAVE TO INCORPORATE INTERDISCIPLINARY STUDY IN MUSIC CURRICULUM

In this section, the benefit of including interdisciplinary study in the Indonesian arts curriculum is discussed. The previous chapter demonstrated this kind of study has several benefits related to 21st century skills. Firstly, an interdisciplinary approach in the classroom supports the development of critical thinking, technology and information management, collaboration, cultural awareness and broader artistic communication.

4.1 *Developing critical thinking*

Critical thinking is an ability that is beyond objective thinking. It is the ability to think wisely to view a problem and used higher order thinking skills (HOTS) to study and solve problem (Beers, 2018; Collins, 2014). HOTS is a complex thinking process that needs several skills such as analysis, synthesis, comparison, inference, interpretation, assessment, and reasoning to solve problems (Budsankom, Sawangboon, Damrongpanit & Chuensirimongkol, 2015). Ivie (1998) states that HOTS is a thinking ability that includes three criteria: (a) abstract thinking skills, (b) integrating information, (c) the ability to judge and think logically (Ivie, 1998). Critical thinking could be considered as the foundation for other skills.

Interdisciplinary study can be employed to develop higher order thinking skills. In this method, students are required to relate the knowledge between subjects that have been learned based on the collected data. To do this, students process and analyze data, interpret the data, and integrate the result into the whole. This process is complex and involves analytical skills, assessment, and interpretation towards information and skills to make logical inferences that help students develop their critical thinking skills.

4.2 *Technology and information management*

To perform interdisciplinary study, we need adequate data support since this study involves several disciplines (subject). Therefore, students must deal with relatively large amount of data that should be processed. The advancement of technology enables people to access information easier.

Based on this situation, we assume that this could help students to improve their ability to manage information. This ability includes accessing/gathering information, analyzing information including data/information validity, synthesizing, and data processing (Beers, 2018). In this era, when information are dispersed rapidly, this skill could be considered as basic life skill that everyone should have since not all information are accurate and some even cause dire consequences to society.

4.3 *Collaborative skill*

According to Beers, collaboration is the ability to share and work together effectively to solve problem (Beers, 2018). It can develop learning mechanism in an active way. As mentioned by Child& Saw, collaboration has many advantages for students such as: increasing learning ability, processing information from various sources (expanding perspective) as well as social skill (Child & Shaw, 2015).

Child & Shaw argue that in collaboration, relation between roles and responsibility toward subtask is flexible. This distinguishes collaboration and other teamwork method such as cooperation. It possess six important point: social interdependence, introducing of new idea, cooperation/task division, conflict resolution, sharing of resource, and communicating (Child & Shaw, 2015).

From the given explanation, we can conclude that collaboration skill is more than just working together in a team. There are sharing responsibility, thinking collaboration (integrated thinking between team member), and social development. In socio-historical study, this skill is needed. Wide range of perspective that should be explored is more effective if discussed in a group especially with artful thinking method that will be discussed in the next chapter.

4.4 *Cultural awareness and broader artistic communication*

Cultural Awareness is a state where people are able to understand and respect different culture and also cooperate with people from different cultural background (Beers, 2018). Through interdisciplinary study, students are encouraged to see a problem from various perspectives. This enables students to accept new perspectives particularly through group activities. Avid discussion and cooperation in group helps students to accept and respect the difference between individual members.

How interdisciplinary study can broaden one's artistic ability? As stated by Hauser (2011), work of art is influenced by environment, experience, and knowledge of the artist itself. Interdisciplinary study is one of the approaches that can broaden students' knowledge of art works or cultural phenomenon. Students' increasing knowledge can be a source of inspiration in creating art work not only from the art work itself but also outside it such as the social environment. The knowledge regarding a particular environment is a good new inspiration for students.

5 CONCLUSION

This paper illustrates how interdisciplinary study could help art education as potential subject to promote 21st century skill. It also discusses interdisciplinary study toward Indonesian art curriculum that show that Indonesia curriculum nowadays has promote 21st century skill as well as having potential interdisciplinary study. Establishing interdisciplinary study-based learning requires careful and rigorous planning by considering certain elements such as relevant topic, student's background knowledge or minimum ability, and appropriate activity that enhance 21st century skill as well as interdisciplinary study itself. The analysis and discussion in this paper are based on a literature study. Therefore, future study needs to conduct field study so that more issues can be covered.

REFERENCES

Beers, S. Z. (2011). 21st century skills: Preparing students for their future. (STEM).

Budsankom, P., Sawangboon, T., Damrongpanit, S., & Chuensirimongkol, J. (2015). Factors Affecting Higher Order Thinking Skills of Students: A Meta-Analytic Structural Equation Modeling Study. *Educational Research and Reviews, 10*(19), 2639–2652.

Child. S., & Shaw, S. (2015). Collaboration in the 21st century: Implications for assessment. *Economics, 21*, 2008.

Collins, R. (2014). Skills for the 21st Century: teaching higher-order thinking. *Curriculum & Leadership Journal, 12*(14).

Hauser, A. (2011). *The sociology of art*. Routledge.

Ivie, S. D. (1998). Ausubel's learning theory: An approach to teaching higher order thinking skills. *The High School Journal, 82*(1), 35–42.

Kementrian Pendidikan dan Kebudayaan. (2016) *Salinan lampiran peraturan menteri pendidikan dan kebudayaan nomor 21 tahun 2016 tentang standar isi pendidikan dasar dan menengah* [A copy of the attachment to the minister of education and culture number 21 of 2016 concerning standard content for primary and secondary education].

Kementrian Pendidikan dan Kebudayaan. (2017). Seni budaya: buku guru [*Cultural arts: teacher books*]. Jakarta: Kemendikbud.

Lattuca, L. R. (2001). *Creating interdisciplinarity: Interdisciplinary research and teaching among college and university faculty*. Vanderbilt university press.

Mansilla, V. B. (2016). A Cognitive-Epistemological Foundation. *The Oxford handbook of interdisciplinarity*, 261.
Spelt, E. J. H., Luning, P. A., van Boekel, M. A., & Mulder, M. (2017). A multidimensional approach to examine student interdisciplinary learning in science and engineering in higher education. *European Journal of Engineering Education*, *42*(6), 761-774.
Styron Jr, R. A. (2013). Interdisciplinary Education; A Reflection of the Real World. *Journal of Systemics*, *11*(9), 47–52.
Swanwick, K. (2002). *A basis for music education*. Routledge.
UNESCO. (2006). *Road map for arts education. Retrieved August, 2018 from* http://www.unesco.org/new/en/culture/themes/creativity/arts-education/official-text/road-map/

21st Century Innovation in Music Education – Astuti et al. (eds)

The impact of the UK curriculum on the musicality of Indonesian students

Priyadi & K.S.Astuti
Universitas Negeri Yogyakarta, Yogyakarta, Daerah Istimewa Yogyakarta, Indonesia

ABSTRACT: This study aims to discuss the implementation and impact of the national music curriculum of the United Kingdom (UK), a curriculum that has was designed specifically to stimulate the development of students' musicality in the elementary school of Luqman Al Hakim, Kota Gede, Yogyakarta Territorial Special Regency, Indonesia. The national curriculum emphasises cross-curricular themes, skills and intra-personal learning. In this study, we discuss how the implementation of the UK curriculum in music subjects improves students' musical abilities, particularly the ability of students to apply melody, harmony and rhythm to a song. Through the approach and experimental method, data were obtained by a post-test only control group design. The results showed that through the application of the UK curriculum, students' musicality improved in the areas of applying the right melody, rhythm and harmony to a song.

1 INTRODUCTION

In the outline of the Indonesian elementary school music curriculum, knowledge of traditional music, archipelago music and foreign music is indicated as necessary materials for inclusion in study. The 2013 curriculum has been carried forward into the 2018/19 school-year learning activities, outlining several themes that are to be converted into learning activities that involve all subject areas, including music, dance and theatre arts.

Sekolah Dasar Islam Terpadu International Luqman Al Hakim (SDIT LHI) was founded in 2008 as an extension of the Luqman Al Hakim International Islamic Elementary School. Beauchamp (2010) revealed that before the national curriculum was implemented in England there were no regulations that stipulated music learning requirements for elementary school students. The learning model in SDIT LHI is developed so that all teachers teach their own classes except for a few subjects. According to Biasutti et al. (2015), studying with specialist teachers like this is ideal learning.

SDIT LHI believes that every child has different potential. This school aspires to develop the potential of every child. As stated by Hirzi (2007), the teaching of music is not just the distribution of hobbies and interests of children, it is also an effort to shape the personality of the child. As a medium to realise this, the school adopted curricula from the United Kingdom (UK) (referred to here as the national curriculum) and Project Holistic Integral (PHI), as well as the curriculum from the Indonesian government. These three curricula are combined into one unit of the LHI curriculum.

A connection can be found in the principles of learning and actualising the 2013 Indonesia curriculum, the UK national curriculum and the PHI curriculum. The concept of carrying a theme, for example, means that students are invited to learn to explore a theme that appears in all subjects. Students are thus able to feel the connection of one subject to another. Such a broad learning design allows students to explore a subject in depth. The love of learning is instilled, and learning is not just about obtaining as much information as possible. These three curricula also hone the character of children, developing their curiosity, observation skills, analytical skills and collaboration skills.

Some obstacles were encountered when trying to implement the UK music curriculum. Students in the UK are grouped according to student attainment, while in Indonesia students are grouped according to year of entry. It is certainly easier to teach students with relatively equal abilities rather than a class of students with varying abilities. The next obstacle was the relatively little availability of music materials from the Indonesian government, particularly for one module that required material on music, art, dance and theatre. There was an imbalance when exploring the basic musical capabilities of students in accordance with the national curriculum compared with "local wisdom" teaching. Also, differences in the basic culture of making some material and song references must be adjusted.

The UK curriculum guide lists the target skills students need to achieve at each level: ability and sensitivity in responding rhythmically, melodically and harmoniously, as well as beginning to hone skills to improvise. However, there are difficulties in adjusting to the teaching materials of the UK curriculum. In Western countries, children are used to listening to classical music early on, such as the compositions of Mozart, Beethoven and others, and many children's songs are rooted in classical music, but in Indonesia this is not the case. This provides a challenge when applying materials from the national curriculum because reference songs from this curriculum sound familiar to children. However, some children's songs in Indonesia are structurally dissimilar to reference songs from the UK curriculum.

Schools always try to improve the music curriculum. Some of the steps taken are to adopt several reference songs from abroad and teach them in the classroom or producing teacher's songs to be recorded and documented. One of them is the song 'Hemat Energi', which is the theme song in the learning theme 'Sumber Energi'. Improvements are carried out continually, and more in-depth research is required to realise more ideal learning according to the objectives of SDIT LHI.

The purpose of this article is to discuss the effectiveness of the implementation of the UK curriculum in improving the musicality of students at SDIT LHI. This is so that it can become a reference in the preparation of the music learning curriculum in Indonesia in the future.

2 THEORETICAL FOUNDATION

2.1 *Implementation*

In this study, the UK national curriculum for key stages 1 and 2 was implemented, and its effect on the musicality of the music students of SDIT LHI was measured. Adjustments were made to the curriculum to accommodate the cultural differences between Indonesian children and children from the UK. More local Indonesian songs were included to adhere to one of the aims of art music learning to promote national cultural identity. Thus, local Indonesian songs such as archipelago, regional and national songs were included. Other than that, the framework and targets of the UK curriculum were maintained.

2.2 *Curriculum*

There are two definitions of curriculum. The first is a number of content plans containing a number of learning stages designed for students, with instructions from educational institutions whose contents are static and which must contain dynamic processes and competencies. The second encompasses all experiences under the guidance and direction of educational institutions that provide learning conditions (Ruhimat, 2011).

Important aspects of the music curriculum are the ways in which it contributes to children's personal, social, mental and physical development. Mind and body are coordinated in the act of singing, playing instruments, collaborating and reading all the signs in the notation (O'Callaghan, 2003). In each curriculum issued every certain period, generally the main points emphasised are the activities of students (Ridwan, 2017).

An activity or a project will run well and have clearer direction if it has been designed beforehand. The programme will run more effectively and efficiently. The time needed is shorter and costs are less, but the results are maximal.

Ahid (2006) revealed three concepts about the theory and practice of applying a curriculum:

1) The concept of curriculum as substance: A curriculum is seen as a plan of learning activities for students at school, or as a set of goals to be achieved. It can also refer to a document that contains the formulation, objectives, teaching materials, teaching and learning activities, and evaluation.
2) The concept of curriculum as a system: A curriculum is part of the school system, education system and even the community system. A curriculum system includes a personnel structure and work procedures on how to compile, implement, evaluate and accomplish a curriculum.
3) The concept of curriculum as a field of study: Here, field of study refers to that of curricula experts and education and teaching experts. The aim of this concept is to develop knowledge about curriculum and curriculum systems.

On the basis of these definitions, it can be concluded that broadly speaking the curriculum is a learning design, which includes the structure of subjects that form the substance and education system.

2.3 *Learning*

According to Ruhimat (2011), learning is an intentional activity carried out by individuals so that there is a change in ability by learning among children who were previously unable to do something, or children who were not skilled at something. Ruhimat (2011) also states that learning is an effort made by a teacher or educator to teach students who are learning. In formal education (school), learning is a task that is borne by the teacher, because it is prepared by professional staff. Learning in schools is increasingly developing from traditional teaching methods to modern learning systems. In'am (2011) suggests there are eight measures of the effectiveness of learning among students:

1) Listen and pay attention to the teacher's explanation.
2) Find strategies to understand problems.
3) Find strategies to solve problems.
4) Convey questions and ideas or ask for explanations from friends or teachers.
5) Respond to the teacher or provide assistance to friends.
6) Improve work results or write something new from the results of the discussion.
7) Summarise how to understand problems or how to solve problems.
8) Engage in activities outside the learning process (i.e. activities not related to learning, such as playing, disturbing friends, etc.).

Learning is a process of interaction between various components of the teaching system. These components are students, teaching materials, media, tools, procedures and learning processes.

2.4 *National curriculum (UK)*

Hawley (2018) revealed that the UK national curriculum was introduced by the Thatcher government after the issuance of the 1988 Education Reform Act. A comprehensive record of its initial history can be found in the 2009 Committee, School and Family Reports, National Curriculum.

In 1987, the UK Ministry of Education and Science, led by Kenneth Baker MP, issued a consultation document that established the rationale for the national curriculum. It identified four broad principles and basic intentions: establishing rights on the basis of a broad and balanced curriculum, increasing school accountability, increasing curricular coherence and aiding 'public understanding' of a school. Furthermore, in 1989 the national curriculum was introduced in elementary schools, with implementation continuing into the mid-1990s. Meanwhile, in 1993, Sir Ron Dearing reviewed the national curriculum response to teachers' observations that the curriculum was "too hard". A revised national curriculum was introduced in 1995, with less prescribed content and new changes to test settings.

In 1996, two pilot support projects in counting and literacy were an introduced. This was brought forward as the 'National Literacy and National Calculation Strategy' by the forthcoming Labour government. Moreover, in 1999, a major review of the curriculum supervised by the Qualifications and Curriculum Authority resulted in further slimming of the determined content and introduction to the statement of purpose.

Review of the secondary curriculum occurred in 2005, again with the aim of reducing the content specified and generating more emphasis on cross-curricular themes, personalised skills and learning. Finally, in 2007, a review of "roots and branches" of the main national curriculum was announced by the government, which was then carried out from 2008 by Sir Jim Rose, and findings were published in April 2009.

3 RESEARCH METHOD

This research was conducted at SDIT LHI, Yogyakarta. The subjects of this study were all students of Grade IV odd semester 2018/19, with a total of 51 students distributed across two classes. The approach in this study uses the post-test only control group design experiment. Class IVa received learning treatment with the UK curriculum, while Class IVb was a control group with conventional learning methods. This study aims to look at the influence of art music learning on student musicality when the local/original curriculum is combined with the UK curriculum. Data were obtained from the post-test assessment carried out at the end of the learning period.

The section of the national curriculum used was Unit 9 of the 1999 UK curriculum music learning guide, themed "Animal Magic", for third- or fourth-year students. In this unit, students and teachers continue to develop the abilities honed in the previous unit, including singing, listening and responding to music; learning to recognise rhythm patterns by moving, stepping or jumping; and experience combining sounds (tones). In this unit, children can use words and phrases related to duration, tempo and notation.

The learning aims for this unit are for students to understand how musical elements can be used together and recognise some descriptive elements such as lyrics based on animal sounds. The song chosen in this study was 'Burung kakaktua'. This song has a relatively simple melody according to the level of students' abilities. Structurally and thematically, it also represents the design of the government curriculum and the UK curriculum.

In the control group, the material was taught conventionally using the lecture method and the imitation method. In the experimental group, the approach used was to implement the guidelines of the UK national curriculum. This included feeling the rhythm pattern with steps or jumping, expressing the lyrics with hand movements and responding spontaneously to the chosen tone (improvisation).

The results of the pretest were obtained after students received learning materials using conventional methods. After learning took place on three occasions, students were given a test with the review component, testing their melodic abilities, rhythmic abilities and harmonious abilities. The performance test (practice) involved singing the song 'Burung kakaktua' in groups. The practice test instruments have indicators that state students' musical ability. These indicators are the accuracy of melody, rhythm and harmonisation between sound part 1 and sound part 2.

Data were analysed from the results of the post-test score using two average similarity tests. First, the pre-condition test was carried out, that is, the normality test and homogeneity test. Table 1 shows the results of the normality test.

The results of the normality of the musical data test students in Table 1 showed that the post-test data were not normal, whereas in the control group the results were normal. It can

Table 1. Results of the normality of post-test data.

	$X2_{test}$	$X2_{table}$
Experiment (IVa)	8.13	7.81
Control (IVb)	5.78	7.81

be concluded that the data were obtained from populations that are not normally distributed, so testing the homogeneity of variants (negligible) is not required. The next step is testing the hypothesis. For data originating from populations that are abnormally distributed, non-parametric tests, such as the U-test or Mann–Whitney test, were used.

4 RESULT AND DISCUSSION

There were 25 students in the experimental group and 26 in the control group. Testing with different *t*-tests was conducted to see whether there were differences in the results of the singing practice test between the control group and the experimental group. The singing practice test score of the experimental group was higher than that of the control group. Thus, it can be concluded that the UK curriculum can improve the musical abilities of students of SDIT LHI.

The results of the *t*-test using SPSS statistical software showed a value of 1.876. For the real limit of 5% and $df = 28$, the critical point in the *t*-distribution is obtained at 1.657. Because the data point t is greater than the critical point, it can be concluded that the null hypothesis (H_0) is rejected and the musicality test score of the experimental group is higher than that of the control group.

In the experimental group, given learning treatment using the UK method and curriculum, it was found that 22 students received good physical scores. This was proven by the fact that they could sing the song 'Burung kakaktua' with the right melody and rhythm and could divide the group into two voices according to the teacher's direction. This shows that the UK curriculum is able to improve students' musical abilities. To establish the correlation of musicality and scores obtained by students, it is necessary to do an *eta* correlation test. This is because the data on musicality are nominal. Testing is done with musical encoding: good musical number 2, less good musical number 1, and medium musical number 0.

The *eta* correlation test produced a score of μ+0.7439 compared with the standard association between two different scale variables. This number is included in the data group with strong or high associations. This means that there is a strong relationship or association between the musicality of the students and the score.

The significance value of the data tested was $F = 4.0328$. If this significance value is compared with the value in the testing using $\alpha = 0.05$ with *df* for (N–k) of 22, then a value of 3.93 will be obtained. The value in the table is smaller than the calculation value for the data. It can be concluded that H_0 is rejected and there is a significant relationship between student musicality and musical practice test scores.

The results of processing the data above show that there is a difference between the musical ability score of the control group and the experimental group. Based on the SPSS calculation of these differences, it was found that H_0 was rejected, which means that the musicality score in the experimental group was significantly higher than the musicality score in the control group.

Table 2. Test results for student musicality.

	Musicality			Total
	Medium	Less Good	Good	
Number of students	4	3	18	25

5 CONCLUSION

This study shows that there is positive effect in applying the UK curriculum in music teaching on students' musical ability. The score of the experimental group was higher than that of the control group, but the results were not that significant. This is because musical ability is not an aspect that can be greatly improved in a short period of time, and the treatment must be applied several times for any dramatic results to emerge. This research shows that there is a need for this method to be applied regularly and continuously. Therefore, this research is expected to be a contribution to the study of how to improve the musical abilities of students in learning music.

REFERENCES

Ahid, N. (2006). Konsep dan Teori Kurikulum dalam Dunia Pendidikan [Curriculum concepts and theories in education]. *ISLAMICA: Jurnal Studi Keislaman, 1*(1), 12–29.

Beauchamp, G. (2010). Musical elements and subject knowledge in primary school student teachers: Lessons from a five-year longitudinal study. *British Journal of Music Education, 27*(3), 305–318.

Biasutti, M., Hennesy S. & de Vugt-Jansen, E. (2015). Confidence development in non-music specialist trainee primary teachers after an intensive programme. *British Journal of Music Education, 32*(2), 143–161.

Hawley, R. (2018). Academic writing as embedded skills acquisition for transition to higher education: An example from a UK university classics department. *Journal of Academic Writing, 8*(2), 67–77.

Hirzi, A.T. (2007). Mengkomunikasikan Musik kepada Anak [Communicate music to children]. *Mediator: Jurnal Komunikasi, 8*(2), 201–201.

Indonesia, R. (2009). *Committee, School and Family Reports, National Curriculum*. Jakarta: Ministry of Education.

In'am, A. (2011). Peningkatan Kualitas Pembelajaran melalui Lesson Study berbasis Metakonisi [Improving the learning quality through lesson study-based metacognition]. *Jurnal Salam, 12*(1), 123-131.

O'Callaghan, E. (2003). On implementing the new music curriculum in primary schools. *Irish Educational Studies, 22*(3), 45–46.

Ridwan, R. (2017). Pembelajaran Seni Musik Tematik sebagai Implementasi Kurikulum 2013 [Thematic music learning as 2013 curriculum implementation]. *Ritme, 2*(2), 18–28.

Ruhimat, T. (2011). *Kurikulum dan Pembelajaran* [Curriculum and learning]. Jakarta, Indonesia: PT Raja Grafindo Persada.

21st Century Innovation in Music Education – Astuti et al. (eds)
© 2020 Taylor & Francis Group, London, ISBN 978-0-367-11122-9

Increasing sight-reading ability through implementation of the Super Sight-Reading Secrets programme

Z.A.W. Putra & K.S. Astuti
Universitas Negeri Yogyakarta, Yogyakarta, Daerah Istimewa Yogyakarta, Indonesia

ABSTRACT: The objective of this research was to investigate and analyse a programme, called the Super Sight-Reading Secrets Programme by Howard Richman, to improve the sight-reading abilities of students in the PIM Piano IV class at the Department of Music Education, Universitas Negeri Yogyakarta. This research involved Classroom Action Research (CAR) that was conducted in two cycles using the model of Kemmis and McTaggart (2002). The subject of this research was the PIM Piano IV students, consisting of 17 students. The data collection techniques used were participatory observation, documentation and testing. The data analysis techniques used were qualitative and quantitative approach. The results show that there is an improvement in the sight-reading skills of the students through the use of the Super Sight-Reading Secrets Programme by Howard Richman. The results show progressive increases in the mean score from the pre-cycle at 68.12 to cycle I at 77.26, followed by cycle II at 78.44.

1 INTRODUCTION

In this modern era, music is in demand and favoured by the wider community. This is shown in terms of its meanings. Music is one of the sciences and one of the fields in art, in the form of sounds that combine in an organised sequence of rhythms, melodies and harmonies that reflect the creator's feelings and emotions. In addition, the wider community is not concerned with age. Learning music can be started at all ages, from children to the elderly. In learning music an institution is needed to learn well and correctly. One of the ways to study well and correctly is by learning music via three education tracks. Education in Indonesia is divided into three tracks, namely formal, informal and non-formal. To learn music well and correctly, one institution for learning music is the Department of Music Education, Faculty of Language and Arts at Yogyakarta State University. In the Department of Music Education, courses are carried out both in theoretical and practical terms. One of the practical subjects taught is Major Instrument Practice (PIM). PIM is a practical course consisting of several instruments selected according to the talents and interests of the students. One instrument that students can choose in the course is the Major Instrument Practice (PIM) Piano. The PIM Piano course is divided into four parts. One of the objectives is that students are able to play or sing musical works from one era by paying attention to music techniques, practising, and interpretation and expression of several composers in one era. According to a preliminary study conducted through interviews with students of PIM Piano IV, when students began learning they were informed that they should have the ability to read beam notation material precisely and accurately. However, students still had deficiencies in mastering reading beam notation. In addition, students also lacked mastery in tempo and dynamics when they played free songs in PIM Piano learning. PIM Piano learning is held once a week in a face-to-face session and students are required to master techniques by perform songs. However, limited face-to-face learning time is one of the factors that prevents students from mastering techniques, practising and performing the music and song. Besides limited face-to-face sessions, PIM Piano learning deficiencies are due to the lack of time devoted to the practice of reading beam notation.

Thus, the ability to read beam notation directly is important for PIM Piano IV students because PIM Piano IV learning materials use beam notation. By instilling the habit of reading beam notation, the ability to read beam notation directly by PIM Piano IV students can increase their reading abilities. Therefore, with the implementation of the Super Sight-Reading Secrets Programme developed by Howard Richman for students of PIM Piano IV, it was expected that the programme would help students to overcome the problems that arose and that reading beam notation directly or sight-reading skills would improve.

2 LITERATURE REVIEW

2.1 *Sight-Reading*

Skills are identical to an activity. Syah (2015, p. 121) said that that skills are defined as activities related to nerves and muscles that appear in physical activities such as writing, typing, and exercise.

Green (2011, p. 4) said that skill is one of the words in social science, which is expressed in language that has meaning and synonyms that are truly like abilities, skills, and talents. Furthermore, Soemarjadi and Zahri (1991, p. 2) stated that skills are behaviours obtained through certain learning stages. It can be said that skills are abilities, talents in doing something or tasks in the form of verbal actions, thoughts of data, people, or objects that require coordination of muscle movements or psychomotor actions. Sight-reading is also known as *prima vista.* Richman (1986) said that sight-reading is the skill of reading music notation directly. Maydwell (2007, p. 2) stated that sight-reading is the ability to display music when reading scores first. This is in line with Pace (1999, p. 1) who said that sight-reading is the ability to read and play music for the first time, for example without preparing to learn the work being played. In addition, sight-reading is a very important provision for musicians. Moreover, Wristen (2005, p. 44) stated that sight-reading is an important and inseparable part of the musical experience of every musician. Meanwhile, Lehmann and Kopiez (2009, p. 344) define sight-reading as a skill in appearance, both vocal and instrumental or *prima vista.*

Based on these opinions, it can be said that sight-reading is the ability to read music notation that has not been seen before. In other words, sight-reading is reading music notation directly.

2.2 *Beam notation*

The beam notation comes from the Dutch language, namely the expression *noten balk*, which means music notation which is in five horizontal lines that can be filled by pitch. This is similar to what was expressed by Kamien (1980, p. 41), who stated that beam notation is a system of music that is written so that tone and rhythmic accuracy can be related. Beam notation is a form of musical writing with five straight lines, in order to indicate high and low tones. In line with Kamien, Harder (1995, p. 17) stated that music notation is the writing of symbols which describe the tone. Meanwhile, Sukohardi (2015, p. 5) said that the notation of the beam indicated by the value of notation is the low height and the length of the short note indicated by the sign.

According to the statement, it can be said that the beam notation is a music system which is depicted by a sign which is arranged into five lines which are called pitches, and can be placed vertically or horizontally so that the tone and rhythmic tone can be interconnected.

The beam notation is divided into various components. The components include: 1) notation value and rest sign; 2) staves; 3) key (clef); 4) barline; 5) dynamic; 6) tempo.

3 MATERIAL & METHODOLOGY

The method used in this study is Classroom Action Research (CAR). According to Hendriana (2014, p. 31), CAR is a systematic study conducted by teachers in an effort to improve practices in education, by taking practical action and reflection of these actions. In addition, according to McNiff (in Kusumah & Dwitagama, 2010, p. 8), looking at the nature of CAR is a form of reflective research conducted by the teacher themselves, whose results are used as tools for developing teaching skills.

This research was conducted at the Department of Music Education, Language and Arts Faculty, Universitas Negeri Yogyakarta, Indonesia. The reason that this research place was selected is because the facilities in the Department of Music Education are quite adequate. In addition, PIM Piano IV students in the Department of Music Education still lacked sight-reading skills, thus hampering the smooth playing of the piano in PIM Piano learning.

In this study, the subjects were 17 students enrolled in PIM Piano IV courses – 12 male students and five female students. PIM Piano IV students were selected because these students did not major in sight-reading training and had difficulties in sight-reading skills.

This research was designed in two cycles. Each cycle consisted of four stages: 1) planning; 2) action; 3) observation; 4) reflection/evaluation. The second cycle was a continuation of the first cycle phase, because in the first cycle there were still some obstacles and shortcomings. This activity was carried out with the aim to get better results in the second cycle compared to the first cycle, and success was determined in accordance with the indicators of research success. The data collection techniques used in this study included: 1) participatory observation; 2) documentation; 3) testing. The research instrument was an assessment test sheet. Practice test sheets were given to the students in the form of assessment aspects, such as running, rhythm and melody. The data analysis technique in this study involved qualitative descriptive analysis. Qualitative descriptive analysis describes the success of actions that have been carried out from the pre-cycle to cycle I and through cycle II, followed by evaluation. In addition to qualitative descriptive data analysis, quantitatively descriptive analysis from observation results were processed by giving a value to each observed aspect using a scoring range of 81–100 (very good), 61-–80 (good), 41–60 (adequate), 21–40 (poor), and 1–20 (very poor).

4 RESULTS AND DISCUSSION

4.1 *Results*

The research began with pre-cycle activities. Pre-cycle activities were undertaken with the aim to determine the level of sight-reading skills of PIM Piano IV students before further learning actions were taken. Pre-cycle activities were held once, on 20 March 2017. From the results of the pre-cycle action activities, five of the 17 students met the success criteria. The average result of sight-reading skills in the pre-cycle action was 68.13. This result shows that students had not met the success criteria within the sufficient value range (71–80).

The implementation of cycle I actions was the beginning of the classroom action research. The implementation of cycle I actions included planning, action, observation and reflection. The implementation of the first cycle was conducted four times, on 27 March, 3 April, 10 April and 17 April 2017. The sight-reading materials in each meeting were different. For the first to fourth meetings, materials from G. Schafer op. 45 nos. 25, 22, 19 and 23 were given. During the action stages in this study, before playing, sight-reading materials were provided using steps from Howard Richman's Super Sight-Reading Secrets Programme. In addition, another method that was used was the lecture and demonstration method. Based on the observation results and reflections on the actions of cycle I, among all of the students, seven students did not sufficiently meet the standard criteria of (71–80). The results of the average value in the first cycle action was 77.26. As a percentage increase from the pre-cycle, the average value increased by 13.2% in cycle I. The implementation of cycle II actions was a follow-up action from cycle I. The second cycle action was carried out as an effort to improve the cycle I action. The implementation of cycle II actions was generally similar to cycle I, which included planning, action, observation and reflection. The difference is that the steps taken were adjusted according to the input and suggestions from collaborators. In addition, by reflecting upon the first cycle, the average value did not meet the criteria for action success. The implementation of cycle I was carried out three times, on 15 May, 22 May and 29 May 2017. Sight-reading material in each meeting was different. For the first to fourth meetings, material from Czerny op. 599 vol. 146 nos. 21, 24 and 25 were given. The second cycle planning stage made improvement possible from cycle I, so that the students' sight-reading skills increased. The emphasis on the material given in cycle I was on the melody aspect, while in cycle II the chord aspect was emphasised. During the action stage in this study, before playing sight-reading material the students were given steps from the Super Sight-Reading Secrets Programme. In addition, the lecture and demonstration method was used. Based on the results of observations and reflections on the second cycle action, all students met the standard criteria sufficiently (71–80). The average value in the second cycle action was 77.26. As a percentage change from cycle I, the average value of students increased by 1.5%. The assessment results are shown in Figure 1.

4.2 *Discussion*

Based on the results of the research, the implementation of Howard Richman's Super Sight-Reading Secrets Programme did improve the sight-reading skills of PIM Piano IV students at the Department of Music Education, Universitas Negeri Yogyakarta. The improvement can be seen from the results of student scores in sight-reading skills, from pre-cycle activities through every meeting in the first cycle and second cycle actions, and at the end of the first cycle and second cycle. The average score from the pre-cycle was 68.12 and the first cycle was 77.26, so that the increase from the pre-cycle to the first cycle was 9.13 or a 13.2% increase. The average value of the first cycle was 77.26 followed by the second cycle at 78.44, showing an increase from cycle I to cycle II of 1.17 or a 1.5% increase. Furthermore, from the results of the average score in the pre-cycle to the cycle II action there was an increase of 10.31 or a 15.1% increase. The results of the study show that there was an increase in sight-reading skills of PIM Piano IV students. This is evident in the success of actions taken in cycle I and cycle II, by applying the Super Sight-Reading Secrets Programme. Therefore, the application of the Super Sight-Reading Secrets Programme successfully improved sight-reading skills of PIM Piano IV students at the Department of Music Education, Universitas Negeri Yogyakarta. The researchers suggest that students can use Howard Richman's programme, not only for PIM Piano learning but also for learning other instruments and improving sight-reading skills.

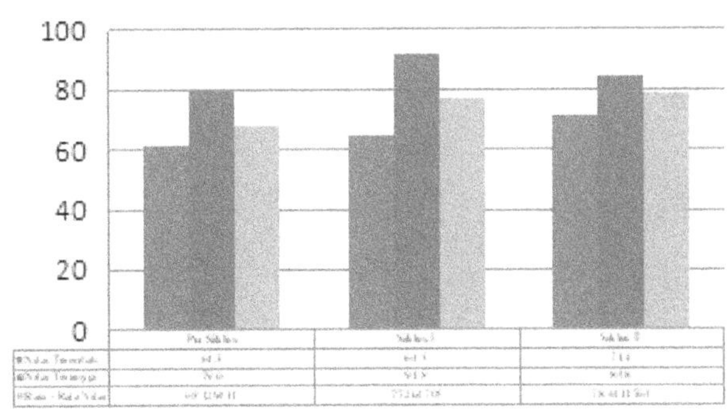

Figure 1. Graph of sight-reading assessment results for PIM Piano IV students: pre-cycle, cycle I and cycle II.

5 CONCLUSION

Based on the results of the research that have been obtained, it can be concluded that the sight-reading skills of PIM Piano IV students improved through the implementation of the Super Sight-Reading Secrets Programme by Howard Richman. This is evidenced by the data results obtained in each cycle. In the first cycle, of the 17 students, ten students met the standard of research success with a score of ≥71. The first cycle showed an increase in average value to 77.26 or 13.2%. In the second cycle, all 17 students met the standard value of the research success, all scoring above 71. Cycle II showed an increase in the average score, rising to 78.44. From cycle I to cycle II there was an increase in sight-reading skills of 1.5%. When viewed as a whole from pre-cycle to cycle II, an increase of 15.1% was obtained. Therefore, PIM Piano IV students met the success criteria for the action taken in cycle II.

REFERENCES

Green, F. (2011). *What is skill? An inter-disciplinary synthesis.* LLAKES Research Paper 20. London, UK: Centre for Learning and Life Changes in Knowledge Economies and Societies.

Harder, P.G. (1995). *Basic material in music theory: A programmed course.* Boston, MA: Allyn & Bacon.

Hendriana, H. (2014). *Panduan Bagi Guru Penelitian Tindakan Kelas Suatu Karya Tulis Ilmiah* [Teachers' guide to classroom action research]. Bandung, Indonesia: PT Refika Aditama.

Kamien, R. (1980). *Music: An appreciation.* New York, NY: McGraw-Hill.

Kemmis, S., & Taggart, M. (2002). R. 1988. *The action research planner.*

Kusumah, W., & Dwitagama, D. (2010). Mengenal penelitian tindakan kelas[Classroom Action Research: an Intriduction]. *Jakarta: PT Indeks.*

Lehmann, A.C. & Kopiez, R. (2009). Sight-reading. In S. Hallam, I. Cross & M. Tahut (Eds.), *The Oxford handbook of music psychology.* Oxford, UK: Oxford University Press.

Maydwell, F. (2007). *Sight reading skills: A guide for sight reading piano music accurately and expressively.* Perth, Australia: New Arts Press of Perth.

Pace, R. (1999). *The essentials of keyboard pedagogy: Sight-reading and musical literacy* (p. 1). East Greenbush, NY: Lee Roberts Music Publications. Retrieved from http://www.iptfonline.org/mono-1.pdf.

Richman, H. (1986). *Super sight-reading secrets.* Reseda, CA: Sound Feelings Publishing.

Soemarjadi, M.R. & Zahri, W. (1991). *Pendidikan keterampilan* [Skills education]. Jakarta, Indonesia: Depdikbud.

Sukohardi, A. (2015). *Teori Musik Umum*[Music Theory] (Revised ed.). Yogyakarta, Indonesia: Pusat Musik Liturgi.

Syah, M. (2015). *Psikologi Belajar*[Psychology of Learning]. Jakarta, Indonesia: Rajawali Pers.

Wristen, B. (2005). Cognition and motor execution in piano sight-reading: A review of literature. *Applications of Research in Music Education, 24*(1), 44–56.

21st Century Innovation in Music Education – Astuti et al. (eds)

The effect of collaborative learning on the *Rinding Gumbeng* music skills and appreciation

V.E. Dewi
Universitas Negeri Surabaya, Surabaya, West Java, Indonesia

ABSTRACT: The objective of this study was to discover the significant effect of the *Rinding Gumbeng* music skills and musical appreciation at SMK Negeri 2 Wonosari Gunungkidul. This study used a quasi-experimental method with a pretest/post-test control group design. The data were collected by way of a test and questionnaire. The data were analysed by using a Multivariate Analysis of Variance (MANOVA) test. The results showed that there were significant differences of the *Rinding Gumbeng* music skills and musical appreciation between students who were taught with collaborative learning and those who were taught with personal learning, at SMK Negeri 2 Wonosari Gunungkidul. This was indicated by the post-test results of the MANOVA test with the Hotelling's trace value of 0.000. It can be concluded, therefore, that there is a significant effect of the implementation collaborative learning on the students' *Rinding Gumbeng* music skills and their musical appreciation.

1 INTRODUCTION

Art education is one of the fields of science in education related to aesthetic education. It is said so because the scope of art education has several important aspects. They include creativity, skills, and sensitivity. The implementation of art education in Indonesia still looks minimal and apprehensive. This can be seen from the many schools and educational institutions in Indonesia that have not incorporated the activities of art into the learning process.

The implementation of art education in Indonesia is still not being given much attention by the government and society. This is due to the tendency of most Indonesians to consider that art education is one part of education that is not so beneficial for the development of children. It is the tendency of most Indonesian society to perceive that exact-science education is far more important for the development of children.

Based on the present condition of the implementation of art education in Indonesia, there should be an understanding of the community in relation to the insights and knowledge of the benefits of art education, which need to be embedded and applied through the development of formal, non-formal, and informal education. Art education in this context is an education oriented towards the growth and development of cultural awareness. The implementation of art education can use literature, dance, art, theatre, or music.

Music education through formal educational institutions lies in the sphere of art and culture subjects and extracurricular activities in schools. Based on the syllabus of art and culture subjects from basic education to higher education and vocational education, the function of art and culture subjects in school is to develop the potentials, skills, and creativities in the arts and cultural arts of the students. Related to the process of music teaching and learning activities in school, music learning activities can be a means of incorporating traditional music and modern musical.That is what has happened in the learning activities of *Rinding Gumbeng* music at SMK Negeri 2 Wonosari Gunungkidul.

SMK Negeri 2 Wonosari Gunungkidul is one of the formal education units under the auspices of general and vocational education. Although this vocational school is based on an engineering programme, art and culture subjects are still included in the learning–teaching process. It is appropriate for students of SMK Negeri 2 Wonosari Gunungkidul to be familiar with the diversity of arts and cultures through the learning of arts, dances, theatre arts, and music arts. The inclusion of arts and culture subjects in the teaching and learning activities of arts can be seen from the learning activities of *Rinding Gumbeng* music in this school.

SMK Negeri 2 Wonosari Gunungkidul is one of the vocational schools in formal education institutions in Gunungkidul applying *Rinding Gumbeng* music in the process of music learning activities in art and culture subjects. There are some factors that cause *Rinding Gumbeng* music to be applied in the process of learning activities of art and culture subjects. One of the factors is the implementation of *Rinding Gumbeng* music is based on the obligation of the syllabus and existing curriculum that students be required to appreciate, demonstrate, and express local music. Another factor is because the music of *Rinding Gumbeng* is one of the original traditional musical streams of Gunungkidul, which it is necessary to preserve.

Rinding Gumbeng music learning activities in SMK Negeri 2 Wonosari Gunungkidul in arts and culture subjects have been conducted since 2014. This has been quite a long time; however, the associated learning processes still face some problems and obstacles.

Based on the previous study conducted between 7 to 14 January 2016, which was done by observing the condition of the teaching and learning process of *Rinding Gumbeng* music, it can be seen that there are still many obstacles and problems. These include the lack of variation and innovation in the model of learning applied by teachers. Another obstacle is the fact that the students' interest and appreciation of *Rinding Gumbeng* music are still very low. It is shown from the attitudes of students who do not appreciate *Rinding Gumbeng* music, and in the lack of enthusiasm and proactiveness of students in attending the teaching and learning activities.

Another obstacle is the lack of the students' appreciation of *Rinding Gumbeng* music in the process of teaching and learning activities in art and culture subjects. This is shown by the number of students who have an *a priori* thought that *Rinding Gumbeng* music is an ancient music that is played only by parents and elderly people. The condition causes most students to be more interested in learning modern music than traditional music.

Problems can also be highlighted from students' report cards and interviews with teachers of art and culture subjects during the implementation of preliminary studies. The scores of the students skills in playing *Rinding Gumbeng* music have not significantly increased from year to year during the past two years. One of the factors that causes this tendency is the lack of the students' appreciation of the music. This is indicated by the condition of the students who are not serious and not interested in learning and playing *Rinding Gumbeng* music. Indirectly, these conditions cause the students' skills in playing *Rinding Gumbeng* music not to increase significantly.

Obstacles and problems that exist in the process of teaching and learning activities of these arts and culture subjects have a great influence on the students' development. These influences include the lack of students developing the abilities and potentials that exist, the lack of the students' knowledge of traditional music, and the lack of interest in and appreciation to traditional music. The present study tries to present a treatment through an alternative learning model in SMK Negeri 2 Wonosari Gunungkidul, which is the application of collaborative learning. Based on the description, this research is intended to observe and find out the effectiveness of the use of collaborative learning of *Rinding Gumbeng* music in the subjects of art and culture in SMK Negeri 2 Wonosari Gunungkidul.

2 METHOD

This study used the quantitative method of the quasi-experiment type. The experimental design used in this study was a pretest/post-test control group design (Campbell & Stanley, 2015, p. 25). The subjects were all 75 students of class X at SMK Negeri 2 Wonosari Gunungkidul. The sample consisted of 63 students, 32 students in the experimental group and 31 in the control group. Sampling used a simple random sampling technique; meanwhile, the sample size was determined by the Slovin technique. The research data were collected through test and questionnaire techniques. The research instruments were a test of music-playing skills and a music-appreciation questionnaire. The validity of the instruments in this study was shown by content and construct validity types. The content validity used Aiken's V validity index, while the construct validity used an exploratory factor analysis. Meanwhile, the reliability estimate was shown by the Cronbach's alpha score and the inter-rater reliability of Cohen's kappa technique.

The validity of the experiments is expected to maximise the research and avoid the existence of internal and external influencing factors. Factors affecting the internal and external validity of the study need to be controlled so that the results of the dependent variable are really the impact of treatment, and can therefore be generalised. The eight internal validity factors (Campbell & Stanley, 2015, p. 8) are maturation, history, regression, selection, testing, instrumentation, mortality, and interaction of selection and maturation. Meanwhile, the five factors that interfere with external validity (Campbell & Stanley, 2015, p. 8) are the interaction of testing and X (interaction between manipulation and testing), the interaction of selection and X (interaction between manipulation and selection), reactive arrangements, multiple-X interference, and multiple treatment interfaces.

3 RESULTS

3.1 *Description of research results*

3.1.1 *Descriptive statistics analysis of music-appreciation questionnaire*

The average pretest score of the experimental group was 111.59, while that of the control group was 114.87. The average post-test score of the experimental group was 115.5, while the control group's average post-test score was 94.29. The experimental group experienced an increase of 3.91 (115.5 − 111.59). Meanwhile, the control group experienced a decrease of 20.58 (114.87 − 94.29). The increase of musical appreciation of the experimental group was shown by the percentage of the criterion A of 59.38% with the frequency of 19 students, rising to 78.12% with a frequency of 25 students. Meanwhile, the decrease of the students' musical appreciation in the control group was shown by the percentage of the criterion A of 77.42% with the frequency of 24 students, dropping to 12.90% with the frequency of 4 students.

3.1.2 *Descriptive statistics analysis of the music-skill test*

The average pretest score of the experimental group was 24.9271, while the average pretest score of the control group was 24.4728. The average post-test score of the experimental group was 37.2083 and that of the control group was 26.1290. The experimental group experienced an increase of 12.2812 (37.2083 − 24.9271). The control group experienced an increase of 1.6562 (26.1290 − 24.4728) at the time of pretest, which rose to 26.1290 at the time of post-test. The improvement of the students' music-playing skills in the experimental group was also shown by the percentage of the criterion A of 0% with the frequency of zero students, rising to 100%. Meanwhile, the improvement of the students' music-playing skills in the control group was seen from the data in the difference in the increase of student numbers on the criterion B by the percentage of 0% with the frequency of zero students, rising to 19.35% with a frequency of 6 students.

3.2 *Test prerequisite analysis*

3.2.1 *Multivariate normality test results*

The normality test was performed to determine whether or not the data had a normal distribution. The multivariate normality test used was the Mahalanobis distance test. The decision criterion (Johnson & Wichern, 2002, pp. 182–183) was based on the condition when 50% of the data range has *dj2* ≤ X2 (0.5). The multivariate normality test results showed that the data were close to the 50% range; it can therefore be stated that the data of both groups had a normal distribution. This is shown by the percentages of 43.75% and 51.61% at the time of pretest, and the percentages of 53.13% and 51.61% at the time of post-test.

3.2.2 *Multivariate homogeneity test results*

Multivariate homogeneity tests were used to determine whether or not the two groups had the same data variance. The multivariate homogeneity test was done using Box's M test by looking at the significance value from the Box's test table. The decision criterion (Pallant, 2007, p. 286) was based on the assumption of the significance value of >0.001. Multivariate homogeneity test results with the significance value of >0.001 from the Box's test table show that the control group and the experimental group at pretest and post-test had homogeneous data variance matrixes. This is indicated by a significance value of 0.076 at the time of pretest and a significance value of 0.363 at the time of post-test.

3.3 *MANOVA test results*

The hypothesis testing was done on the basis that the data had normal distributions and were homogeneous. The multivariate test of the hypothesis used a Multivariate Analysis of Variance (MANOVA) test with the following Hotelling's trace with the help of the SPSS 23.0 statistical software program. The decision criterion for the significance value (Pallant, 2007, p. 287) was <0.05.

Based on the results of the MANOVA test, it was found that there were differences in test results between the pretest and post-test. The difference in results can be seen from the significance value in the column of Hotelling's trace. The MANOVA showed a value of 0.112 on the pretest and 0.000 on the post-test. For the pretest, the MANOVA test result was >0.05, so it can be said that there were no differences in the pretest scores for skills appreciation. From the score results of the pretest, it can be concluded that there was no difference in the students' music-playing skills and musical appreciation between the experimental group and the control group. This means that there were no differences in both aspects between students who were taught using the collaborative method and students who were taught using the personal learning method.

On the other hand, the value of significance in the MANOVA post-est results showed <0.05. It can be concluded that there were differences in the post-test of *Rinding Gumbeng* music skills and musical appreciation between students in the experimental group and those in the control group. Based on these results, it can be concluded that there was a significant difference in music-playing skills of *Rinding Gumbeng* and musical appreciation between students who were taught by using collaborative learning and students who were taught by using personal learning at SMK Negeri 2 Wonosari Gunungkidul.

4 DISCUSSION

As stated in the introduction, the goal to be achieved in this study is understand the significant differences in the skills of playing *Rinding Gumbeng* music and musical appreciation between students taught by using collaborative learning and students taught by using personal learning. Based on the results of the hypothesis testing conducted, it is known that the null hypothesis (H_0) was rejected and H_a was accepted, with a significance value in the column of Hotelling's trace of 0.000. Because the significance value shows <0.05, it can be said that there is a significant difference of music-playing skills of *Rinding Gumbeng* and musical appreciation between students taught by using collaborative learning and students taught by using personal learning at SMK Negeri 2 Wonosari Gunungkidul.

Collaborative learning is one of the efforts being made to improve the benefits of music education in Indonesia. As revealed by Sawyer (2014, p. 185), more than two decades in research have proven that the form of collaborative exercises is uniquely useful for learning in different regions. Another point is also made by Cane (2009, p. 33) that, in a musical perspective, collaboration enriches music education for learners. Opinions and research results are also reinforced by the results of this study, whereby the application of collaborative learning has an influence on the skills of playing *Rinding Gumbeng* music and musical appreciation of students. This is shown through the MANOVA test results. The increase of the average score is one of the implications in this research with the application of collaborative learning. Thus, it can be concluded that collaborative learning is one of the alternative learning modes that can enrich music education for students.

The functions and benefits derived from the application of collaborative learning for the development of students associated with this research is enriching the insights and knowledge of students through improving the students' skills in playing traditional *Rinding Gumbeng* music and increasing the students' appreciation of this traditional music. The other benefits obtained by students include the existence of social attitudes such as cooperation, cohesiveness, togetherness, and tolerance of students, all visible through the complex interaction and communication among students during the running of the collaborative learning. As expressed by Sato (2014, p. 28), collaborative learning is focused on cultural practices (cognitive activities with cultural contents) and attaches importance to the social aspect of learning as the formation of meaning and relationships.

5 CONCLUSION

Based on the results of the study, it can be concluded that there are significant differences in the skills of playing *Rinding Gumbeng* music and the musical appreciation between students taught by using collaborative learning and students taught by using personal learning at SMK Negeri 2 Wonosari Gunungkidul. This is evident from test results of the MANOVA post-test showing a Hotelling's trace value of 0.000, which is lower than 0.05. Thus, it can be concluded that there is an influence in the application of collaborative learning to the skills of playing *Rinding Gumbeng* music and to the musical appreciation of students of SMK Negeri 2 Wonosari Gunungkidul.

REFERENCES

Campbell, D.T. & Stanley, J.C. (2015). *Experimental and quasi-experimental designs for research*. Boston, MA: Ravenio Books.

Cane, S. (2009). Collaboration with music: A noteworthy endeavor. *Music Educators Journal, 96*(1), 33–39.

Johnson, R.A. & Wichern, D.W. (2002). *Applied multivariate statistical analysis* (5th ed.). Upper Saddle River, NJ: Prentice Hall.

Johnson, R.A. & Wichern, D.W. (2007). *Applied multivariate statistical analysis* (6th ed.). New York, NY: Prentice Hall.

Pallant, J. (2007). *SPSS survival manual*. New York, NY: McGraw-Hill.

Sato, M. (2014). *Mereformasi Sekolah: konsep dan praktek komunitas Belajar* [Reforming Schools: Learning community concepts and practices]. Tokyo, Japan: JICA Publication.

Sawyer, R.K. (2014). *Group creativity: Music, theater, collaboration*. Hove, UK: Psychology Press.

21st Century Innovation in Music Education – Astuti et al. (eds)
© 2020 Taylor & Francis Group, London, ISBN 978-0-367-11122-9

Vocal learning efforts to increase music teachers' abilities in *Panti Dewi* kindergarten

P.P Rusdewanti, D. Kristianingsih & D. Jogjaningrum
Yogyakarta State University, Yogyakarta, Daerah Istimewa Yogyakarta, Indonesia

ABSTRACT: This study provides an approach for developing the musical ability of teachers in Panti Dewi Berbah Sleman kindergarten through vocal learning. The research used the Classroom Action Research method during July 2018. The results show that providing vocal learning approaches, increasing the musical ability. Based on the assessment results for each basic indicator that includes basic abilities (tone, rhythm and tempo), singing ability (sound production, articulation, intonation, frequency), and additional abilities (expression, interpretation and appreciation) teachers obtained an average score of 73.7, which is in the medium category while the learning process is in the good category. Thus, the action research with the help of good learning media that was conducted by training teachers to understand singing well, to be more sensitive to tone, rhythm and tempo and how to better express, interpret and appreciate songs, succeeded in helping develop the musical ability of teachers in Panti Dewi Kindergarten.

1 INTRODUCTION

Kindergarten or Early Childhood Education (ECE) teachers are the most important education figures for the early stages of after-child education(Undang-undang RI, 2005). While being a kindergarten/ECE teacher is not an easy profession(Sanjaya, 2013), people often simply see kindergarten/ECE teachers teaching only singing, storytelling, colouring, drawing and games. The ability or competence possessed by a kindergarten/ECE teacher looks only "half-way" or sober. For example, kindergarten/ECE teachers must be able to sing, but whether the singing is good or not good is not considered at all. What often happens is that the kindergarten/ECE teacher invites students to sing but the teacher's tone is "false". There are also kindergarten/ECE teachers who can accompany songs using a keyboard, but the accompaniment played is not in accordance with the proper rhythm or harmony of the song being played.

Panti Dewi Kindergarten/ECE which is located at Komplek Tanjung TRP Tanjungtirto Berbah Sleman, consists of four kindergarten classes and one ECE class. There are 10 teachers in for these classes and all of them are women. There are three learning materials or activities related to music—singing, drumming and dancing. The singing material is held at every flag ceremony on Monday and at the beginning of learning sessions every day. Drum band and dancing material is given on certain days according to the schedule of each class. In each of these activities, all 10 teachers were involved. However, only a few teachers teach dancing.

Judging by what was provided as learning materials and activities, the teachers' musical ability in kindergarten/ECE was still not optimal before the study began. The teachers gave singing examples, and yet they could not feel that the tone that was sung was not correct. The teachers sang and clapping hands, but the tempo became too fast. Sometimes there was too much enthusiasm so that teachers sang songs by shouting and that were not pitched, because shouting as singing was considered an example of singing passionately to students. Thus, the rules of singing were neglected. Likewise, when a teacher accompanies Indonesian Raya, Mengheningkan Cipta songs and selected songs for ceremony, the song's rhythm and harmonization are less harmonious. There are other ways that become advantages. For example, some teachers can make and create songs that are sung once with the students. Among the teachers themselves, the songs were already created before being given to the students. If there are certain activities such as a visit to a place, some teachers spotlessly make songs related to the activity. Uniquely, the songs are only memorised by poetry and tune but there is no written one in accordance with the rules of music. In drum band learning, the teacher participates in giving commands to train band students rhythmically and melodically. However, sometimes the marching is not in accordance with what is given by the actual drum band coach. Likewise during dancing or exercising in the morning, the movements are not in accordance with the count or beats of the music accompaniment.

Through observation, the research was carried out to study the development of teacher musicality in Panti Dewi Kindergarten/ECE through vocal learning and in accordance with the professional field and competence of kindergarten teachers.

2 METHOD

This study uses a classroom action research approach (Madya, 1994). The choice of this approach is practical and directly relevant to actual situations. Additionally, action research is flexible and adaptable by allowing changes to occur to develop new skills or new approaches for solving problems (Suryabrata, 2013,:94). In action research, researchers work directly with communities, involve themselves during the process of developing musicality activities and are followed continuously.

The study was conducted in Panti Dewi Kindergarten Komplek Tanjung Tirto Berbah Sleman, starting in April 2018 for pre-field activities in the form of observations. The study continued with the process of action and assessment until July 2018. Out of 10 teachers who became subjects, there were nine teachers in the field.

3 RESEARCH STAGE

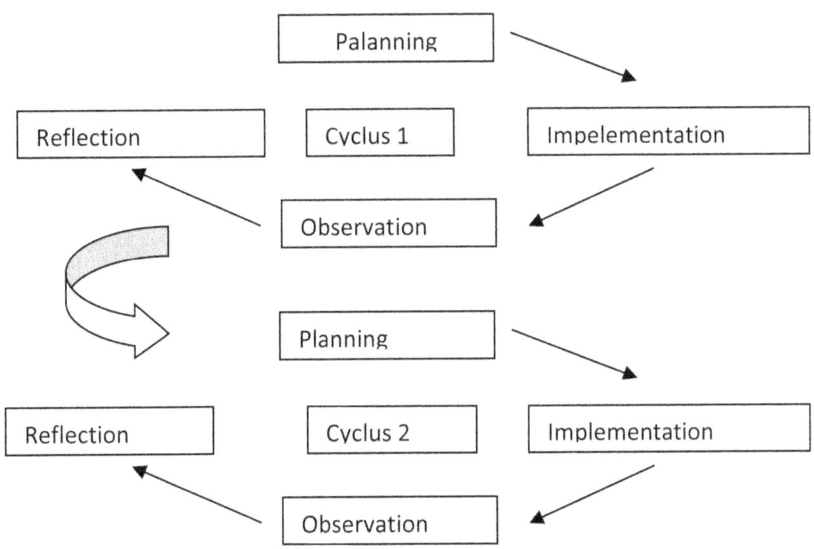

Figure 1. Acting Stages.

The study was conducted in two cycles beginning with pre-cycle activities. The stages of research following the model proposed by Kemmis and Taggart in Suharsimi (2006:16) were carried out in two cycles as shown above.

3.1 *First cycle*

1. Planning consists of activities that determine the time for conducting research with the research team as well as with collaborators and the Principal of Panti Dewi Kindergarten, together preparing the teachers to be observed, preparing vocal learning modules and preparing observation and assessment instruments.
2. The implementation of the action that consists of activities conveying material objectives, explaining vocal learning material, demonstrating vocal learning material, practicing vocal techniques of the teachers, observing the teache''s training process activities during learning. The teachers display the results of the training both in the classroom and independently and provide an assessment of the process and results of vocal learning.
3. Observations involve noting the situation of learning activities, activeness, the participation of teachers and the ability of teachers to complete tasks in groups and independently.
4. The result of reflection is that the teacher can know and understand goals, materials and demonstrations, so that during the learning activities the teacher can practice and display the results of the training both independently and in vocal learning groups.

3.2 *Second cycle*

1. The researcher makes a learning plan based on the results of reflection during the first cycle.
2. The implementation of vocal learning research is conducted based on the learning plan resulting from reflection in the first cycle.
3. Observation of vocal learning activities.
4. Reflection is carried out on the implementation of the second cycle and analysing and drawing conclusions on the vocal learning process as an effort to develop the teachers' musical abilities.

4 RESULTS OF RESEARCH AND DISCUSSION

4.1 *Data description*

4.1.1 *Pre-cycle Stage*

In the pre-cycle stage, the kindergarten/ECE teacher learning process usually took place in two meetings. At the first meeting, nothing was specifically planned to carry out the action process. The teacher sang with the students during the flag ceremony on the first Monday of the research period. The songs that were sung were "Indonesia Raya", "Mengheningkan Cipta" and "Ibu Kartini". At the next meeting, the teacher sang with the students in the class according to the learning theme. In class A, English learning took place and the teacher taught English songs. In class B, singing teachers were assisted by audio visual media using cell phones. At that time, researchers only observed how the teacher sang.

By observing the aforementioned situation, the planning began for cycle phase I. The results of the assessment can be seen in Table 1 below:

Table 1. Results of average values of Pre-Cycle singing assessment.

No	Teacher's Name	Basic Ability	Singing Ability	Additional Ability
1	Pr	61	61	62
2	Ka	61	62	62
3	Ru	67	63	61
4	Sr	64	64	67
5	Ch	70	70	71
6	Es	62	62	70
7	Fe	67	65	70
8	De	60	61	61
9	Er	70	70	70
10	Mi	56	54	50
	Sum	638	632	644
	Average	63,8	63,2	64,4
	Total Average Value	63,8		

4.1.2 *Cycle 1 (Four Meetings)*

The first cycle was carried out in four stages:

1. Planning
   a. Determine that the time for conducting research for the first meeting would begin in July after students return from school holidays.
   b. Prepare for classes in kindergarten, keyboard, speakers, and for the 10 teachers who will be trained.
   c. Determine vocal learning material in the form of modules and audio media.
   d. Prepare observation and assessment instruments.
2. Implement actions
   a. Deliver material goals
   Submission of this goal is given only at the first meeting. Then during the second and third meetings explanations are no longer given. The purpose of vocal learning is to develop the musical ability of the teacher.
   b. Demonstrate vocal material.
   The researcher demonstrated breathing material. At the beginning of the meeting, the teacher was equipped with a vocal learning module. There are eight main subjects that were delivered for the learning: posture in singing, breathing techniques, articulation, resonance, relaxation, how to produce sounds and basic vocal techniques. The researcher gave an example, and then each teacher attempted to do what the researchers demonstrated. This activity was observed by other research members while at the same time making improvements when an error occured. Vocal training is useful for increasing teacher sensitivity to tone, rhythm and tempo.
3. Observation
   The results of observing vocal learning were generally good and the learning went smoothly. The teachers listened to the material provided by the researcher. The module supported the teacher in understanding the terms contained in vocal techniques. Illustration examples using pictures that are in the module also supported teachers to learn independently. When one teacher asked a question, the other teachers listened well, so that when the practice was carried out, the teacher understood the steps to be taken. There are times when the teacher asked about other things such as the meaning of song lyrics, but these kinds of questions were in line with the purpose of giving vocal learning material and research objectives.
4. Reflection
   After the first cycle phase was implemented, there were issues that become deficiencies during implementation. These were:

a. There continued to be problems with posture and standing position. In a standing position, the teacher often does not believe in herself and often moves erratically. In a sitting position, because the bench used is the same low seat that is used by children of kindergarten, sitting became uncomfortable.
b. When asked to sing, teachers still tended to shout. There were times when the targeted tone was not right, so it sounded "false".
c. Breathing training for singing is the most difficult exercise, because the teacher has not been able to distinguish between normal breathing, talking and breathing while singing.
d. The teachers' lack of musical experience influenced the vocal learning process, especially during vocalisation exercises with the scales shifting and following the introduction of the song.
e. There was one teacher who did not fully take part in the process because she was assigned to a workshop at the Ministry of Education for two weeks that took her away from the school. But this teacher's absence did not affect the overall average value of the teachers' abilities obtained at the first evaluation.

4.1.3 *Cycle II (Four Meetings)*
The second cycle was carried out through four stages:

Table 2. Results of average cycle I singing values.

No	Teacher's Name	Basic Ability	Singing Ability	Additional Ability
1	Pr	65	65	68
2	Ka	67	67	68
3	Ru	70	65	65
4	Sr	70	68	70
5	Ch	72	72	75
6	Es	65	68	70
7	Fe	-	-	-
8	De	65	65	65
9	Er	70	72	70
10	Mi	60	60	60
	Sum	604	602	611
	Average	67,1	66,8	67,8
	Total Average Value	67,2		

1. Planning
   a. Re-established implementation time.
   b. Prepared classes in kindergarten, keyboard media, speakers and 10 teachers who will be trained.
   c. Determined vocal learning material in the form of modules and audio media.
   d. Prepared observation and assessment instruments.

2. Implementation of actions
   a. Delivered material goals
   Submitted learning objectives based on the results of reflection in cycle I.
   b. Demonstrated vocal material.
   The researcher demonstrated the material based on the module by covering basic abilities, singing abilities and additional abilities. In this case, researchers increasingly focused on teachers who in the first cycle, still earned low scores. Vocabulary material is more varied so that the teacher could also improve the musical experience. Before attempting the song material, the teacher must first perform further vocalisation exercises. When the vocal material was finished, the action continued with the children's

song material starting with listening to the introduction, entering the song, the interlude and the closing. Vocals were assisted by keyboard and children's songs were sung using audio media.

3. Observations
The results of the vocal learning observation in the second cycle was generally good and went smoothly. The teachers listened well to the material provided by the researcher. During practice, the teachers began to understand the shortcomings of what is found in cycle I. The teachers assessed each other's shortcomings and attempted to discuss all of the steps to improve. The suggestions given by the researchers were carried out as well as possible. The results in cycle II were even better than in cycle I.
4. Reflection
Although cycle II was not quite reached, there were matters relating to technical implementation:
   a. a. Two teachers arrived late, so they did not have time to listen to the vocal materials. Thus, the researchers interrupted the planned time sequence by having to again explain the material and adjusted the advanced material.
   b. b. There were two unhealthy teachers. Thus, the learning focus was somewhat disturbed. However, during the time when values for the three abilities were taken, the two teachers sang according to the direction of the researcher as well as possible so that they could then further develop musicality.

Table 3. Results of average cycle II singing rating.

No	Teacher's Name	Basic Ability	Singing Ability	Additional Ability
1	Pr	72	70	74
2	Ka	71	72	73
3	Ru	71	72	76
4	Sr	73	74	71
5	Ch	80	80	78
6	Es	71	72	74
7	Fe	-	-	-
8	De	71	71	71
9	Er	79	80	80
10	Mi	71	71	73
	Sum	659	662	670
	Average	73,2	73,5	74,4
	Total Average Value	73,7		

4.2 *Discussion*

The process of vocal learning from the beginning until the end went smoothly. This process was supported by the teachers' enthusiasm and the availability of media facilities provided by the researchers and the schools. During each meeting, there appeared to be a development of the teachers' musicality. Although the results of practice that were scored in the form of numbers as obtained by teachers are not much different from the previous research in 2017, the existence of additional material related to music appreciation, interpretation and expression added to musical ability improvements by the teachers.

At the time of the assessment, each teacher practiced independently. There were times when a teacher felt that the time for preparation was lacking because there were other teacher assignments that also needed to be carried out. Nevertheless, it turns out that the teachers stated that singing is important and since the teachers would teach students, they would do her best to practice so that their students would also benefit. The results of the questionnaire showed that teachers very much benefited from this vocal learning process. The Principal and the teacher also asked permission to use audio media (Minus-One CD) that was already being used for learning at the school. The researcher responded very well to this and responded that if there was something wrong with the audio media, the teacher was asked to contact the researcher again so that the media used for learning could be used properly.

The results of classroom action research conducted at Panti Dewi Kindergarten/ECE teachers not only measured abilities via the final values obtained during the evaluation. The vocal learning process specifically added to the teachers' musical experience. Based on the results of the research questionnaires and interviews, the teachers felt that there was a difference between the study that began in 2017 when the teachers were still unfamiliar with beam notation, and the 2018 study where the teachers increasingly understood the "what's" and "how's" of beam notation. The most obvious difference between 2017 and 2018 is the further development of singing. The teachers developed more sensitivity to tone, rhythm and tempo. In the previous year, the teachers just sang the important aspects correctly. But after learning good and correct vocal training techniques, the teachers became more confident to teach the students. The teachers at Panti Dewi Kindergarten also sang as a choir at the IGTK to show the results of the vocal learning process. Growing confidence is what increasingly enhances the results and development of musicality.

Looking at the results of evaluation and reflection at the last meeting of the study, the average value obtained is 73.7, showing that the learning process and the materials provided by the researcher are good. The purpose of research through teacher vocal learning is to develop musicality optimally and that can be achieved. The most important thing is that the teachers gained musical experience that became an asset that the teachers could then impart to their students.

5 CONCLUSION

The results of this study show that the musical abilities of teachers at the Panti Dewi Kindergarten grew. This growth can be seen in the assessment results that include attained levels of basic abilities (tone, rhythmic and tempo), singing abilities (sound production, articulation, intonation and phrasing) and additional abilities (expression, interpretation and appreciation) The obtained average value is 73.7 which in the good category. Thus, the action research that was carried out by training teachers through vocal learning and with the support of helpful learning media succeeded in supporting the teachers at the Panti Dewi Kindergarten in developing and enhancing their musical abilities.

REFERENCES

Amir, M. Taufiq. (2009)*Inovasi pendidikan melalui problem based learning*.[Educational Innovation through Problem Based Learning: How Educators Empower Learners in the Age of Science] *Jakarta: Kencana Prenada Media Group* (2009).

Madya, S. (1994). *Panduan penelitian tindakan*. [Series of Research Methodology. Action research guide] *Yogyakarta: Lembaga Penelitian IKIP Yogyakarta*, 92.

Sanjaya, Wina. 2013. *Learning Strategy*. Jakarta: PT Raja Grafindo Persada.

Suharsimi, A. (2006). *Prosedur penelitian suatu pendekatan praktik*.[Research Procedure: a Practice Approach] *Jakarta: Rineka Cipta*.

Suryabrata, S. (2013. *Metodelogi penelitian*. [Research Method]*Jakarta: Raja Grafindo Persada*.

Undang-Undang RI 2005 No. 14, *Teachers and Lecturers Law*. Jakarta: RI.

21st Century Innovation in Music Education – Astuti et al. (eds)
© 2020 Taylor & Francis Group, London, ISBN 978-0-367-11122-9

Comparison of cognitive and improvisational learning models on music students' learning

Y Widiyanto & K.S. Astuti
Universitas Negeri Yogyakarta, Yogyakarta, Indonesia

ABSTRACT: The cognitive learning model is a form of learning theory called a conceptual model. Meanwhile, improvisation is an activity that encourages the generation of spontaneous responses in a variety of situations. The purpose of this study is to investigate a better learning model used between the cognitive learning model and improvisational learning model. The research methods included a quasi-experimental approach with non-randomised control group pretest and post-test design. The sample in this research are 115 students. Data was analysed using *t*-tests. The results of this study are: (1) learning music with a cognitive learning model demonstrates effective use; (2) learning music using an improvisational model also demonstrates effective use; (3) there is a significant difference in student improvement between the cognitive and the improvisational learning models; (4) the improvisational learning model is more effective than the cognitive learning model for improving the outcome results of music students.

1 INTRODUCTION

Education is one of the most important fields to note because education is the basis for the formation of civilised human beings. Education also has an important role to play in the progress and development of a nation, such that education outcomes serve as benchmarks in evaluating the progress of a nation. To achieve the appropriate learning outcomes of a standard education, what is required are appropriate learning strategies. The effective teaching of learning strategies helps teachers provide assistance so that students can achieve learning goals. Thus, it is necessary to develop the quality of education and learning processes in a sustainable manner in all aspects of learning. If the learning process is not optimal then the quality of education and learning will decrease.

Art education is actually a part of basic education along with language. If art is one form of culture, then art should be taught in ways that align with exact teaching and language learning methods. At present, art education in Indonesia is still limited in its role as being complementary to the required curricula. Art education is not prioritised compared to exact or language learning, thus impacting on the number of teaching hours devoted to art and culture, which are very minimal in the schools. This indirectly impacts on the quality of the students, and art appreciation by the students is very lacking. Less teaching hours in art education also results in a less effective delivery of teaching and learning activities. Many of the factors that contribute to low levels of teaching music in schools are not profound. Teachers lack sufficient numbers of music teaching hours, lack school facilities to teach music and some teachers lack creativity in delivering music materials. The subject of music consists of theory and practice, so a model is required that combines both practice and theory with efficient time for both kinds of endeavours. The model of cognitive learning emphasises the process of thinking and can encourage mastering theories that will be applied to the practice of playing music. Students can apply theories related to the material used for music practice. Students can be freed to look for tools that support the material. Thus, when the practice material is completed students are also equipped with theories that support the practice, and as expected, students will achieve good learning outcomes.

In addition to the cognitive learning model, one of the other effective learning models for music education for students to achieve good learning outcomes is the improvisational learning model. The improvisational model is a model that emphasises practice. Students are given practical materials first. Thus, the students are expected to understand theories applied to practice materials, with teachers as facilitators of student practice using an improvisational model. In this model, teachers are expected to master the material and musical instruments used in practice sessions that would hopefully produce maximum learning outcomes. This research was conducted to compare the cognitive learning model and the improvisational model in music subjects and to determine which model is better in improving learning outcomes in music education. This research is conducted in SMA N 1 Way Jepara, East Lampung regency.

2 RELATED WORKS/LITERATURE REVIEW

The term 'cognitive' comes from the word 'cognition', which means 'understand' or 'understanding'. Broad understanding is the defender, the arrangement and the use of knowledge that contains the process of thinking (Anwar, 2017). Awareness is a general concept that includes all concepts. These concepts include seeing, watching, giving, thinking, reading, estimating, considering, speaking and judging. The term 'cognitive' is related to the ability to think, including the ability to memorise. Cognitive abilities comprehend appearances that can be used as results, or refers to the process of learning knowledge through one's own experience. The cognitive learning model is oriented towards thinking ability, which includes intellectual abilities, remembering, and problem-solving skills that require students to connect and combine several ideas, methods or procedures to solve problems. Lemut et al. (2013) say that conceptual strategies in the cognitive domain include behavioural solutions to problems, analysis, structuring, concepts and solutions.

The cognitive learning model is a form of learning theory called a conceptual model. The cognitive learning model is a learning model that emphasises the very important internal processes, the denominations of behaviour, including learning that is always occurring or the actions taken as part of behavioural learning. In the student cognitive learning model there is a reflection on the meaning of information they receive by using their own words. As described by Leighton and Gierl (2012), cognitive models are simple descriptions of human problems, taken on by tasks that fulfil educational standards and that help to characterise knowledge and skills at different levels. Ultimately, the cognitive learning model prioritises the development of shared knowledge and values in each student.

Bhinnety (2008) explained that human memory has at least three components, namely the sensory register, working memory (short-term) and long-term memory. Cowan (2014) argues that the development of learning and cognitive education is based on memory. In the cognitive learning model, the teacher provides the basics of the material that is taught for student development and the continuation is left to the students to remember the material. The teacher's duty is to monitor, organise and explain the flow of materials that has been given. In this way, the cognitive learning model helps students to maximise their memories to remember the materials that have been given. To be able to understand and always remember the concepts given to students, the teacher should give emphasis to repeating or reviewing past materials that have been given so that what is received by students can enter into the long-term memory as a model of cognitive learning.

Improvisation is an activity that encourages the generation of spontaneous responses in a variety of situations. Improvisation is a mindset for organisational analysis and in general improvisation works to foster active, initiative, creative and innovative power. Improvisation can also sharpen creativity, imagination and skill in playing arts in a spontaneous way. Djohan (2009) says that the essence of music is good hearing and educational experiences that include a series of performance activities to strengthen and facilitate the wide development and understanding of various musical skills. In Indonesia, Curriculum 13 emphasises integrated thematic learning. Astuti et al. (2018) suggest that there are three important aspects in the development of integrated thematic teaching. First, there must be integration of subject matter and character education. Second, there must also be integration of theory and practice. Finally, the teaching process should be fun.

The improvisational model is very supportive of Curriculum 13 because there is character education and there is the integration between theory and practice. Theory and practice are applied into a more easily understood way, so the students do not run with their own practice and theories themselves. Improvisation is also certainly fun because students are free to express what is on their minds and is certainly different from memorising and reading. In improvisation there must be creativity because cultural improvisation is related to creativity. According to Beckstead (2001), improvisation is a far more creative attempt to see how far the creative part of the brain is accessed. If the goal is to involve students in accessing and training the 'creative' part of the brain then there is no need to worry about how good the students' improvisational skills are. Improvisation can also increase cohesiveness and cooperation, innovation and performance in teams.

Music improvisation means playing music without preparation, and can teach students how to adjust to changing circumstances spontaneously by hearing and playing music. Djohan (2009) says that learning music through the ears (hearing) and improvisation are important to achieving success. Playing music spontaneously is one of the learning processes. Azzara (1999) says that the best way to learn music and to further know about music is by playing it.

In the model of improvisational learning, the students are given practice materials first. Thus, the students are expected to understand the theories used in the practice materials. In the improvisational model the teacher is a facilitator and guides student practice. In this model, the teacher is expected to help students master the material and musical instruments that will be used, hopefully producing maximum learning outcomes.

3 MATERIAL & METHODOLOGY

3.1 *Data*

This research was conducted on the students of class X SMA N 1 Way Jepara. The sample population in this research are 115 students. Data collection techniques include using a comparative approach with paired samples (sample related). Paired samples are samples that are subjected to pretest and post-test evaluations. In this study, data collection is divided into two groups, namely one student class that uses a cognitive learning model and another class that uses an improvisational learning model. Both classes were pretested to measure the students' prior abilities and then subjected to a post-test to measure the outcomes of the research. The data collection instrument was a multiple-choice question-set test instrument with four possible answers prepared, based on a grid of questions that adhere to the content standards in the 2013 curriculum (K13). This test was used to evaluate student learning outcomes before and after the applied models of cognitive learning and improvisation.

3.1.1 *Normality Test*

As a condition of data analysis using a *t*-test, the data must have normal distribution and be a homogeneous variant. Test normality was determined using the Kolmogorov-Smirnov test because the number (n) is greater than 30 (large sample). The data is considered to be normally distributed if the significance is greater than 0.05. The research data that was produced was conducted via a normality test:

a) data for learning outcomes based on the cognitive learning model.
b) data for learning outcomes based on the improvisational learning model.

The results of the normality tests can be seen in Tables 1 and 2.

Table 1. Results of normality test for the cognitive learning model.

	Cognitive learning model	Kolmogorov-Smirnov		
		Statistic	df	Sig.
Cognitive	Pretest	.096	55	.200
	Post-test	.110	55	.092

Table 2. Results of normality test for the improvisational learning model.

	improvisational model	Kolmogorov-Smirnov		
		Statistic	df	Sig.
Improvisational	Pretest	.091	60	.200
	Post-test	.095	60	.200

From the results of data analysis, the data has met the normal assumption of pretest value of 0.200 and the post-test value of 0.092. The significance values of 0.200 (>0.05) for pretest and 0.092 (>0.05) for post-test means it can be tested further.

From these results, the data is considered normal. Pretest and post-test significance data is 0.200. Because the significance value of 0.200 is above 0.05, the data is eligible for further testing.

3.1.2 *Homogeneity Test*

The homogeneity test is used to test the similarity of variance between the two groups. The homogeneity test uses the F or Levene's test. For homogeneity analysis, the data is said to be homogeneous when the significance value (Sig.) is more than 0.05. The homogeneity test results can be seen in Tables 3 and 4.

The homogeneity test of cognitive learning produced a significance value equal to 0.801 (>0.05). Thus, the cognitive class data is considered homogeneous.

The homogeneity test for improvisational learning generated a significance value of 0.781 (>0.05). Thus, the data for the improvisational class is considered homogeneous.

Table 3. Homogeneity test for cognitive learning.

Levene statistic	df1	df2	Sig.
.064	1	108	.801

Table 4. Homogeneity test for improvisational learning.

Levene statistic	df1	df2	Sig.
.077	1	118	.781

3.2 *Method*

This research used a quantitative approach with quasi-experimental methods. The selection of quasi-experimental research was made in order to reveal a comparison of learning methods. What is compared are the cognitive learning model and the improvisational learning model.

This research was conducted by using two groups. One group of 55 students received the cognitive learning model and the second group of 60 students received a learning model of improvisation. Before being subjected to the two learning models, both groups were subjected to pretesting. After both groups were exposed to the learning materials, the post-test was administered in order to establish and compare the results of learning method treatments given to the two groups. Figure 1 shows the research design.

As a quasi-experimental research, the researcher determined the two classes as the object of research. The experimental class was given the improvisational learning model and the control class was given the cognitive model.

The assessment of these two learning models was made possible by the pretest + treatment + post-test design. The assessment aimed to find a significant difference between the students' learning outcomes prior to the application of each learning model via pretesting, and with a post-test after the treatment of cognitive learning models (experimental classes) and improvisational learning models (control classes). The learning results in both classes were then processed with statistical techniques with the help of SPSS 16 for Windows, to generate the difference in gain (gain).

The gain is based on the acquisition of the average pretest value and the mean value of the post-test for the two classes, and then calculated so that the results obtained for research can be used to answer the research hypothesis proposed in this study.

3.3 *Testing the research hypothesis*

After administering the hypothesis prerequisite test, that are normality and homogeneity tests, hypothesis testing is performed. From the results of the analysis that has been done, the data that was tested was shown to have normal and homogeneous distributions. Therefore, the researchers proceeded with parametric statistics. The hypothesis test used in this research was a *t*-test using SPSS 16 for Windows. The *t*-test states that if the value of significance is below 0.05 then an independent variable has an effect on a dependent variable. Whereas, if the value of significance is above 0.05 then an independent variable has no effect on a dependent variable. The *t*-tests used are a paired sample *t*-test and independent sample *t*-test.

The purpose of this hypothesis testing is to compare which of the two models is a better learning model for musical arts learning outcomes: the cognitive learning model or the improvisational learning model. The data used in this hypothesis testing is the gain values of the cognitive learning model and the improvisational learning model. The test that was carried out paired sample tests of each group: the group using the cognitive learning model and the group using the improvisational learning model. The hypothesis testing was performed, providing a collection of data before and after the learning model treatments in different time with the independent sample *t*-test administered to know the mean values of two different

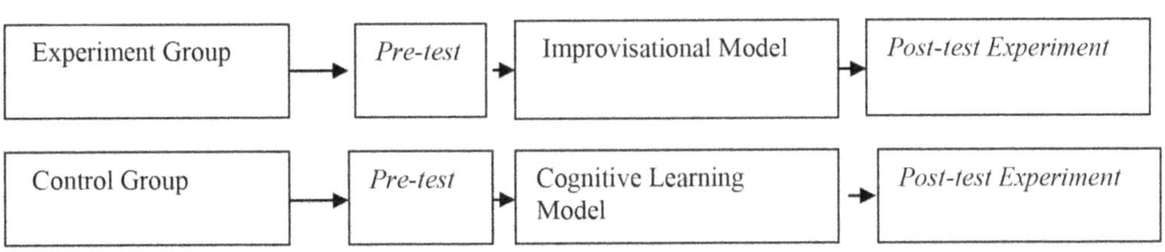

Figure 1. Non-randomised control group pretest/post-test design.

groups. Decision criteria in paired sample tests are not accepted if the value of significance is smaller than the real level used (0.05) or Sig. <α.

4 RESULTS & DISCUSSION

4.1 *Results*

This study aimed to compare the cognitive learning model with the improvisational learning model for efficacy on learning outcomes in musical arts. From the results of descriptive analysis, the average pretest value on the cognitive learning model was 48.53, while the post-test value on the cognitive learning model was 63.22. For the improvisational model, the average pretest value was 50.60 while the average post-test value was 67.47. Each model showed an increase in value from the pretest to post-test averages. The cognitive learning model obtained an increase of 14.69 while the improvisational learning model showed an increase of 16.87.

The results of the paired sample *t*-test analysis with SPSS 16 for Windows statistical software showed the cognitive learning model had a significance value of 0.001 while the improvisational learning model also had a significance value of 0.001.

To find the difference in the improvement of learning outcomes between the cognitive learning model and the improvisational learning model an independent sample *t*-test was used. From the results, the significance value is 0.03. Hypothesis testing was also conducted to find out which model is better. From the results obtained, the mean value of cognitive learning model was 0.2813 while the improvisational learning model obtained a mean of 0.3382.

4.2 *Statement of results*

The results of data calculations can be seen in Tables 5, 6, 7, 8 and 9.

Table 5. Calculation results for both learning models.

	Cognitive		Improvisational	
Group	Pretest	Post-test	Pretest	Post-test
Average	48.53	63.22	50.60	67.47
Max	67	80	67	83
Min	30	47	30	50
SD	8.56	8.35	8.49	8.21
Var	73.25	69.69	72.13	67.48

Table 6. Paired sample *t*-test for the cognitive learning model.

		Sig. (2-tailed)
Pair 1	Pretest_cognitive Post-test cognitive	.000

Table 7. Paired sample *t*-test for the improvisational learning model.

		Sig. (2-tailed)
Pair 1	Pretest_improvisation Post-test_improvisation	.000

Table 8. Independent test sample *t*-test.

		Sig. (2-tailed)
Gain	Equal variances assumed	.030
	Equal variances not assumed	.030

Table 9. Model comparison.

	Model	N	Mean
Gain	Cognitive	55	.2813
	Improvisational	60	.3382

4.3 *Explanation*

Table 5 shows the calculation results for both models. The average pretest and post-test averages of both models are obtained. The average value obtained from pretest and post-test for both models improved. Likewise, the minimum value and the maximum value obtained between the pretest and post-test values increased. The provision of learning method treatment after the pretests in this study show improved learning outcomes.

The decision criteria in the paired sample test is rejected if the significance value is smaller than the real level used (0.05) or Sig. $<\alpha$. The data shows that the significance value is 0.001. The decision criteria in the paired sample test is not acceptable if the value of significance is smaller than 0.05. The paired test sample *t*-test on the cognitive learning model is shown in Table 6. The data shows a significance of 0.001 (<0.005), so it is concluded that there is a difference between the pretest value and post-test values for the cognitive learning model. Table 7 shows the same value of 0.001 with a significance of 0.001 (<0.005), so it is concluded that there is a difference between the pretest and post-test values in the cognitive learning model.

In Table 8, the decision criteria on the independent sample tests is rejected if the value of significance is smaller than the real level used (0.05) or Sig. $<\alpha$. The data shows the significance value is 0.03, meaning that there is a difference in the increase of learning outcomes by using the cognitive model and the improvisational model. To see which model is better, the mean values of Table 9 are used. The data shows that learning outcomes were higher in the improvisational method than the cognitive method. Therefore, the researchers concluded that the improvisational model is better to be used than the cognitive model in improving the learning outcomes for music education.

4.4 *Discussion*

The results of this study are in accordance with the expectations of researchers, that teaching with the improvisational learning model is better for improving learning outcomes compared with the cognitive learning model. The materials for the cognitive learning model were delivered by emphasising the understanding of theory, by giving the materials and then the students were given the freedom to process information, imagine, estimate and solve problems posed in the material being studied. The improvisational learning method was delivered by teaching the practice material in advance to stimulate students to know the meaning of the theory to be applied. Having the practice material in advance made it possible for students to know the theory material via practice. Thus, the improvisational learning model is better because concepts and theories were obtained directly through practice.

5 CONCLUSION

This study presented a comparison of the cognitive learning model and improvisational learning model. The researchers conclude that: 1) the improvisational learning model is better used to improve learning outcomes in music education; 2) although the improvisational learning model is better, this does not mean that the cognitive learning model is not as useful. the cognitive approach needs to be further developed to get maximum results; 3) this learning model may be less useful if applied to subjects other than music; 4) this model of learning can be an alternative approach for teachers in delivering learning materials; 5) it is expected that similar research will be carried out in the future to improve the learning outcomes of music education.

REFERENCES

Anwar, C. (2017). *Teori-Teori Pendidikan Klasik Hingga Kontemporer* [Theories of classical to contemporary education]. Yogyakarta, Indonesia: IRCiSoD.

Astuti, K.S., Barensend, M., Widyantoro, A. & Boel, C. (2018). *Effect of improvisation learning based on chord progression methods on Indonesian and Dutch musical skill students.* Singapore: International Congress for School Effectiveness and Improvement (ICSEI).

Azzara, C.D. (1999). An aural approach to improvisation. *Music Educators Journal, 86*(3), 21–25.

Beckstead, D. (2001). Will technology transform music education? *Music Educators Journal, 87*(6), 44–49.

Bhinnety, M. (2008). Struktur dan proses memori [Structure and process of memory]. *Buletin Psikologi, 16*(2).

Cowan, N. (2014) Working memory underpins cognitive development, learning, and education. *Educational Psychology Review, 26*(2), 197–223.

Djohan. (2009). *Psikologi musik* [Music psychology] (3rd ed.). Yogyakarta, Indonesia: Galangpress.

Leighton, J.P. & Gierl, M.J. (2011). *The learning sciences in educational assessment: The role of cognitive models.* New York, NY: Cambridge University Press.

Lemut, E., DuBoulay, B. & Dettori, G. (Eds.). (2013). *Cognitive models and intelligent environments for learning programming.* Berlin, Germany: Springer Science & Business Media.

21st Century Innovation in Music Education – Astuti et al. (eds)

The effectiveness of vocal learning using the drill learning method

N.R.D. Arini & H.S. Mudjilah
Universitas Negeri Yogyakarta, Yogyakarta, Indonesia

ABSTRACT: This research focuses on the effectiveness of vocal learning using the drill method to improve the musical ability of students in Yogyakarta State University. The research aims to determine the effectiveness of the drill method in improving 23 vocal students' musical ability by using an experimental research method. The result shows a significant difference between the experimental class and the control class. The experimental class, which was treated with the drill method, improved its musical ability, while the control class showed no improvement. The result of the multivariate homogeneity test in the experimental group and the control group in the post-test with Box's M reached a value of 3.308, an F value of 1.063, and a significance value of 0.363, which is above 0.001.

1 INTRODUCTION

From Greek mythology, *the muses* are the nine goddesses of art and science. The word *muse* can be understood, along with *polyhimnia*, as meaning composers of different types of songs. More specifically, the word *muse* means a musician's art or technique of playing (Kaufman, 2005). Music (Kaufman, 2005) is a combination of elements such as tone, melody, harmony, rhythm, structure, sound quality (timbre), articulation, and dynamics.

The term 'musical' in this research has a different meaning from what has been understood as "good in playing musical instruments or singing" only. It emphasises the sensitivity of sounds, such as high-low tone (pitch), rhythm, and melody. This sensitivity of sound is related to one's sensitivity to something that happens in the surrounding environment. A child who has a good level of musical ability does not necessarily have good playing skills. Basic musical abilities can be achieved by conducting training on high/low tone (pitch), rhythm, and melody. Exercises should be carried out regularly in order to achieve a good musical skill. Hence, it can provide a sense of deep musical ability for students. In addition, regular exercise that is often done can activate the memory for remembering tones and sounds. A student's sensitivity can also be honed by the dynamic signs that are heard and played with vocals and musical instruments. Routine exercises can also develop hearing sensitivity. It also will increase the knowledge of the musical terms and symbols.

Routine exercises can also allow students to imagine the music being played. It enables them to achieve the appropriate tones, melodies, rhythm, accuracy, intervals, and dynamics. Thus, if students can master these abilities, they will not experience difficulties in playing the material or musical scores given by the teacher. Besides the ability to hear, play a melody, and to play notation on a musical instrument, is the basic and comprehensive ability to master singing or playing music. Music education has the ability to improve musical ability. In addition to musical ability, in learning music, students are expected to master *Solfeggio*. *Solfeggio* consists of ear training, sight-singing, and sight-reading abilities. Ear training is the ability to hear or listen. Sight-singing is the ability to sing by reading song scores. Sight-reading is the ability to read and analyse tone, melody, or rhythm on sight of a score. Students can be trained in *Solfeggio* through regular practice reading melodies, tones, and rhythms in the form of song scores, both in number and beam notation.

Music education activities within the scope of educational institutions can be found in non-formal and formal institutions. Music learning activities in formal institutions are the scope of teaching and learning activities in art and culture subjects. Music education activities can also be found at college level.

Based on the results of an interview with one of the lecturers of PIM I (*Praktik Instrumen Major I*), vocal lessons, it is known that PIM 1 Vocal learning material is derived from various books. These books are *Caesari* vocal books, *Etude Panofka* nos. 1–3, *Sieber* nos. 16–17, and vocal lessons 2–4. *Caesari* and *Etude Panofka* contain melodies of songs in beam notation as vocal warmups. *Sieber* and *Vaccai* are collections of songs (*etudes*) accompanied by basic classical song lyrics in beam notation. The *etudes* are used to practice singing after vocal warming. Meanwhile, the song materials used for PIM 1 Vocal are Indonesian and Western classical songs. The Indonesian classical songs are 'Bahagia' and 'Sandiwara', while the classical Western songs are 'Sulmar Vieni' and 'Passing B'.

Based on the results of observation, it is known that PIM 1 Vocal students were full of enthusiasm when attending the lecture. However, there are still many obstacles and problems in the learning process. These constraints include students who still have difficulty in singing melodies and reading scores of beam notation. Therefore, there are still many PIM 1 Vocal students who have not mastered the practical knowledge of music and basic singing techniques, the techniques of breathing, intonation, articulation, and phrasing. In addition, theoretically, students have also not yet mastered the knowledge of music including the basic theory of singing, such as *Solfeggio* (ear training, sight-singing, and sight-reading).

The other problem is the students' difficulties in targeting tones and singing the tones correctly. Another obstacle is that some students who come from high schools or vocational high schools which do not provide music lessons or music-based extracurricular activities do not have knowledge of music theory and music practice. Based on the constraints and the problems that have been described, a special learning method is needed to improve the musical ability of PIM 1 Vocal students in listening, writing and re-singing a melody. Thus, the researcher used the drill learning method. The method was applied to determine the effectiveness of vocal learning for improving the musical ability of PIM 1 Vocal students to hear, write and re-sing melodies.

The identified problems are: first, the students still have difficulty in singing the melodies of the song and reading the notes on the score notation; second, the students have not yet mastered the knowledge, theoretically or practically, as the problems occur in terms of basic singing techniques – breathing, intonation, articulation, and also the technique of phrasing; third, the students have not yet mastered the basic music theory of singing and Solfeggio which consists of ear training, sight-singing and sight-reading; fourth, the students have difficulty in targeting the tone correctly; fifth, the students have difficulty in reading rhythms and tapping the songs; finally, they do not know the effectiveness of vocal learning using drill learning methods to improve their musical ability to hear, write, and re-sing the melody of the song. The purpose of this research is to determine the level of effectiveness of vocal learning by using drill learning methods to improve the musical ability of PIM 1 Vocal students to hear, write and re-sing a melody.

2 LITERATURE REVIEW

2.1 *The nature of teaching and learning*

2.1.1 *Definition of learning*

Learning is a process carried out by someone to obtain a change in a whole, as a result of his or her own experience in the interaction with his or her environment (Slameto, 1995, p. 2). The characteristics of behaviour change in terms of learning are: (1) changes occur consciously, (2) changes in learning are continuous and functional, (3) changes in learning are positive and active. Learning (Witherington & Burton, 1982, p. 7) is a change in personality that is a new pattern of reactions in the form of conversation, attitudes, habits, and personality. Burton (1962, p. 13) also defines learning as follows: 'learning is change in the individual, due to the interaction of individual and his environment, which fills a need and makes him more capable of dealing adequately with his environment'.

Learning is a change that occurs in a person forming a new response or giving a meaningful new one (Woolfok & Nicolich, 1984, p. 161). The learning process occurs from experience that shows there are changes or modifications in the pattern of adjustment. This change is an interaction between a person and his environment so that the needs can be fulfilled as the person can easily adjust. Next, Chance (2013, p. 36) states that "learning is a change in behaviour that is due to experience". Learning must involve students both intellectually and physically (Campbell, 1996, p. 37). Children should be active, using their basic knowledge and experience in new situations and continue to grow.

Based on the opinions above, it can be concluded that learning is a process of change the behaviour, both physically and mentally, obtained by training or adaptation to the environment. Thus, learning is an active process, the process of reacting to all situations around individuals. The process directed to the goal, the process of acting through various experiences and the processes of seeing, observing, and understanding something that is learned. Changes in learning behaviour are not caused by physiological growth or maturity processes.

2.2 *The definition of teaching and learning*

Teaching learning is a set of actions designed to support student learning by taking into account extreme events that play a role in a series of internal events experienced by students (Winkel, 1999). The equivalent term for 'Teaching learning' that can be found in foreign literature is 'instruction' (Winataputra, 1995, p. 2). According to Romiszowski (1981, p. 4), the term instruction refers to the teaching process centred on goals, or goal-directed teaching processes that can be re-planned. Therefore, the term instruction is often interpreted as a learning process. It is also interpreted as the process of making people learn according to design.

Djaafar (2001, p. 2) said that teaching learning is called instructional activity, which is an effort to manage the environment intentionally so that someone learns to behave in a certain way. Furthermore, teaching and learning is nothing but the activity of teachers helping students in learning. It works to help students learn, to create a learning environment, to motivate students, and to control discipline and atmosphere. This activity includes providing learning resources, designing activities that must be carried out by students, arranging time allocation, providing a place to research, and managing the classroom. Meanwhile, Hamalik (2003, p. 57) stated that teaching and learning is a combination of human, material, facilities, equipment, and procedures that influence each other to achieve the learning goals. Gagne and Briggs (1979, p. 19) state that "Instruction is the means employed by teachers, designers of materials, specialists curriculum, and others whose purpose is to develop an organised plan to promote learning", which implies that teaching and learning is not something that happens by accident, but rather through the teacher's knowledge of the basics of good teaching.

Based on the various opinions stated above, it can be said that teaching and learning is a set of actions designed to support student learning processes carried out by the teacher. It is work to help students learn, to create a learning environment, and to motivate students. This activity includes providing learning resources, designing activities that must be carried out by students, arranging time allocation, providing a place to research, and arranging classroom management to achieve the learning goals.

2.3 *Vocal learning*

Vocal elements are very important, especially in the relation to singing. One way to express music (Mahmud, 1995, p. 39) is through vocals expressed by singing. The level of ability to reveal the content and meaning of the singing depends on how to maintain the voice. The vocal sounds being produced are a variety of vowel sounds, double vowels, and consonants. There are basically five kinds of vowel sounds, which are A, E, I, O and U. Duplicate vowels are two vowels combined, such as ai, au, and oi. The examples of consonants are l, m, n, r, z, and w (Jamalus, 1988, pp. 55–57). From the definition above, it can be concluded that a vocal is an expression of contents and meanings with the mouthpiece, which then produces various vowel sounds, double vowels and consonants.

2.4 *The human voice and its classification*

The human voice has volume (hard and soft), pitch (tone accuracy), quality (quality), and duration (length of time). Hence, a good voice is a voice which can adjust the softness of its sound. The human voice involves the production of sound through the mouth coming from the throat. The low part of the throat will produce loud sounds or high or low sounds, while the bright or gloomy sounds come from the lungs (Magill & Jacobson, 1978). There are two kinds of classification of human voice, based on male and female voices: (1) male voices are categorised as bass (low voice), baritone (medium voice), and tenor (high voice); (2) women's voices are alto (low noise), mezzosoprano (moderate), and soprano (high voice) (Magill & Jacobson, 1978).

2.5 *The basics of singing techniques*

The basics of singing techniques include posture, breathing, and sound or vocal formation. Posture is a musical instrument of a singer. A good way to stand is to slightly rotate the thigh, knee, and ankle joints outward so that the two heel muscles behind the thighs must be tightened. Jamalus (1988, p. 49) says that the best technique for singing is diaphragmatic breathing, as when we breathe, the lungs increase in size so added space is needed, so the surface of the convex diaphragm moves down almost horizontally increasing the circumference of the diaphragm, and the surrounding body part increases.

Sound or vocal formations, consists of live vowels and dead vowels. Live vowels are A, E, I, O and U, dead vowels are usually at the end of words ending with L, NG, R, B, and others. People's voices have various qualities. A good voice will depend on the shape of the mouth. The sound made through the mouth that is opened wide and rounded will sound full, round and pleasant (Jamalus, 1988, pp. 52–53).

2.6 *Musical ability*

The term 'musical' in this research does not only mean 'good' at playing an instrument or singing, but also emphasises sensitivity to the sound of music, be it high or low tone (pitch), rhythm, or melody. George and Hodges (in Salim, 2009, p. 53) stated that musical ability is a sensitivity to respond music or a sensitivity to musical stimuli which include appreciation and understanding of music without having to have the skills to play any musical instruments. Another opinion expressed by Hallam (2006, p. 425) is that musical ability is considered to be related to rhythmic sensitivity, followed by the ability to understand and interpret music, thoughts and feelings through tone expression, being able to communicate through sound, motivation to engage with music, and being able to engage in music with others.

The difference between musical ability and musical talent is that musical ability leads to an understanding of the ability to accept musical stimuli, which is more related to sensitivity, feelings, and appreciation of music, while musical talent leads to performance abilities in music, such as the ability of musical expression through playing musical instruments. On the contrary, Sumaryanto (2000, p. 3) defines musical ability as things related to the concept of musical thinking and memory, tone composition, rhythm, emotional appreciation, the quality of songs, hearing, and range of sound which all lead to knowledge, potential, and attitudes that are reciprocal to the music itself. Based on some experts' opinions, it can be said that musical ability is a sensitivity about music or that is musical in nature without having to have musical skills.

2.6.1 *Aspects of musical ability*

Musical ability refers to the innate ability inherent in individuals in responding to musical elements (Sumaryanto, 2000, p. 4). According to "Measure of Music Audiation" theory

(Hallam, 2006, p. 426), the musical abilities of children aged 5–8 years can be measured through three aspects, namely tonal imagery (melody and harmony), rhythm imagery (tempo and metre), and musical sensitivity (phrasing, balance, and style).

The aspects of musical ability according to Dyson and Gabriel (in Sumaryanto, 2000, p. 3) are stated in three stages. Those are: First, pitch recognition factors, including recognition ability, pitch addition, and tonal memory. Second, the ability to perceive rhythmic movements of the tone, the tonal memory, and the tempo. Third, harmony factors, including memory of chords, chord analysis, and tone. Fourth, experience factors, including tonal movements and rhythm recognition. Fifth, musical assessment factors, including phrase recognition (melody) and musical style. Sixth, intensity factors, including the ability to respond to the quality of the tone and the colour of the tone.

2.6.2 *Factors influencing musical ability*

Every ability possessed by an individual certainly has things that affect its development, as well as the development of musicality in children. According to Lumbantouran (2012, p. 28), the background experience of musicality refers to the direction of musical experience that forms an awareness of bringing a person closer to music activities in general, especially singing as one of the main branches of artistic ability.

Furthermore, Salim (2009, p. 76) mentions five factors that influence musicality, namely intelligence, auditory acuity, gender, race, and cultural background. From both experts' opinions, it can be concluded that the factors that influence musicality are musical experience, intelligence, listening ability, reproductive ability, creativity ability, gender, cultural background, and race.

2.6.3 *Solfeggio ability*

Sight-singing is singing practice by reading the scores of songs, either individually or in groups (Henry & Mobberley, 1986, p. 1). Sight-singing can also be called the ability to read a musical work. Sight-singing is also the practice of singing tones according to the melody. The ability to sing a tone (sight-singing) is a student's level of fluency in changing the form of the notation into a sound or vocal without prior preparation. At the implementation of vocal learning in PIM 1 Vocal, sight-singing and ear training are suitably applied, according to the grade level of the students.

Ear training is a practice using the ear to listen to and understand music. It is almost similar to an audio recording. The practice of singing by listening is so-called because the singer must prioritise their own hearing or ears to control whether their own voice is good or not. Music hearing practice is usually in the form of dictation in the form of tones sung then mimicked, preceded by auditory and memory exercises. The dictation is in the form of a melody, chord, and rhythm. This hearing practice requires serious concentration so that the effect of the music can be understood.

Florentinus (1997, p. 62) divides the ability to hear notes (ear training) into three indicators of ability, which are: (1) the ability to hear and remember rhythm, write it down and voice it again; (2) the ability to hear and remember melodies or a series of notes, write them down, and reproduce them (3) the ability to hear and remember chords or the harmony of the combined notes. According to Benward, quoted by Sumaryanto (2000, p. 35), auditory ability is a combination of factors of habit and disposition. Habitual factors can be developed through regular practice, while pure innate factors come from a person's inbuilt abilities in the form of musical talent.

Sight-reading, according to Kodijat (2004, p. 10), is "the ability to simultaneously read and play songs that have never been known before". According to Kodijat (2004, p. 81), sight-reading is playing without prior preparation, at first glance. According to Banoe (2003, p. 379), sight-reading is "playing or singing (sight-singing) in one (first) reading".

2.7 *Drill method*

In this research, the researchers investigating drill learning theorised that learning takes place in two stages (Spence, 1956). These two stages are: (1) phase response learning or a learning response that is learning to respond to things in depth (during this stage, early learning errors occur more often; this stage requires practice so that the students are able to master the material); (2) the associative stage, which involves connecting responses to correct stimuli. During this stage, errors become more common.

2.8 *Hypothesis of the research*

Hypotheses were constructed to explain the cause and effect of the research and support a clear indication of the generalisation of the research. Specifically, the hypotheses result in a range of results that can be limited and other modifying factors influencing the research can be reduced. According to the formulation, the research hypothesis is: Ear training and sight-singing methods are effective for improving the musical ability of PIM 1 Vocal students.

4 RESEARCH METHOD

4.1 *Type of research*

This research uses a quantitative research method with an experimental and quasi-experimental design. Sugiyono (2013, p. 114) stated that quasi-experimental design has a control group, but it cannot function fully to control external variables that affect the implementation of the experiment.

McMillan and Schumacher (2010, pp. 258–259) state that there are six characteristics of research methods, which are: (1) hypotheses are built from theory (construct); (2) statistics on the comparison of treatment classes to control classes; (3) all control variables and bound variables are applied to the subject evenly; (4) every independent and bound variable can be measured; (5) research uses inferential statistics; (6) all research variables can be controlled. The quasi-experimental design was chosen because it was adapted to the circumstances of the research area. In accordance with the opinion of Kirk (1995, p. 6), who states that quasi-experimental research is based on practical and ethical reasons, it is sometimes not possible to put the subjects into groups.

In this research, the experimental design used was a pretest/post-test control group design. There were two groups in this design, which were selected randomly, then given a pretest to establish their initial state and whether there was a difference between the experimental group and the control group. The pretest results can be said to be good if the values for the experimental group and the control group are not significantly different.

4.2 *Research variables*

The variables in this research are experimental variables that consist of the independent variable and the dependent variable. The independent variable in this research is the drill method itself (X). The dependent variable in this research is the musical ability to listen to the song melody (ear training) (Y1), the musical ability to rewrite the song melody (sight-reading) (Y2), and the musical ability to re-sing the song melody (sight-singing) (Y3).

4.3 *Data collection*

This research was conducted in Yogyakarta State University PIM 1 Vocal class 2017 in the 2017/2018 school year. The stages of the research are the preparation stage (including submission of titles, making proposals, surveys and observations at the relevant university, permission for the research, and preparation of instruments), the implementation stage (the activities that take place in the field which includes testing instruments and collecting data), and the last stage which is data processing and the preparation of the research reports. The data collection technique used in this research was a test. The data needed in this research is of the musical ability of students aged 19–21 years. The research used an oral test (individuals and groups). The process for developing the test was formulating objectives, compiling outlines, making specification tables, selecting items, compiling items, composition of the instructions, developing answer sheets, making answer keys, carrying out tests, and revising tests. Observation was carried out during the research process and was conducted during the pretest, at the time of the treatment, and post-test. The observation was to observe the attitudes of students during the research.

The musical ability test was done to measure the students' musical ability at the age of 19–21. This is a test to measure the level of musical ability based on the aspects seen. Respondents were given directions for taking this musical ability test, so that at the time of the assessment the respondents only followed the instructions from the researcher.

The instrument for measuring musical ability consists of two indicators and five sub-indicators, each of which consists of five items so that the total number of items for the test is 20.

a. Ear training ability treatment

An oral test, a practice test, and observation were used to measure the ability of ear training of PIM 1 Vocal students aged 19–21. Each student listened to the tone and then had to mimic it. Students who have good ear training skills can mimic the tone correctly, while students who do not imitated the wrong tone. Major scales and minor scales were the tones imitated by the students.

b. Sight-singing ability treatment

Oral tests, practical tests, and observations were used to measure sight-singing ability of the vocal students aged 19–21. Each student was given scores to read and sing. Students who had mastered sight-singing could read the scores properly, while students who had not mastered it read the scores poorly.

4.4 *Instrument validity and reliability*

The validity of the construct used in this research can be seen through trials of research instruments at universities that are not the subject of the research. Validation of the construct is needed in this research in order to identify and pair the items to determine the levels of cognitive and psychomotor aspects. Meanwhile, the validation of the content of this research used expert judgement techniques. For expert judgement of the musicality test instruments, the experts consulted were Prof. Djohan Salim, M. Si and Dr. Kun Setyaning Astuti, M. Pd. As in the content validity, to determine the level of construct validity, the preparation of items can be done based on the instrument grid. Meanwhile, to estimate the reliability of the instrument, the techniques used were Cronbach's alpha and inter-rater reliability using Cohen's kappa.

4.5 *Population and sample*

The population of this research was 23 PIM 1 Vocal students. The students were divided into two groups, namely, an experimental group and a control group. The experimental group consisted of 15 students while the control class consisted of eight. The sampling in this research used a saturated sampling technique where all members of the population were sampled. The samples for the control group and the experimental group were randomly selected.

4.6 *Data analysis technique*

The steps taken in describing the data are by making a distribution of pretest and post-test data, which are obtained from descriptive statistical results. The summary is in the form of mean, median, mode, standard deviation, variance, minimum score, and maximum score using SPSS 16 for Windows statistical software.

5 RESULTS

This research began with the initial test at the first meeting with the control and experimental groups. The test was a musicality listening, writing, and singing test. The initial test was used to determine the initial levels of the students' musicality. At the second meeting, the control group and the experimental group were given a musicality test. The research was continued by giving repeated treatments to the experimental group from the third meeting until the ninth meeting. After the experimental group had been given the treatment, the research continued at the tenth meeting with the final test (post-test). The final test was in the form of a musicality test involving listening to the melody (ear training), rewriting the melody (sight-reading), and singing the melody (sight-singing). The final test was used to determine the final condition of the musicality test. After the musicality test was completed, at the eleventh meeting, the experimental group was given a drill method to improve musical ability, while the control class was not.

Pretest scores were used to determine the experimental group's status and their initial level of musical ability before being given a treatment. In addition, the pretest scores were also used to determine musicality tests in the control group that was not given the treatment. Besides this, the post-test score was used to determine the experimental group's final level of musical ability after having been given the treatment. The post-test score was also used to determine the status and final level of musicality in the control group that was not given a treatment. Descriptions of each set of test data are described in what follows.

The musicality test data consisted of the results of the pretest and post-test in the form of quantitative data. The musicality test was given to the experimental group in two meetings, before and after the treatments. The treatments were given during the seven meetings. In addition, musicality tests were also given twice to the control group that was not given any treatment. After the data were obtained, the next step was to analyse the data using descriptive statistics.

The variance homogeneity test was conducted to test the variance similarity of the experimental group and the control group for the pretests and post-tests. This homogeneity test was conducted on the experimental group that was given a treatment and the control group that was not given a treatment. This homogeneity test was also conducted on all dependent variables of the two groups to determine the similarity of the covariance variance matrix of the dependent variables simultaneously. In addition, the multivariate homogeneity test was also used to determine whether the two groups have the same data variance or not.

The multivariate homogeneity test was tested using the Box's M test by looking at the significance values of the Box's test table. The decision criteria (Pallant, 2007, p. 286) are the data that meet the assumption of multivariate homogeneity if the significance value is > 0.001. This multivariate homogeneity test was carried out by using SPSS 23.0 statistical software.

The results of the multivariate homogeneity test in the Box's test table indicates that the control and the experimental groups at the pretest and post-test stages have a homogeneous data variance matrix. This is indicated by a significance value of 0.076 at the pretest and a significance value of 0.363 at the post-test.

6 CONCLUSION

The results of this study show a significant difference between the experimental class and the control class. The experimental class, which was treated by the drill method, improved its musical ability, while the control class showed no improvement in musical ability. The results of the research use test scores to show improvement in the experimental group's musicality. This is indicated by the higher post-test scores of the experimental group's musicality test, which are higher than the average students' scores. The improvement in the average scores of the experimental group is indicated by the score of 3.91. Meanwhile, there is a decrease in the control group's musicality score. This is indicated by the students' average post-test scores, which are lower than the average scores. The decrease in the control group is shown by the score of 20.58. The results of the multivariate homogeneity test in the experimental group and the control group at the post-test with Box's M produced a value of 3.308, an F value of 1.063, and a significance value of 0.363, which is above 0.001. Thus, it can be said that the drill learning method is effective for improving musical ability.

REFERENCES

Banoe, P. (2003). *Kamus music* [Music Dictionary]. Yogyakarta: Kanisius.

Burton, W.H. (1962). *The guidance of learning activities*. New York, NY: Appleton-Century-Crofts.

Campbell, N.A. (1996). *Biology*. Menlo Park, CA: Benjamin.

Chance, P. (2013). *Learning and behavior*. Scotland,UK: Nelson Education.

Djaafar, Z. (2001). *Kontribusi strategi pembelajaran terhadap hasil belajar* [Contribution of learning strategies to learning outcomes]. Fakultas Ilmu Pendidikan, Universitas Negeri Padang.

Florentinus, T. S. (1997). *Pengembangan Instrumen Pengukuran Kemampuan Solfeggio* [Development of Solfeggio Capability Measurement Instruments]. *Jakarta: IKIP Jakarta.*

Gagne, R.M. & Briggs, L.J. (1979). *Principles of instructional design*. New York, NY: Holt.

Hallam, S. (2006). *Music psychology in education*. London: University of London.

Hamalik, O. (2003). *Proses belajar mengajar*. Jakarta: Bumi Aksara.

Henry, E. (1997). *Sight singing*. New Jersey: Prentice Hall.

Henry, E. & Mobberley, J. (1986). *Musicianship: Ear training, rhythmic reading, and sight singing* (vol. 1). California: ICON Group International.

Jamalus, D. (1988). *Pengajaran Musik Melalui Pengalaman Musik* [Music Teaching through Music Experience]. Jakarta, Indonesia: Department of Education.

Kaufman, J. C., & Baer, J. (Eds.). (2005). *Creativity across domains: Faces of the muse*. New York: Psychology Press.

Kirk, R. E. (1974). *Experimental design: Procedures for the behavioral sciences*. Pacific Grove, CA: Brooks/Cole.Spence, K. W. (1956). *Behavior theory and conditioning* (Vol. 35). New Haven: yale university Press.

Kodijat-Marzoeki, L. (2004). *Istilah-istilah musik* [Music Terms]. Yogyakarta: Djambatan.

Magill, P. C., & Jacobson, L. (1978). A comparison of the singing formant in the voices of professional and student singers. *Journal of Research in Music Education, 26*(4), 456-469.

Mahmud, A.T. (1995). *Musik dan Anak* [Music and children]. Jakarta, Indonesia: Departemen Pendidikan dan Kebudayaan.

McMillan, J.H. & Schumacher, S. (2010). *Research in e: Evidence-based Inquiry, MyEducationLab Series*. New Jersey: Pearson.

Pallant, J., & Manual, S. S. (2007). A step by step guide to data analysis using SPSS for windows. In *SPSS Survival manual*. Berkshire UK: Open University Press.

Romiszowski, A.J. (2013). *Developing auto-instructional materials*. London: Routledge.

Salim, D. (2009). *Psikologi Musik* [Music psychology]. Yogyakarta, Indonesia: Galangpress.

Slameto. (1955). *Belajar dan Faktor-faktor yang mempengaruhinnya* [Learning and the factors that influence it] (Rev. ed.). Jakarta, Indonesia: Rineka Cipta.

Sugiyono, P.D. (2013). *Metode penelitian kuantitatif dan kualitatif dan R&D* [Quantitative and qualitative and R & D research methods]. Bandung, Indonesia: Alfabeta.

Sumaryanto, F.T. (2000). *Kemampuan Musikal dan Pengaruhnya terhadap Prestasi Belajar Musik* [Musical ability and its influence on the achievement of learning harmony music]. *Harmonia: Journal of Arts Research and Education, 1*(1).

Winataputra, U.S. (2014). *Teori belajar dan pembelajaran* [Learning and learning theory]. Bandung, Indonesia: Rosda Karya.

Winkel, W.S. (1999). *Psikologi Pengajaran* [Teaching psychology]. Jakarta, Indonesia: Grasindo.

Witherington, H.C. & Burton, W.H. (1982). *Teknik-teknik belajar dan mengajar* [Learning and teaching techniques]. Jakarta, Indonesia: Grasindo.

21st Century Innovation in Music Education – Astuti et al. (eds)
© 2020 Taylor & Francis Group, London, ISBN 978-0-367-11122-9

Early inculcation of character values through children's songs

H.Y.Wicaksono, Agustianto & Y.J.Fitria
Universitas Negeri Yogyakarta, Indonesia

ABSTRACT: Character education is an educational movement that supports the social, emotional and ethical development of students. It involves proactive effort on the part of schools, districts, and states to instill in students core ethical values such as caring, honesty, diligence, fairness, fortitude, responsibility, and respect for self and others. Character education provides long-term solutions to moral, ethical, and academic issues that are of growing concern in our society and our schools. Character education teaches students how to be their best selves and how to do their best work. This paper describes how teachers use children's songs as a medium to instill character values in early childhood at a preschool in Yogyakarta, Indonesia. Interviews and observations were used to collect data on the classroom learning activities in which songs were used to teach character values.

1 INTRODUCTION

Character education is an educational movement that supports the social, emotional and ethical development of students. Character education teaches students how to be their best selves and how to do their best work. It is important that children receive character education as early as possible to help them become responsible adults and prepare them for life's challenges and possibilities. However, good character is not formed automatically; it is developed over time through a sustained process of teaching, example setting, learning and practice. This is the purpose of character education. The intentional teaching of good character is particularly important in today's society since our youth face many opportunities and risks unknown to earlier generations. They are bombarded with many more negative influences through the media and other external sources prevalent in today's culture (Pala, 2011). Thus, teachers in schools need to provide learning materials that can develop students' character. As Kaelan (2007:180) has observed, character education is directed at providing advice, standards, and a set of oral and written rules about how humans must live and act in order to become good human beings.

Dewey (in Noddings, 2002:166) identifies two purposes of character education: to impart the principles of good character to students and to produce citizens of good character through education.

Ki Hadjar Dewantara has long thought in terms of character education. It is important to sharpen intelligence in order to build a strong culture through which people can realize personality (*persoonlijkhheid*) and characterThis allows people to overcome primal such as passions and traits such as ruthlessness, anger, and greed) (in Siswa,1977, 24). In other words, character education is education that teaches good values that provide an important foundation for human integrity and personality.

Many people believe that character education needs to be included in the school curriculum. Children who are taught good behaviour and values in schools develop respect for others and for the law. Schools are responsible for guiding children's development in a step-by-step process, and character development or learning ethical values is one step in this broader development process. One emerging approach to inculcating character values in schools, including in early childhood, is the use of songs about virtues such as love, mutual respect, and purity. This was also done by teachers at the Nurul Dzikri preschool in Yogyakarta.

In a preliminary study conducted in 2018, one of the teachers was interviewed about the role of songs in instilling character values in children. The particular songs were called “One-by-one dear mother”, which teaches love of family, and “Thank you”, in which children express their gratitude to the teachers who have guided them.

At very young ages, the desired message is more easily conveyed through song than through words, since the sound of musical instruments and singing voices can evoke emotional responses in listeners, including children, making them feel happy, sad, calm, comfortable, or peaceful (Salim, 2005).

According to Kusumawati (2013), songs that are usually sung by children are cheerful and reflect noble ethics. They refer to simple activities associated with childhood and often tell about love for God, parents, brothers and sisters, and others, the beauty of nature in simple language that can be easily understood.

In summary, for children are simple, have a cheerful rhythm and reflect the age of the child. The lyrics often have an educational function, hence the songs can be used as a medium for teaching character.

2 RESEARCH METHOD

The study employed qualitative research methods to collect rich data in a natural setting (Soejono, 1999). Such descriptive data can be analyzed to generate in-depth understanding of the phenomenon under investigation (Creswell, 2013)—in this case, the process of instilling character values in early childhood through songs (Creswell, 2013)

The research was conducted in a natural setting, namely, in the classroom and in the school yard of PAUD (early childhood education) Nurul Dzikri Yogyakarta and in classrooms when the pupils were taught character-forming songs.

2.1 *Research subjects (participant)*

The research participants were five teachers at the school who taught character education through songs and the head master, who participated in singing in class as a way to teach character education.

2.2 *Procedures*

Data were collected through observation and interviews. The first stage involved observing the classroom as the teachers taught the song ‘Wake Up’. The children’s responses were video recorded and extensive field notes were taken. In this observation, it was seen that children were very happy to see the expressions and gestures presented by the teacher. This can be seen from several children following the movements carried out by the teacher. Photographs of the teaching process were also taken as an evidence. These multiple data gathering methods were used as triangulation.

In the next stage, the teachers and principal were interviewed to obtain their perspectives on the success of the use of children’s songs to instill character values. It is known that the songs that are used to instill these character values are more themed love, discipline, and responsibility. The interviews were audio recorded and transcribed.

The data were analysed using the interactive model developed by Miles and Huberman (1994) and comprising reduction, data display, and conclusion/verification.

3 RESULTS

In interview, the headmaster strongly supported the value of instilling character values in children through songs, which are easily understandable: “*The teachers here teach character values with songs. Songs are enjoyable for children happy, and make it easier for them to understand the material.*"

Teacher supported this view. In relation to instilling hygiene values, for instance: *"So, if you want to introduce any character values, for example about cleanliness, you can do it with songs. With songs, they are easier to accept. When we replay the song, the students were reminded of the value."*

Another teacher, Ms Ny, also commented on the role of body movements in reinforcing the message of the song. Because children like to play, the teacher accompanied the singing with body movements that reflect the theme being taught as she also played the samba. *"Usually in singing, the teacher not only sings, but also uses body language, and while playing, because what is spoken with the movement is faster, and with body language (movements that match the song), the message is more easily conveyed".* The observations showed that the teacher modified her expression in accordance with the lyrics, and that the children happily followed The lyrics of the song "Wake Up", for instance, indirectly convey the character values of discipline and cleanliness, telling children that they should take a bath after their wake up and help their mothers to make their beds.

4 DISCUSSION

Character education needs to be provided at all educational levels, from preschool to university, since today's young people are exposed to numerous negative influences, especially through the media. Therefore, it is critical to create schools that simultaneously foster character development and promote learning. Children's songs are an important means through which schools can developing character, especially in early childhood education.

Nurul Dzikri Yogyakarta is one of the early childhood schools in Indonesia that has adopted this approach. Through songs, children easily understand the various character values that their teachers are seeking to impart, since children naturally like to play. In this study, the principal emphasised the importance of instilling character values early.

The literature supports the view that there are important benefits in introducing character education from an early age (Pala, 2011). Students more readily develop important human qualities and understand why it is important to live by them. Character education develops a positive moral climate by engaging the participation of students, teachers and staff, parents, and communities. It creates an integrated culture of character that supports and challenges students and adults to strive for excellence, and teaches adults and students to understand, engage in, care about, and act on core ethical values such as respect, justice, citizenship, fairness, and responsibility for self and others in school and as part of a larger community.

Each lesson at Nurul Dzikri preschool begins with a song intended to teach character values. . If, for example, the desired values are are discipline, cleanliness, responsibility, respect, and mutual help, the teacher sings the song "Wake Up", using body movements and expressions while playing. Then the children imitate the teacher's singing, movements and expressions.

The use of songs to instill character values helps children to grasp the message more quickly and effectively than is possible through words alone.

At Nurul Dzikri, these songs are sung every day, at the beginning of the lesson, and while the children are resting. At rest, children are played with songs that contain character values, such as the song "Thank You", "Dear Teacher", and "First Dear Mother".

Character education can be initiated at any grade level. It is important to set a strong foundation during the earlier grades and to reinforce and build upon that foundation during the later grades. To be effective, character education must include the entire school community and must be infused throughout the entire school curriculum and culture (Pala, 2011).

5 CONCLUSION

Character education is an important component of education that should begin as early as possible. This is as expressed by Gunawan (2017) who stated that character education needs to be applied from an early age, as early childhood (between from 2 to 7 years) is a time when people begin to learn about the ways of life and socializing.

In t Nurul Dzikri preschool in Yogyakarta, Indonesia, character values are instilled through children's songs in several ways. At the beginning of each class, the teacher sings the song, using body language and expressions that reflect the lyrics. Then the children sing the song, imitating the teacher's expressions and body movements. Finally, they all sing the song together.

Schools can play a significant role in imparting character values, and there should be ongoing development of such programs. Nonetheless it is vital that this process also involves families and other community members.

REFERENCES

Creswell, J.W., &Creswell, J.D. (2017). *Research Design: Qualitative, Quantitative, and Mixed Methods Approaches*. Sage Publications.

Gunawan, R. (2017, May). The Role of Character Education for Early Children in Early Childhood Education Programs in Happy Kids Bogor Indonesia. In *1st Yogyakarta International Conference on Educational Management/Administration and Pedagogy (YICEMAP 2017)*. Atlantis Press.

Kaelan, A.Z. (2007). Pendidikan Kewarganegaraan[Civic Education]. *Paradigma, Yogyakarta.*

Kusumawati, H. (2013). Pendidikan Karakter Melalui Lagu Anak-anak [Values Education through Children's Songs]. *Imaji, 11*(2).

Miles, M.B. & Huberman, A.M. (1994). *Qualitative Data Analysis*. California: Sage Publications, Inc.

Noddings, N. (2002). *Educating moral people: A caring alternative to character education*. Teachers College Press, PO Box 20, Williston, VT 05495-0020.

Pala, A. (2011). The need for character education. *International Journal of Social Sciences and Humanity Studies, 3*(2), pp. 23-32.

Salim, D. (2005). Respons Emosi Musikal dalam Gamelan Jawa.[Musical Response in Javanese Gamelan]. *PSIKOLOGIA 2*, pp.63-75.

Siswa, M.L.P.T. (1977). Karya Ki Hadjar Dewantara. *Bagian pertama: Pendidikan. Yogyakarta: Majelis Luhur Persatuan Taman Siswa.*

Soejono, A. (1999). *Metode penelitian: Suatu pemikiran dan penerapannya*. [Research methods: Theory and Implementation]. Jakarta: Reneka Cipta.

21st Century Innovation in Music Education – Astuti et al. (eds)
 ISBN 978-0-367-11122-9

Effect of a backing track on improvement of students' electric guitar playing skills

A. Suprayogi & K.S. Astuti
Universitas Negeri Yogyakarta, Indonesia

ABSTRACT: This research aims to describe the effects of audio backing tracks on the improvement of electric guitar playing skills of learners at the Ahmad Dhani School of Rock, Magelang Branch. The research approach is quantitative with an experimental method. The data analysis used a multivariate analysis of variance followed by a Dunnett's test. The results of the research show that audio backing tracks are effective in improving aspects of posture by as much as 21.6%, basic technique mastery by 19.7%, perfect pitch by 19.9%, phrasing/timing by 2.1%, intonation by 19.4%, and song interpretation by 9.9%.

1 INTRODUCTION

The development of music education in Indonesia is advancing rapidly in both formal and informal music institutions (Bernard, 2007, p. 12). The Ahmad Dhani School of Rock (ADSOR), Magelang Branch, uses an innovative teaching strategy involving audio backing track media to support teaching. This research reveals the musical elements that can be developed using that media.

2 LITERATURE REVIEW

Audio backing tracks are the audio or Musical Instrument Digital Interface (MIDI) files recorded to be played along with musicians when they play a musical instrument, sing a song or to fill or add the missing parts of music instruments. According to Kai Yang and Xi Zhou, as cited in Subiakto (2012, pp. 1–12):

> ...MIDI (Musical Instrument Digital Interface) is a digital communications language and compatible hardware specification that allows multiple electronic instruments, performer controllers, computers and other related devices to communicate with each other within a connected network. Standard MIDI files are a popular source of music, and MIDI music is widely accepted by musicians, composers, music lovers, teachers, mobile phone users and game makers. Especially, MIDI has demonstrated a very broad application prospect in music education....

Based on the above description, MIDI is a digital communication language. It is more well-known as a music source and widely used by musicians, composers, music lovers, teachers and game makers. Specifically, MIDI is also used widely in music teaching (Case, 2007, p. 225).

Govan (2002, p. 108) states that a band or solo musician also often uses backing tracks or "minus-one" tracks to fill the empty part of music, as well as to help improve sound and replicate sound which is closer to the original sound. Minus-one is a track whose application is not accompanied by a voice or an instrument, whose function is the same as that of an audio backing track. Audio backing track media are expected to play a role as a supporting media facility for learners in learning some of the aspects of musicality, including posture, basic technique mastery, perfect pitch, timing/phrasing, intonation and song interpretation.

3 METHOD

This research is quantitative, and utilises an experimental method (Pallant, 2010, p. 283). The research design used is a pretest/post-test control-group design. The sample consists of 20 learners divided into two classes: the experimental and control classes (Glass & Hopkins, 1984, p. 128). Before receiving treatments, the two classes were given a pretest (X1). Then a treatment (X2) was given to the experimental class, comprising electric guitar instruction using the audio backing track.

The evaluation was done by using a performance test in the form of the skill in playing an electric guitar, performing the song 'Sweet Child O' Mine'. The test indicators consist of six aspects, including posture, basic technique mastery, perfect pitch, timing/phrasing, intonation and song interpretation.

The treatment given to the experimental class used the audio backing track media in four class meetings in February 2017. The data obtained were in the form of the scores of the learners' skill test.

3.1 *Analysis*

This research has one independent variable and more than two dependent variables and therefore, it uses a Multivariate Analysis Of Variance (MANOVA). The requirement for the MANOVA test is that the distribution must be normal and homogeneous. The following is the result of the analysis of the requirements (Setyanto, 2015, p. 31).

3.2 *Requirements analysis*

a. Normality test
   The result of the normality test of the pretest and post-test shows significant values in the aspects of posture (experimental class = 0.834, control class = 0.992), mastery of basic technique (experimental class = 0.941, control class = 0.126), perfect pitch (experimental class = 0.603, control class = 0.527), phrasing/timing (experimental class = 0.436, control class = 0.302), intonation (experimental class = 0.560, control class = 0.307) and song interpretation (experimental class = 0.503, control class = 0.139). Every aspect has p-value higher than 0.05 (or $p > 0.05$), and therefore it can be concluded that the research data in the pretest and post-test of meetings I, II and III are normally distributed.
b. Homogeneity test
   The result of the homogeneity test shows that the significance value of the aspect of posture is 0.058, mastery of basic technique is 0.125, perfect pitch is 0.424, timing/phrasing is 0.870, intonation is 0.171 and song interpretation is 0.209. Every aspect has $p > 0.05$ and, therefore, it can be concluded that the two groups of data have the same variants and are homogeneous.

The treatment was conducted on 19 February 2017. The meeting was at the experimental class, consisting of the pretest class and the post-test class. The result of the research with the experimental class is presented as follows.

a. Pretest of the experimental class

The data of the pretest consisted of six aspects, including posture, basic technique mastery, perfect pitch, timing/phrasing, intonation and song interpretation, with the respondents consisting of eight learners.

Based on the data on the process of electric guitar training of the students at ADSOR before they were given the treatment, the lowest score for the *posture* aspect is 17.00, the highest score is 26.00, mean 21.62, mode 23.00, median 22.50, standard deviation 3.07. The lowest score for the *basic technique mastery* aspect is 15.00, the highest 30.00, mean 22.87, mode 24.00, median 23.50, standard deviation 4.29. The lowest score for the *perfect pitch* aspect is 15.00, the highest 27.00, mean 21.75, mode 21.00, median 21.50, standard deviation 3.80. The lowest score for the *timing/phrasing* aspect is 16.00, the highest 22.00, mean 19.00, mode 16.00 and 21.00 (this is called a bimodal mode because of two in-orderly modes), median 19.50, standard deviation 2.39. The lowest score for the *intonation* aspect is 16.00, the highest 24.00, mean 20.75, mode 22.00, median 22.00, standard deviation 2.87. The lowest score for the *song interpretation* aspect is 16.00, the highest 22.00, mean 19.00, mode 19.00, median 19.00, standard deviation 2.00.

The categorisation of the data of the effect of audio backing track training on the improvement of the electric guitar skill of the students at ADSOR used the formula based on the research observation sheet shown in Table 1.

The results are grouped into five categories by using the interval system in Table 1. The result of the data analysis follow below.

Table 2 shows that before the treatment in the form of using audio backing tracks, no learners are in the Excellent or Good categories (0.00%), one learner is in the Sufficient category (12.50%), five learners are in the Poor category (62.50%), and two learners are in the Very poor category (25.00%). Therefore, it can be concluded that the tendency of the electric guitar skill of students at ADSOR without treatment using the audio backing tracks is in the Poor category (62.50%).

b. Post-test of experimental class

The post-test result of the experimental class is the data captured after they were given the treatment in the form of training by using the audio backing track media.

Based on this data, the lowest score for the *posture* aspect is 22.00, the highest score is 32.00, mean 26.87, mode 25.00 and 32 (this is called a bimodal mode because of two in-orderly modes), median 26, standard deviation 3.83. The lowest score for the *basic technique mastery* aspect is 22, the highest 36.00, mean 29.75, no mode, median 30.50, standard deviation 5.26. The lowest score for the *perfect pitch* aspect is 24.00, the highest 33.00, mean 27.75, mode 24.00, median 26.50, standard deviation 3.69; The lowest score for the *timing/phrasing* aspect is 23.00, the highest 30.00, mean 25.87, mode 24.00, median 25.50, standard deviation 2.64. The lowest score for the *intonation* aspect is 23.00, the highest 30.00, mean 26.12, mode 23.00 and 25 (there are two in-orderly bimodal modes of 23 and 25), median 25.50, standard deviation 2.87. The lowest score for the *song interpretation* aspect is 22.00, the highest 31.00, mean 25.37, mode 26.00, median 25.00, standard deviation 2.82.

Table 1. Electric guitar playing skill categorisation.

Interval	Category
35–40	Excellent
29–34	Good
23–28	Sufficient
17–22	Poor
10–16	Very poor

Table 2. Results of the mean category of the experimental class pretest.

No.	Interval category	Frequency	Percentage (%)
1	35–40	0	0.00
2	29–34	0	0.00
3	23–28	1	12.50
4	17–22	5	62.50
5	10–16	2	25.00
Total		8	100.00

Table 3. Hypothesis testing: results of MANOVA.

	Model	Sum of squares	df	Mean square	F	p
Posture	Regression	1.581	1	1.581	9.147	.009
	Residual	2.419	14	.173		
	Total	4.000	15			
Basic technique	Regression	1.590	2	.795	4.287	.037
	Residual	2.410	13	.185		
	Total	4.000	15			
Pitch	Regression	1.812	3	.604	3.311	.057
	Residual	2.188	12	.182		
	Total	4.000	15			
Phrasing	Regression	2.908	4	.727	7.326	.004
	Residual	1.092	11	.099		
	Total	4.000	15			
Intonation	Regression	2.931	5	.586	5.484	.011
	Residual	1.069	10	.107		
	Total	4.000	15			
Interpretation	Regression	3.002	6	.500	4.513	.022
	Residual	.998	9	.111		
	Total	4.000	15			

The results of the MANOVA test (Table 3) show that the posture aspect has a regression value of 1.581 with a significance of 0.009; the mastery of basic technique aspect has a regression value of 1.590 with a significance of 0.037; the pitch aspect has a regression value of 1.812 with a significance of 0.057; the phrasing aspect has a regression value of 2.908 with a significance of 0.004; the intonation aspect has a regression value of 2.931 with a significance of 0.011; and the song interpretation aspect has a regression value of 3.002 with a significance of 0.022.

4 RESEARCH FINDINGS

This experiment was conducted over a period of two months, from February to March 2017. The materials taught involved basic techniques for electric guitar, including bending, slurring, trilling, sliding, vibrating, tapping, alternative picking, harmonising, sweep picking, arpeggiating and barre. An explanation of musical scales was given, including major, minor natural, minor harmonic, minor melodic and pentatonic scales, and the learners were given the song 'Sweet Child O' Mine' by Guns N' Roses to play. The song was used as a measure of the application of basic techniques of playing the electric guitar and the understanding of musical scales.

The results are classified into five categories using the intervals shown in Table 1. The results of the mean category of the experimental class post-test (after the learners were given treatment in the form of training in the use of audio backing track media) were as shown in Table 4.

Table 4. Results of the mean category of the experimental class post-test.

No.	Interval category	Frequency	Percentage (%)
1	35–40	0	0.00
2	29–34	2	25.00
3	23–28	6	75.00
4	17–22	0	0.00
5	10–16	0	0.00
Total		8	100.00

Table 4 shows that after the treatment in the form of using audio backing tracks, no students were in the Excellent, Poor or Very poor categories (0.00%), two learners were in the Good category (25%) and six learners were in the Sufficient category (75%). Therefore, it can be concluded that the tendency for improvement in the skill in playing electric guitar of students at ADSOR of Magelang Branch after treatment is in the Sufficient category (75%).

5 CONCLUSION

There is a positive effect of audio backing tracks on the improvement of electric guitar playing, and it contributes to the process of teaching electric guitar, especially in the aspects of posture (21.6%), mastery of basic technique (19.7%), perfect pitch (19.9%), and intonation (19.4%), while for the aspect of song interpretation, the contribution is 9.9% and for phrasing/timing is 2.1% and the significance level of 0.005/6 aspects = 0.0083. Therefore, it can be concluded that backing tracks play an effective role in helping the process of teaching the electric guitar, especially in the aspects of posture, mastery of basic technique, perfect pitch and intonation.

There may be other applications for the use of backing tracks for training in other instruments, such as drums, piano/keyboard, bass, saxophone, and so on. There are also possibilities for research and development of products that can test the effectiveness of the use of backing tracks in musical training.

REFERENCES

Bernard, S.L. (2007). *Les Musik Untuk Anak Anda (*Music Tutoring For Your Children*)*. Jakarta, Indonesia: PT Gramedia Pustaka.

Case, A.U. (2007). *Sound FX*. Oxford. UK: Focal Press.

Govan, G. (2002). *Creative guitar 1*. London, UK: MGP Books Bodmin.

Glass, G.V. & Hopkins, K.D. (1984). *Statistical methods in education and psychology* (2nd ed.). Englewood Cliffs, NJ: Prentice Hall.

Pallant, J. (2010). *SPSS survival manual*. Sydney, Australia: Allen & Unwin.

Setyanto, A.E. (2015). Memperkenalkan Kembali Metode Eksperimen dalam Kajian Komunikasi (Reintroducing the Experiment Method in the Communication Study). *Ilmu Komunikasi* (Communication Study*)*, *3*(1), 37–48.

Subiakto, T. (2012). *Peran Teknologi Komputer dalam Proses Kreatif Musik* (The Role of Computer Technology in the Creative Process of Music). Yogyakarta: UGM Press.

Sugiyono. (2009). *Metode Penelitian Kuantitatif Kualitatif Dan R & D* [Quantitative, qualitative and R&D research methods]. Bandung, Indonesia: Alfabeta.

21st Century Innovation in Music Education – Astuti et al. (eds)
© 2020 Taylor & Francis Group, London, ISBN 978-0-367-11122-9

How music education can contribute to moral development and citizenship?

W. Veugelers & J. Terlingen
University of Humanistic Studies, Utrecht, The Netherlands

1 INTRODUCTION

We like music, and we listen a lot to all kind of music. Ranging from traditional music from all over the world to rock-music, in particular the music we grow up with, the soul and folk-rock of the sixties and seventies. Music is part of culture and it can be created actively and consumed passively. However this consumption is often not considered by the listener as passive, it is engaged and part of his or her personal identity. According to sociologists of generation your period of adolescents is very crucial for your identity development, and the culture and subculture in which you are growing up influences your identity and stay relevant the rest of your life.

We can conclude that music is very important for the identity development of people, in particular adolescents. Music contributes to their moral identity and citizenship. There are many different kinds of genres in music, for example traditional folk music, religious music, formal national music, children songs, songs from specific social and cultural groups, rock, disco, techno, etc. Music can be more traditional, contemporary or experimental. Music can be more local inspired or by global international trends.

Often there is a fusion of styles. There are many examples of crossover cultural politics that became very successful. Michael Jackson became a world star when he combined soul with rock on his album Thriller. Youssou N'Dour combined African rhythms with disco. And nowadays you hear in Indonesia crossovers of gamelan music with electronic dance. Such crossovers are expressions of feelings and contribute to new cultural identities that show a mix – a 'bricolage' - of inspirations.

Music is often part of a broader cultural style with own words, own clothes, and own attributes: either in a traditional style, in specific cultures or religions, or in a more modern or global style. Such a global style is often in fact quite Western oriented. Music is really part of social, cultural, moral and civic developments. It is part of life and creates a lifestyle.

Music education can support youngsters in their moral and citizenship development. Music education can focus on:

- Technical aspects of rhythm and playing instruments, including singing
- Focus on understanding cultural meanings in music
- Creating new meanings by playing music

For the relationship of moral development and music education in particular the non-technical cultural understanding and creating new meanings are important. However technical music skills can help in giving meaning and in understanding music.

2 MORAL VALUES: ADAPTATION, AUTONOMY AND SOCIAL CONCERN

We often speak of moral values. But what kind of moral values are we referring to? In our theoretical and empirical research, three moral values are pivotal: *adaptation, autonomy, and social concern*. We find them in our empirical studies, of both quantitative and qualitative nature. These values are also very central to our theoretical analyses:

1. *Adaptation* is at the heart of Durkheim's (1913) sociology of socialisation and in traditional forms of character education, as in the work by Lickona (2004). It is about following norms and values: hence, adaptation.
2. *Autonomy* is a central concept in Kohlberg's moral development theory, and in moral reasoning and moral identity. Autonomy is a very important concept in moral psychology. Within sociology, Giddens combined structure with agency. Introducing agency in sociological theory has made societal development more dynamic and gives room for a personal articulation in identity development, and so contributes to personal empowerment.
3. The *social* moral dimension can we find in a range of theories; from psychological emphases on empathy (Selman, 2004) and care (e.g. Noddings, 2002) to more transformative political articulations in solidarity and social justice (e.g. Freire, 1982; Veugelers, 2017). The social orientation, in particular in more political and social justice orientations, can contribute to collective empowerment.

In the moral development of each human being there is a specific combination – 'bricolage' – of these three moral values. Also in educational activities, visions and policies these three moral values are embedded in specific combinations.

We also can see these moral values even in football. The Barcelona football team is a good example of this. As in society, these three moral values are relevant in football. Football needs adaptation: perform your task – assigned by the coach – in a disciplined way. But also social concern: help each other, play as a team. The combination of adaptation and social concern can lead to quite defensive play, therefore football needs some autonomy: players like Messi who dribble and have an individual action. But too much autonomy and neglecting the social orientation and adaptation leads to losing the ball and chaos in the team.

In empirical educational research we constantly find these three moral values. The statistical analyses of a survey among a representative sample of Dutch teachers revealed three different types of citizenship that made different combinations of the three moral values (Leenders, Veugelers & De Kat, 2008). An adapted type of citizenship emphasizes adaptation (discipline) and social orientation, in particular the own community. The second type is the individualised citizenship with a strong focus on autonomy and also to some extent adaptation, but with a weak social orientation. The third type, which we call a critical-democratic citizenship, combines a focus on autonomy and on social concern. We call it democratic because of its search for balance between the person and the other, the common good and a democratic way of life and politics. We add critical to stress the dynamic character of democracy and the possibility for individuals to influence societal developments.

We think contemporary society and the future of the world need such critical-democratic citizenship, but we are also aware of the fact that this is a normative choice. A choice based on political ideas about a good life, a human world and a just society. We found these different types of citizenship in several studies among teachers, students, and parents. We could also distinguish different educational practices related to these types of citizenship and citizenship education (traditional teacher-directed education, individualised education, and cooperative and inquiry-oriented education). However, we also found large disparities between goals and actual practices. At the level of goals the critical-democratic type is quite strong in the Netherlands, but in practice the adaptive dominates.

The individualised citizenship type is partly 'hidden', or implicit, in educational practice: in assessing students, in competition and selection, and in child-centred pedagogies. Child-centred pedagogies neglect the influence of society and express what Aloni (2007) called a naturalistic-romantic view of human development. The combination of a market-oriented neo-liberal education strategy and a focus on a child-centred view on the unique individual strongly emphasises an individualised citizenship.

3 MORAL VALUES AND THE POLITICAL

Linking these moral values and citizenship and citizenship education already showed how moral values are tied up with social and political relationships, and how the societal context influences the articulation of moral values. Moral values never appear in a vacuum. Perhaps it does in abstract theoretical thinking: but then moral thinking is more a kind of preaching; a repeating of mantras of nice though very abstract values.

Our research on citizenship education, but also research by Haste and Westheimer and Kahne, clearly shows how moral values are embedded in societal and political contexts and relationships. Different societal and political orientations emphasise different moral values and give different articulations to moral values. This linking of the moral and the political does not devaluate the relevance of moral values; instead and quite conversely, it shows how important moral values are in daily life. Moral values constitute and frame human relationships. Research on moral values will increase its validity by analysing moral values in their social, cultural and political contexts.

4 MUSIC EDUCATION AND MORAL AND CITIZENSHIP EDUCATION

One of our students Jos Terlingen did a very interesting study on music education. He studied how music education contributes to citizenship education in secondary schools in the Netherlands. He used the above presented theoretical framework and observed classroom practices and interviewed teachers. We will present here his conclusions. We honour his work by making him co-author. Jos is musician himself.

4.1 *Democracy*

Music education in secondary education can contribute to democratic skills and attitudes by focusing on making together music, in particular by composing together with improvisation as part of the creative process. Students start working together to make something beautiful. In this process of co-creation they learn to cooperate, to attune to each other, to empathy with each other, to give space and to take space and to solve problems together. Making music and in particular signing is an exciting activity for youngsters. Teachers need to pay a lot of attention to make a safe climate, clear agreements and a respectful interaction of all involved.

Youngsters learn in the music lessons to take together responsibility for each other. Another aspect of how music education can contribute to democratic education is the capability to receive critics and to give feedback to each other. This can be done in the process of co-creation, in composing themselves, but also as part of performing. Students can show their art in the classroom. Other students can get the task to criticize in a respectful way and to give feedback.

In this way music education can be a playground for citizenship. Making music together, present this to each other, and react to it can all contribute to democratic skills, but also to democratic attitudes as respect for each other, belonging, connectedness and social concern of the students to each other.

4.2 *Participation*

Participation is another aspect of citizenship. It is the willingness and the capacity to participate in society and, based on a feeling of responsibility, to get actively engaged in society. Music education can contribute to this in different ways. By paying attention to the rich societal significance of music and to get into dialogues with students about it, the music teacher can enhance the societal consciousness en the engagement to society of the students.

Music education can also go one step further and indeed stimulating students to cultural participation outside the classroom. This can be in out-of-school activities like a musical, but also by giving students the task to attend a concert and to write a reflective report about it.

Music education can play a larger role in the public domain by giving performances in local theaters, in the street, in the community, or in residential homes for elderly people. For participating trust in one own's capacity is important. Efficacy is a crucial concept in this. Music education can contribute to trust by honoring in particular the effort of a student instead of the result. Many subjects in secondary education emphasize strongly assessment. In music education this is less the case. Music education seems in this way less enforcing a competitive attitude between students and is in essence more focused on the participation of youngsters.

In music education students learn in particular to present themselves. Making music, performing and in particular singing is exciting for youngsters. They master themselves if they perform music. Finally, music education is a place where youngsters who are less visible in more cognitive oriented subjects can show their co-students what they are able to in this area. In this way music education can empower students, and enlarge the possibilities of societal participation of all students.

4.3 *Identity*

For many youngsters music is strongly linked with their identity. With their musical preferences youngsters show who they are. Often youngsters strongly identify with musicians, in particular pop musicians. By paying attention in schools to the personal and cultural meaning of music a dialogue about their personal and cultural identity can be started.

Music has a strong emotional component. By analysing their musical preferences youngsters can learn to reflect on their moral values, attitudes and feelings. When students get the task to present their favorite music in the classroom, and to tell what this music means for them students learn to acknowledge each other better and can develop more empathy for and understanding of each other.

5 HOW THE CONTRIBUTION OF MUSIC EDUCATION TO CITIZENSHIP DEVELOPMENT CAN BE ENLARGED?

The interviews with the music teachers show that these teachers consider citizenship education as a normal task of education. In particular they refer to democratic skills and attitudes. Attention for participation in society and identity development is less in their picture. This study shows that music education contributes to all the three types of citizenship that has been explained in the beginning of the chapter. The adapted and the individualistic citizenship are dominant. The challenge for music education is to improve the focus on a critical-democratic citizenship education.

Music education delivers a strong contribution to nearly all citizenship skills and attitudes. For improving citizenship knowledge music education is less the significant subject. Social studies is more the subject for these knowledge goals. For improving really the contribution of music education to citizenship development teacher education should pay more attention to the link between music education and moral and citizenship development. All student teachers should learn about citizenship education and how they can include this in their teaching. They should learn to reflect on processes of meaning giving in music education, on what they think is good citizenship and how they can work on it in their lessons. This gives changes for a future music education in which moral development and citizenship is strongly embedded. In teacher education student teachers should learn to explore possibilities for enhancing civic participation, dialogues about students' identity and contributing to a critical-democratic citizenship.

REFERENCES

Dewey, J. (1923). *Democracy and education.* New York, NY: Macmillan.
Durkheim, E. (1971[1923]). *Moral education.* New York, NY: Free Press.

Freire, P. (1985). *The politics of education: Culture, power and liberation.* South Hadley, MA: Bergin & Garvey.
Elliott, D. (2012). *Another perspective: Music Education as/for Artistic Citizenship.* Music Educators Journal 99 (1), 21–28.
Giddens, A. (1990). *The consequences of modernity.* Cambridge, UK: Polity.
Haste, H. (2004). Constructing the citizen. *Political Psychology*, 25 (3), 413–440.
International Civic and Citizenship Education Study (2010). *ICCS 2009 International Report.* Amsterdam: IEA.
Leenders, H., Veugelers, W., & De Kat, E. (2008a). *Teachers' views on citizenship in secondary education in the Netherlands.* Cambridge Journal of Education, 38 (2), 155–170.
Leenders, H., Veugelers, W., & De Kat, E. (2008b). Moral education and citizenship education at pre-university schools. In F. K. Oser & W. Veugelers (Eds.), *Getting involved: Global citizenship development and sources of moral values* (pp. 57–74). Rotter dam, Netherlands: Sense Publishers.
Lickona, T. (1991). *Educating for character.* New York, NY: Bantam Books.
Mouffe, C. (2005). *On the political.* London, UK: Routledge.
Noddings, N. (2002). *Educating moral people.* New York, NY: Teachers College Press.
Nussbaum, M. C. (1997). *Cultivating humanity.* Cambridge, MA: Harvard University Press.
Parker, W. (2004). *Teaching democracy.* New York, NY: Teachers College Press.
Power, F. C., Higgins, A., & Kohlberg, L. (1989). *Lawrence Kohlberg's approach to moral education.* New York, NY: Columbia University Press.
Regelski, T. (2005a). Music and Music Education: Theory and Praxis for 'making a difference'. In: Lines, D. (ed.) *Music Education for the New Millennium: Theory and Practice Futures for Music Teaching and Learning.* Oxford: Blackwell.
Regelski, T. (2005b). Curriculum: Implications of Aesthetic versus Praxial Philosophies. In: *Elliott, David J. (ed.) Praxial music education: reflections and dialogues.* Oxford: Oxford University Press.
Terlingen, J. (2013). *Muzikaal burgerschap (musical citizenship).* Utrecht: University of Humanistic Studies Masterthesis.
Veugelers, W. (2007). Creating critical-democratic citizenship education: Empowering humanity and democracy in *Dutch education. Compare*, 37 (1), 105–119.
Veugelers, W. (Ed.) (2011a). *Education and humanism.* Rotterdam, Netherlands: Sense Publishers.
Veugelers, W. (2011b). The moral and the political in global citizenship education: Appreciating differences in education. *Globalisation, Societies and Education*, 9 (3-4), 473–485.
Veugelers, W., De Groot, I, & Stolk, V. (2017). *Research for Cult Committee – Teaching Common Values in Europe.* Brussels: European Parliament, Policy Department for Structural and Cohesion Policy.
Westheimer, J., & Kahne, J. (2004). What kind of citizen? The politics of educating for democracy. *American Educational Research Journal*, 41 (2), 237–269.

21st Century Innovation in Music Education – Astuti et al. (eds)
© 2020 Taylor & Francis Group, London, ISBN 978-0-367-11122-9

Musical ability effect on interpersonal intelligence, empathy and self-confidence

C.H.K.Adi & Sutiyono
Yogyakarta State University, Yogyakarta, Indonesia

ABSTRACT: This study aims to reveal the effect of: (1) musical ability on the interpersonal intelligence; (2) musical ability on empathy; (3) musical ability on the self-confidence of grade 11 students of SMAN (State Senior High School) 1 Bulus, Sukoharjo. The population of this ex post facto research was all 320 students of grade 11. The data collection used practice tests and questionnaires. The data were validated using content and construct validity, and their reliability was tested using Cronbach's alpha formula. The data were analysed using the simple regression analysis. The results of the research are as follows: (1) there was an effect of the musical ability on the interpersonal intelligence of the participants,; (2) there was an effect of the musical ability on the empathy of the participants; and (3) there was an effect of musical ability on the confidence of the participants.

1 INTRODUCTION

Musical arts education is one of the arts educations in public schools that aims to develop students' personality, values of beauty and aesthetic feeling. Aesthetic feeling is a feeling used to distinguish the good from the bad or the beautiful from the ugly (Ahmadi, 2005: 98). To develop that aesthetic feeling, students should be introduced to music-related activities. The introduction of music to students since an early age is closely related to the development of students' future characters and intelligence. In addition, music-related activities can also establish inner calmness among students.

Interpersonal intelligence is the ability to understand and cooperate with others (Amstrong, 2002: 4). This intelligence requires the ability to absorb and respond to moods, temperaments, intentions, and desires of others. Interpersonal intelligence shows one's ability to relate to others.

High interpersonal intelligence allows people to cooperate with other people and achieve a synergy to produce positive results (Lie, 2003: 8). Children with high interpersonal intelligence will be able to establish effective communication, develop a deep empathy, build harmonious relationships, and carry out group work with others. Interpersonal intelligence can also be considered as social intelligence, which is defined as one's ability and skills to create, build, and maintain social relations based on mutual benefits (Safaria, 2005: 23).

Children develop musical competence more quickly as a result of the learning process because it involves interaction with the environment. For children, besides the ability to establish close friendships, social intelligence can also help them to lead, organize, and handle disputes between friends, and gain sympathy from other children. They often show different social reactions. If children find it difficult to socialize, they will not develop well in the future since in any situation, they will be required to relate to others. This ability becomes important because humans cannot escape from social interaction with one another.

Empathy is an effective response that comes from understanding the emotional state of others or the ability to feel what other people feel. Empathy is an integral tool to know and relate to others and also increase the quality of life and wealth of social interaction. Empathy has an important role in the development of social understanding and positive social behavior and serves as a relationships foundation and becomes the basis for coping with stress and conflict resolution (Barr and Higgins, 2009: 5).

This is in line with Republika.co.id's post on September 19, 2016, describing the study conducted by Finnish researchers published in the Frontiers in Psychology. The experiment involved 102 volunteers who were asked to listen to instrumental music with a sad tone. Instrumental music was deliberately chosen to eliminate specific factors such as lyrics that could recall certain memories. Then, the research team evaluated the participants' emotional responses after examining their heart rate, blood pressure, and other things that could affect their reaction toward music. As a result, there were three categories of response types in different sadness intensity. Those three classifications were 'peaceful sadness', 'anxious sadness', and 'intense and transcendental sadness'. The pattern revealed from the study also showed the relationship between empathy and emotional response. Those who have high levels of empathy were more likely to provide intense emotional responses to sadness.

Related to these findings, Goleman (2005: 53) explained that an empathy that was constantly honed would lead to moral development. A teenager who had high empathy would be easier to feel what others felt and responsive to the social environment so that they could control their behaviour (Rachmat, 2008).

Empathy is the ability to understand the consequences of human behaviours on other people's feelings and to share other people's feelings. People who are reluctant to share will grow into individualistic and egocentric individuals while those who understand others' difficulties since the childhood will commonly be more sensitive and willing to help others.

Self-confidence is also important to be examined in this research. Self-confidence is one of the elements of personality that play an important role in human life. Many experts recognize that self-confidence is an important factor in determining one's success. Surya (2007: 31) stated that self-confidence is an important part of one's personality development. It determines how a person behaves and acts. It is obvious that to achieve something in life, one needs self-confidence. The problem is that many people do not have self-confidence even though they are academically intelligent. This is because this self-confidence is not something that can grow and exist in a person by itself.

High school students are children who are entering adolescence. These characteristics make them inseparable from the characteristics of teenagers who find themselves in difficult times, and have to adapt to the various changes in themselves. Hurlock (2000: 23) said that adolescence is a period characterized by physical and psychological changes that begin with physiological changes such as emotional irritability, turbulence and changeability. These changes make teenagers feel dissatisfied with their condition sometimes and often cause them to fall into a state/condition of self-doubt. As said by Martanian and Afiatin (1998), self-confidence develops through the interaction of individuals with their environment. A conducive psychological and sociological environment will foster and increase one's self-confidence.

Based on the explanation above, it can be said that one of the ways to form students' interpersonal intelligence, empathy, and self-confidence is to have musical abilities. Gardner (2003: 4) argued that musical ability is a potential ability that grows earliest and emerges unpredictably if it is compared to other fields of human intelligence. This ability involves sensitivity to sound, tone, rhythm, melody, harmony and notation. This musical ability is located in the right part of the brain. Adiningsih (2008: 7) stated that musical ability is the ability to process or utilize something related to rhythm, tone and sound including sounds produced by natural sources.

Music is much related to intelligence. Widayati and Utami (2008: 163) found that listening to music or learning music is useful for improving speech, hearing, self-confidence, coordination skills, and optimizing intelligence of children. However, the position of music in the education process is not always considered. Teachers and parents sometimes underestimate the musical abilities or intelligence that children have, so that their abilities do not develop.

Based on the description above, this research investigated the affective domain related to students' interpersonal intelligence, empathy, and self-confidence. The research was conducted at SMAN 1 Bulu. The reason why researcher chose SMAN 1 Bulu as the location of the research was because this school was the favorite school in the Sukoharjo area. From the observation conducted by the researcher on January 12, 2018, the researcher found an extracurricular activity, especially in the music field, in the form of a group of marching bands.

The marching band group consisted of all students of grade 10 and 11 who had certain musical abilities such as piano, violin, guitar, bass, and other musical abilities accompanied by one conductor. The majority of this marching band group was dominated by the grade 11 students. This was because grade 12 students were focused on preparing themselves for national exams.

The result of the observations at SMAN 1 Bulu showed that grade 11 students who were members of the school marching band group tended to find it easy to socialize with their peers and actively participate in student activities held by the school. This was because the grade 11 students who were members of the marching band are calm, adaptable, fun, communicative, confident, and able to interact well with their environment.

Besides that, the grade 11 students who were involved in marching band activities had different musical skills from other students. Different things were shown by children who lacked of musical skills. During the observation, the grade 11 students who were not involved in the marching band activities seemed to be less able to adjust themselves so they tended to be passive, reserved, not confident, and selective in choosing friends for playing.

As mentioned earlier, introducing music to children can also help children become independent individuals, improve motoric control, improve language skills and speech, and control their emotional and social development. Students' interpersonal intelligence, empathy, and self-confidence are very necessary in social interaction so that students are able to socialize well with their environment, convey constraints, conduct consultations, deliver answers, work in team and interact with others in supporting school activities. Therefore, research in these areas are needed. Moreover, there is also lack of understanding of the relationship among musical abilities, interpersonal intelligence, empathy, and self-confidence of the grade 11 students at SMAN I Bulu, Sukoharjo.

This research was aimed to test, analyse, and obtain empirical evidences about (1) musical ability effect toward interpersonal intelligence; (2) musical ability effect toward empathy; and (3) musical ability effect toward self-confidence of the grade 11 students at SMAN I Bulu, Sukoharjo.

Musical ability is achieved through effort done in training or outside the training to recognize, analyse, and feel the high and the low of the pitch by the hearing sense (and other senses) to analyse and compose music elements in the form of melody, harmony, and rhythm that form a single piece of recognizable and understandable song. Musical ability was measured by using a practical score of 20 assessment items.

Interpersonal intelligence is the ability to understand the intentions and feelings of others and create a harmonious relationship with others. Interpersonal intelligence was measured by using a questionnaire with an ordinal measurement scale of 20 statement items. Interpersonal intelligence indicators were measured based on Yaumi's (2013) theory of reading social cues, expressing emotions in their place, and controlling emotions.

Empathy is feeling and thinking about the conditions that others feel without actually having to involve themselves in the feelings or responses of others. Empathy was measured by using a questionnaire with an ordinal measurement scale of 20 statement items. The indicator used to measure empathy according to Nashroni (2008) was perspective taking, fantasy, empathic concern, and personal distress.

Self-confidence is a person's mental or psychological condition, where an individual can positively evaluate him or herself which subsequently create positive feelings in the form of confidence and trust in his or her abilities and potential, which create a sense that she or he will be able to do all the tasks well and achieve life goals. Self-confidence was measured by using a questionnaire with an ordinal measurement scale of 20 statement items. The indicator used to measure students' score according to Martaniah and Afiatin (1998) was the ability to interact with the environment, responsibility, willingness to ask questions and express opinions.

2 THEORITICAL REVIEW

2.1 *Musical ability*

Musical ability is achieved through effort done in training or outside training to recognize, analyse, and feel the high and the low of the pitch by the hearing sense (and other senses) to analyse and compose music elements in the form of melody, harmony and rhythm that form a single piece of recognizable and understandable song. Musical ability can be measured based on musical achievements gained through effort done in training or outside training to recognize, analyse, and feel the high and the low of the pitch by the hearing sense (and other senses) to analyse and compose music elements in the form of melody, harmony and rhythm that form a single piece of recognizable and understandable song. Indicators used for children's musical ability(Lwin, May & Khoo,2008). Those indicators for measuring musical ability were accuracy in reading rhythmic notation, imitating the rhythmic sounds, distinguishing major from minor chords, and imitating melody by the correct rhythmic and melody.

2.2 *Interpersonal intelligence*

Interpersonal intelligence is the ability to understand the intentions and feelings of others to create a harmonious relationship with others. Interpersonal intelligence is important in human life because a human essentially cannot live by him or herself. Many activities in human life are related to others. Likewise, a child also needs the support of the people around him. Child social skills are intertwined through relationships with peers. According to Yaumi (2013: 45) there were three important elements of interpersonal intelligence in building communication. These three important elements were: reading social cues, controlling emotions, and expressing emotions in their place. Interpersonal intelligence indicators in this research were as follows:

1. Reading Social Cues
   Paying full attention on how other people communicate, and understanding verbal and nonverbal communication used in interacting. Reading social cues were measured by:
   a. Starting a new relationship with others
   b. Showing an openness in relationship with others
   c. Showing a trust in others to share feelings
2. Controlling Emotions
   If you felt tense about the topic being discussed, you should step back a little to calm, then continued the conversation. Controlling emotions were measured based on:
   a. Appreciating the differences of others
   b. Thinking positively toward others
   c. Not overly suspicious
3. Expressing Emotions in Their Place
   Knowing when to express compassion and love, emotional connection, or positive emotions. Expressing emotions in their place were measured based on:
   a. Giving support to friends
   b. Giving appreciation to others
   c. Spontaneity
   d. Putting self on a par with others
   e. Recognizing the importance of the presence of others
   f. Two-way Communication

2.3 *Empathy*

Empathy is a feeling and thinking about the conditions that others feel without having to actually involve themselves in the feelings or responses of others. An emphatic individual does not only experience an event as experienced and felt by others, but also understand events from others' point of view. Experts distinguished empathy responses into two components (Fuad, 2008: 32), namely:

a. The cognitive component in empathy is an intellectual process to understand others' perspective appropriately. In this cognitive component, a person is expected to be able to distinguish others' emotions and accept their views.
b. The affective component is a person's tendency to experience the emotional feelings of others.

Furthermore, each of those two empathy components has two aspects, namely the cognitive aspect, which includes perspective taking and fantasy; and the affection aspect, which includes empathic concern and personal distress.

a. Perspective taking is to imagine a person's condition and feeling in mind by putting views and minds on the others' position that causes an individual to become more aware of and paying attention to other people's opinions about him or herself.
b. Fantasy is a person's ability to change himself imaginatively to experience feelings and actions in imaginary characters in books, movies or plays that is read or watched.

a. Empathic concern is sympathy that is oriented to others and attention to the misfortune of others. Empathic concern reflects feelings and sympathy that are closely related to sensitivity and concern for others.
b. Personal distress emphasizes self-oriented personal worry and anxiety in dealing with unpleasant interpersonal situations. The high personal distress shows a lack of ability to socialize. This can cause internal rewards such as patience and love. Individual with high personal distress is the main consideration about helpful behavior. Thus, the higher the personal distress, the lower the prosocial and external reward behavior such as popularity, the compliment from others.

2.4 *Self-confidence*

Self-confidence is a person's mental or psychological condition, where an individual can positively evaluate him or herself which subsequently creates positive feelings in the form of confidence/trust in his or her abilities and potential, and perceived ability to do all his or her tasks well and achieve his or her life goals with these abilities and potentials. Martaniah and Afiatin (1998) formulated several aspects of the characteristics and indicators of self-confidence. They are as follows:

1. Being able to interact with the environment
2. Having responsibility
3. Daring to ask questions and express opinions

3 RESEARCH METHODOLOGY

This study employed ex post facto research. Sugiyono (2010: 7) stated that "ex post facto research is a research conducted to examine the events that have occurred and then look back to find out the factors that can cause those events". The population in this research were all 320 students of grade 11 in SMA 1 Bulu. The sampling technique used proportional random sampling based on Krecjie and Morgan table at a significant level of 5%. The sample in this research were 175 students of grade 11 in SMA 1 Bulu. The data were collected using questionnaires. The data were then analyzed using simple regression analysis.

4 RESEARCH FINDINGS AND DISCUSSIONS

4.1 *Research findings*

4.1.1 *Musical ability*

The results of data analysis revealed that there were 54 students (30.86%) in the high musical ability category, 98 students (56.00%) in the medium musical ability category, and 23 students (13.14%) in the low musical ability category. Thus, the majority of the musical ability tendency of the grade 11 students at SMAN 1 Bulu, Sukoharjo was in the medium category with the percentage of 56.00% (98 students).

4.1.2 *Interpersonal intelligence*

The results of data analysis revealed that there were 43 students (24.57%) in the high interpersonal intelligence category, 132 students (75.43%) in the interpersonal intelligence category, and no students in the low interpersonal intelligence category (0.00%). So, the majority of interpersonal intelligence tendency of the grade 11 students at SMAN 1 Bulu, Sukoharjo was in the medium category, with the percentage of 75.43% (132 students).

4.1.3 *Emphaty*

The results of data analysis revealed that there were 33 students (18.86%) in the high empathy category 142 students (81.14%) in the medium empathy category, and no students in the low empathy category (0.00%). So, the majority of empathy tendency of the grade 11 students at SMAN I Bulu, Sukoharjo was in the medium category with the percentage of 81.14% (142 students).

4.1.4 *Self-confidence*

The results of data analysis revealed that there were 49 students (28.00%) in the high self-confidence category, 126 students (72.00%) in the medium self-confidence category, and no students in the low self-confidence category (0.00%). So, the majority of self-confidence tendency of the grade 11 students at SMAN I Bulu, Sukoharjo was in the medium category with the percentage of 72.00% (126 students).

4.1.5 *Hypothesis testing*

The results of the t-test statistic for the motivation of members showed a significance value of 0.000, where the significance value was less than 0.05 (0.000<0.05), and the regression coefficient had a positive value of 0.447. Thus, the hypothesis which stated that "there is an influence of musical ability toward interpersonal intelligence on the grade 11 students at SMAN 1 Bulu, Sukoharjo" was accepted. Therefore, it could be concluded that there was a positive and significant influence of musical ability toward interpersonal intelligence on the grade 11 students at SMAN 1 Bulu, Sukoharjo.

The results of the t-test statistic for the motivation of members showed a significance value of 0.000, where the significance value was less than 0.05 (0.000 <0.05), and the regression coefficient had a positive value of 0.265. Thus, the hypothesis which stated that "there is an influence of musical ability toward empathy on the grade 11 students at SMAN 1 Bulu, Sukoharjo" was accepted. Thereofore, it could be concluded that there was a positive and significant influence of musical ability toward empathy on the grade 11 students at SMAN 1 Bulu, Sukoharjo.

The results of the t-test statistic for the motivation of members showed a significance value of 0.000, where the significance value was less than 0.05 (0.000<0.05), and the regression coefficient had a positive value of 0.507. Thus, the hypothesis which stated that "there is an influence of musical ability toward self-confidence on the 11 grade students at SMAN 1 Bulu, Sukoharjo" was accepted. Therefore, it could be concluded that there was a positive and significant influence of musical ability toward self-confidence of the grade 11 students at SMAN 1 Bulu, Sukoharjo.

4.2 *Discussions*

4.2.1 *The Effect of Musical Ability toward Interpersonal Intelligence of the Grade 11 Students at SMAN 1 Bulu, Sukoharjo*

Based on the results of data analysis, there was an effect of musical ability toward interpersonal intelligence of the grade 11 students at SMAN 1 Bulu, Sukoharjo. This was shown from the significance value that was less than 0.05 (0.003<0.05), and the regression coefficient that had a positive value of 0.447.

Interpersonal intelligence was the ability to understand the intentions and feelings of others to create a harmonious relationship with others. Interpersonal intelligence was important in human life because a human essentially cannot live by him or herself. Many activities in human life were related to others. Likewise, a child also needs the support from the people around him or her.

A child's social skills were learnt through the influences of their peers. One way to form students' interpersonal intelligence was by having musical abilities. Musical ability is achieved through effort done in training or outside the training to recognize, analyse, and feel the high and the low of the pitch by the hearing sense (and other senses) to analyse and compose music elements in the form of melody, harmony, and rhythm that form a single piece of recognizable and understandable song.

Musical ability is the expression of the thought through melody, rhythm, and harmony with supporting elements such as shapes, characteristics, and sound colours. In line with Safaria (2005:23) If students had musical ability, so, indirectly, the students had interpersonal intelligence capable of creating, building, and maintaining their social relations based on mutual benefits.

4.2.2 *The effect of musical ability toward empathy on the grade 11 students at SMAN I Bulu, Sukoharjo*

The results of the data analysis showed that there was an effect of musical ability toward interpersonal intelligence of the grade 11 students at SMAN 1 Bulu, Sukoharjo. This was shown from the significance value that was less than 0.05 (0.000<0.05), and the regression coefficient that had a positive value of 0.265.

Empathy was feeling and thinking about what others feel without actually having to involve themselves in the feelings or responses of others. An empathic individual did not only experience an event that was experienced and felt by others alone, but also understand events from others' point of view. One way to form one's empathy could be done by increasing musical ability. Bernstein & Picker (1972) said that music is sounds that are organized in time, have artistic value and can be used as a tool to express the ideas and emotions of the composer to the audiences.

The musical ability relevance was considered appropriate in shaping someone's empathy because by having ability to play a musical work, a person would have the ability to read a musical work, and the ability to hear or analyse and express these abilities by creating and or playing a musical work.

This was in line with Howe (2012) which explained if empathy is a process of symbolizing the conscious personal experience supported by the empathic response in him or herself. It is an ability to express an implicit experience into a clearer picture to understand what is experienced by others. Therefore, it would automatically change confusion that was exist before into a clarity and calmness.

4.2.3 *The effect of musical ability toward self-confidence on the grade 11 students at SMAN 1 Bulu, Sukoharjo*

Based on the results of the data analysis, it was known that there was an effect of musical ability toward self-confidence on the grade 11 students at SMAN 1 Bulu, Sukoharjo. This was shown from the significance value that was less than 0.05 (0.000<0.05), and the regression coefficient that had a positive value of 0.507.

Self-confidence is a person's mental or psychological condition, where an individual can positively evaluate him or herself which subsequently creates positive feelings in the form of confidence/trust in his or her abilities and potential, and perceived ability to do all his or her tasks well and achieve his or her life goals with these abilities and potentials.

One of the abilities that could shape self-confidence was the musical ability of each student. This was in line with Syukur (2005: 54) which explained that music is the art of carefully arranging sound into a regular and melodic pattern created from musical instruments or human sounds. Music usually contains elements of rhythm, melody, harmony, and sound color.

Besides that, Angelis (2003: 10) stated that self-confidence is a belief in the abilities possessed, a purpose or objective in life and the ability of a character to carry out what is desired, planned and expected. Due to those explanations, it becomes clearer that the musical ability can form students' self-confidence.

5 CONCLUSIONS

5.1 *Conclusions*

The results of this research showed that:

1. There was an effect of musical ability toward interpersonal intelligence on the grade 11 students at SMAN 1 Bulu, Sukoharjo as shown from the significance value that was less than 0.05 (0.003<0.05), and the regression coefficient that had a positive value of 0.447.
2. There was an effect of musical ability toward interpersonal intelligence on the grade 11 students at SMAN 1 Bulu, Sukoharjo as shown from the significance value that was less than 0.05 (0.000<0.05), and the regression coefficient that had a positive value of 0.265.
3. There was an effect of musical ability toward self-confidence on the grade 11 students at SMAN 1 Bulu, Sukoharjo as shown from the significance value that was less than 0.05 (0.000<0.05), and the regression coefficient that had a positive value of 0.507.

5.2 *Suggestions*

1. In this research, the school is advised to be able to explore the students' potential so that students could have interpersonal intelligence, empathy, and self-confidence in any case.
2. Students should be liberated to explore new things held by the school so they can recognize their abilities and thus they can hone and improve their interpersonal intelligence, empathy, and self-confidence.

REFERENCES

Adiningsih, N.U. 2008. Permainan kreatif asah kecerdasan musik balita [*Creative Games to hone Toddler Music Intelligence*]. Bandung:Karya Kita.

Martaniah, S. M., & Afiatin, T. (1998). *Peningkatan Kepercayaan Siri Remaja Melalui Konseling Kelompok.* [Enhancing Youth Confidence Through Group Counseling] *Psikologika Jurnal Pemikiran dan Penelitian Psikologi Nomor VI tahun III.*

Ahmadi. 2005. *Strategi Belajar Mengajar.* [Strategy of Teaching and Learning Interactions]. Bandung: Pustaka Setia.

Amstrong, Thomas. (2002) 7 Kinds of Smart. *Menemukan dan Meningkatkan Kecerdasan Anda Berdasarkan Teori Multiple Intelligence.* [7 Kinds off Smart: Finding and Enhancing Your Intelligence Based on Multiple Intelligence Theory]. *Jakarta: Gramedia Pustaka Utama.*

Angelis, B. D. (2003). Confidence: *percaya diri sumber sukses dan* kemandirian.[*Confidence: Source of Success and Independence*] *Jakarta: PT. Gramedia Pustaka.*

Barr, J. J., & Higgins-D'Alessandro, A. (2009). How adolescent empathy and prosocial behavior change in the context of school culture: A two-year longitudinal study. *Adolescence*, *44*(176), 751-773. Academic research Library.

Gardner, Howard. (2003). Multiple Intelligences: *Teori dan Praktek*.[Multiple Intelligence: Theory and Practice]. Batam Centre: Interaksara.
Goleman, D. (2005). Working with emotional intelligence: *kecerdasan emosi untuk mencapai puncak prestasi*. [Working with emotional intelligence: emotional intelligence to reach highest competency] *Jakarta: PT Gramedia Pustaka Utama*.
Howe, David. (2012). *Empathy: What it is and why it matters*. Macmillan International Higher Education.
Hurlock, Elizabeth. B. 2000. *Psikologi perkembangan anak*. [Developmental Psychology: An Approach throughout the Life]. Jakarta: Erlangga.
Lie, Anita. 2003. *101 Cara untuk menumbuhkan kecerdasan anak*. [101 Ways to Develop Children's Intelligence]. Jakarta: PT Elek Media Komputindo.
Lwin, May & Adam Khoo, dkk. 2008. *Cara Mengembangkan Berbagai Komponen Kecerdasan [How to Develop Various Components of Intelligence]*. Yogyakarta: Indeks.
Fuad, Nashori. 2008. *Psikologi Sosial Islami*. [Islamic Social Psychology]. Bandung: PT. Refika Aditama.
Safaria, Triantoro. (2005). *Interpersonal intelligence: Metode pengembangan kecerdasan interpersonal anak*.[Interpersonal Intelligence: Child Interpersonal Intelligence Development Method]. *Yogyakarta: Amara Books*.
Sugiyono. 2010. *Metode Penelitian Kuantitatif, Kualitatif, dan R&D*.[Educational Research Methods: Quantitative, Qualitative, and R&D Approaches]. Bandung: Alfabeta.
Surya, H. (2007). *Percaya diri itu Penting*.[Confidence is Important]. Jakarta: Elex Media Komputindo.
Syukur, Abdul. (2005). *Ensiklopedi Umum untuk Pelajar*. [General Encyclopedia for Students]. Jakarta: Ichtiar Baru van Hoeve.
Widayati, Sri., & Widijati, Utami. (2008). *Mengoptimalkan 9 zona kecerdasan majemuk anak*.[Optimizing 9 Child Multiple Intelligence Zones]. *Yogyakarta: Luna Publisher*.
Yaumi, Muhammad. (2016). *Prinsip-prinsip desain pembelajaran*.[Principles of Learning Design]. Jakarta: Fajar Interpratama Mandiri.

21st Century Innovation in Music Education – Astuti et al. (eds)

A revitalization of Lubuklinggau folk songs to engraft the patriotism

A. Fadlilah & K.S. Astuti
Yogyakarta State University, Yogyakarta, Indonesia

ABSTRACT: This research aims to revitalize Lubuklinggau folk songs for the students of the Lubuklinggau Junior High School 1 and to engraft character building with the spirit of love for country, or patriotism, by learning Lubuklinggau folk songs. This research used a mixed method combination of sequential exploratory models. In the first stage, qualitative data was obtained in the form of an indicator value of the patriotism spirit in folk songs that will be used as a treatment at the experimental stage. The data in this research was obtained based on the main informant and documents. The data collection techniques were conducting interviews, completing documentation and then giving a questionnaire. The results show that there were differences in values between the experimental class and the control class with a significance value of 0.03.

1 INTRODUCTION

Indonesia has various arts and cultures that are different in each region. Arts and cultural differences are what become character reflecting the characteristics and background of a society. Art is part of culture that intertwines with society. One of the works of art is folk songs that are also a part of Indonesian society.

Stated that folk songs are a manifestation of a culture used as a communication medium in the form of human expression, as a means to invite sympathy for a reality and for imaginative stories, and as a medium to express someone's emotions and feelings with the aim of engrafting attitudes or values in an educative culture (p. 2). In line with Batubara, Djohan (2006) states that every culture in the world has music that is specifically heard or played based on historical events in a society's life journey. Music is also a major practice for completing and refining various forms of art in various cultures (p. 23).

Overall, Lubuklinggau culture has a close connection to Malay culture and is inseparable from the majority culture that developed in Sumatra island. Malay culture has a big influence on the development of music in Lubuklinggau both as musical accompaniment and as entertainment. Based on the interviews with artists conducted on 3 July 2017, Lubuklinggau artistry was robust during the 1980s during the time when there were many active artists in Lubuklinggau who created folk songs and dances. In the 2000s, a VCD album of Lubuklinggau folk songs was presented accompanied by Malay genre songs and visual style dances. The album was quite popular in Lubuklinggau society and schools until 2008. The songs have very positive moral messages conveyed in the song lyrics for the young generation of Lubuklinggau, especially introducing them to the local wisdom culture and character building. In line with this, character building is a deliberate effort to help people understand, care about and act upon core ethical values (Istiningsih, 2016).

There are many factors that make Lubuklinggau folk songs less popular. These factors include the lack of artists' creativity and lack of societal awareness and not knowing that their own folk songs are still lacking. In particular, the teachers in the schools also lack understanding of their own folk songs. Thus, teachers have many limitations that mitigate providing folk song materials in school. Technological developments are inevitable and cannot be avoided.

New technology provides many choices for listening to music from local to foreign music. However, the use of new technology does not align with the development of local music as it is still rare to find Lubuklinggau folk songs online and on music player media, such as joox, spotify and so on.

In line with new developments, technological progress has also changed the way people listen to music (Purwanto, 2017). Although there are people who are still interested in folk songs, conventionally listening to music with a cassette player or compact disc (CD), these tools are now beginning to be abandoned. Technology makes it easier for music lovers to listen to their favourite music through the digital music sold in various applications and music streaming that can be an option. At this juncture, people are getting used to listening to music through music streaming applications. Moreover, the interest of people today to listen to folk songs has become rare. This change was based on the interview conducted on 13 February 2018. Additionally, there are some students who often choose foreign music to listen to.

Based on the previous discussion above, this research departed from problems found in the field and instead conducted research to provide solutions to problems such as lack of interest in local music by providing local song learning materials based on local wisdom. What is thus anticipated is finding a way to engraft the spirit of patriotism in students at the Lubuklinggau Junior High School 1.

1.1 *The structure*

Student awareness of local music is still lacking. Many students choose Western music to listen to, and mention reasons such as, "to be looked more awesome and modern". This can happen because the local music itself, especially Lubuklinggau songs, have not become the host music in its own place/region. The simplest way to learn art and culture through Indonesian folk songs is through school. However, school teachers have even less knowledge of folk songs. Many folk songs actually invite listeners to get to know the potential of the region. However, the lack of interest in local folk songs along with societal conditions reduce the sense of belonging while at the same time maintaining folk songs become on the wane.

In building students' spirit of love for country, the act of revitalizing Lubuklinggau folk songs is the right decision to make. Through learning Lubuklinggau folk songs proper conditions will be created for students to know their own folk songs, understand the meanings of the lyrics and know the potential of their own region. Thus, the effort to engraft the spirit of love for country, or patriotism, can consistently be a part of the students.

2 RELATED WORKS

The work of Desyandri (2016) titled, *Revitalization of Educative Values of Minang Songs to Build Students' Character* revealed and described educative values in Minangkabau songs and their implementation in the scope of basic education in Padang, Sumatra, Indonesia. Desyandri's work aimed to build student character and was an effort to revitalize Minangkabau songs. Both the previous research by Desyandri and the present research use folk songs as the object of analysis, aiming to reveal the educative values of the spirit of love for country. In the present research, the folk songs of Lubuklinggau were implemented as part of the process of learning folk songs at the Lubuklinggau Junior High School 1. This research brings novelty in terms of using quantitative methods to evaluate the implementation of folk songs to engraft the spirit of love for country.

Another previous publication was made possible through the research conducted by Abi Karoma Batubara (2017) titled, *Revitalizing Educational Values of Rejang Cultural Character through Rejang Folk Song and Its Implementation in Senior High School 4 Curup, Rejang Lebong Regency, Bengkulu Province*. The research aimed at the preservation of the arts using the tools of observing, interpreting and reflecting the results of the analysis of cultural values of Rejang folk songs toward character building. The study revealed that a folk song contains strong moral values to be implemented in the world of education. Both the previous researchand the present research use the same object of study, namely, folk songs in order to reveal the values of character building. The difference in analytical method is that the present research used the mix method to provide more detailed results.

3 MATERIAL & METHODOLOGY

3.1 *Data*

Obtaining the data for this research was carried out in two stages. The first stage was the qualitative stage where data was derived from the main informants and documents. The data was collected through documentation and interviews with the artists and teachers with interview topics related to the problems of sustaining Lubuklinggau folk songs. The second stage was a quantitative stage where the data was obtained by distributing a questionnaire to students at Lubuklinggau Junior High School 1 with questions about the learning of Lubuklinggau folk songs and the aim to engraft the spirit love for country. This research also used a school guidebook about the development of education in national cultural and character (Manullang, 2013), and that uses indicators for the spirit of love for country. These indicators were then used to develop sub-indicators based on the analysis of the values of spirit love to country in Lubuklinggau folk songs.

3.2 *Method*

This research was conducted in Lubuklinggau city, at the Lubuklinggau Junior High School 1 in South Sumatra Province, Indonesia. This junior high school was chosen because it was the first school that implemented the 2013 curriculum at grade levels VII - VIII in 2015. The first research (qualitative) focused on identifying Lubuklingau folk songs based on interviews for selecting songs as the object of this research and that were analysed with the artists and teachers. The songs were analysed and used as a sub-indicator of the spirit of love for country and in reference to the 2010 character-building school guidelines. While two stages were necessary, the second stage (quantitative) results of song analysis is used as the basis for making the questionnaire items. The questionnaire was used as a tool to measure engrafting the spirit of love for country through learning Lubuklinggau folk songs.

Mixed method means that qualitative and quantitative data were merged at one end of continuum while kept separated at the other end of the continuum, or combined in several other ways. These two data can be written separately, but both remain implicitly connected to each other (Creswell & Plano Clark, 2010, p. 311). In this research, the exploratory sequential combination became the focus of the method which was used as the reference as it was in line with the stages of the steps used (Creswell, 2014, p. 301). An exploratory partial mixed method is a design by which researchers started by exploring qualitative data to make an analysis and then used findings in the second, quantitative phase (Creswell, 2016).

4 RESULTS AND DISCUSSION

4.1 *Identification and song analysis of the Lubuklinggau region*

The results of the interviews with teachers from the Lubuklinggau Junior High School 1 were obtained via the identification of six Lubuklinggau folk songs: "Jengan Harak", "Mangun Linggau", "Bumi Silampari", "Ragam Beregam", "Ayo Ngelong ke Lubuklinggau" and "Kesenia Lame".

Of the six songs identified, the songs that were chosen as the object of this research were "Ayo Ngelong Ke Lubuklinggau", "Kesenia Lame" and "Mangun Linggau". To analyse these three folk songs, an in-person interview with each songwriter was conducted. "Kesenian Lame" and "Mangun Linggau" were created by Azman Bainuri, an active artist in Lubuklinggau since the 1980s until the present who began creating songs starting in 1993 into the 2000s.

The interview results related to the "Mangun Linggau" folk song revealed that the song is the result of a tone adaptation of the "Mangun Dusun" folk song. "Mangun Dusun" is an unknown song and the name of the composer is not known but the song was very popular in the 1990s and used Lubuklinggau local language. "Mangun Linggau" or "Mangun Dusun" both have the same meaning and values inviting members of all ages in Lubuklinggau society to jointly build and love the potential of their own region.

The results of the "Kesenian Lame" song interview revealed that the song contains meanings inviting members of local society to jointly play Lubuklinggau *berejung* original artistry, *bepantun* and also recognize *kaen* dance as one of Lubuklinggau's original dances. "Ayo Ngelong" in Lubuklinggau has the same meaning as the two previous songs that also invite members of Lubuklinggau society to come on a tour and to know Lubuklinggau city. The song invites listeners to get to know the folklore of *Dayang Torek* and *Bujang Kurap*. Also, the songs mention Lubuklinggau potential tourism locations such as Bukit Sulap, Taman Kurma As Salam, Bendungan Watervang, Temam Waterfall, Sando Waterfall, Takli Waterfall, and the Subkos Garuda Museum.

The Ministry of National Education Guidelines for Character Building (2010) mention that the spirit love for country is a way of thinking, behaving, and acting that shows loyalty, caring, and high appreciation for the language and physical, social, cultural, economic and political environment of the nation (p. 43). The results of the three Lubuklinggau folk songs analysis show that the songs contain educational value for the purposes of character building and loving the nation. This educational value is reflected in the songs' meanings as described previously, that Lubuklinggau folk songs appeal to ideas of bringing society together and to know Lubuklinggau city through the potential of its local wisdom.

Hash believes that teaching the values of morality and citizenship will make it possible for people to be able to live their best lives regardless of their social class (Hash, 2006). As Lubuklinggau folk songs contain values that encourage the spirit of love for country, if the songs are implemented early in the schools, the song will have a good impact on the preservation of local and national culture.

Table 1. Result indicator and sub-indicator.

Aspect	Indicator	Sub-Indicator	The sum of the items	
			No Item	The sum
The spirit of love for country	Love for the geograpical wealth and prosperity of the Indonesian's land	Knowing the geograpical area of Lubuklinggau city	22,28	2
		Knowing the tourist attraction places in Lubuklinggau	23,27,30	3
		The prosperity of Lubuklinggau city.	24,25	2
	Love for the diversity of culture and arts in Indonesia	Knowing the original culture of Lubuklinggau	15,16,17,26	4
		Knowing Lubuklinggau folk songs	18.19,20	3
		Knowing Lubuklinggau folklore	21,29	2
	Love for the diversity of tribes and local languages in Indonesia	Using Lubuklinggau local language	4,5,6,14	4
		Love for Lubuklinggau local language		
	Adoration and love for the industrial and technological products of the nation	Love for Lubuklinggau folk songs	2,3,9,10,12	5
		Singing Lubuklinggau folk songs	1,7,8,11,13	5
Total				30

Table 2. Independent samples test.

		t-test for Equality of Means		
		T	df	Sig. (2-tailed)
Score	Equal variances assumed	2.131	67	.037
	Equal variances not assumed	2.138	66.998	.036

The following table shows the analysis results for Lubuklinggau songs related to indicators of the spirit of love for country that is in the Ministry of National Education Guidelines for Character Building (2010, p. 43).

4.2 *Implementation*

To answer the research questions, class samples from the student population at Lubuklinggau Junior High School 1 were randomly taken. Class VIII was chosen as the research sample because the class learning materials were in accordance with the object of the research- Indonesian folk songs. The continuation process involved treatment in the experimental class as class A and the control class as Class B.

Treatment in the experimental class obtained significant results. The significance shows that 0.03 <0.05, meaning that there is a difference between the value of students receiving treatment in the form of Lubuklinggau folk songs compared to the value of students who were not given the same treatment, who were not given Lubuklinggau folk songs.

HO: There is no difference between the experimental class and the control class
HA: There is a difference between the experimental class and the control class

The significance value is 0.03 <0.05 meaning that HO is rejected, and it also means that there is a difference between Class A and Class B. The results of this research show that delivering Lubuklinggau folk songs can engraft students into understanding more deeply about their own region.

5 CONCLUSION

Based on the results of the research and the discussion, the following conclusions can be drawn. Lubuklinggau folk songs have educative values that promote character building and the spirit of love for country, also known as patriotism. The indicators and the sub-indicators that were adopted from the meaning of Lubuklinggau folk songs include: knowing the geograpical area of Lubuklinggau city, knowing the tourist attraction places in Lubuklinggau, appreciating the prosperity of Lubuklinggau city, knowing the original culture of Lubuklinggau, knowing Lubuklinggau folk songs, knowing Lubuklinggau folklore, using Lubuklinggau local language, love for Lubuklinggau local language, love for Lubuklinggau folk songs and singing Lubuklinggau folk songs. Learning in the simplest way by introducing folk songs can engraft the spirit of love for country. Thus, appropriate ways of learning folk songs can result in character building values that will shape students' spirit of love, whole-heartedly caring about and loving the culture.

REFERENCES

Creswell, J. W. (2010). *Research design pendekatan kualitatif, kuantitatif, dan mixed* [Research design approaches are qualitative, quantitative, and mixed]. Yogyakarta: Pustaka Pelajar.

Creswell, J. (2016). *Research Design.: Qualitative, Quantitative, Mixed Methods Approaches*. University of Nebraska-Lincoln.

Desyandri, D. (2016). *NILAI-NILAI EDUKATIF LAGU-LAGU MINANG UNTUK MEMBANGUN KARAKTER PESERTA DIDIK* [Educative Values of Minang Songs to Build Student Character],. *Jurnal Pembangunan Pendidikan: Fondasi dan Aplikasi*, *3*(2), 126–141.

Djohan. (2006). *Terapi musik: teori dan aplikasi* [Music therapy: theory and application]. Galangpress.

Hash, P. (2006). Character development and social reconstruction in music education at the turn of the twentieth century.

Howard, K., Campbell, P. S., Schippers, H., Drummond, J., Dunbar-Hall, P., & Wiggins, T. (2005). *Cultural Diversity in Music Education: Directions and challenges for the 21st century*. Australian Academic Press.

Istiningsih, I. (2016). Character education in the most developed nations of ASEAN. *Journal of Education and e-learning*, *3*(1), 32–37.

Manullang, B. (2013). Grand Desain Pendidikan Karakter Generasi Emas 2045. *Jurnal Pendidikan Karakter*, (1).

21st Century Innovation in Music Education – Astuti et al. (eds)
© 2020 Taylor & Francis Group, London, ISBN 978-0-367-11122-9

Gandang Tambua & Tasa: Elements of music and relevance to character education

M. Afdhal & A.N. Machfauzia
Universitas Negeri Yogyakarta, Yogyakarta, Daerah Istimewa Yogyakarta, Indonesia

ABSTRACT: The aim of this study was to apprehend the relevance of Gandang Tambua and Tasa arts music with character education using an ethnographic, qualitative approach. The data was collected through observation, interview and documentation techniques. Elements of the music were categorised, classified and analysed using semiotics. Therefore, the data was interpreted by understanding the signs found and associated with character values. Data validity testing was undertaken via triangulation. The results of the study show that: a. Elements of music contained in the art music of Gandang Tambua and Tasa are the main elements—elements of expression and game techniques; and b. the relevance of character values in Gandang Tambua and Tasa arts are expressed as discipline values, democratic values, love of country, friendly or communicative values, the value of responsibility and creative values.

1 INTRODUCTION

Art is a form of culture in human civilization. Art activities are expected to be able to preserve the nation's culture through its role in social growth developing values and morals that can be used as guidelines in life. In addition to the aesthetic basis, there are also ethical and moral elements in an art form (Suryadi in Ratnasari, 2013, p. 1).

Art is one manifestation of local wisdom that must be preserved (Kuntoro, 2012, p. 6). Local wisdom is used to show that there are several components in the region such as nobility, values, truth, kindness and beauty that can be used as guidelines for making patterns of relationships among people in a community as a basis for a vision of life and can be the basis of character education in schools. Rosala (2016) suggests that in the context of art education, art and culture can be used as learning materials to be lived and analysed bringing forth examples of wisdom for living (p. 25). Therefore, arts education is an effective means of fostering the moral character of students.

Music as a form of art is a work of sound in the form of song and instrumental composition. Jamalus (1998) states that music is a work of art in the form of songs or musical compositions that expresses the thoughts and feelings of the creator through elements of music such as rhythm, melody, harmony, form, song structure and expression as a whole (p. 1). Elliott (1997) says that the sound of instrumental music acts as a physical stimulant that penetrates directly through the ears to the soul to induce pleasure and arouses sadness, happiness and other special emotions (p. 771). Musical sounds contain a kind of message, moral lesson or veiled analogy intuitively detected by listeners. Lee (2014) also states that there is a change when music activities are integrated into a school curriculum that increases character development and the values of cooperation, responsibility and honesty (p. 10). Hash (2008) furthermore asserts that in general music in schools is considered an effective means to develop morale and character, especially through strengthening the abilities of students to participate in religious activities and to develop discipline and a sense of nationalism (p. 4).

Cooperation, responsibility, honesty, a sense of nationalism, religious attitudes and discipline are some of the values developed in character education. Character education is an innovation toward creating schools that foster ethics, a sense of responsibility and caring for people through emphasis on universal values. Pala (2011) suggests that character education is an old program which at the present must now be integrated in each school subject (p. 25).

Gandang Tambua and Tasa are musical arts that originate from the city of Pariaman, West Sumatra, Indonesia. This art-music consists of rhythmic percussion instruments consisting of six double-headed drums and one Tasa drum which is a single-headed drum. The playing of this music is an art game built from the rhythmic patterns of drums and Tasa but without using instruments that function as melodies. The musical characteristics of Gandang Tambua and Tasa are energetic and loud. The drumming presentation attracts a lot of attention such that this music dominates processions and esplanade events. Gandang Tambua's performances have heroic sounds, are vibrant and tend to encourage the audience's imagination to move along with the rhythms according to Wahyuni and Indrayuda (2014, p. 135). Tambua drum instruments and Tasa were used as stimuli to stimulate emotions and instigate audience supporters' spirit during the show that may be considered as 'violent'. Asril (2015) notes that drum and Tasa rhythms played in a fast and dynamic hard tempo produce a sound that is "explosive" and triggers the emotions and enthusiasm of show supporters (p. 71).

The Gandang Tambua and Tasa arts have spread throughout the province of West Sumatra. The existence of this art can still be felt even though the influence of modern music development among the people in West Sumatra is appreciable. This art must be taught in extracurricular activities from elementary to secondary school levels. The uniqueness and existence of Gandang Tambua and Tasa art should be a concern of arts education advocates especially for the purpose of character education that is strongly connected to arts subjects. Not much is known about the art of Gandang Tambua and Tasa, especially the musical elements and character values contained in it, so that the elements of music in Gandang Tambua and Tasa's art need to be identified and their relevance to the values of character education.

2 RELATED WORKS/LITERATURE REVIEW

The form and structure of the presentation of Gandang Tambua and reveal the AMA process as a musical process in the presentation of Gandang Tambua in Koto Malintang Nagari, Tanjuang Raya Agam District, West Sumatra. The rhythms played through the drums and Tasa made the atmosphere tense and engendered explosive emotions. This situation is portrayed visually with the appearance of an actor who performs as a knight using a sword cutting down a banana stem at the location of the ceremony. Asril (2015) explained that the role of Gandang Tasa is to build enthusiasm and the emotional atmosphere at the Tabuik show in Pariaman. The study's findings show that AMA arises because of personal, situational and musical factors. Tambua *malingka* is a unit in a situational context with a repertoire of songs of complexity performed by a totality of players playing together with a serious approach within a personal context.

This study used a qualitative method extracting data directly from resource persons using observation, survey and interview techniques to describe the structure and role of Gandang Tambua in the Tabuik performance in Pariaman. While this study uses qualitative methods, the analysis relies on a semiotic approach to identify character values contained in Gandang Tambua and Tasa art and the musical elements contained therein. Semiotics is the study of signs looking for and examining signs in the elements of Gandang Tambua and Tasa art, then connecting the characteristics of the signs with the meanings of character values contained in the music.

3 MATERIAL & METHODOLOGY

This study uses ethnographic and qualitative approaches. The Gandang Tambua game is explored objectively by presenting cultural facts from the field. The data is then analysed with musical elements described and further explained in terms of relevance to character education.

Data observation and collection involved the process of training and performances from Gandang Tambua and Tasa arts and working with several informants as resource persons. Data was collected through observation, interviews and documentation. The data collected was categorised and classified into elements of music contained in the art of Gandang Tambua and Tasa. The data was then analysed using semiotic data analysis and interpreted by understanding the signs found and the meanings contained within the signs. Testing the validity of the data was conducted by triangulation. In order to analyse the relationships between musical elements contained in the art of Gandang Tambua and Tasa, that process included coding the data, sorting, classifying and connecting value aspects of character education based on theoretical studies.

4 RESULTS AND DISCUSSION

4.1 *The musical elements of Gandang Tambua and Tasa art in the Darak Badarak studio*

The musical elements in the art of Gandang Tambua consist of the main musical elements, elements of expression and game techniques. The main elements describe the form of melody, rhythm and harmony found in Gandang Tambua and Tasa. Gandang Tambua and Tasa are ensembles of two types of rhythmic musical instruments without any support or role from the melodic or harmonious musical instruments in the art of Gandang Tambua and Tasa.

On Gandang Tambua and Tasa art, harmony is created from two different musical instrument characters. The combination of drum Tambua characters and Tasa drums does not only play the same motif simultaneously. Musical patterns are comprised of a division of roles referred to as children, basic and acting that signal a sound be played together or interlocking.

The elements of expression in the game Gandang Tambua and Tasa illustrate the nuances of tempo, dynamics, sound color and style of playing. Gandang Tambua and Tasa arts are found in Pariaman area and are played with a fast tempo and hard strikes to the drums. Such games present elements of emotions such as joy creating fun and entertainment. The musical elements of happiness and enthusiasm are in accordance with the character of the Pariaman people who live on the Sumatran coast. The very fast tempo played in Gandang Tambua and Tasa arts ranges from 140 to 170 Lpm (steps per minute). In certain songs such as the Kureta Mandaki song, there is a transition to 110 to 130 Lpm.

Dynamic elements are very thick in the art of Gandang Tambua and Tasa. This is because Gandang Tambua and Tasa consist of only two types of rhythmic instruments that are thick in the use of dynamic techniques. The changes in dynamics in this art function as tempo transitions and song changes. For example, in the song Kureta Mandaki, the transition to the *sosoh* parts uses dynamic techniques.

Dynamic control in the game Gandang Tambua and Tasa is also based on Tasa drum players. Tasa drumming begins to change by reducing the dynamics of the song via weakening the strength of the drum strike. The Tambua drum player will also follow this change by lessening the strength of his strike. Almost every song in the game Gandang Tambua and Tasa is actually found using dynamic techniques in order to spur the spirit of the players and listeners.

Gandang Tambua and Tasa are rhythmic musical ensembles consisting of two types of drums. One type is the two-headed Tambua drum with a body made of wood and drum heads made of goat skin or cow leather. The shape of the Tambua drum can be seen in Figure 3.

Figure 1. The song score of Gandang Tambua and Tasa.

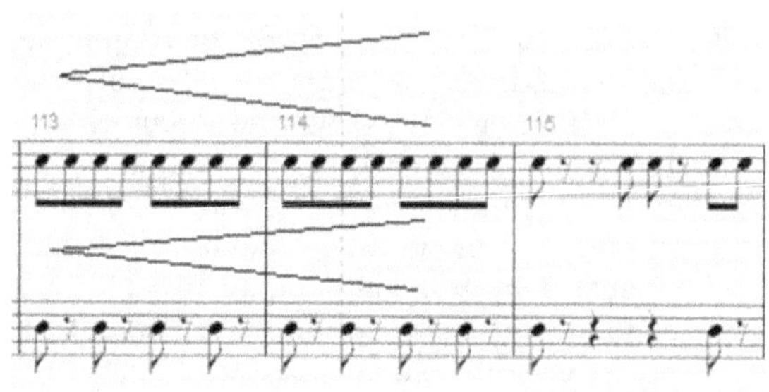

Figure 2. The song score of Kureta Mandaki.

Figure 3. Tambua drums.

The single-headed Tasa drum has a semi-circular shape made of fiber and a membrane for its face. The shape of the Tasa drum can be seen in Figure 4.

Gandang Tambua and Tasa have only two sound characters that are high and low. The sound produced from Gandang Tambua and Tasa comes from the sound of the wooden bat or *pangguguah* (the term used by the Minang people) upon striking the head of Gandang Tambua or Tasa. Gandang Tasa has a high sound that is dominant and piercing.

The sound produced from this instrument has three sound characters. The first is the sound of "tang" with a high character produced by the Tasa drum; the second is the sound "dum" characterised as low sounding produced by the drum membrane Tambua; and the third is the sound "tak" a high character sound produced by pounding the *peluguah* or wooden bat on the body or the periphery of Tambua drum membrane.

The formation of each drums in a Gandang Tambua and Tasa show are arranged in a circular position. Six Tambua drum players surround one Tasa drum player in the centre aiming to make it easier for players to communicate. A circle is formed around one player by the other drummers, so that they can listen to the game playing of each Tambua or Tasa drum player. The drum players formation can be seen in Figure 5.

The performance is a game that has a nuanced spirit. A style emerges during the playing of Gandang Tambua and Tasa. Traditionally, the choreography depends on the players' feelings, not only emotionally, as the atmosphere of the game also deeply influences the choreography

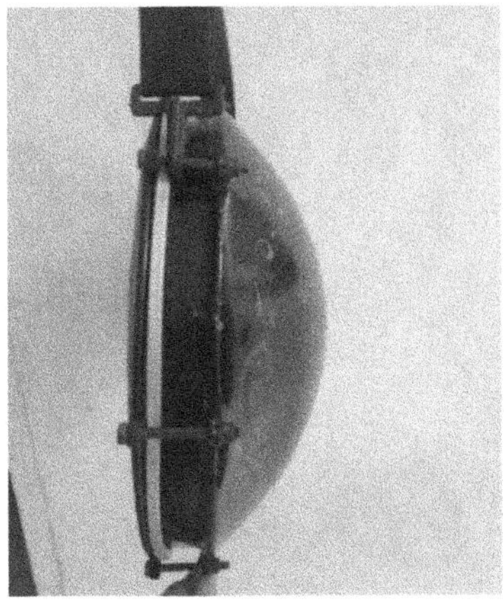

Figure 4. Tasa drums.

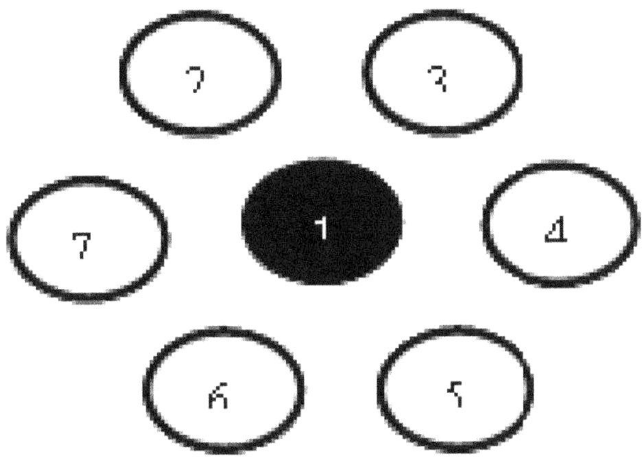

Figure 5. Formation of Gandang Tambua and Tasa players.

that emerges from the players. The choreography that arises is not taught but nearly all Gandang Tambua and Tasa players in Pariaman create a choreography and players can freely express themselves without relying on a set cue. The choreography is almost the same among the Gandang Tambua and Tasa players in Pariaman. This can be called the habit of the Gandang Tambua players and the Tasa.

Gandang Tambua and Tasa art classifications are included in the percussion category of music arts. The game Gandang Tambua and Tasa are played by beating the drums with the *pangguguah.* The Tambua drum is struck using a stick made of wood and the Tasa drum is struck with a rattan stick. The game techniques used in Gandang Tambua and Tasa are described in terms of the technique of how to hold the bat and how to strike the drum. The technique of holding the Gandang Tambua bat and Tasa is shown in Figure 6.

Traditionally, the way to hold a stick in order to strike the Gandang Tambua is such that the position of the right hand uses a matched grip and the left hand using a traditional grip. The hand and stick positions aim to make it easier for the Gandang Tambua players to play by having the stick held in a tilted position to the right, so the right hand can freely use a matched grip to hit the head of Gandang Tambua that is tilted upwards. The left hand holds the stick with the traditional grip technique along with the support of the weight of the

Figure 6. Rule to hold a stick of Tambua drum.

Figure 7. Rule to hold a Tasa stick.

Gandang Tambua drum, make it easier for players to strike the head of the Gandang Tambua that tilts downward. As for how to hold the Tasa bat, this can be seen in Figure 7.

Figure 7 shows how to hold the beater for the Tasa drum using both hands with a matched grip with the tip of the bat pointing towards the bottom of the drum body. This position makes it easier for players to grab the head of Tasa drum which is worn close to the player's body.

The role of the Tasa drummer is as a leader directing the Gandang Tambua and Tasa arts performance. The Gendang Tasa acts as a giver of cues to a row of Tambua drum players for transitions from one song to another song and initiates the exchange of motives or rhythm patterns and tempo shifts.

Each player must control all the motives in the song being played. Without the use of verbal signals, when the Tasa drum player has played a certain pattern such as a pattern called *sosoh* that signals the ending of each song, the Tasa drum players understand that they have entered into the last part of the song.

4.2 *The character values in Gandang Tambua and Tasa arts*

Based on the results of the analysis, there are six character values found in the Gandang Tambua and Tasa arts, as described in the following:

4.2.1 *Discipline*

Discipline is a reflection of having an orderly attitude and adheres to applicable regulations. There is an orderly or regular, constant meaning in the game Gandang Tambua and Tasa. The Tambua drum must be played with a regular and constant tempo following the tempo played by the Tasa producing a synchronous tempo. During the training process, Gandang Tambua and Tasa players are required to be disciplined with the applicable regulations such as starting the training on time and adhering to the process of playing Gandang Tambua and Tasa arts. Players are required to be organized and constant in how they use the tools they play with. What is very significant in maintaining the stability of the tempo and dynamics in the Tambua drumming game is the physical stamina of each player. This art game is played over a long duration, and players must continue to hold a fairly heavy musical instrument while managing the Tambua drum that is of sizeable dimensions. Thus, a disciplined attitude is necessary to maintain a constant tempo and dynamic that is possible by the players' strong physical condition. Hash (2008) mentions that in general, music arts are considered as an effective means of developing disciplinary abilities (p. 4).

4.2.2 *Democratic*

In Gandang Tambua and Tasa arts, there is a distribution of game patterns that are referred to as children, basic and acting. These patterns delineate the division of tasks among the players in what each drummer plays. Children, basic and acting have different rhythmic patterns but are played with the same tempo. As democracy is a reflection of thinking and acting with the same obligations as others, Tambua drum players must understand and carry out their roles as children, basic and acting by producing the same and regular sounds in order to produce rhythmic patterns creating melodies while playing Gandang Tambua and Tasa.

The sound colours produced in the game Gandang Tambua and Tasa are low and high. The same sound colour produced from Gandang Tambua comes from stick holding techniques and hitting methods which must be used by the players. This shows that there are democratic values in the Gandang Tambua and Tasa game techniques.

4.2.3 *Love of country*

Love for the homeland is a value for preserving national culture. This love and preservation can be seen through the participation of the community in an artistic activity and teaching the art to the next generation. Gandang Tambua and Tasa art games are capable of raise a sense of love for the homeland and is strengthened by an increasing number of enthusiasts for this art.

Every education unit in the Pariaman has applied this art to extracurricular activities. Students in extracurricular activities can already play the basic pattern of the Gandang Tambua and Tasa game motif without having been taught prior to the school activity. This has been made possible because there are many Gandang Tambua and Tasa arts ensembles found playing at wedding parties and traditional parties such as *batagak datuak*. Gandang Tambua and Tasa arts have become icons and are very popular traditions for the people of Pariaman in the Tabuik party. In addition, each village in the Pariaman have the music tools and the players to propagate Gandang Tambua and Tasa arts.

Finding players for Gandang Tambua and Tasa arts in Pariaman is not difficult because the regeneration of Gandang Tambua and Tasa players is always there. For example, in the Darak Badarak studio, there are 10 generations of Gandang Tambua and Tasa players. This certainly shows that this art can increase the interest of the community, especially the younger generation, to participate in and learn about the art. The interview results found that Gandang Tambua and Tasa play in the *kureta mandaki* song was very fun. *Kureta mandaki* is the most legendary song and is fun when played especially due to the use of its dynamic technique. The feeling of pride for the arts is so strong that interviewees desire to introduce Gandang Tambua art to friends outside of West Sumatra. Hash (2008) found that music is a way of developing morals to inspire a sense of nationalism in young people (p. 4).

4.2.4 *Friendly/Communicative*

Collaboration in the art of Gandang Tambua and Tasa is related to the "feeling" of the players. Every player involved in Gandang Tambua and Tasa arts must memorize all the patterns in each song and must memorize the parts according to his role as children, basic or acting. This is performed successfully without a verbal signal whereby the players must be able to feel and understand what they have to do to adjust the tempo and dynamic of the song being played. This of course must be done with good cooperation between the players, communication through the "taste" that is felt by all players who must understand each other so as to produce the same expression. A study found that children felt very pleasant cooperation while playing music and that the value of cooperation is fostered by and in children who were interested in the various sounds and rhythms of the song, the lyrics or the role of each instrument and that cooperation was needed for the successful playing of music (Lee, 2014, p. 10).

4.2.5 *Responsibility*

The Gandang Tambua and Tasa game is composed of a variety of motifs and patterns that are played simultaneously and that produce mutually cohesive sounds that are in harmony. This requires a high level of responsibility by the drummers playing Gandang Tambua and Tasa arts. If there is a mistake during the game, this can disrupt the harmony created by various Tambua drum patterns and Tasa drums. Therefore, this game requires a cooperation by all players in performing each of their respective roles with a sense of responsibility by each player towards other players. Lee (2014) also states that there is a change when music activities are integrated into the school curriculum that increases character development and the value of responsibility (p. 10).

4.2.6 *Creative*

The Gandang Tambua and Tasa arts have not have developed new patterns for each song played. However, the studio groups in the Pariman made innovations by collaborating and introducing other Minangkabau musical instruments such as *talempong* and *bansi* into Gandang Tambua and Tasa.

The Gandang Tambua and Tasa arts that have been innovated upon are no longer called the Gandang Tambua and Tasa art that is completely based on tradition. Instead, the Pariaman community named the new form as Gandang Tambua Creation. In Gandang Tambua Creation, players can develop new patterns freely at the Tambua and Tasa drums. This was stated by one of the Gandang Tambua and Tasa players in the Darak Badarak studio who he himself has often made new patterns in the song. This shows that the gatekeepers of Tambua creation accept creativity by producing new patterns in an established song compostion. Sulistyowati (2012) asserts that one character value in art is creativity (pp. 77-78).

5 CONCLUSION

Based on this paper's discussion, the are two conclusions have been reached. First, the musical elements contained in the art of Gandang Tambua and Tasa are: 1) Main elements that include harmonisation; 2) Elements of expression that include: tempo, dynamics, sound color and playing style; and 3) Game technique that involves prescribed ways of striking drums.

Second, There are six character values found in the Gandang Tambua and Tasa arts. These values are: discipline; democratic; love of country; friendly or communicative value; responsibility and creativity.

Based on the conclusions of this study, recommendations include that educational units integrate the Gandang Tambua and Tasa arts not only into extracurricular activities but also into the arts education curriculum in order to instill character values that have been identified in Gandang Tambua and Tasa music art.

REFERENCES

Asril, (2015). Gandang Tasa's Role in Building the Spirit and Atmosphere at Tabuik Show in Pariaman. *Journal Humaniora*, (27), 067–080.

Elliott, D.J. (1995). "*Music Matters: A New Philosophy of Music Education*", New York: Oxford University Press.

Hash, P.M. (2008). Character Development and Social Reconstruction in Music Education at The Turn of The Twentieth Century, *Faculty Publications - Music* 1–30.

Jamalus. (1998). *Belajar Musik Melalui Pengalaman*[Teaching Music Through Music Experience] Jakarta: Departemen Pendidikan dan Kebudayaan.

Kuntoro, S.A. (2012). *The Concept of Education Based on Local Wisdom as The Basis of Character Formation of The Nation*, In the Proceedings of the National Seminar on Education.

Lee, A., (2014). Implementing Character Education Program through Music and Intergrated Activities in Early Childhood Settings in Taiwan. *Journal Music Education* 1–12.

Pala, A., (2012). The Need for Character Education. *International Journal of Social Sciences and Humanity Studies*, (3), 1309–8063.

Ratnasari, S.D., (2016). *Nilai Pendidikan Dalam Pertunjukan Kesenian Kethek Ogleng Pacitan*[*Educational Values in the Ogleng Pacitan Kethek Show*]. Journal Culture 3, 1.

Rosala, D., (2016). *Pembelajaran Seni Budaya Berbasis Kearifan Lokal Dalam Upaya Membangun Pendidikan Karakter Siswa di Sekolah Dasar*[Learning of Cultural Wisdom Based on Local Wisdom as an Effort to Build Students Character Education in Primary Schools]. *Journal Ritme 1.*

Sulistyowati, E. (2012). *"Implementation of character education curriculum"*, Yogyakarta: PT. Citra Aji Parama.

Wahyuni, I., & Indrayuda. (2014). Gandang Tambua's Structure As an Embodiment of Local Wisdom on the Cubadak Aia Village, North Pariaman District, Pariaman., *Journal Humanus* 135 (13).

21st Century Innovation in Music Education – Astuti et al. (eds)
© 2020 Taylor & Francis Group, London, ISBN 978-0-367-11122-9

Auditive stimulation of dance accompaniment in the *Sekar Kinanti* art studio as a method of increasing kinesthetic intelligence

G.K.W. Anggraheni, K.S. Astuti, A. Husna & W.O. Wiko
State University of Yogyakarta, Yogyakarta, Daerah Istimewa Yogyakarta, Indonesia

ABSTRACT: Auditive stimuli focus works through using sounds or voices to stimulate creativity. Auditive stimulation is one of the initial stimuli in dance while auditive stimuli is one of the new methods to develop one's creativity in dance. The Sekar Kinanti Art Studio (SSKS) in Yogyakarta, Indonesia uses auditive stimulation methods to support the learning process. This research is a descriptive, qualitative study that reveals the progression of events that occur in the SSKS, and in particular investigates the learning methods used in the studio. As part of the learning process in SSKS auditive stimulation methods successfully shape the studio students' kinaesthetic abilities. Kinaesthetic abilities (motion and the sense of motion) is observed in SSKS graduates who have passed the learning process after a period of two to five years in the SSKS. Auditive stimulation is one method implemented to help form one's kinesthetic abilities.

1 INTRODUCTION

The essence of music is sound. Sound can come from the natural voice. Humans as part of nature, have feelings (senses) toward rhythm. Millions of sounds in nature are processed by humans into music. Psychologically, music is a product of the mind. Basically, music elements such as vibration, frequency and duration cannot be said to be "music" if these elements have not been transformed neurologically and then interpreted by the brain into pitch, timbre, tempo, and dynamics (Djohan, 2009, p. 32).

In its development, music has two main functions: as a means of nemesis and as catharsis (soul purification). Music as a media nemesis is shown in sixteenth century opera performances where music can be an emotional reflection of the human condition. The function of music as a means of catharsis is believed to be a medium for expressing oneself. Music helps to shape, organise and deliver emotions while media trains inspiration. In a specific cultural context, music has more complex functions. Music is present in every aspect of people's lives such as for ceremonies related to marriage, death and birth, as well as religious ceremonies, rituals, traditional activities and dances (Djohan, 2009, pp. 87-88). Music and dance are two sides of a whole that are difficult to separate from each other. In dance performances, music is one of the elements that must exist. Music in dance functions to regulate emotions, motion, space and time, and inspiration in the dance. Hidajat (2008) states that music has three functions in choreography: 1) music as motion accompaniment; 2) music as an affirmation of motion; and 3) music as illustration.

Music as illustration presents or creates nuances underlying the events in a dance work, building the audience's perception to enter into the work, and is used by dancers as a medium that helps determine the movement and deepens the inspiration of the work.

Sensitivity to music is not immediately owned by dancers. Musical sensitivity can be trained using a variety of methods including stimulation. Auditive stimuli as one stimulus can present and increase kinaesthetic awareness (motion and sense of motion). As a basic concept, humans have a physical response while listening to various kinds of sounds. This basic concept was thus taken and applied by the Kinanti Sekar Art Studio (SSKS) as one of the studio's dance learning methods.

SSKS is one of dozens of dance studios in the Special Region of Yogyakarta that along with other studios, care about sustainability and the future life of art especially traditional arts. The SSKS offers three kinds of dance classes: classical dance class, creative dance class and children's dance class. The learning system in SSKS is new and unique because the system presents a method of responding to initial stimuli. One of the methods involves using audio stimulus to train the studio students' musical sensitivity (*sedulur*). Musical sensitivity is needed because the music or dance accompaniment used in the SSKS uses traditional collaborative music and modern EDM (Electronic Dance Music) in the MIDI (Musical Instrument Digital Interface) format. The transition between each bar in music is difficult for beginners to understand, so the ability to hear transitions needs to be trained early.

Understanding music is considered important for Kinanti Sekar Rahina, the owner of the Kinanti Sekar Art Studio. As music plays an important role in, for example, being able to count from one to eight in a dance work. During the learning process each facilitator (teacher) at the SSKS is encouraged to know and understand every detail of dance accompaniment music. During the process of delivering the dance material, the facilitator minimizes counting one to eight, and replaces this traditional way of teaching by instead imitating the music. In certain classes the facilitator will play music and the students are given the freedom to respond to the music they hear.

Based on the above-mentioned learning events that occurred at the SSKS, the researcher was interested in examining the role of auditive stimuli upon dance accompaniment at the Kinanti Sekar Art Studio as a method of increasing kinesthetic abilities. The accompaniment music that was studied are called motion dance music and songs for the *Sekar Taji* Mask dance in the child category called *Kenes Gandhes* dance music for new dance in the creative class category.

2 METHODOLOGY

This research is a descriptive, qualitative method study. Descriptive qualitative methods were chosen to explain the overall object of study in the research while the analysis refers to relevant theories. The data was collected by the researchers as key instruments for the investigation. The data was collected by observing and seeing the facts in the field, in various settings and with various methods and sources (Sugiyono, 2014, pp. 62-63). In this study, the data was obtained through observation, interviews, documentation and triangulation techniques.

The research took place at the the Kinanti Sekar Art Studio in Yogyakarta, Indonesia. The data was obtained by observing the learning process at the SSKS, by interviewing the studio owner, the management personnel, the dance class facilitator and others in the SSKS. Documentation studies are also part of research techniques that made it possible to read student documents from the beginning when the SSKS was first established and until this research was conducted.

3 DISCUSSION

3.1 *Auditive stimulation*

Stimulation, often referred to as stimulus, is a term used in psychology to explain something that provokes a certain response. Stimulation can trigger the appearance of behaviours or attitudes. Auditive stimulation is one of the elements found in the initial stimulus. Auditive stimuli include sounds that stimulate creativity. These sounds can come from musical instruments and music tapes. As for the sounds of nature and the environment, such as the sounds of waves, animals as well as various, frenzied sounds generated by human activities (Susanti, 201, p. 46).

In dance, auditive stimuli have a very big role. At the present, many dancers are inspired to move after listening to certain music, sounds or voices. In creating dance works, choreographers can use auditive stimuli to stimulate the emergence of new movements in accordance with the theme of the choreographer's work. In modern dance learning, auditive stimuli are used as a method to hone one's kinesthetic creativity.

3.2 *Kinanti Sekar Art Studio*

The Kinanti Sekar Art Studio was founded on 24 July 2015. The SSKS has four kinds of classes: 1) classic dance class consisting of the *Nawung Sekar* dance, the *Golek Ayun-Ayun* dance and the *Golek Kenya Tinembe* dance; 2) creative dance class consisting of the *Kenes Gandhes* dance, the *Gelegar Nusantara* dance, the *Sonteng* dance and the *Pendet* dance; 3) children's dance class consisting of dance and song, the *Kwek-Kwek* dance, the *Nyawiji* dance, the *Padhang Bulan* dance and the *Nusantara Anak* dance; and 4) *Kinari* class, an advanced class for children that is specifically open to advanced level students. Additionally, there is a forum for studio alumni called the Kinanti Dance Laboratory.

Dance material is delivered using the methods of imitating and responding. The imitation method is applied to all classes especially classical dance classes. In the children's classes and creative classes mimicking and responding methods are used simultaneously. Students in the *Sedulur* studio are trained to respond to initial stimuli such as kinaesthetic, touching, visual and auditive in order to cultivate kinaesthetic and auditive abilities that students possess. The focus of delivering dance material during children's classes and creative classes is to introduce and guide the development of sensitivity to dance accompaniment as well as understanding the dance itself. This sensitivity is needed because the music that accompanies dances taught at the studio is a combination of modern and traditional MIDI format. The modern music is EDM.

> One of the most distinctive characteristics of EDM is the way in which it is produced- namely, through the use of electronic technologies such as synthesizers, drum machine, sequencers, and samplers. Although increasingly common in popular music in general in recent years, these technologies have always formed the backbone of musical creation in EDM, in which traditional instrument or live vocal is the exception rather than the rule. (Butler, 2006, p. 33)

For the process of making music with EDM, there are features that can be used to create traditional instrument sounds or other desired sounds. However, it is a difficult transition to distinguish sounds in EDM music in MIDI form. For beginners, the transition between cultivation in each bar of music is difficult to understand.

The style of music produced at the SSKS is different from music in other studios. Lay people (outside the SSKS) may not be able to dance with the music from the SSKS because the music has its own characteristics. In fact, students at the first level need time to adapt and get to know the dance accompaniment (Kinanti, personal communication, 3 September 2018). The absence of special cues for the transition of motion such as cues found in traditional music (*gamelan*), is an obstacle for beginning dancers making it challenging to determine the movements or changes in dance movements based on the SSKS music.

The children's classes use music that comes from children's songs such as the "Gugur Gunung" song for the *Nyawiji* dance, the "Padhang Bulan" song for the *Padhang Bulan* dance and a MIDI medley of *nusantara* music for the *Nusantara Anak* dance. The "Dolanan Anak" song was chosen as an arrangement material because the song is familiar to children and helps children to remember the music used in the dance.

For the learning process in creative classes body formation and musical sensitivity are prioritised. The class facilitator provides the basics of dance at the beginning of the class session and gradually over time trains the students' musical sensitivity. The method used is listening to the selected piece of music repeatedly and students are also asked to move according to what they hear. Repeating the material up to seven times in one class session is one way to make students listen to and recognize dance music.

In addition to regular classes, the SSKS has natural class programs as a dance learning support program. The natural class vision and mission is to invite all students at the SSKS to learn and interact with nature. The concept of auditive stimuli is actually implemented during this activity where students will explore what they hear and visualize what they hear through motion.

3.3 *Kinaesthetic intelligence*

Intelligence can be interpreted as an expression of ways of thinking that are used as modalities in learning. A theorist on intelligence named Gardner (1993), describes nine intelligences possessed by humans: linguistic intelligence, mathematical logic intelligence, spatial intelligence, musical intelligence, interpersonal intelligence, intra personal intelligence, naturalist intelligence, spiritual intelligence and physical or kinaesthetic intelligence (Khasanah, 2016). Gardner (1993) argues that kinaesthetic intelligence is the ability of a person to use his or her entire body or a part of the body to solve problems.

Kinesthetic intelligence can also be interpreted as possessing good coordination between the nerves, mind and limbs.Kinaesthetic intelligence or physical intelligence is the ability for a person to use his or her limbs to organise movements into activities such as running, dancing, building something, doing art activities or cubits works

The core component of kinaesthetic intelligence is the ability to use specific physical abilities of coordination, balance, skills, strength, flexibility, speed and ability to receive stimuli (proprioceptive) and is also related to tactile and haptic abilities or touch (Yuningsih, 2015).

3.4 *Auditive stimuli at the Kinanti Sekar Art Studio: Dance accompaniment music as a method of improving kinaesthetic intelligence*

The method of imitating and responding used a the SSKS is used to train young people to not only know how to dance but to also have good kinaesthetic abilities by responding to kinaesthetic stimuli, visual stimuli, touch stimuli and auditive stimuli. The body's responses to these stimuli is extended through exploration and presentation of motion.

Auditive stimulation is one method that is always used at the SSKS. Auditive stimulation is a process of searching for motion through audio. In dance this concept is used by choreographers to create works by referring to what is heard, producing work patterns by considering auditive aspects related to the body's ability to visualise what it hears (Hidajat, 2011, p. 95). Responses to various stimuli is a skill that is very important to introduce to student early in order to train creativity and the body.

By observing SSKS graduates researchers found that the SSKS implemented the following processes:

a. Systematic Learning System
   The learning process at the SSKS is carried out by using the quarterly chess system. Competency testing is done at the end of the fourth month. The class facilitator manages 16 face-to-face sessions with the students to complete one dance material.

   Based on observations made by the researchers since January 2016, the SSKS has graduated ten students from the creative dance class, three students from the classical dance class and one student from the *Tembang* class. Additionally, fourteen students from the children's *Kinari* class have also completed the qualification exam at the SSKS (Andi, personal communication, 30 July 2018).
b. Learning Support program
   The learning support program at the SSKS consists of the nature class and the studio staff. The head of the studio, Andi Wicaksono, further described the two programs:

   > As the name implies, natural class is a visit to nature whether it's to the beach, mountains or rice fields. The purpose of the existence of a natural class is to introduce a studio in the surrounding environment that brings positive energy in our lives (humans). *Sedulur* is invited to unite with nature, learn in the open to accept whatever nature gives. Besides

that, SSKS has a studio program. *Sowan* in Javanese means visiting; *Sowan sanggar* means visiting other studios to establish friendship and sharing knowledge between studios. The hope of this activity is to know about the learning activities in other studios, and to exchange experiences in the arts and culture. (Andi, personal communication, 3 March 2018).

The research yielded results such as discovering interesting phenomena related to the dance learning system at the SSKS. During the last two years, the researchers observed five children's dance classes and five creative classes. The basic class for children is coded 'a', the first batch is coded A, the second batch is B, the third class is C, the fourth class is D and code E for the *Kinari* class. The first class of the creative category is coded as 1, the second generation is 2, the third class is 3, the fourth class is 4 and the fifth class is coded as 5.

The children's class with code 'a' is a dance class for 3 to 6-year olds. This class was only opened in the tenth class of the SSKS in August 2018. The material that is given is about motion and songs using the method of playing and moving. The application of the auditive stimulation method starts at this level where students are introduced to various kinds of dance accompaniment music, development music, children's songs and dancing with simple movements. The motion material that is given includes basic movements such as heating, that are arranged into a series of interesting movements according to the music that is used. The purpose of the movement and song class is for children to enjoy the learning process and assume that 'dancing is fun'.

The children's class that is coded A uses the dance material called *Kwek-Kwek*. This class is classified as a basis class for children who are 7 years old. Students are introduced to the basics of dance to be used at the next level. The movement in this dance aims to train the students' physical strength. Running, jumping and moving are the hallmark movements in this class. The facilitator has the responsibility to introduce complete traditional music by calculating each transfer. The facilitator uses the auditive stimulation method by repeatedly playing dance accompaniment while asking students to move to demonstrate the dance movements. If the motion carried out is not in accordance with the accompaniment or the student loses the count in the music, the facilitator repeats the activity from the beginning and is repeated until the student understands the key to transferring the music. In general, the students' kinesthetic abilities at this level was not observed. However, by the end and after the material has been given, there is an increase in dance skills by the students in class A.

The next level, called the *Nyawiji* class, is coded as B. At this level the students are given more complex dance material, both in terms of motion and music. Dance movements in the *Nyawiji* class are prioritized in a way so as to form and bring out the character of each student's body. During the learning process, the facilitator is the model for the students and occasionally the students alternately become models for each other. Realizing that music that has a high degree of difficulty for children, the facilitator uses the method of repeated listening so that the students can better know the accompaniment music. Improvements in the students' motion and musical sensitivity begins to appear at this level and the students tend to grasp material given by the facilitator more quickly. Each student's dance movements were observed as more organized at this level.

Next, students enter the *Padhang Bulan* class which is coded C. The music that is used has a faster tempo than the previous class, namely moderato (100-103). This music is intentionally used in class C to train the students' sensitivity to be able to respond to music with a fast tempo by adjusting to the music using existing dance moves. Based on observations, class C students have developed higher levels of musical and body sensitivity compared to the previous class.

The final level in the children's classes is called *Nusantara Anak* coded as D. This class is a fairly complex class such that students are introduced to dance forms from several regions in Indonesia. The dance accompaniment music is a medley of music consisting of *nusantara* songs that represent the area chosen by using EDM. Musical concentration and sensitivity are necessary to be able to understand the movements of dance accompaniment music. Based on the researchers' observations, these students developed stronger dancing skills. The student's motion techniques experienced a significant increase; the form of motion began to look like 'waking' and organized. The ability of students to respond to the auditive stimuli of dance accompaniment music also showed a significant increase. Students were able to determine when to move based on the music they heard. Students found the 'code' for the movement of music so that dance movements were based on 'taste' in a simple sense. The ability to memorize dance movements and dance accompaniment music also increased compared to before. At level D, the level of kinesthetic abilities increased from class C.

Similar achievements were also seen in adult class students in the creative class. In the first stage in class 1 with *Kenes Gandhes* dance material, many students were getting to know dance for the first time in their lives. The dance material was given with a time duration that is not too long so that it would be easier for these students to memorise the dance movements and the music accompaniment. The music that is used is gamelan and EDM collaborative music. The *Kenes Gandhes* dance music is difficult because from the beginning to the end of the music there is no significant change in cultivation patterns. What is needed is repetition and that the intensity is not small to get into the accompaniment of dance music. For this reason, the class facilitators often use auditive stimulation methods to hone and develop the musical sensitivity of studio students.

The Level 2 goals for the *Gelegar Nusantara* dance material is to strengthen the musical and kinesthetic sensitivity of the studio. At this level, the material has a duration that is longer than Level 1. The purpose of this class is to introduce students to dance forms and movement of the Indonesian archipelago. Dance formation is emphasized at this level and indirectly, some students begin to see the forms of specific dances. The music in this class has a fairly high level of difficulty in grasping movement/music signs that serves as markers for certain dance movements and that require precision and high concentration. During the observation process, students found it easier to receive the material from the facilitator so that there were no special difficulties in delivering the dance material.

At Level 3, the students are exposed to complete dances from different regions. Examples of dances include the *Pendet* dance and the *Sonteng* dance. At this level 3 class, physical capabilities and the ability to identify music increase compared to levels 1 and 2 students. Students in level 3 more easily receive dance material because the ability to receive increases significantly. Likewise, musical sensitivity is at higher level when students are given the material. Generally speaking, the kinetic abilities of the SSKS students increased from one class level to the next as seen from the improvements in shape, technique, ability to respond to music creatively using the body while dancing.

For level 1 to 3 students, the graduation test is carried out by using dance material from levels 1 to 3. One requirement that must be fulfilled by all SSKS students in the creative class is to create a dance work that departs from the dance materials given in the classes. Students who have passed the graduation test are coded 4 and 5. Based on the results of the study, the researchers observed that students in levels 4 and 5 had good body movements and could be said are not inferior to dancers from official dance institutions such as schools and dance arts universities. Student excellence at this level is the ability to create dance works. Creating dance works is not an easy thing to do, but the SSKS students are able to create dance works at a satisfactory level meeting the criteria for graduation from the studio.

Level 4 students create works that prioritise high sensitivity to music. Semi-contemporary dance is chosen as a foothold in making dance works. The *Prasapa* dance work was created based on a variety of Yogyakarta-style movements and contemporary movements. The movements are those that are discovered by students while creating works in response to the four stimuli: kinesthetic, touch, visual and auditive. The results of this exploration then become the power of motion in the work that is being made. The music used to accompany this dance is made with the FL Studio application. Music to accompany dance works is made with the same pattern from beginning to end but occasionally uses tempo variations. The auditive excitement concept is seen in this dance accompaniment and is inspired by the learning process at the SSKS that uses auditive stimulation methods in creating dance movements.

To fulfill the graduation test, level 5 students also create dance works armed with the knowledge and experience gained during their studies at the SSKS. The work created is called *Kekepetan* dance that is made by developing motives and motion techniques acquired in the previous class. The music used in this work is EDM music.

In addition to level 4 and 5 students, there is another class where dance progress seems to increase. This is the *Kinari* class with code E where students have received all the material in the creative class and children's class. *Kinari* students are familiar with the learning methods carried out by Kinanti Sekar Rahina in the SSKS. The intensity of listening to the accompaniment music of a typical SSKS dance makes it possible for students to possess a high level of musical sensitivity. This sensitivity was observed in the *Sekar Taji* mask dance. The *Kinari* children's kinesthetic abilities was observed when responding to the music, which if heard does not have clear displacement or can be said to be contemporary music. A high level of kinesthetic intelligence makes it possible to perceive and interpret the beat of music and motion simultaneously even though the count is not clearly heard. This process is called 'ngrasakke', or feeling the meaning of the dance while performing, not just memorising the movements of the dance.

After reaching this stage, there were a lot of positive changes observed in the *Kinari* children. The biggest change was the growth in kinesthetic abilities. The children not only danced but were able to convey the meaning of the dance and delivered the meaning to the audience. To be categorized as a professional dancer is a bonus for *Kinari* students earned through the learning process. However, the main benefits are that children build a sense of solidarity, tolerance, responsibility, honesty, sportsmanship, family and togetherness.

4 CONCLUSIONS

The present era brings changes to dance learning systems. The method of delivering dance material is always developing every day. One development is responding to kinaesthetic, visual, touch and auditive stimuli.

The application of new methods to develop the ability to respond to stimuli is carried out to train the SSKS students' kinaesthetic abilities. In natural class programs, auditive stimuli are used as the main dance learning method where students are directed in such as way so as to be able to respond to the sounds of nature. Additionally, students are directed to create a simple work by responding to kinaesthetic stimuli, auditive stimuli and visual stimuli that exist in nature. This method is also used in regular classroom learning processes.

The use of this method at the SSKS produces students who have good bodily acumen, high musical sensitivity and optimal kinaesthetic abilities. As observed by the researchers, the SSKS graduates who have gone through the learning process for two years possess a form of dance that can be said to be equal to dancers graduating from formal institutions. The success of using the auditive stimulation method as a way to increase the level of kinaesthetic abilities can be seen in *sedulur* with codes C, D, E for child classes and codes 3, 4, and 5 for creative classes.

Thus, the learning method using auditive stimuli increase the level of kinaesthetic abilities. Auditive stimuli helps students develop kinaesthetic potential (motion and sense of motion) that is inherent in children. Using auditive stimuli makes the process easier for students to explore how to use their bodies. The ability to develop greater sensitivity and inspiration is enhanced using auditive stimuli.

REFERENCES

Butler, M.J. (2006). *Unlocking the Groove: Rhythm, Meter, and Musical Design in Electronic Dance Music*. Bloomington: Indiana University Press.

Djohan. (2009). *Music Psikology*. Yogyakarta: Best Publisher.

Gardner, H. (1993). *Multiple Intelligences: The Theory in Practice*. New York: Basic.

Hidajat, R. (2011). *Choreography and Creativity*. Yogyakarta: Media Kendil.
Khasanah, I. (2016). Enhancing Child Kinesthetic Intelligence through Angguk Traditional Dance at TK Melati II Glagah. *Journal of Early Childhood Education*, 5(*3*), 293.
Sudarsono. (1977). *Indonesian Dances I*. Jakarta: Proyek Pengembangan Media Kebudayaan Direktorat Jendral Kebudayaan, Departemen Pendidikan dan Kebudayaan.
Sugiyono. (2014). *Understanding Qualitative Research*. Bandung: Alfabeta.
Susanti, D. (2015). Application of Alma Hawkins Creation Method in Gundah Kancah Dance. *Jurnal Ekspresi Seni*. Institut Seni Isi Padangpanjang.
Yuningsih, R. (2015). Kinesthetic Intelligence Enhancement through Minang Dance Basic Motion Learning. *Journal of Early Childhood Education*, 9(2).

21st Century Innovation in Music Education – Astuti et al. (eds)
© 2020 Taylor & Francis Group, London, ISBN 978-0-367-11122-9

Cultivating the spirit of nationalism through the song 'Manuk Dadali'

S.R. Mahmudah & Sumaryadi
Universitas Negeri Yogyakarta, Yogyakarta, Indonesia

ABSTRACT: The purpose of this study is to foster a sense of nationalism in children by introducing the local song 'Manuk Dadali' at school. This is very important because children are more interested in playing games than studying or knowing their Indonesian culture. Studying the song 'Manuk Dadali' in *Dadali Campeurnik* dance can foster a sense of nationalism in children. Besides, by learning the song 'Manuk Dadali', students can recognise and understand the art and culture of Indonesia in their own regional characteristics. When teaching and learning activities take place, the role-playing learning model is very helpful for achieving the expected goals. In the learning process, students are directed to learn the song 'Manuk Dadali' in *Dadali Campeurnik* dance, and understand the meaning contained in the lyrics of the song 'Manuk Dadali'.

1 INTRODUCTION

Elementary school-aged children (5–12 years old) are at the stage of physical and psychological growth. During this age, the child's world is very pleasant and full of imagination (Priyanto, 2014). It is at this age, or before they reach adolescence, that the nationalistic values need to be cultivated. In the digital age, however, children are more interested in playing games than studying or knowing their Indonesian culture. This poor sense of nationalism is largely influenced by the external culture which is considered to be more attractive. This situation greatly affects the habits of children, making them lazy about learning their own culture.

One way to foster a sense of nationalism in children is by introducing local songs at school. Local songs contain the ethics and moral values of the Indonesian people that need to be introduced from an early age, which subsequently improve Indonesia's character education (Mulyono, 2012). The nationalistic values contained in these are very essential for achieving better character education because characters are often identified with morals.

Through learning local songs, it is expected to see changes in the children and their behaviour towards a better, nationalistic attitude and love for their homeland. Sambas Mangundikarta created 'Manuk Dadali'. He was a reporter and journalist at Radio Republic of Indonesia (RRI). The song grew out of his love for Indonesia, a love that begs hopes for the presence of a strong and great country. This song could inspire nationalism.

2 LITERATURE REVIEW

2.1 *Learning components*

For every child's education to be implemented optimally and have a clear learning plan to achieve its aims, it has to have certain necessary components. These components include: learning methods/models/strategies, learning media, learning materials/content and learning evaluation. These components include objective components, content components, method components, and evaluation components.

2.2 *The purpose of learning the song 'Manuk Dadali' in Dadali Campeurnik dance*

Growing the value of nationalism in students, can be aligned to the applicable curriculum through the learning of the song 'Manuk Dadali' in *Dadali Campeurnik* dance in accordance with the values of the nation's cultural education and character. These values include being religious, honest, tolerant, disciplined, hard-working, creative, independent, democratic, curious, having national spirit, having love of the homeland, respecting achievement, being friendly, loving peace, loving reading, caring for the environment, being socially aware, and being responsible.

2.3 *Learning methods/models*

One of the learning models used in teaching learning is the role-playing model. A role-playing model is a way of mastering learning materials through developing students' imaginations and appreciation. Development of the imagination and appreciation is carried out by students by acting as a living figure or inanimate object. This game is generally carried out by more than one person, depending on what is played. Blatner (in Baroroh, 2011) stated that role playing, a derivative of a sociodrama, is a method for exploring the issues involved in complex social situations.

This model is used when the lesson is intended to: (a) explain an event in which there are many people, is based on didactic considerations and it is better to dramatise it than to tell it, because it will be clearer and can be experienced by the children; (b) train children so that they are able to solve social-psychological problems; (c) train children so they can get along and provide possibilities for understanding other people and their problems.

The role-playing method is used according to the time of the lesson, which is in the hours after the break when students are already saturated with the previous lesson and the weather is also hotter than in the morning. So, this role-playing method has the opportunity to rebuild the mood of students to learn. While the teaching methods used are the lecture method, the demonstration method, and the question and answer method, the capacity of the demonstration method is greater than the lecture method and the question and answer method.

3 DISCUSSION

3.1 *The spirit of the song 'Manuk Dadali'*

Character building in children from an early age through the song 'Manuk Dadali' in *Dadali Campeurnik* dance helps instil nationalistic values and the song is expected to have a big influence on improving a child's character and developing their nationalistic spirit, whose characteristics are:

1. Loving the nation and homeland of Indonesia
2. Willing to sacrifice for the sake of the nation and state
3. Proud of the nation and land of Indonesia
4. Placing the interests of the nation and state above personal and group interests.

Some examples of nationalistic behaviour or attitude in daily lives that can be applied to early childhood education is to preserve Indonesian culture and uphold the values of national unity. As for the characteristics of nationalism itself, one of them is love for the homeland, race, language or shared cultural history. Recognising that the sense of nationalism among teenagers and children has begun to fade, teachers will do no harm if they instil nationalism into students because at this age it is not uncommon that many students do not care about their own friends and tend to be individualistic. Individualism has made people not care about their nation. In addition, our society, especially young people who forget about their identity as Indonesian people, prefer and imitate western culture and lifestyle rather than Indonesian culture and lifestyle.

Character development in human life is very important and strategic because characters are often identified with morals (Faturrohman et al., 2013). Cultivating nationalistic values through the song 'Manuk Dadali' in *Dadali Campeurnik* dance might provide an insight into ethnic diversity in Indonesia. It is very appropriate to instil nationalistic values in children in elementary schools.

Local songs that dominate the dance work of children created by Indonesian choreographers should get the support and attention from the government because indirectly they have helped implement character education through dance movements, local songs in dance, traditional music (which is often used as accompaniment in child dance) make-up, and fashion that resemble Indonesian culture. The nationalistic value which contains the teachings of manners and love of the Indonesian homeland needs to be taught to elementary school-aged children because they need to know the Indonesian identity which will form their behaviours, attitudes, and characters. Dewantara (1967) said that cultivating nationalism has to be done in elementary school because children will readily imitate.

3.2 *The relationship between arts and the cultivation of character education*

Art education is one of the subjects given in schools, from early childhood education to high schools. According to Plato, in Jazuli (2008), art education should be the basic education, which means that art education must be the foundation for education. Hikmawaty (2014) reveals that art education teaches children how to become independent in environments such as at home and school. The steps in conducting this study are identifying general ideas, conducting observation, developing general plans, developing the first step of action, implementing the first step, evaluating, and improving common practices.

Character education, as stated, is an important part of the world of education. National education aims to develop and shape a dignified national character and civilisation in order to educate the nation, aimed at developing students' potential to become believers and to fear God Almighty as well as to be noble, healthy, knowledgeable, capable, creative, independent, democratic and responsible citizens.

Efforts to form a sense of nationalism in early childhood is not solely done in the school environment, but can be done in the environment outside of schools as well. For example, in the family environment, play environment, studio environment, and so on.

Nationalism values need to be developed among students from an early age through learning art such as learning the song 'Manuk Dadali' in *Dadali Campeurnik* dance. In addition to the nationalistic value, mental attitude and discipline are also found in art education in both dance and musical arts in schools. This way, art lessons (dance/music art/fine arts) are the media for children to learn character education. Children's dance will influence the sharpness of mind, subtlety of taste, and strength of will and the sense of independence.

3.3 *'Manuk Dadali' regional song learning in Dadali Campeurnik dance*

The song 'Manuk Dadali' has a nationalistic value, as explained in the book entitled *Negara Paripurna* (Plenary State) (Latif, 2013),

- Lyrics of the song 'Manuk Dadali'

Mesat ngapung luhur jauh di awang-awang
Mébérkeun jangjangna bangun taya karingrang
Kukuna ranggoas reujeung pamatukna ngeluk
Ngepak mega bari hiberna tarik nyuruwuk

Saha anu bisa nyusul kana tandangna
Gandang jeung pertentang taya bandingannana

Dipikagimir dipikasérab ku sasama
Taya karempan kasieun lébér wawanénna

Refrain:
Manuk dadali manuk panggagahna
Perlambang sakti Indonesia Jaya
Manuk dadali pangkakoncarana
Resep ngahiji rukun sakabéhna

Hirup sauyunan tara pahiri-hiri
Silih pikanyaah teu inggis béla pati
Manuk dadali ngandung siloka sinatria
Keur sakumna Bangsa di Nagara Indonesia

- Translation of the lyrics of the song 'Manuk Dadali'

Flying high, far away in the air
Stretching his wings, appearing without hesitation
Its legs are long and the beak is curved
Welcome the clouds while flying quickly

Who can rival his courage
Unmatched mighty stout
Respected and loved by others
Without hesitation without fear, it was great

Refrain:
Garuda bird, the most dashing bird
The symbol of Indonesia's magnificent Jaya
Garuda, the most famous
Happy to be united, get along well

Living together without envy
Love each other, do not hesitate to sacrifice lives
Garuda is a chivalry

3.4 *The song 'Manuk Dadali' as a learning material*

The learning media used in the learning process of art and culture is the song 'Manuk Dadali' in *Dadali Campeurnik* dance to foster nationalistic values in students using audio media, visual media, and audiovisual media. The audio media used were the song 'Manuk Dadali', the visual media used was the text image of the song 'Manuk Dadali', while the audiovisual media used was the *Dadali Campeurnik* dance video, complete with costumes and make-up with the theme of the Indonesian emblem, the Garuda bird (*Dadali*).

The song 'Manuk Dadali' originated from the Sunda region (West Java), while the *Dadali Campeurnik* dance, despite having a Sundanese name, has a variety of dance movements like those of Nusantara dance, namely a variation of Sundanese, Kalimantan, Central Java and Balinese dance.

The song 'Manuk Dadali' is an accompaniment for *Dadali Campeurnik* dance, based on creative dance. Students can easily perform *Dadali Campeurnik* dances. It is not uncommon for students to sing the song 'Manuk Dadali'; because of the pleasant accompaniment to be heard and the lyrics that are easily memorised, students can easily learn both. In early childhood, children really like the song 'Manuk Dadali', thus, it will be easy to provide an explanation of the values of character education to students, especially the nationalism values found in the song 'Manuk Dadali' in *Dadali Campeurnik* dance.

4 CONCLUSION

Nationalism needs to be applied as early as possible so that students can learn the diversity of the archipelago (the country of Indonesia), considering that many children are more aware of modern culture and songs than the culture of Indonesia. Studying the song 'Manuk Dadali' in *Dadali Campeurnik* dance can foster a sense of nationalism in children. Besides, by learning the song 'Manuk Dadali', students can recognise and understand the art and culture in Indonesia with their own regional characteristics. When teaching and learning activities take place, the role-playing learning model is very helpful to achieve the expected goals. In the learning process, students are directed to *learn the song 'Manuk Dadali'* in *Dadali Campeurnik* dance, and understand the meanings contained in the lyrics of the song 'Manuk Dadali'.

REFERENCES

Baroroh, K. (2011). Upaya meningkatkan nilai-nilai karakter peserta didik melalui penerapan metode role playing[Efforts to improve the character values of students through the application of role playing methods]. *Jurnal Ekonomi dan Pendidikan*, *8*(2), 149–160.

Dewantara, K.H. (1967). *Kebudayaan Bagian IIA* [Culture part IIA]. Jogyakarta, Indonesia: L Madjelis Luhur Persatuan Taman Siswa.

Fathurrohman, P., Suryana, A.A. & Fatriani, F. (2013). *Pengembangan pendidikan karakter* [Development of character education]. Bandung: PT Refika Aditama.

Hikmawaty, L. (2014). *Penerapan model pembelajaran kritik tari untuk meningkatkan pemahaman multikultur siswa kelas xi sma negeri 7 tangerang* [Application of dance critical learning model to improve multicultural understanding of Class XI students of SMA Negeri 7 Tangerang] (Doctoral dissertation, Universitas Pendidikan, Indonesia).

Latif, Y. (2013). *Negara Paripurna* (Plenary State). Jakarta: Gramedia Pustaka Utama.

Jazuli, M. (2008). *Paradigma Kontekstual Pendidikan Seni* [Contextual paradigm of art education]. Surabaya: Unesa University Press.

Mulyono, U. (2012). Pendidikan nilai luhur melalui tembang (lagu) dolanan anak [Valuable education through children's song]. *Selonding*, *1*(1), 120–127.

Priyanto, A. (2014). Pengembangan Kreativitas Pada Anak Usia Dini Melalui Aktivitas Bermain [Development of creativity for early childhood through games activity]. *Jurnal Ilmiah Guru Caraka Olah Pikir Edukatif*, *2*(1), 67–73.

21st Century Innovation in Music Education – Astuti et al. (eds)
ISBN 978-0-367-11122-9

The value of local wisdom in children's song for character education

M.Y. Putrajip, K.S. Astuti & Y. Fitrianto
Yogyakarta State University, Yogyakarta, Indonesia

ABSTRACT: Children's song has a value of local wisdom that can be integrated into character education. The values of local wisdom in *dolanan* or game songs, such as 'Padhang Bulan', 'Ilir-Ilir', 'Sluku-Sluku Bathok', 'Jaranan', 'Dhondhong Apa Salak', 'Menthok-Menthok' and 'Gundhul-Gundhul Pacul', are the values of life, as in the nine pillars and 18 characteristics of national identity defined by Indonesia's education ministry. This study aims to identify how far the value of Javanesse dolanan songs fot character education. The approach of this research semiotic. The result of the study is game songs for children as education means not only seeing the song as an expression of art but also understanding the rules and philosophical meanings contained in each song's lyrics.

1 INTRODUCTION

The Javanese community has a variety of cultural forms that contain local wisdom values. This local wisdom that is applied to education is meaningfully derived from the nature of knowledge and values of life. The values of local wisdom will be meaningful to social life if they can be reference material for the world of education. The value of local wisdom also has a strategic function to inculcate character. Exploring and reinvesting the value of local wisdom through education can be an effort to build the spirit of nationalism and can be a filter of the impact of globalisation.

One form of Javanese community culture that contains noble local wisdom values and can be implemented in education is the Javanese song. Javanese game songs or children's songs, such as 'Ilir-Ilir', 'Menthok-Menthok', 'Buto Galak', 'Cublak-Cublak Suweng', 'Gugur Gunung', 'Gambang Suling', 'Sluku-Sluku Bathok', 'Jaranan' and 'Kupu Kuwi', have local wisdom values that can be used in the world of education. A Javanese game song is a song for children that has a symbolic meaning that is suitable to give to children to instil character education. Education through art forms as a national identity is an effort to maintain the culture of Indonesian society.

A Javanese game song is a means of communication and socialisation for children with their environment. Through game songs, children can have fun, play and learn. Javanese game songs contain educational moral values, conveyed through simple language, so that they are easily memorised and digested according to the level of psychological maturity or development of the child's soul. This period is also an important period in cognitive, affective and psychomotor development. Children's physical and psychological growth should be used as a foothold in providing education, including instilling character education. Children need good guidance to support success in their growth and development. According to Montessori (in Riyanto, 2009), when educating children what must be considered is that they are unique individuals and will develop according to their own abilities. The value of local wisdom in game songs is conveyed through parables and analogies that are packaged in simple and beautiful language so that they are easily understood by children in learning.

Learning that is in accordance with the child's development stage will affect and determine the personality of the child in adulthood. Success in educating children is also influenced by how well and appropriately parents and educators behave and their attitude toward children. Environmental factors also influence the success of child development. Child development is fast; this is a period that parents and educators must not ignore. In the learning process it is the role of educators to really pay attention to children's development and carry out learning through well-planned activities, so that they can support the physical growth and development of the children's cognitive, affective and psychomotor abilities. This understanding implies that education for early childhood is stated to be successful if it considers the level of development of children in achieving educational goals.

In instilling character education, there are two elements that must be considered, which are learning methods and media. Learning media is a component of learning resources or physical vehicles that contain instructional material in the student environment that can stimulate students to learn (Arsyad, 2007). These two aspects are much related; the selection of one particular teaching method will affect the type of media that will be used during the learning process. However, there are still various aspects that must be considered in choosing media, including learning objectives, types of assignments and responses that children are expected to master after learning takes place, and the learning context includes the characteristics of children. The use of media in the learning process can arouse new desires and interests, generate motivation and stimulation during learning activities and even bring psychological influences to students and teachers (Arsyad, 2007). It means that the main function of learning media is as a teaching aid that is also influenced by the climate, conditions and learning environment that is organised and created by the teacher or educator.

As described by Asmani (2009), in children's education, the use of music is an important part of the educational material. The music includes playing musical instruments and singing songs such as children's songs. Game songs in children's education have two fundamental roles. First, music and songs are used in children's education to optimise children's physical, cognitive and psychomotor development. Second, music and songs are used in children's education to stimulate the development of children's musical intelligence (Asmani, 2009). Javanese game songs are used as a learning media to instil character education for children because the lyrics contained in game songs are more easily understood by children.

Character education aims to instil values in children, so that students are able to develop their characters in full, ensuing that they are integrated and balanced. Students who have ethical values will use all their knowledge, skills and emotions in solving problems they face (Asmani, 2011). Through character education, children are expected to be able to independently improve and use their knowledge, study and apply a noble character so that it manifests in their behaviour.

2 RESEARCH METHODOLOGY

2.1 *Data*

In qualitative research, the research instrument is the researcher herself/himself. The researcher is the key to the validity of the data obtained in the field. Data is collected from various settings, methods and sources (Sugiyono, 2014). In this study data was obtained using observation, interview, documentation and combined techniques (triangulation).

2.2 *Method*

This research is a qualitative research with a semiotic approach. Semiotics is the study of signs, how the function of a sign and the production of the meaning of signs that exist can convey information so that it communicates (Piliang, in Tinarbuka, 2008). A sign is something that represents or describes something else (in the mind of someone who thinks about it).

3 DISCUSSION

Music in Javanese society has been around for a long time. Most of the ancestral cultural heritage is packaged in the form of ballads or songs. One of the cultural heritages that has developed within the Javanese community is the game song. This song is not only functioning as a song sung by children while playing and socialising with their environment. More than that, a game song is a work of art that contains the value of local wisdom and moral messages that can be implemented in character education.

Endraswara (2005) explains that children's songs are sung while playing, or songs sung in certain games. Children's game songs are unique, with meanings different from other forms of Javanese songs. According to Danandjaya (1985), there is a children's game song in oral Javanese that is classified as a folk song. Folklore's important characteristics related to children's game songs are: (1) the language is simple; (2) it uses a simple *cengkok* or melodic style; (3) the number of lines is limited; (4) it contains things that are in harmony with the child's situation and things that are entertaining for a group (Endraswara, 2005). These characteristics are also found in praise verses, the difference being that game songs involve nuances for children, while the usual praise poetry is sung by adults although this does not rule out the involvement of children. The following is a children's song that in the lyrics implies the value of local wisdom:

1) *Jaranan*

Jaranan-jaranan... jarane jaran teji
sing numpak ndara bei, sing ngiring para mantri
jeg jeg nong, jeg jeg gung, prok prok turut lurung
gedebug krincing gedebug krincing, prok prok gedebug jedher

Riding the horse, riding the horse, the horse is *teji* (big and high)
The one who rose was Mr. Bei, who accompanied were the minister
Jeg-jeg nong, jeg-jeg gung, prok prok menyusuri jalanan
Gedebug krincing gedebug krincing, prok prok gedebug jedher

The game song 'Jaranan' contains values for respect and courtesy for superiors, older people or higher ranks. The song also teaches the nature of love, care and cooperation with other people. The lyrics in the song have a message about the importance of togetherness, because humans need each other. People who have a higher position need lower people and vice versa. *Ndara Bei* is a symbol of a high-ranking person and/or rich noble because it is symbolised by having a large, high horse to ride (*jaran teji*) so that the noble must be accompanied by his walking subordinates (ministers).

2) *Padhang Bulan*

Yo prakanca dolanan ing njaba
Padhang mbulan padhangé kaya rina
Rembulané kang ngawé-awé
Ngélikaké aja turu soré-soré

Let's play outside

Bright moonlight

Moon as if waving

Reminds us not to sleep late in the afternoon

The game song entitled 'Padang Bulan' contains the value that a human being must love God and nature and other creatures of God. The meaning contained in it also teaches the nature of compassion, care and togetherness toward fellow human beings. The lyrics in the game song contain messages teaching people to always thank God by enjoying the beauty of their natural creation. To show gratitude we are expected not to spend time only sleeping in the afternoon, but instead we should use the time to worship God.

3) *Ilir-Ilir*

Lir ilir, lir ilir, tanduré wus sumilir
Tak ijo royo-royo tak sengguh temantèn anyar
Cah angon, cah angon, pènèkna blimbing kuwi
Lunyu lunyu yo pènèken kanggo mbasuh dodotiro
Dodotiro, dodotiro, kumitir bedhah ing pinggir
Dondomana j'rumatana kanggo séba mengko soré
Mumpung padhang rembulané, mumpung jembar kalangané.
Yo surako surak hiyo.

Wake up, wake up! Plants have blossomed

So green like a new bride

Shepherd boy, shepherd boy climbs the star fruit tree!

Let it be slippery and difficult, still you climb to wash your clothes

Your clothes, your clothes are torn apart on the side

Sew it, correct it! To meet later in the afternoon

While the moon is shining brightly, while there is plenty of free time

Cheer with cheers Yes!

The game song titled 'Ilir-Ilir' contains religious values, responsibility, discipline, hard work and to never give up. The song has the message that as human beings they must be able to rise from adversity, by having faith and striving to get happiness. The star fruit picked by the shepherd with difficulty is like the command of Allah to carry out the five daily prayers. Even though it was heavy with many obstacles in carrying it out, it was likened to clothes until it was torn apart, but it had to be done. By always obeying God's commands, it open hopes for humanity to improve so that it will be ready when the time comes to face and fulfil the divine call.

4) *Menthok-Menthok*

Menthok-menthok tak kandhani, mung solahmu angisin-isini
Bokya aja ndheprok, ana kandhang wae
Enak-enak ngorok, ora nyambut gawe
Methok-menthok, mung lakumu megal- megol gawe guyu

Menthok-menthok I advised you, your behaviour was shameful
Don't just sit and sit, just in the cage
Good sleep doesn't work
Menthok-menthok, your way to shake your butt makes people laugh

The 'Menthok-Menthok' song implies that humans must have a humble attitude and want to be introspective. As humans, we should not be arrogant and must respect others, because all of God's creations have their own advantages and disadvantages. Like '*menthok*' (wild duck), an animal whose appearance is ugly, unattractive, likes to sleep and is lazy is still useful for others, because it is able to make others laugh at the humour of their behaviour. Therefore, you should not be reluctant to see your own shortcomings and not easily demean others. This song also means that you should not be lazy because it is not a good trait.

5) *Gundhul-Gundhul Pacul*

Gundhul gundhul pacul cul, gemblèlengan
nyunggi nyunggi wakul kul, gemblèlengan
wakul ngglimpang, seganè dadi sak ratan
wakul ngglimpang, seganè dadi sak ratan

Bald head without hair like a hoe, big head (arrogant, haughty)

Bring baskets, with a big head style (arrogant, haughty)

The basket fell, the rice spilled messy on the road

The basket fell, the rice spilled messy on the road

The 'Gundhul-Gundhul Pacul' play song illustrates the character of an ugly, arrogant child that behaves badly. Bad behaviour and attitude make it unable to work properly, and do things that are not useful. The message in the song is to be a person who should not feel themself the smartest, the most powerful, arrogant or careless. Such a nature will only cause failure, because people who are arrogant and careless will not be able to carry out their responsibilities properly.

6) *Sluku-Sluku Bathok*

Sluku-sluku bathok
Bathoke ela-elo
Si Rama menyang Solo
Oleh-olehe payung motha
Mak jenthit lolo lobah
Wong mati ora obah Nek obah medeni bocah
Nek urip goleka dhuwit.

Swing head

His head was shaking

The father went to Solo

Therefore, the mutha umbrella

Suddenly it moved

Dead people do not move

If it moves it scares people

If life is looking for money

The 'Sluku-Sluku Bathok' song contains the value of love for God and a sense of responsibility for the life lived, discipline responsibility and independence in living life. The implicit meaning in the song is that humans always cleanse their minds by praying (*dhikr*), by shaking their heads (*elo*), by saying "*Laa illa ha illallah*" (there is no god but Allah) both in joy and sadness and when getting pleasure and calamity. When still alive, be diligent in worship, because life and death no one knows.

7) *Dhondhong Apa Salak*

Dhondhong apa salak, dhuku cilik-cilik
Andhong apa mbecak, mlaku dimik-dimik

Dhondhong or *salak, dhuku* is small

By carriage or by pedicab, walk slowly

The game song 'Dhondhong Apa Salak' has the meaning to always do good, and not hurt others. It also teaches to have the nature of independence and not be happy to depend on the help of others with all the advantages and disadvantages that are owned.

4 CONCLUSION

Character education is an effort to instil and develop good values based on the virtues that apply in society. The value of virtue that applies in community life in general has been agreed both in writing and unwritten. Character education is an effort to educate children to have a good understanding so that they are able to behave in accordance with the prevailing norms in society. Character education produces individuals who have intact intelligence personality in terms of cognitive, affective and psychomotor, and are able to find solutions to any problems faced.

In realising child character education, a strategy is needed and involves various elements within the community. Formal education institutions and the role of parents are the most effective places to instil character education for children. Character values developed in formal education include the values of honesty, responsibility, healthy living, discipline, hard work, confidence, entrepreneurial spirit, creative thinking, logical, innovative, independent, curious, knowledgeable, polite, tolerant, democratic and nationalist (Asmani, 2011). It has been described that children's songs as education have meaning and are used to instil character for children. Education through children's songs contains moral messages and good behaviour in accordance with the rules of society. Children's songs are implemented in character education in which character values are appreciated, understood and expressed.

Game songs for children as education means not only seeing the song as an expression of art but also understanding the rules and philosophical meanings contained in each song's lyrics. Through this understanding, children will have a direction from the point of view of children's songs as a work of art and song. Children are taught values for life and moral teachings to strengthen character, which is called character education. The values of local wisdom contained in children's songs can be integrated into character education so that the child has a strong identity.

It is necessary to realise that implanting character education through children's songs is not as easy as imagined by the practice of art. The practice of art through children's songs must be designed according to the character of the child and the environmental conditions of the education. Character education through children's songs can be packaged through appreciation, expression and understanding the meaning contained in children's songs.

REFERENCES

Arsyad, A. (2007). *Instructional media* [Instructional media]. Jakarta, Indonesia: Raja Grafindo Persada.

Asmani, J.M. (2011). *Handbook on internalization of character education in schools*. Yogyakarta, Indonesia: Diva Press.

Asmani, J.M. (2009). *Strategic management of early childhood education*. Yogyakarta, Indonesia: Diva Press.

Danandjaya, J. (1985). *Folklore Indonesia, gossip and others*. Jakarta, Indonesia: Pustaka Utama Grafiti.

Endraswara, S. (2005). *Javanese oral tradition*. Yogyakarta, Indonesia: Laksbang Presindo.

Mulyono, U. (2013). Sublime education through song Dolanan Anak. *Selonding*, *1*(1), 121–128.

Riyanto, H.Y. (2009). *Paradigma baru pembelajaran: sebagai referensi bagi guru/pendidik dalam implementasi pembelajaran yang efektif dan berkualitas* [New paradigm of learning: as a reference for teachers/educators in implementing effective and quality learning]. Jakarta: Kencana Prenada Media Group.

Sugiyono. (2008). *Metode penelitian kuantitatif dan kualitatif dan R&D* [Quantitative and qualitative research methods and R & D]. Bandung, Indonesia: Alfabeta.

Tinarbuko, S. (2008). *Semiotika komunikasi visual: metode analisis tanda dan makna pada karya desain komunikasi visual* [Semiotics of visual communication: Methods of sign and meaning in visual communication analysis]. Banten: Jalasutra.

Wang, J.C. (2015). Games unplugged! Dolanan Anak, Traditional Javanese children's singing games in the 21st-century general music classroom. *General Music Today*, *28*(2), 5–12.

Winarti, D., Wijana, I.D.P., Poedjosoedarmo, S. & Ahimsa-Putra, H.S. (2015). Variations of directive speech act in Tembang Dolanan. *Jurnal Humaniora*, *27*(3), 305–316.

Yunita, L.S. (2014). Symbolic forms and functions of Javanese song. *Nosi*, *2*(5), 472–478.

21st Century Innovation in Music Education – Astuti et al. (eds)
 ISBN 978-0-367-11122-9

Gamelan music therapy to train the inner sensation to get the feeling of the dance

D.A. Fajar & R. Wiharsih
Universitas Negeri Yogyakarta, Yogyakarta, Indonesia

ABSTRACT: The process of learning through the body gets less attention in research. Students usually only use self-conditioning exercises until they are able to understand the feeling. In this writing, the formation of "taste" appears within the period of time of at least three years where new learning can be mastered, and only if the skills are honed continuously. This research can be an alternative method of teaching inner sensation using gamelan music therapy. The case study method is used along with referring to literature studies in order to find supporting data that might be able to overcome problems that exist in the field. This research is only preliminary and is expected to become research practice in the field.

1 INTRODUCTION

The brain is an organ that plays an important role in life. Research on the brain has a lot to do with various things. One such research deals with music and the learning process. Some studies use music to influence the brain to improve concentration and learning processes. Music has a strong influence on the learning environment. Research shows that learning is easier and faster if students are relaxed and receptive. The heart rate of people in this state is 60 to 80 times per minute. In this state, the brain enters the alpha wave (8-12 hz), which is a relaxed but alert brain condition so that part of the brain, namely the hippocampus and somatosensory, can work optimally. Gamelan music can affect the electrophysiology of the brain which has been reported in many studies.

The use of gamelan music therapy is expected to accelerate the process of learning the feeling and self-conditioning process of dance that accompanies music. A systematic sense-of-training method will be obtained and is expected to become a learning model that will be used in departments that require sense-training. However, there are few theories and methods that discuss the practice of inner sensation.

2 RELATED WORKS

Previous research that has been carried out include: the effects of non-invasive Javanese gamelan music therapy on the hemodynamic status of patients with acute miocardial infarction in the Intensive Care Units in Panembahan Senopati Bantul hospital and Wates By Parlina (2015); and brain, music and learning research carried out by Supradewi (2010). This research differs from previous research as it applies the use of gamelan music as therapy in students who are just learning to dance, so that the gamelan music patterns can be captured by the brain and subconsciously shape the dancers' mental environments and habits.

3 METHOD

The research method is carried out by researchers through the collection of information or data and investigating the data that has been obtained. This research includes a literature study of theories that are relevant to cases or problems found. Theoretical references are used as the basic foundation and the main tools for research practice in the field.

This research begins with a study of a number of literatures that support the analytical approach including: a study of theories dealing with the relationship between music and brain waves, the function of gamelan music and the formation of taste through gamelan music. The source of the study is in the form of reference books, statutory provisions and related regulations, published scientific journals and references on websites accessed via the internet.

4 DISCUSSION

4.1 *Olah tubuh (body work)*

The Body is known in various branches of scientific discipline that make the body as a medium, for example in theater, dance, martial arts, etc. Santoso (1986) found *olah tubuh* is to train physical endurance, endurance of muscles, bones and breath. Breath coordination exercises and movements for dancers are useful for balancing the two. The balance of breath and motion produces strong expression and strength. However, the quantity and type of training will be different according to its purpose. Experts also express the meaning of *olah tubuh.* Sardono W Kusuma (Santoso, 1986) say the body work (*olah tubuh)* is a preparation so that the body condition can reach states such as *sareh* (peaceful), *sumeleh* (relax) and is done with *pranayama* (deep breath). Meanwhile Suprapto Suryadarmo(Santoso. 1986) say body work is a unit of life between the body, god and nature through conscious improvisation.

It can be concluded that the body deals with physical processing activities, used in order to improve one's abilities with expected goals that include dealing with the concept of the body and soul as a whole. "Processing" means that actions are done consciously in a systematic and planned manner. Every human body has several different physical conditions where different habits can affect different anatomical structures such as bone abnormalities, imbalanced muscle capacity, incorrect body postures and so on. Therefore, it is necessary to empty the body, which means to restore the anatomy of body to ideal proportions before doing any formation or enhancement.

The term "body" does not mean the body in opposition to the soul, but refers to the body with all its constituents, as well as the soul as its personality. This is what makes people whole, with their expressions, emotions and mentality. In a dance, the movements of a dancer must have certain demands, even though the principle is to move the body's segments that is different from everyday movements. In dance, aesthetic elements are needed. Discussing the aesthetic value of the body as an abstract object (matter) is difficult. To convey the aesthetic quality of a motion requires feeling or expression. The soul or sense of being also needs to be processed in addition to the body (material) which is the starting point as a means of delivering the meaning.

4.2 *Inner sensation*

In the Surakarta dance style there are three main aspects, namely *wiraga*, *wirasa* and *wirama*. Feeling in Javanese is called *wirasa*, and is the appreciation and inspiration in dance, such as firm, soft, happy and sad that are expressed through movements and facial expressions that give birth to beauty. Sense in intention is seen in body movement so that the message in dance can be conveyed. The inner sensing is almost the same as inner beauty, like the charisma that radiates from the body. For example, two characters such as a king and a guard dance together. Even though the characters perform with the same movements the audience can distinguish between the king and the guard through inner sensation of the dancers performing.

In conventional methods commonly used by artist, practicing daily influences the performance characters, so that unconsciously the character will become a "habit". However, it takes a long time to practice in order to be able to master one's thoughts and feelings.

4.3 *Music therapy and brain waves*

The human body has a basic vibration pattern. Pinel (2009) say the music vibration is closely related to the body's basic frequency and vibrating pattern, which has a very good healing effect for the whole body, mind and soul that can cause changes to emotions, organs, enzymes, cells and atoms.

Sounds with a high frequency of 3000 - 8000 Hz will generally vibrate in the brain and can affect cognitive functions such as thinking, spatial perception and memory. The sounds that has a medium frequency of 750-3000 Hz tends to stimulate the heart, lungs and emotions. While for sounds that have a low frequency of 125 - 750 Hz affect physical movements. High frequencies are more than 100 Hz, and low frequencies are below 100 Hz.

Pinel (2009) also found the human brain network produces fluctuating (up and down) electrical waves. Fluctuating electrical waves are called brain waves In 1929, Hans Berger, a German psychiatrist, discovered electroencephalography (EEG). The EEG is a device that can be used to measure electric waves produced by the brain.

The frequency of the human brain is different for each of the following phases: conscious, relaxed, light sleep, deep sleep, trance (unconsciousness), panic and so on. Mustajib (2010) found through lengthy research, neurologists agree that brain waves are related to mind conditions. Certain types of brain wave frequencies affect human brain conditions as in the following:

a. *Gamma (16 Hz -100 Hz)*
   Gamma is the brain wave that occurs when a person experiences very high mental activity, for example being during competition, a championship, appearing in public and when one experiences panic or fear. Therefore, gamma describes conditions of someone in full awareness.
b. *Beta (12 Hz - 19Hz)*
   Beta is when someone experiences mental activity while fully awake, for example when doing daily activities and interacting with other people.
c. *Sensory Motor Rhytm (12 hz - 16 hz)*
   Sensory motor rhythm commonly called SMR is still included in the low beta vibration group and is a wave that can make people focus or concentrate. If someone does not produce this wave, he or she will automatically not be able to concentrate, for example, epileptic sufferers, and those diagnosed with ADHD (Attention Deficit and Hyperactivity Disorder) and autism.
d. *Alfa (8 hz - 12 hz)*
   Alfa occurs when someone experiences relaxation. Alpha waves reflect relaxed alertness or sometimes is also called relaxed awareness. People who started meditation also produce alpha waves. .Mustajib (2010) says meditation sometimes also called *semadi*, is the practice of relaxation that involves releasing the mind from all things that are interesting, burdensome, or worrying in our daily lives. The literal meaning of meditation is the activity of flipping through the mind, thinking, pondering Alpha frequency is also the frequency controlling and connecting the conscious mind and subconscious
e. *Theta (4 hz - 8 hz)*
   Brain waves that occur when someone experiences light sleep or is very sleepy are called theta waves that are usually marked or characterised by slow and deep breathing conditions. In addition to being asleep, some people can also produce this condition under certain conditions, for example, when meditating deeply, praying or undergoing religious rituals solemnly. In addition, people who are able to drain *chi*, *prana* or inner energy can also produce theta waves while they exercise or channel energy to others.
f. *Delta (0.5 hz - 4 hz)*
   Delta is a brain wave with a large amplitude (deviation furthest from the point of balance on vibration) and low frequency that is below 3 Hz. If someone sleeps without dreaming,

the brain will produce this wave. The delta phase is also called the resting phase for the body and mind. Because while asleep, the body will undergo the process of self-healing and improve nerves and produce new cells.

In music therapy, there are five important elements that must be considered; namely pitch (frequency), volume (intensity), tone color (timbre), interval and rhythm (tempo or duration). For example, a high pitch with a fast rhythm and a hard volume can cause muscle tension or an uncomfortable feeling. Whereas, a low pitch with a slow rhythm with a low volume will cause a relaxing effect. Slow tempo may decrease respiratory rate and pulse rate compatible with the rhythm of the music. Pitch and rhythm will affect the brain system which can affect emotions.

Music therapy with a frequency of 40-52 Hz can reduce pain. Music therapy can begin with a frequency of 40 Hz, which is the basic frequency in the thalamus, so that vibration stimulation with the same frequency initiates cognitive effects for therapy. Music with a frequency of 40 to 60 Hz is proven to reduce muscle tension, reduce pain, and provide a relaxed feeling. The volume unit for listening to sound vibrations is decibel (dB). Individuals who listen to music use stereo handsfree (headset), usually listening with a volume of 70 - 90 dB. The volume said to be comfortable is 70 dB.

The volume that can have therapeutic effects is 40 to 60 dB. The recommended therapy for having a maximum therapeutic effect is 60 dB for 20 - 60 minutes in one session. This can be done when going to bed for 45 minutes to get the maximum relaxation effect. Music must be heard for at least 15 minutes to get a therapeutic effect. Individuals will tend to feel relaxed if the music used has a stable musical element. But if the music that is heard varies every time and causes the individual to feel a sudden change, it will increase the level of excitement due to stimulation.

a. Potential stimulation elements

1) Unpredictable tempo changes,
2) Sudden changes to volume, rhythm, timbre, pitch and harmony,
3) Variety of music structures,
4) Unexpected dissonance (uncomfortable combination of sounds),
5) Unexpected accents (pressure on tone),
6) Rough Timbre (tone colour),
7) Lack of structure and form of music,
8) Quicker, slowing down, hardening, and softening suddenly, and
9) Stopping unexpectedly.

b. Potential elements of relaxation

1) Tempo is stable,
2) Stability or gradual change in volume, rhythm, timbre, pitch and harmony,
3) Consistent texture,
4) Predictable harmony modulation,
5) The right cadence. cadence is a melody or harmony configuration that gives the impression of calmness or resolution. cadence are generally found at the end of the phrase, period, or end of a composition in Western tonal music,
6) Selected melody lines, and
7) Repetition of material.

Husada (2015) found if music can produce stimulus with waves that are delivered through the ossicles to the middle ear through cochlear fluid then to the auditory nerve and to the area of the autonomic nervous system. Then the auditory nerve delivers the resulting signal to the auditory cortex in the temporal lobe. Stimulation of the music will produce endorphins. Endorphins have a function that can provide a sense of relaxation in the human body. Music can have the effect of stimulating the sympathetic nerves to produce a reaction in the form of relaxation. Effects that arise from relaxation include reducing muscle tension and increasing the threshold of consciousness. Indicators that can be measured from the relaxation reaction generated from music, show reduction in blood pressure, heart rate and breathing.

4.4 *Gamelan javanese music*

The gamelan music orchestra is usually called *karawitan* derived from the word "*rawit*" which means complicated, smooth, small. Gamelan comes from the word *nggamel* (in Javanese)/gamel which means to hit or beat, followed by the suffix "an" which makes it a noun, ferdiansyah (2010). The term "gamelan" refers to a group of musical instruments played together. Gamelan music is one type of traditional music which consists of several instruments, namely the *bonang*, *kenong*, *saron*, *peking*, *kendhang*, *rebab* and *gong*. All components are played classically and together. Each instrument is played in different ways. For example, the *bonang*, *kenong*, *saron* and *peking* are played by being hit with a layered stick called *tabuh*. *Kendhang* is sounded with a beat using a hand without any other tools. The *trigon* is played by swiping the threads and the gong is played by hitting the protruding knob using a layered bat, Salim (2005).

Each component of the gamelan produces different sounds. The sound produced by playing gamelan instruments in harmony produces a unique music. The unique sound produced by the gamelan is a characteristic of gamelan music with sounds that cannot be produced by other musical instruments. Djohan (2006) explains that Javanese gamelan is pentatonic as seen in the following:

> One complete gamelan game consists of two rounds, namely *slendro* and *pelog*. *Slendro* has 5 tones per octave, 1 2 3 5 6 [CD E + GA] with small interval differences. *Pelog* has 7 tones per octave, namely 1 2 3 4 5 6 7 [C + D E-F # G # AB] with a large interval difference. Gamelan music composition is created with several rules, consisting of several turns and *pathet*, *Pathet* is a regulation of gamelan tones or traditional Javanese music. Pathet applies in the tone and notes pattern of the *pelog* and *slendro* gamelan. *Pathet* gives the beauty and harmonization of gamelan blows. Each *pathet* on the diatonic scale is also called the basic tone. Restricted by one gong and the melody created in a unit consisting of 4 tones. The notes on the gamelan if measured as frequencies, are not the same as the Western musical tones, Djohan (2006).

Gamelan can be played by several people and cannot be played by a solo performer. Gamelan requires togetherness and cooperation when playing a set of gamelan instruments. Gamelan music tends to sound melodious and seductive, calm and pleasant to listen to, and the longer it is heard the music will make the listener feel sleepy because of the harmonious tones produced by musicians playing a set of gamelan instruments. Javanese gamelan can produce high beta waves through the theta wave depending on the gamelan music pattern that is played.

Table 1. Gamelan waves.

No.	wave	Patern	Tempo
1.	High beta	*Sampak*	
		Srepeg	
2.	Beta/SMR	*Ladrang*	
		Lancaran	
3.	Alpha	*Ketawang*	
		Irama tanggung	1/2*
4.	Theta	*Irama dados*	1/4*
		Irama wiled	1/8*
		Irama rangkep	1/16*

* For gamelan tempo, maybe not same as regular music

4.5 *Gamelan music therapy method*

Nationalism is growing among students in accordance with what has been stated. Learning is a mental activity that takes place in active interaction with the environment that produces changes in knowledge, understanding, skills, values and attitudes, Winkel (1996). In order to get maximum learning results, it is necessary to support an effective learning process. Pasiak (2007) also suggests that there is a relationship between emotional involvement, learning and long-term memory. Without emotional involvement, brain nerve activity is less than what is needed to "attach" lessons in memory.

Music rhythms, beats and harmony affect human physiology especially brain waves and heart rate, in addition to arousing feelings and memories. Ostrader (1994) found that baroque music aligned the body and brain. Baroque music can unlock the emotional key to super memory, the limbic system of the brain. This system not only processes emotions, but also connects the conscious part of brain with the subconscious brain.

Music has a strong influence on the learning and study environment. Research shows that learning is easier and faster if students are relaxed and receptive. The heart rate of people in this state is 60 to 80 times per minute. Most baroque music is in accordance with the conditions of a relaxed human heartbeat in optimal learning conditions. In this situation the brain enters alpha waves (8- 12 Hz), brain waves that occur when a person experiences relaxation [11]. Alpha waves are "relaxed awareness" or "relaxed alertness". The brain's alpha rhythm is a relaxed but alert brain condition so that part of the brain, the hippocampus and somatosensory, can work with optimal results, Pasiak (2007).

When we use music from brain wave therapy, our brain functions according to the audio we are listening to (the synchronisation process). This condition is referred to as the "state of meditation". When we are in this state of meditation the benefits include:

- Increased emotional stability, psychological balance and mental harmony,
- Increased motivation. Improved mental processing ability and brain synchronization,
- Improved memory and capacity for learning, concentration and problem solving,
- Increased feeling of happiness, confidence and feeling good, improve sleep quality. Makes deep sleep, and is a powerful antioxidant,
- Increased energy and creativity, and recovery from mental fatigue,
- Eliminate uncontrolled bad habits, fears and phobias, reduce cortisol levels,
- Reduction of stress, worry, anxiety, irritability, depression, and moodiness. Sufficient amounts of seratonin are needed to overcome depression, insomnia, and other problems,
- Reduction of 'mental chat'. Eliminate unwanted patterns of personality/personality, and overcome past traumas,
- Stabilizes the mood. Reduces anxiety, addiction, panic, insomnia and epilepsy,
- Improves memory and improves feelings of well-being, and
- Internal pain killers, lowers blood pressure and increases feelings of happiness.

By familiarising oneself with the tempo and dynamics of gamelan music, the activities carried out in therapy can process inner sensing. The function of exercise on the other hand will build spiritual honesty and spiritual liberation from things that are binding and limiting. Character dance is also included in acting, which shows the beauty and skill of a person in realizing various thoughts, emotions and feelings in the role played according to the specific character.

A dancer must have the ability to be someone who he or she is not. Of course, this can happen if you are able to concentrate on processing feelings and emotions. For this reason, a dancer needs to practice concentration, feelings and emotions by the feeling exercise. Furthermore, the imposition is expected to help develop feeling in order to give birth to ideas, inspiration and creativity.

The exercises that should be done include body building exercises and emotional formation will be accompanied by music to shape the soul and strengthen the feeling of music. Some of the rhythmic structure of *gamelan* music used is *sampak, srepeg, ladrang, lancaran, ketawang, irama tanggung, irama dados, irama wiled, irama rangkep.* The explanation of the rhythmic as follows:

1. *Sampak* is the simplest form of rhythmic structure. This structure serves as a beginning or end of every musical or puppet show beginning with *kendhang, t*ones played in only one tone, this music played with hard and quickly, the characters of music are usually dashing/ enthusiasm.

2. *srepeg* is almost the same as *sampak*, but the game pattern is 1/2 times slower and the tone is more varied. showing a sense of enthusiasm is usually used to describe an agile but regular movement
3. *ladrang* contains 32 beats in 1 round of *gamelan* (*gong* punch), meaning that the tempo is 1/16 slower than sampak. giving great nuance and feeling of calm and dashing impression. usually played at a moderate time, not fast or slow
4. *lancaran* the tones used in *lancaran* are usually rather complicated because of the many variations. in one round of the game there are 16 pitch punches. but the tempo produced is the same as the tempo of the *ladrang* so that it produces a more complex atmosphere. the pattern of fluency is widely used in describing the atmosphere but with a rather fast tempo
5. *ketawang* contains 16 punch of tone in 1 round of play but the tempo is 2 times slower than the pattern of *lancaran*. so the music feels smoother and softer.
6. *Irama tanggung* pattern is like a *ketawang* but the tempo is 2 times slower. the impression is made more deeply, the atmosphere produced will feel calmer.
7. *Irama dados* pattern is like a *ketawang* but the tempo is 4 times slower. the impression is made more deeply, the atmosphere produced will feel calmer.
8. *Irama wiled* pattern is like a *ketawang* but the tempo is 8 times slower. the impression is made more deeply, the atmosphere produced will feel calmer.
9. *Irama rangkep* pattern is like a *ketawang* but the tempo is 16 times slower. the impression is made more deeply, the atmosphere produced will feel calmer.

The steps to practice is by concentration, sensing, sense sensitivity, and imagination. Concentration is the 'gate' that determines the continuity of regulating and controlling a psychological phenomenon in mastering roles. In this section, an actor will attempt to alienate himself from the life he lives every day to stimulate creativity, inner sensing and feeling in one center of attention. Concentration training is the practice focusing our thoughts on an object in accordance with the purposes. Basically, the teaching of concentration is the doctrine of mastery and self-control. We concentrate our mind and spirit into what we do and within the time that we need. Important elements of the psychological phenomena of concentration include: impression of self-control, honesty and surrender of the heart, sense sensitivity, readiness and mental strength, concentration of mind and attention. The exercises by: emptying the mind, concentrating on an object and concentrating on certain events imaginatively. Gamelan *irama dados*, *wiled* and *rangkep* increase full concentration and helps a person enter into the meditation mode. The second stage is sensing. The body responds or react to various things, especially those things related to hearing. Activity trains sensitivity to music, and of the ears and body. As an exercise one can feel how music will form an atmosphere that lets the body: move freely following the music in accordance with the atmosphere created by gamelan music and use gamelan *ladrang lancaran* because in this pattern there are many images of taste in the song. Gamelan musical patterns leave great impressions when the tempo is not too fast or not too slow. The stage of sense sensitivity learning section is the main focus of the sense in practice. Since the beginning stages of concentration, meditation and sensing, it is hoped that participants can have a sense of *sukma* or sense, or inner appreciation that is able to develop the skills of controlling emotions in a play whenever needed. Sense is the inner strength of the actor that is then shown to the audience through media such as mime, expression, gesture (body motion), dramatic action and character. Feelings are encouraged and included in emotional sensitivity training sadness, fear, anger, joy and hatred. By using gamelan *lancaran*, *irama tanggung* to feel music with a slower rhythm can bring more experience to exploring the characters and explore roles. The last stage is Imagination. The ability to create imaginary power as a result of the sensitivity of emotional memories from everyday life, using metaphors related to animals, plants and other natural elements or via deep reflection that can present positive fantasies. This exercise can stimulate your imagination as if an event happened now and you feel it. This can be done individually or as a shared imagination by playing a game using imagination. For example, you imagine going on an adventure into the wilderness, climbing a high peak, going down a steep ravine and meeting a variety of beasts that are both tame and savage. One can also imagine various natural scenes such as a refreshing waterfall, fallen trees, rain or even feel a mountain that will erupt. One can imagine being the wind, water, sound, the sun, moon, stars, trees and birds and use a variant of gamelan pattern in this practice combining various types of patterns ranging from the fastest and containing passionate emotions, then changing to a pattern that is slows and permeates every detail of the music. This practice can bring insight into the variety of flavours that are included in making the music.

A dancer must have a concentration like a Balinese dancer. It can be said that as a professional dancer her or his eyes can imply the meaning of the dance. The eye as the center of expression can be said to be alive when the dancer pours the inner sensing into his or her eyes. The greater the feeling that comes out of the dancer's body, the stronger the hypnotic power of the dancers to the audience as if the audience entered the dance. As a professional dancer, he or she must be able to live a true and imaginary life. This abstract life is what concerns us but it is not easy because it is very fragile.

5 CONCLUSION

The results of theoretical studies that have been carried out by researchers show that music can indeed affect brain waves, the physiological neuro-human body that when used in the learning process brings positive results. Music influences the learning process. Music can organize moods, change a person's mental state and support the learning environment. The making of this gamelan music therapy method originated from the problem of the lack of effectiveness of inner sensing processing. It is expected that this method can be one of the alternatives to overcome problems that occur in inner sensing processing.

REFERENCES

Djohan. (2006). *Terapi musik: teori dan aplikasi* [Music Therapy Theory and Application]. Galangpress.

Ferdiansyah, F. (2010). *Kesenian Karawitan-Gamelan Jawa* [Karawitan Arts - Javanese Gamelan]. Yogyakarta: Garailmu.

Husada, s. T. I. K. K. P*emberian terapi musik terhadap status hemodinamika pasien koma pada asuhan keperawatan tn. s dengan stroke hemoragik di ruang High Care Unit (HCU) Anggrek II RSUD* [Giving Music Therapy to Hemodynamic Status in Coma Patients in Nursing Care S With Stoke Hemor-rhagic at the High Place Care Unite (HCU) Anggrek II].

Mustajib, A. (2010). *Rahasia Dahsyat Terapi Otak*[The Big Secret of Brain Therapy] . Jakarta: PT. Wahyu Media.

Pasiak, H. T. (2007). Brain Management for Self Improvement. Mizan Pustaka.

Parlina, J., & Yani, J. A. P*engaruh terapi musik gamelan jawa terhadap status hemodinamik non invasive pada pasien acute miocardial infarction*[the effects of non-invasive javanese gamelan music therapy on the hemodynamic status of patients with acute miocardial infarction]. DI ICU RSUD, P.S.B, Wates,.

Pinel, J. P. (2009). *Biopsikologi* [Biopsikology]. Yogyakarta: Pustaka Pelajar.

Salim, D. (2005). *Respons emosi musikal dalam gamelan jawa* [Response to Musical Emotions in Javanese Gamelan].

Santoso, S. (1986). [principle of body exercise] *Prinsi-rinsi Latihan Olah Tubuh.*

Supradewi, R. (2010). *Otak, Musik, dan Proses Belajar*[Brain, Music and Learning]. Buletin Psikologi, 18(2).

Winkel, W. S. (1996). *Psikologi Pengajaran Edisi Revisi* [teaching psychology]. Jakarta: pT. Gramedia.

21st Century Innovation in Music Education – Astuti et al. (eds)
© 2020 Taylor & Francis Group, London, ISBN 978-0-367-11122-9

Educational values in the *Erpangir Ku Lau* ritual

A.P.H. Purba & K.S. Astuti
Universitas Negeri Yogyakarta, Yogyakarta, Indonesia

ABSTRACT: This study aims to investigate the ritual process in terms of meanings and values of the *Erpangir Ku Lau* ritual ceremony and the presentation in the context of the Batak Karo ethnic music of *Gendang Lima Sendalanen*. This was a qualitative descriptive study using an ethnographic method. The research setting was Semangat Gunung village, Karo Regency, North Sumatra Province, Indonesia. The informants were Batak Karo artists and lecturers of traditional Batak Karo music at the State University of Medan. The results of the study show that *Erpangir Ku Lau* contains meanings and values of social, cultural, and multicultural education, reflected in the lyrics of oral literature from the spells of the ritual ceremony chanted in ritual songs such as 'Simalungun rayat', 'odak-odak', and 'Patam-patam'.

1 INTRODUCTION

All aspects of art, including art exhibition groups that developed in certain ethnics can be examined in two aspects, namely the aesthetic aspect and the functional aspect. The aesthetic aspect focuses on several things about the beauty of a work, including form (proportion), subtlety, colour, harmony, etc. related to the aesthetic concept. On the other hand, the function of art is related to how far the art can serve people. Art has also a role as a media for various rituals that are usually in line with a religious system or a prevailing custom system.

The Batak Karo community living to the north of Toba Lake, which is called *Taneh Karo* in the Batak language, has land of 5000 km^2. Tanah Karo, geographically, covers most of Dairi, Deli Serdang, and Langkat Regencies. In the north, Tanah Karo borders Aceh Province.

In Batak Karo culture, one of the most important aspects as the guiding point for social integration and ethnic identity for the Batak Karo people is *merga* (clan). The Batak Karo people use a patriarchal system, meaning that the lineage and inheritance rights of each clan is handled by the males. Batak Karo people from the same *merga* (clan) like to live in groups, based on the same *merga* (clan). Since they have one ancestor, even if it is not by blood, they are considered as family since they come from the same clan.

Additionally, the Batak Karo people also have an art tradition supported by music, dance, performance, fine art, etc. There are two types of ensemble in Batak Karo musical culture, that is, *Gendang Lima Sendalanen* and *Gendang Telu Sendalanen*. *Gendang Lima Sendalanen* is an ensemble consisting of five types of musical instruments, namely: a *sarune* (double reed aerophone), which is a double-tongue wind instrument; a *gendang singindungi* (single head conical drum), which has a membrane at one end shaped like a cone; a *gendang singanaki* (*single head conical drum*), which has one-base membrane shaped like a cone; a *gung* (gongs), which is a gong with a *pencu*; *penganak* (gongs), which is also a gong with a *pencu*. Meanwhile, *Gendang Telu Sendalanen* consists of three types of musical instrument, namely: a *kulcapi* (long neck lute, two-stringed), which is a two-stringed musical instrument with a neck; a *keteng-keteng* (two-stringed idiochord tube zither), which is a tube-shaped musical instrument with two strings on the body of the instrument; a *mangkuk mbentar/mangkuk putih* (Tarigan, 2004).

The Batak Karo ethnic people have a distinctive religious system which believes in the creator and ruler of the universe having infinite power. Traditionally, in the Karo ethnic origin, there is a religion combining its belief system of 'all-spirit' and gods which is harmonious and complementary. In the concept of dynamism and animism, Karo ethnic people believe that their lives are surrounded by a cosmic power, applying myths to understand life and their environment.

Such a religion is called *Pemena* or *Perbegu*. Related to the people's belief of the concept of 'all-spirit', their activities are automatically characterised by rituals to worship the spirits considered as the ruler of the universe and to bring luck as well as anything related to human life. One of the rituals performed by the Batak Karo people that will be discussed in this study is *Erpangir Ku Lau*, although there are other rituals based on Batak belief.

Considering the complex cultural diversity in the ritual ceremonies of the Batak Karo people, and related to the traditional art in the form of music, the author would like to analyse educational values in the social structure of Batak Karo people, that is, *rakut sitelu*, through *Gendang Lima Sendalanen* in the *Erpangir Ku Lau* ritual.

2 LITERATURE REVIEW

Etymologically, the word 'culture' comes from the Latin 'colera' which means maintaining, prioritising, and upholding. As time passed, the word 'colera' changed into 'cultura', which means human mind (Dewantara, 1967, p. 72). Generally, culture is the root of the human mind containing the nature of nobility, tenderness, ethic, and aesthetic, either physical or spiritual aspects of human life (Dewantara, 1967, p. 77). According to Koentjaraningrat (1990: 180), it is stated that culture is a system in which its entirety contains ideas, actions, and results that are useful for the society, which has to preserve it in the future.

According to Rostiyati (1995, p. 74), the educational value of culture is something considered to be good and valuable by the ethnic group or tribe, yet it is not always considered to be good by other ethnic groups or tribes since a cultural value limits and gives characteristics to a certain community and its culture. Cultural value is the most-abstract level of custom and life, rooted in people's minds, so that it is hard to be replaced by another culture in a short time. Cultural values in human life exist because humans interpret space and time. The interpretation will be inter-subjective because it is grown individually, but it is shared, accepted, and approved by the community, being an integrated background for the phenomenon described. The system of cultural values is the core of the culture that will influence and manage elements existing in the surface structure of human life, including behaviour as the unity of symptoms and objects as material unity. Based on the opinion above, it can be said that the cultural-educational value system has a central and important position in the cultural framework which is abstract and can only be expressed or explained through observation of the more real symptoms and material objects as the result of expressing value concepts through patterned-behaviour.

According to Evans-Pritchard (2013, p. 154), every community, in spite of its very simple form, will have a form of family life, recognition of family ties, economic and political systems, social status, religion, and conflict solutions as well as punishment for criminals, and others in addition to material culture, which is a collection of knowledge of the universe, techniques, and tradition.

The term *Gendang Lima Sendalanen* in the Karo language is basically defined as music. Additionally, *gendang* also has some definitions based on the word following it, such as *gendang* as a music instrument (*Gendang Anakna/Singanaki, Gendang Indungna/Singindungi*), *gendang* as an ensemble (*Gendang Sarune/Gendang Lima Sendalanen*), *gendang* as a repertoire (*Gendang Guru*), *gendang* as a composition (*Gendang Simalungen Rayat*), and *gendang* as a ceremony (*Gendang Serayaan*).

Gendang Lima Sendalanen in this research refers to *gendang* as composition, that is, *Gendang Simalungen Rayat*, *Gendang Odak-Odak*, and so on, and *gendang* as ensemble, that is, *Gendang Sarune/Gendang Lima Sendalanen. Dalan gendang* is a term from the Karo language consisting of two words, that is, *dalan* and *gendang*.

3 RESEARCH METHOD

The method used in this research is the qualitative method. Qualitative research is research aiming to describe and analyse phenomena, events, social activities, belief behaviour, perception, and notions of people as individuals or a group (Nana, 2007, p. 60).

Koentjaraningrat (2011, p. 329) regards qualitative research as ethnography, describing culture of a nation through an anthropological approach. This is confirmed by Fathoni (2006, p. 98); because the material of the cultural unity of an ethnic group in a certain place becomes the main description of an ethnographic work, it is divided into chapters concerning cultural elements based on the standard order.

In this research, data analysis will be conducted by dividing the whole *Erpangir Ku Lau* ritual through the context of the Batak ethnic music of *Gendang Lima Sendalanen*, and then it is analysed using Saussure's semiotic theory. Saussure's theory focuses more on the way signs (in this case, words) relate to the research objects.

This research uses Saussure's semiotic theory (Gaines, 2007), in which there are three elements, namely: signifier, signified, and signification. Those three elements will be separated and facilitated by the researcher to interpret the *Erpangir Ku Lau* ritual in the context of the Batak ethnic music of *Gendang Lima Sendalanen*. Separation of the *Erpangir Ku Lau* ritual will guide the researcher in interpreting the *Erpangir Ku Lau* ritual in the context of the Batak ethnic music of *Gendang Lima Sendalanen*. To analyse a text based on Saussure's theory, there are some steps that can be used to interpret the *Erpangir Ku Lau* ritual in the context of the Batak ethnic music of *Gendang Lima Sendalanen*. These steps are as follows:

a. Signifier
   This includes the material aspect of language: what is said, what is heard, and what is read. The signifier can also be defined as sound or writing with meaning. In this research, the signifier is the *Erpangir Ku Lau* ritual in the context of the Batak ethnic music of *Gendang Lima Sendalanen*.
b. Signified
   This is the description of a signifier; an interpretation phase of the text of the research object. In this research, it is the result of interpretation of the context of the Batak ethnic music of *Gendang Lima Sendalanen*, which is related to educational values.
c. Signification
   This is a process of signification after the interpretation step of the *Erpangir Ku Lau* ritual in the context of the Batak ethnic music of *Gendang Lima Sendalanen*. The researcher will relate the context of the music to educational values. In this research, the signification is performed by connecting the *Erpangir Ku Lau* ritual in the context of the Batak ethnic music of *Gendang Lima Sendalanen* with educational values.

4 RESULT AND DISCUSSION

4.1 *Cultural area of the Karo people*

The Karo tribe is one of the Batak sub-tribes in North Sumatera, Indonesia, so it is often called Batak Karo. Karo, besides being the name of a Batak sub-tribe, is also the name of the administrative city, that is, the Karo regency area that covers the entire Karo highlands. The description of the Karo people's domicile can be seen and understood through Neuman's book about the Karo people's cultural life (Tarigan, 2009, p. 36), where it is stated that the area inhabited by the Karo tribe is bounded, in the east, by the road separating the highlands from Serdang. In the south, it is bounded by the Biang river (which is named as the Wampu river if it enters Langkat); in the west it is bounded by Sinabung mountain, and in the north, it extends to the lowlands of Deli and Serdang.

Based on the description of the area of the Karo people's domicile, there are undoubtedly some groups of people living in the beach area who are neighbours of Malays, and the two tribes gradually blended and acculturated with each other. Thus, the Karo people spread and acculturated with other tribes, resulting in different names for their community areas, namely, Karo Kenjulu, Karo Teluh Dereng, Karo Singalor Lau, Karo Baluren, Karo Langkat, Karo Timur, and Karo Dusun.

4.2 *The process of the erpangir Ku Lau ritual in the tourist attraction of Lau Sidebuk-debuk bathing in Semangat Gunung Village, Karo regency*

Batak Karo people in the tourist attraction of Lau Sidebuk-debuk bathing in Semangat Gunung village, Karo regency, still preserve traditional culture, which is the *Erpangir Ku Lau* ritual, and they still uphold the tribe's belief, called 'Pemena' in Batak Karo language. This belief still holds that supernatural power and ancestors' spirits exist in the world, guarding their descendants from distress. In the concepts of dynamism and animism, the Batak Karo people still believe that their lives are surrounded by cosmic power. They also use myths to understand the rite of the *Erpangir Ku Lau* ritual of the Batak Karo people living in Semangat Gunung village, Lau Sidebuk-debuk, Karo regency.

The Erpangir Ku Lau ritual is carried out by Culture Department of Karo regency to maintain Batak Karo ethnic culture. The reasons for the Batak Karo people in Semangat Gunung village, Lau Sidebuk-debuk bathing still performing the *Erpangir Ku Lau* ritual are as follows:

a. The *Erpangir Ku Lau* ritual shows gratitude for ancestors' spirits blessing their descendants who now get good luck, are kept away from danger, get an abundant harvest, and recover from any diseases.
b. Batak Karo people perform the *Erpangir Ku Lau* ritual to keep their descendants away from misfortune, and their ancestors will save them by giving warnings through dreams.

4.3 *The Sendalanen ensemble interpreting the steps of the Erpangir Ku Lau ritual*

Considering the function of each musical instrument integrated in *Gendang Lima Sendalanen*, *sarune* is the carrier of the main melody and *gendang singanaki* is the accompaniment that produces the rhythm or a certain rhythmic pattern playing repeatedly in one or more compositions. At the beginning of each composition, *gendang singindungi* introduces various rhythmic patterns, which are different from those brought by *gendang singanakin,* but they will follow the rhythmic pattern of *gendang singanaki* repeatedly, so they seem mono-rhythmic. *Penganak* and *gung* function as accompaniments that generate a repetitive beat pattern and as tempo/speed regulators to make each composition stable or constant.

4.4 *Creative multicultural educational values (culture and community)*

Based on the *Mangmang* and *Enjujungken Mayang* music described above, the *Erpangir Ku Lau* tradition has existed in the Karo community giving a sense of past times reflected in the form of culture, and it has been inherited through generations. In this case, symbols and behaviour rules selected through acculturation become heritage in the form of sociocultural continuity from the past preserved until now.

As time goes by, the tradition changes through the acculturation process in terms of living place, from an individual to another as well as from one generation to another generation. It makes the living culture become the vital element as the basic principles in forming a cultural landscape. Thus, culture becomes something that has been created by the Karo people as a place to live in for people running their activities. These cultural spaces fill the spatial parts of rural areas, and possibly urban areas, as a place to organise their collective activities, along with cultural values and the architectural design that have been put in their type of building as a place to live in with their family and community. In totality, the values contained in an environment where people live creates a special place for them. In their daily activities, symbols of social status can be realised in the architectural form or style appearing in the *Erpangir Ku Lau* tradition of the Karo people, along with its social strata.

5 CONCLUSION

Erpangir Ku Lau is a ritual ceremony of Batak Karo ethnic music contains meanings and values of social, cultural, and multicultural education. It can be used as a reference to internalise the educational values of the community's life. It is a way to stimulate people to understand and to identify the meaning as well as the values in the *Erpangir Ku Lau* ritual, which is possible to be implemented in other communities.

Considering the benefits of examining and understanding the *Erpangir Ku Lau* ritual in redeveloping internal values, there are some suggestions to be addressed:

1. Teachers teach about cultural values contained in the *Erpangir Ku Lau* ritual as a way to instil educational values in the next generation of the Karo community.
2. Karo people are expected to not merely follow the *Erpangir Ku Lau* ritual but also understand its educational values as well as know the values of the ritual as a basic educational value.
3. The regional government should pay more attention more to the local culture and preserve the *Erpangir Ku Lau* ritual by creating other types of music supporting the ritual in order to preserve the values of the *Erpangir Ku Lau* ritual.

REFERENCES

Dewantara, K.H. (1967). *Karya Ki Hadjar Dewantara bagian kedua A (Kebudayaan)* [Art Work of Part II Culture]. Yogyakarta, Indonesia: Majlis Luhur Persatuan Tamansiswa.

Evans-Pritchard, E.E. (2013). *Social anthropology*. London: Routledge.

Fathoni, A. (2006). *Antropologi sosial budaya: suatu pengantar* [Sociocultural anthropology]. Jakarta, Indonesia: PT Rineka Cipta.

Gaines, E. (2007). The semiotic analysis of myth: A proposal for an applied methodology. *American Journal of Semiotics*, *17*(2), 311–327.

Koentjaraningrat, R.M. (1990). *Pokok-pokok Antropologi Sosial* [The core of social anthropology]. Jakarta, Indonesia: PT Dian Rakyat.

Koentjaraningrat. (2011). *Pengantar antropologi* [Introduction to anthropology]. Jakarta, Indonesia: PT Rineka Cipta.

Nana, S.S. (2007). *Metode Penelitian Pendidikan* [Educational research methods]. Bandung, Indonesia: PT Remaja.

Rostiyati, A. (Ed.) (1995). *Fungsi Upacara Tradisional pada Masyarakat Pendukungnya Masa Kini* [The function of traditional ritual for the community at present]. Yogyakarta, Indonesia: UNY Press.

Tarigan, H.G. (1990). *Percikan budaya Karo: kumpulan karangan mengenai adat istiadat Merga Silima* [Karo cultural spark: A collection of essays about Merga Silima customs]. Bali: Yayasan Merga Silima.

Tarigan, P. (2004). *Musik Tradisional Karo dalam Ben* Pasaribu. *Pluralitas Musik Etnik Batak Toba, Mandailing, Melayu, Pakpak-Dairi, Angkola, Karo, Simalungun* [Karo traditional music in Ben Pasaribu. Plurality of Batak ethnic music: Toba, Mandailing, Melayu, Pakpak Dairi, Angkola, Karo, Simalungun]. Medan, Indonesia: Pusat Dokumentasi dan Pengkajian Kebudayaan Batak, Universitas HKBP Nommensen.

Tarigan, S. (2009). *Lentera kehidupan orang karo dalam berbudaya* [The lantern of the Karo people in culture]. Medan, Indonesia: University of Northen Sumatra.

21st Century Innovation in Music Education – Astuti et al. (eds)
© 2020 Taylor & Francis Group, London, ISBN 978-0-367-11122-9

Moral teaching in children's religious songs

G. Dwiyantoro & H. Pamadhi
Universitas Negeri Yogyakarta, Yogyakarta, Indonesia

ABSTRACT: This article examines a number of children's religious songs which contain essential moral values for children's education. The songs that will be studied include "*Allah Maha Tahu*", "*Belajar Puasa*", "*Berani Jujur*", "*Muliakan Tetangga*", and "*Rumahku Syurgaku*". The song entitled "*Allah Maha Tahu*" teaches us that Allah knows everything we do and invites us to be good in every activity. "*Belajar Puasa*" song invites children to learn to fast. The song entitled "*Berani Jujur*" tells children to always say something good and right. The song "*Muliakan Tetangga*" invites children to keep in touch with their neighbours so that they are always in love. Finally, the song entitled "*Rumahku Syurgaku*" is about a home, a place where we live happily with our parents. These five songs will be discussed further in this article.

The approach of the research is qualitative , the method of analyse use descriptive, and the validity use semantic.

1 INTRODUCTION

Education is a very important and emotional process towards nature and other humans (Dewey, 1982: 9). Education usually starts very early since childhood. This is the period that determines where someone starts learning. One means of education for developing children's positive personality is through art education. There are some typical kinds of art studied in schools such as dance, fine arts, theatre arts, and craft arts. This article will discuss music art education through religious songs whose lyrics contain essential moral values for developing children's behaviour.

According to Asmani (2009), the use of music is important in children's education as it becomes a part of educational materials. Here, the term music encompasses playing musical instruments and singing songs. Music and songs in children's education have two fundamental roles. First, they are used in children's education to support the optimisation of children's physical, cognitive, and psychomotor development. Second, music and songs are used in children's education to stimulate the development of children's musical intelligence (Asmani, 2009: 153, 159, 163-168). This view is consistent with the fact that children's songs are commonly used as a part of the children's learning process in formal, non-formal, and informal education.

Value or *valere* (Latin) means useful, capable, and valuable while a strong value indicates the quality of something that makes it high-quality, useful and valuable. A quality or appreciation of what is achieved that underlies a person's behaviour is called value (Muchson and Samsuri, 2013: 21). Meanwhile, according to Kartono and Dali (2014: 14), values are things that are considered important and good, such as a person's beliefs about what should or should not be done (e.g., honest, sincere) or the ideals that someone wants to achieve e.g., happiness, freedom). From the opinions of these experts, it can be concluded that value is the quality or appreciation of something which is considered important and good that determines the behaviour or ideals someone wants to achieve.

Moral is manifested in the form of child behaviour. There are some moral values that we know in our daily lives such as self-related values, which include honesty, hard work, firmness, patience, resilience, cheerfulness, openness, visionary, autonomy, rigidity, courage, reflectiveness, responsibility, and discipline. There are also other-related values such as helpfulness, tolerance, friendliness, generosity, cooperativeness, communicativeness, *amar ma'ruf* (enjoining good), *nahi munkar* (preventing evil), care (to human and nature) and fairness. There are also divine values such as *ikhlas* (sincerity), *ikhsan* (right virtue), *iman* (faith), and *taqwa* (piety). From this description, it can be said that fostering children's moral values is an effort to shape and direct children's behaviour toward a better direction where they will eventually be able to think logically and be responsible for themselves and others. In addition, moral values can be interpreted as values that are related to the concept of whether something someone is doing is good or bad and a means to measure the goodness of one's actions, the consequences of a person's actions, and the moral consequences they cause.

Moral teaching is advice and a mandate regarding the goodness of human attitudes in communities (Nurgiyantoro, 2012: 320). This teaching is useful for humans who live in a society in learning good attitudes (Hadiatmaja and Endah, 2008: 56). Therefore, moral teachings are important for the community, especially for children because moral values must grow early on.

Song lyrics are one of the literary works, so it is safe to say that song lyrics also contain moral teachings (Amirul et al., 2017: 2). Through music, children can increase their creativity, strengthen their sensitivity to feelings and emotions, reduce their aggressive behaviour, and build their mentality to love harmony and beauty. Appreciation of music can also fulfil some social needs, emotional needs and social roles that are important in subsequent developments.

Song lyrics can be divided based on the types of moral teachings they contain. Thus, we can find the core teachings that humans can digest more when applying in their lives. Moral education is expected to produce personally and socially competent students as an effort to build good citizens. The moral education policy has to be directed towards the development of children which is in accordance with the standard of democratic, anti-violent, virtuous, moral, and participatory society, the foundation of legal certainty, the priority towards egalitarian values and justice, the respect towards human rights, the law enforcement, and the respect towards differences in SARA (ethnicity, religion, race, and class) within the frame of national unity. In addition, we must also uphold religious values in accordance with the implementation of *Pancasila* (five pillars). Both are objectively and subjectively treated as the paradigms. Moral education must be a part of everyday life because it will greatly support a conducive atmosphere that is important for the implementation of moral education and the development of an ideal society especially for children. In this case, moral teachings will be conveyed through children's religious songs.

Children's songs are usually aesthetic, musical and cultural. In terms of musicality, the form and rhythm of the songs are adjusted to the development of the child's musical sensitivity. Culturally speaking, Javanese *Dolanan* songs can teach children how to be disciplined and maintain harmonious relationships with the nature, parents and other people. Teaching *dolanan* songs is an alternative to overcome modernisation which generally keeps children from having good morals (Triyono, 2000: 1-12).

Songs are a series of words made with tones and accompanied by musical instruments. Songs are very influential on children's behaviour. With songs, children can learn a lot of things only if they listening to them. Songs can be used as a good medium of teaching children about behaviour. Islamic-themed songs can influence children and lead them to the right path in accordance with the rules of the religion.

2 METHODOLOGY

This study used a qualitative method. A qualitative method in literary research is carried out by using descriptive forms (Ratna, 2007: 46). The data in this study were in the form of words, phrases, and words. The purpose of this study was to find out the moral values in religious songs for children created by Ms. Fiani Gee, S.Pd. This study focused on analysing the lyrics of children's religious songs in relation to moral values. The moral values in question are those related to the behaviour of the children themselves, to others, and to God.

The first stage of data collection in this study was to read the lyrics of the children's religious songs several times. The next step was to look for types, forms, and ways to explain moral teachings. The collected data was then linked to the purpose of this study. Then, the data were divided into words and lines in children's religious song lyrics. This study used data reduction to eliminate data that had no relationship with the purpose of this study. In this way, it was expected that the data found was accurate.

The data in this study were validated using semantic validity carried out through expert judgment. Semantic validity measures the sensitivity level of symbolic meaning related to the context (Endraswara, 2013: 164). Semantic validity was done by connecting symbolic meaning with the context of moral teaching data in children's religious song lyrics. Linking symbolic meaning to context was done so that the data were valid. Expert judgment was done by consulting with lecturers who were experts in their fields. The reliability of the data was shown by means of accuracy, intra-rater, and inter-rater. According to Endraswara (2013: 164), accuracy lies in the relationship between the results of the research and the literature review formulated. Intra-rater reliability was done by reading the lyrics multiple times and then understanding them carefully. Inter-rater reliability was carried out by discussing the results of research with friends who understood the topics and shared similar problems regarding moral teachings.

3 DISCUSSING AND FINDING

Before discussing the lyrics, this article will first discuss the music, especially children's music. Music can be interpreted as sounds arranged in such a way that they become rhythms, melodies, harmonies, shapes, structures, and expressions that can make people happy and comfortable. Music does not only entertain its listeners, but it can also be a means of sending moral messages (Ainoer et al., 2017: 1).

To develop children's morals, it is necessary to put a struggle using structural and cultural strategies simultaneously. The most effective structural strategy in a political sense is through the children's education curriculum in schools. Through formal education institutions, community aspirations about morals can be channeled and moral values can be advocated in the form of input from the public to the government, especially the Ministry of Education and Culture. Input from the public to the government will be elaborated in the form of policies or laws that require the implementation of moral education for children.

To socialise moral values, teachers are needed as they are moral fighters who have never flinched, despaired nor frustrated despite obstacles, unfavourable environments, and destructive conditions. They are not tired of socialising moral education to realise universal moral values that respect others.

One part of how to examine moral teachings can be seen from the types of moral teachings. The types of moral teachings can be freely included in events. These types of moral teachings can enter into life and life events, things that are related to human worth and dignity (Nurgiyantoro, 2007: 323).

Children's songs are a means of educating children because children's songs can send messages (about divinity, natural beauty, self, family, social environment, and nationalism) to children in a simple and fun way (Johar, A, 2015: 1). Children's songs according to Endraswara (2013: 4) are songs that are typically cheerful in nature and reflect noble ethics. Children's songs are songs that are usually sung by children and contain simple things that are usually done by children. Children's songs usually tell stories on love for others, parents, sisters/brothers, beauty of nature, and greatness of God written in a simple language in accordance with the nature of children's thinking.

Seen from the explanation above, it can be concluded that the types of moral teachings can be divided into four: the moral teachings of human-God relations, the moral teachings of human-human relations, the moral teachings of human relations with themselves, and the moral teachings of human-nature relations. This article will analyze five children's religious songs. The songs that will be analysed include "*Allah Maha Tahu*", "*Belajar Puasa*", "*Berani Jujur*", "*Muliakan Tetangga*", and "*Rumahku Syurgaku*". These songs have not been so popular in the world of children's music because they were created and recorded in 2017 by Fiani Gee, an early childhood education teacher, assisted by Ginanjar Dwiyantoro, a graduate of elementary school teacher education.

Following are song lyrics along with the descriptions of their moral values.

a. Allah Maha Tahu
Song by: Fiani Gee and Ginanjar Dwiyantoro

Allah melihatmu teman, teman
Berbuatlah kebaikan
Dalam terang, dalam gelap
Allah melihatmu teman, teman

Allah mendengarmu teman, teman
Berkatalah kebaikan
Dalam sepi dalam ramai
Allah mendengarmu teman, teman

Dia maha tau segala sesuatu
Dia maha penyayang kepada semua orang
Dia maha pengampun dosa yang tak terhitung
Dialah Allah, Laa ilaaha illallah

This song contains a message about the greatness of Allah who knows all about our daily lives in the world, all see and hear, all merciful, forgiving his sinful servant, and the Almighty God. The repetition of these words in the lyrics of this song is because the creator of the song wants to emphasise to the listeners that God is omnipresent and omnipotent. The message conveyed in this song is very clear, that is, to invite everyone to do good because every bad deed will be seen and heard by Allah.

b. Belajar Puasa
Songs by: Fiani Gee dan Ginanjar Dwiyantoro

Duh. Adikku yang lucu
Sini kakak beritahu
Besok itu bulan yang berbeda
Besok itu bulan puasa

Nanti kita belajar bersama
Menahan lapar di siang harinya
Makan dan minum ketika berbuka
Menunggu Maghrib dengan bahagia

Duh.. Adikku yang manis
Nanti lapar tak boleh menangis
Boleh bermain tetap senyum manis
Hingga senja telah habis

Nanti kita bersabar bersama
Sambil mengaji dengan Ayah Bunda
Juga berdzikir menyebut AsmaNya
Semoga Allaah terus jaga kita

Eh Ramadhan datang
Kita sambut dengan senang
Eh Ramadhan akan datang
Sudah masuk bulan Sya'ban
Eh Ramadhan datang
Ayo semua kita siapkan
Eh Ramadhan akan datang
Semoga Allaah makin saying

This song tells a story of an older brother who always gives a good example to his younger brother. The older brother told him that the month of Ramadan will come soon and they have to welcome the month with pleasure, learn to fast since childhood, learn to be patient in the month of Ramadan, and always worship Allah. The message we get in this song is about an invitation to learn and worship since childhood.

c. Berani Jujur
Songs By: Fiani Gee dan Ginanjar Dwiyantoro

Hai teman jangan bohong
Bicara terus terang
Ayo berkata benar
Yang jujur anak pintar

Bohong.. Bohong tidak berguna
Bohong.. Bohong tidak disuka
Ayo teman jujur saja
Katakan yang sebenarnya
Katakan yang sebenarnya

Ikuti cara Nabi
Berkata hati-hati
Salahpun diakui
Yang jujur anak brani

Bohong.. Bohong tidak berguna
Bohong.. Bohong tidak disuka
Ayo teman jujur saja
Katakan yang sebenarnya

This song tells us that saying good and being honest are the right things to do and lying is useless. What is conveyed on this song is very clear by looking at the lyrics and understanding its contents on high moral elements. It is about an invitation to always be honest because telling lies is useless and God does not like it, advice to follow the prophet's way of speaking and the message that an honest child is a brave child.

d. Muliakan Tetangga
Songs by: Fiani Gee dan Ginanjar Dwiyantoro

Ada mereka yang tinggal
Di depan dan blakang rumah kita
Ada mereka yang tinggal
Di kiri dan kanan rumah kita
Mereka... membantu kita
Dalam suka dan duka
Dalam tangis... dan canda
Seperti layaknya saudara

Pesan Nabi kita
Muliakanlah mereka
Agar kita dan tetangga
Tetap saling cinta

This song tells us about our neighbours. We do not live alone in this world. There are many neighbours around our house. They are the ones who will help us when we are in trouble and when we need something. This song also includes the words conveyed by the prophet, namely respecting neighbours so that we and our neighbours still love each other and avoid any conflict so that our environment is maintained and full of intimacy with our neighbours.

e. Rumahku Syurgaku
Songs by: Fiani Gee dan Ginanjar Dwiyantoro

Rumahku indah
Buatku betah
Bersama ibu dan ayah

Rumah tempat gembira
Rumah tempat bahagia
Rumahku Surga di dunia

Alhamdulillah
Ku senang di rumah
Terimakasih Allah

The last song tells about the situation at home. Home is a place for us to take shelter, to gather with family, and to learn from our parents.

4 CONCLUSION

Moral education will succeed if the teacher gives a stimulus to students so that they can respond according to their wishes. With the stimulus and response, students will get a classical conditioning which will create a more conducive learning condition. In order for the purpose of moral education to be achieved, the teacher can give rewards to students who are successful and impose punishments for those who fail. However, this punishment must still be in the corridor of humanity. The process of stimulus and response in moral education must be given continuously and programmed so that children will have a habitus (education that changes behaviour) to realise morals of Indonesian people.

A songwriter can be considered as a person who struggles in education if he uses his space and time to educate children through a song. The songs that we wrote all are religious children's songs, whose lyrics contain messages on how to do good to family, neighbours, friends, and even people whom we do not know at all. In addition, these songs also invite us to always worship the almighty since childhood, so that when we grow older our religious knowledge is stronger.

The song "Allah Maha Tahu" gives us a message that Allah indeed knows everything we do as long as we live in this world. So, the core message is to do good so that we are not hated by Allah. The song entitled "Belajar Berpuasa" gives us a message to learn to fast while in childhood in the month of Ramadan. The third song entitled "Berani Jujur" gives us a message to imitate the honesty of the prophet because lying is not good and to realise that honest people are brave people. The fourth song entitled "Muliakan Tetangga" gives us a message about the importance of respecting neighbours to maintain mutual love and avoid conflict, according to what the prophet taught us. The last song entitled "Rumahku Syurgaku" gives us a message that home is where we find happiness with our parents.

REFERENCES

Ainoer, R, Ikhwanul, Q, & Gatut, R, (2017). Music and Songs Media in the Learning Process. Banyuwangi: *Journal of the Basic Education Postgraduate Program*, Universitas PGRI. Retrived from http://journal.stkipsingkawang.ac.id/index.php/JPDI/article/view/330/393.

Amirul, N, W, & Kundharu, S, (2017). Moral Teachings in Song Lyrics of Dolanan Anak. Surakarta: *Journal of the Postgraduate Program*, Universitas Sebelas Maret. Retrived from http://jurnal.isi-dps.ac.id/index.php/mudra/article/view/107/50.

Asmani, Jamal M'mur. (2009), *Strategic Management of Early Childhood Education*. Yogyakarta: Diva Press.

Dewey, John. (1982) *Fundamentals of Education*. Bandung, Angkasa.

Endraswara, Suwardi. (2013). *Literary Research Methodology*. Yogyakarta: CAPS.

Hadiatmaja lan Endah. (2008), Javanese Philosophy, Yogyakarta, Kanwa Publisher.

Johar, A (2015). Children's songs as one of the means to educate children. Majenang: *Journal of Basic Education* Retrived from https://scholar.google.co.id/citations?user=jg2GIgkAAAAJ&hl=en

Kartono, Kartini dan Dali Gulo, (2014) Kamus Psikologi, Bandung: Pionir Jaya.

Muchson dan Samsuri. (2013). *Basics of Moral Education Base on Character Education Development*. Yogyakarta: Ombak.

Nurgiyantoro, Burhan. (2007), *Fiction Study Theory*, Yogyakarta, Gajah Mada University Press.

Ratna, Nyoman Kutha. (2007), *Theory, Methods, and Literary Research Techniques*, Yogyakarta, Pustaka Pelajar.

Triyono, Bramantyo. (2000), *Song Dolanan Anak*, Yogyakarta, Tarawang Press.

21st Century Innovation in Music Education – Astuti et al. (eds)
 ISBN 978-0-367-11122-9

Developing character values through Javanese *gamelan*

A. Pranata & S. Rudiyati
Universitas Negeri Yogyakarta, Yogyakarta, Indonesia

ABSTRACT: This study aims to describe how character values are developed through the application of character education through the art of Javanese *gamelan* music. This research is a qualitative descriptive study with research subjects that are principals, teachers and students. The object of this research is the character values in the Javanese *gamelan* culture. This study uses data collection techniques in the form of observation, interviews and documentation. The results of the study show that the character values that exist in Javanese *gamelan* are very influential to those who play this music. Javanese *gamelan* is an alternative teaching material for applying character values. Javanese *gamelan* art unconsciously fosters good character and cooperation in the singing of *gamelan* music. Javanese *gamelan* art culture contains historical, moral and philosophical values that are conveyed to students during the learning process.

Keywords: character values, cultural arts, Javanese *gamelan*

1 INTRODUCTION

According to Law No. 20 of 2003 (Indonesia, 2003) concerning the National Education system, education is a conscious and planned effort to create a learning atmosphere and learning process for students to actively develop their potential to have religious spiritual power, self-control, personality, noble intelligence and the skills needed by their selves, society, and the nation (Wibowo, 2012). The current trend of globalisation means that character education must be taken up and faced by schools. However, globalisation causes a cultural shift that reduces the value of national character. Character education is integrated into learning in each subject, including the study of cultural arts, one of which is the Javanese *gamelan* culture. The Javanese *gamelan* culture is one of the cultural arts inherited from Indonesian ancestors and is still practised today. In Javanese *gamelan*, one type of *gamelan* music includes the *kendang* (drums), *rebab* (fiddle), *gambang* (xylophone) and gong. Each instrument has its own function in *gamelan* music. *Gending* in Javanese *gamelan* is a harmony used as a way to regulate emotional balance, and organise pitch and harmonious behaviour. This is one of the mediums from which character values can be derived.

Culture-based education is a mechanism that provides opportunities for everyone to enrich science and technology. Education is a 'process' (culture) that humans use to develop in a better direction. In order to support culture-loving students, the government in *Perda* (or Local Regulation) No 5 of 2011 (Indonesia, 2011) states that the implementation of culture-based education is to be carried out at the levels of basic education to higher education, both in formal education and non-formal education. *Perda* No 5 Tahun 2011 article 2 verse 2 (Indonesia, 2011) states that the noble cultural values that need to be developed are: "Honesty, humility, order, discipline, politeness, patience, and cooperation in tolerance, responsibility, justice, caring, self-knowledge, integrity, hard work, leadership and toughness". Character values have not been socialised, even though character values are the main key to building a quality young generation by applying culture-based character education. The increasingly sophisticated flow of globalisation causes students to forget the order of values of national culture and local culture.

Javanese *gamelan* is one of the cultural arts of Indonesian ancestral heritage, which is still much favoured by the community, proving that Javanese *gamelan* as *Adiluhung*, or beauty sublime, is a cultural value that has been upheld from one generation to the next. Javanese *gamelan* uses a set of instruments called *karawitan*. The Javanese *gamelan* contains historical and philosophical values because the Javanese *gamelan* is one of the inherited cultural arts which is still popular and sought after (Haryono, 2001).

2 LITERATURE REVIEW

A study conducted by Widiadnyana et al. (2014) showed that the model of culture-based learning can be well received, despite the low level of insights and knowledge of the subjects. The lack of insight/knowledge of students towards the culture-based science learning model lies in the inability of students to look for examples of events that contain the values of local wisdom integrated into the syllabus or plan for implementing learning. It is evident from *Rencana Pelaksanaan Pembelajaran* (RPP), or learning implementation plan, that the teacher composes very little material in relation to local culture. Likewise, in direct observation during the learning process, the teacher still seems hesitant and awkward in connecting with local culture. The research that refers to the culture-based learning model in SMP 2 Denpasar is done with the students' knowledge of the culture-based learning model while still seeing the wisdom of local culture. The research that the authors conducted on the culture-based character education strategy discussed how student character education was made into a culture-based school.

The development of Javanese *karawitan* musical aesthetic art in the community shows that by learning Javanese *karawitan* art, artistic values and aesthetics that exist in society can be learnt. The development of the musical aesthetics of Javanese music in society influences various aspects of society, including economic, social and cultural factors. The relationship or mutually beneficial relationship between *pengrawit*, or the *gamelan* musician, and *pandhemen*, or lay music listeners, with the existence of Javanese musical is seen in the horizontal system. This connection is a manifestation of mutual symbiotic relations between the *pengrawit* and *pandhemen* of Javanese music.

3 MATERIAL & METHODOLOGY

The research used in this study is a qualitative approach, and uses observations of *gamelan* culture games that have character values that can be applied. The purpose of the research is to find character education contained in the Javanese *gamelan*. Qualitative research is defined by Moleong (2007, p. 6) as research that intends to understand the phenomenon of what is experienced by research subjects such as behaviour, perception, motivation, actions, and others holistically, with descriptions in the form of words and language, in a special natural context and by utilising various scientific methods Sugiyono (2010). This research includes descriptive qualitative research, and this study aims to describe information about the data obtained from the field in the form of field observations, and written data and interviews of the subjects and objects studied during the implementation of the study. Techniques for collecting data include observation, interviews and documentation. Data analysis techniques used are data reduction, data presentation and conclusion. Triangulation techniques are used to test the credibility of data by checking the data from the same source or informant with different data collection techniques. This research includes descriptive qualitative research. This study aims to describe information about the data obtained from the field in the form of field observations, and written data and interviews of the subjects and objects studied during the implementation of the study.

4 RESULTS AND DISCUSSION

4.1 *Character building*

Character education in schools is indispensable. Character education is carried out by exploring local cultural values, one of which is the Javanese *gamelan* culture, which is the study of *gamelan* as a musical instrument that gives a sense of harmony and humanity, imparting the values of togetherness, empathy and tolerance. By studying this *gamelan* culture there is harmony between students that influences the quality of the music produced, as playing the *gamelan* requires a high sense of empathy and tolerance. Character education does not only teach right and wrong but is also responsible for instilling good habits into the child's cognitive and psychometric personality. Character education needs to be done in order to realise character, which is the ultimate goal of the education process. Culture is very influential in character education because by building up a student's understanding of their own culture, you can accelerate the realisation of the expected character. Education is defined as the process of changing one's attitudes and behaviour in an effort to develop children's potential.

Lickona (2009) said that the noble character includes knowledge of goodness, then raising commitments of intentions towards goodness, and finally, truly doing good. Characters refer to a series of knowledge cognitive attitudes and motivations, behaviours and skills. Character education is the deliberate effort to help people understand, care about, and act upon core ethical values. Character education is the deliberate effort to cultivate virtue, which is objectively good human qualities that are good for the individual person and good for the whole society Zamroni (2011: 111). School culture is a pattern of values, principles, traditions and habits formed in the long journey of the school, developed by the school for a long period of time and becomes a handle, and is believed by all school members, thus encouraging the emergence of attitudes and behaviour of school residents.

Character crises that occur in students are characterised by the rise of promiscuity, cultural influences that become social problems in society. Building national character is not easy. The nation is currently losing local wisdom as the national cultural character. One example is the rise of student fights that show a crisis of character in youth. Thus, the formation of a national character that is a character that upholds morals, nobility and nationalism are vital. (Wibowo, 2012).

Karawitan is related to the formation of student character through culture as character values can be developed unconsciously, and as students understand music better, the sound of *gamelan*, and the strains of the *gamelan* make the heart become calmer and more peaceful. In studying *karawitan* there is a component of character education that is obtained, namely moral knowing, feeling and action, because when studying *gamelan* music a curiosity is needed in the form of self-awareness, moral feeling is used when playing *gamelan*, and moral action is an action forming character through culture. School-based arts and cultural programmes have character values that can be obtained by students as a habit in character-building through culture.

4.2 *Javanese gamelan culture*

From the results of the study, it was found that the supporting factors for culture-based character education in schools were seen from internal factors such as the response of the school community regarding culture-based character education, school policy, personnel instructors who are experts in their fields and cultural development, and external factors such as adequate school facilities and supporting environmental conditioning.

4.3 *Character values in Javanese gamelan culture*

The Javanese *gamelan* culture is an alternative method of character education to cultivate students with habits in thinking, behaving and working together in the school environment and society. The art of the Javanese *gamelan* culture is related to social, moral and spiritual values.

Character education includes tolerance, discipline, hard work, creativity, and responsibility. Javanese *gamelan* music as an alternative that can shape character education through traditional culture. For students, this is felt after studying the *gamelan*, as follows:

a. In the drumming game, the Javanese *gamelan* fosters a sense of togetherness, eliminating the sense of individualism that exists in students.
b. Singing the Javanese *gamelan* has an impact on the peace of mind and the concentration of children at school.
c. Students have a greater sense of tolerance and responsibility.
d. Preserving the arts, especially the Javanese *gamelan* culture by playing musical instruments combined with instruments that can foster a harmonious sense when listening to the chants from Javanese *gamelan*.

5 CONCLUSION

Based on the results of the research into the character values conveyed in the art of the Javanese *gamelan* culture, it can be concluded that the values contained therein can subconsciously influence the character traits of *gamelan* players, to develop greater tolerance, as well as hard work and responsibility. Integrating character education into subjects and self-development in schools as a step in developing character education is a strategy that is customary and well-developed in schools. Javanese *gamelan*, as one of the alternative teaching materials for character education for students, can explore Javanese culture while promoting it in character education. Javanese *gamelan* has very strong historical, moral, and philosophical values, developed for the learning process of students with character education through the Javanese *gamelan* culture. The art of Javanese *gamelan* can be chanted and studied diligently. This *gamelan* art can become a territory in which to continue cultivating Indonesian art through the sounds and chanting of Javanese *gamelan* music.

REFERENCE.

Haryono, T. (2001). *Sejarah dan Makna Game* [Game history and meanings]. Yogyakarta, Indonesia: UGM Press.

Indonesia, R. (2003). *Undang-undang Republik Indonesia Nomor 20 Tahun 2003 tentang Sistem Pendidikan Nasional.* [Republic of Indonesia Constitution Number 20 of 2003 Concerning the National Education System]. Jakarta: Government of the Republic of Indonesia.

Indonesia, R. (2011). *Republic of Indonesia Local Regulation Number 05 of 2011 Concerning the Culture-based Education.* Jakarta: Government of the Republic of Indonesia.

Lickona, T. (2009). *Educating for character: How our schools can teach respect and responsibility.* New York: Bantam Books.

Moleong, L.J. (2007). *Metodologi Penelitian Kualitatif* [Qualitative research methodology]. Bandung, Indonesia: PT Remaja Rosdakarya.

Mulyana. (2018). *Kearifan Lokal Indonesia* [Indonesian local wisdom]. Yogyakarta, Indonesia: Tiara Wacana.

Sugiyono. (2010). *Educational research methods: Quantitative, qualitative and R&D approaches.* Bandung, Indonesia: Alfabeta.

Wibowo, A. (2012). *Pendidikan karakter: Strategi membangun karakter bangsa berperadaban* [Character education: Strategy to build a character of a civilised nation]. Jakarta: Student Library.

Widiadnyana, I.W., Sadia, I.W. & Suastra, I.W. (2014). *Pengaruh model discovery learning terhadap pemahaman konsep IPA dan sikap ilmiah siswa SMP* [the effect of discovery learning model on understanding of science concepts and scientific attitudes of middle school student]. *Jurnal Pendidikan dan Pembelajaran IPA Indonesia, 4*(1), 123-129.

Zamroni., M., (2011). *Dinamika Peningkatan Mutu* [Dynamics of quality improvement]. Yogyakarta, Indonesia: Galvin Kalam Utama.

21st Century Innovation in Music Education – Astuti et al. (eds)
© 2020 Taylor & Francis Group, London, ISBN 978-0-367-11122-9

Tolerance education through a choral lesson in schools

O. Pratama & K.S. Astuti
Universitas Negeri Yogyakarta, Yogyakarta, Daerah Istimewa Yogyakarta, Indonesia

ABSTRACT: The aim of the study is to instill tolerance through a choral lesson which focuses on *Nusantara* songs. Indonesia's diversity fosters differences of opinions, habits, and perspectives. Therefore, we need to understand how to respond to these differences which can be accomplished through the process of understanding *Nusantara* songs originated from diverse regions in the form of school choirs. The research is experimental by using pretest post test design. Students are invited to examine and experience *Nusantara* songs from Sumatra, Java, to Papua so that students are able to recognize and understand the diversity of Indonesia as reflected in the *Nusantara* songs.

1 INTRODUCTION

Indonesia has a diversity of ethnicities, languages, religions, and cultures in each region. According to the Centre of Statistics and Data for Education (PDSPK) of the Indonesian Ministry of Education and Culture, Indonesia has six official religions, 250 ethnicities, and 250 regional languages that make up various cultures spread across 34 provinces and 16,056 islands. These differences make Indonesian people more vulnerable to various problems and conflicts. Some national news and surveys inform that various conflicts between communities are due to differences.

The Indonesian Ministry of Religion survey states that Indonesia has various conflicts that arise because the people involved have used their respective religious and affirmed their religious interests toward others. Therefore, actions to prevent conflict in all aspects such as in formal education are urgently needed. Promotion of character education values such as tolerance values needs to be held. Choirs, which are part of extracurricular activities in schools, can be a medium for teaching and applying tolerance values.

2 CHOIRS

The result of the choir is the combination of many sounds into one colour of sound, namely the colour of the choir (clear and balanced sound) which is a unified whole. A Choir is the uniformity of a group of people who sing together on the same basic note.

In addition, there are various types of voices and groups based on gender in choirs. Soeharto (1992: 15) said that generally a choir presents several parts of voices that consist of Sopran, Mezzosopran, and Alto (for women) and Tenor, Bariton, and Bass (for men). Human voices are divided into female voices, male voices, and children's voices. Differences in voice timbre in men, women, and children will provide more colors to choir groups.

The *Tim Pusat Musik Liturgi* (Liturgi Music Centre Team) (2011: 13) explained the formation and tone colour of choirs in which they stated that the ideal number of members in the choirs is between 40-50. If the number is too small, it will be difficult because the sound produced will be soft. Meanwhile, the large number of members results in the difficulty of maintaining group cohesiveness. Choirs involve many interactions and social processes in its formation both in terms of singing ability and attitudes in learning choirs in school.

However, *Tim Pusat Musik Liturgi* (Liturgi Music Centre Team) (2011: 14) in its discussion of the standard number of members and the arrangement of songs performed in the choir also said that choirs can have members between 25-30 people with the characteristic of 2 or 3 sounds. That formation will also result in a beautiful performance even without musical accompaniment if they sing softly.

According to various opinions above, it can be concluded that a choir is a group of 25 people or more. However, if the number of choir members ranges from 40 to 50, the volume of sound formation can be maximized. In general, a choir can fulfil the choir requirements if there are 25 people in the group. If it is less than that, then the group can be considered as a vocal group. Integrating voices in choir groups consisting of many people cause various events to occur in the process of social interaction between students and teachers. These interactions provide opportunities to experience the process of understanding each other related to the process of tolerance education.

3 TOLERANCE EDUCATION VALUES

3.1 *Definition of values*

Value is a scale to measure something. Scheler (Wahono & Hum, 2014) argues that value is a quality that does not depend on its bearer; something that has been felt by humans without going through worldly experience first. Value is something that human beings, according to their essence and characteristics, believe as important or useful for humanity. The value in this study is not defined as a measured value in units of numbers but an explanation of events based on a description of the facts in people's interaction in a choral lesson.

3.2 *Education*

Education is very important for the growth and development of human life. Soehardjo (2005: 1), referring to the Indonesian Law No. 2 of 1989 concerning the national education system, said that education is a conscious effort to prepare students through guidance, teaching, and/or training activities for their role in the future. Education has a role that is always related to human benefit. Yaumi (2014: 6) says that only through education, all the potentials possessed by humans are developed to become fully human. Children with their capacities have to be directed towards education to make them able to develop their thoughts, feelings, psychotropic, and, more importantly, their souls as a source of spirit in life. Likewise, Arslan (2013: 4115) said that education is the most effective process in forming, developing, guiding and preparing healthy individuals and communities for their future. Thus, education must continue to be inherent in human life so that humans develop into complete human beings and are useful in all ages.

3.3 *Tolerance*

In the table of values of character and national culture, Yaumi (2014: 83) includes tolerance into groups of descriptions of character education values. Tolerance is an attitude and action that respects differences in religions, ethnicities, opinions, attitudes, and actions toward others which are different from us. Differences encountered by children must be framed using tolerance values so that they respect one another.

Yaumi (2014: 91) says that tolerance is the attitude of accepting differences in others, not imposing beliefs on others, not disliking others because they do not share the same beliefs or understandings, and not judging others based on their background, appearance, or habits. Therefore, tolerance is needed for character building and education in the midst of Indonesia's cultural diversity. The value of tolerance can be one of the pillars in making Indonesia a unified whole.

Judiani (2010: 284) says that tolerance is an attitude and action that respects differences in religions, tribes, ethnicities, opinions, attitudes, and actions of others. Therefore, tolerance is needed in dealing with various differences in the learning process of students who have various ethnic, cultural, and religious backgrounds.

From the theories above, it can be concluded that tolerance values play an important role in people's education. This value is found in educational interactions related to guidance and training that are useful for optimizing the full nature of human beings. Tolerance is one of the most needed attitudes, especially in multicultural Indonesia. The value of tolerance education is closely related to various activities in people's daily lives. One of them is in the choir whose process of achieving sound integration involves many students with various attitudes and behaviours.

4 TOLERANCE EDUCATION THROUGH *NUSANTARA* SONGS

Indonesian regions are rich in culture and language. Therefore, Indonesia has a variety of regional songs called *Nusantara* songs. *Nusantara* can literally be translated as 'archipelago'. However, this paper emphasizes the meaning of *Nusantara* as the combination of *Nusa* (island) and *Antara* (among). *Nusantara* songs in this paper refer to the songs that are developed among all the islands in Indonesia. Therefore, the authors will persistently use the term *Nusantara* without being translated to 'archipelago'. Moreover, *Nusantara* songs show or accentuate the characteristics of Indonesian-ness, both in terms of language and melody style.

All 34 provinces in Indonesia provide a diversity of regional songs. This diversity is a tool for carrying out tolerance education in choirs. Darrow (2017: 20) says that teaching vocal music is related to lyrics in a unit and instrumental compositions written for the purpose of promoting unity by highlighting social justice in music classrooms. The strength of the meaning of song lyrics can be a tool and material for tolerance lessons. Various styles of melody, distinctive regional tones, and the language and meaning of songs can be a learning tool in the process of tolerating various differences.

For example, Sumatra regional songs have a distinctive style of Malay music which is dominated by Malay *cengkok* (melodic style), *gambus* musical instruments, *rampak rebana*, and *pantun* (a traditional form of repetitive prose) contest. The Java region has a gamelan which is a distinctive tone in the song (pentatonic tone) and are rich in meanings inherited from ancestors. Meanwhile, the Papua region tends to describe the strength of nature in its regional songs. Therefore, students will be taught how to deal with and behave on different cultures reflected in *Nusantara* songs. Students learn how to understand the contents of regional songs that do not originate from the regions where they live and accept difficulties in various regional songs both in terms of song lyrics, tones, melodies, and interpretations. *Nusantara* songs become objects in a choral lesson for students under the teachers' guidance. The teachers can apply the value of tolerance education in the learning process. The habitual process to understand diversity can be done through *Nusantara* songs.

5 TEACHERS' ROLES

The learning process in choral classes which focuses on the value of tolerance education certainly requires a figure of teachers who are professional and organized in regulating classrooms. Kuyumcu (2012: 479) says that suitable teachers in the art education environment are organizers who are able to teach how to use materials, talk about what students are doing, organize art education in such a way that shows and preserves things, follows their application, and provides necessary guidance.

Teachers have a very important role in managing and achieving the success of the class they are teaching. This role needs special attention, especially when applying tolerance education to choral classes. This special case is a challenge for the future of music education, especially choir, amidst various problems in Indonesia.

The application of tolerance education values certainly requires critical thinking skills guided by the teacher. Music lessons also aim at developing social character, problem solving skills, cognitive abilities, critical thinking dispositions, and academic achievement. Tolerance is a social character that requires a critical mindset, so children are able to think actively, especially when applying tolerance.

Observing music education in schools will offer more or less adequate information about what music education is like today that can be used to build music education in the future. Music educators must be able to see the development of music along with the development of students and their development for music education, especially choirs that are able to develop educational values of tolerance.

Current technological developments help teachers to get several examples of cultural diversity in Indonesia. Documentation in the form of photos and videos can be easily accessed by the teacher to be used as references by students in recognizing various cultural diversity of the *Nusantara*. These references can strengthen and explore appreciation when singing *Nusantara* songs through choirs.

For example, the sound of Gondang Batak (percussion instrument in the region of North Sumatra) imitated in choir arrangements can be more alive when students see and listen to the sounds of Gondang Batak through video. Thus, the role of teachers who are active, innovative and able to adapt to changing times is needed in facilitating tolerance lessons in school choirs. Learning is not only limited to the learning process of singing but also recognizes indirectly the diversity of the *Nusantara* and applies tolerance values through maximizing the use of technology.

6 CONCLUSION

Character education has actually been a part of education so far. However, the educational tools in the process of implementing character education in Indonesia must be developed and maximized according to the potential of a region. Indonesia with various conflicts caused by differences (lack of understanding of the importance of tolerance), requires various media so that tolerance education can be effectively applied in the learning process.

Choir is an alternative topic in applying tolerance education processes. The number of members which generally ranges from 30-40 people offers an opportunity for social interaction (which is important in implementing tolerance education). In addition to musical interactions in choirs, diverse Indonesian songs in Indonesia can be a strategic medium in implementing tolerance education. The learning process is certainly fully held by the teacher. Therefore, teachers must be able to develop materials for choral lessons in schools by looking at the opportunities of Indonesia's cultural diversity. By learning choirs and *Nusantara* songs, students are directed to understand the importance of tolerance since school age.

REFERENCES

Ahmad, H. A. (Ed.). (2013). *Survei nasional kerukunan umat beragama di Indonesia.*[*National Survey of Religious Harmony in Indonesia*] Kementerian Agama RI, Badan Litbang dan Diklat, Puslitbang Kehidupan Keagamaan.

Arslan, A. A. (2014). A study into the effects of art education on children at the socialisation process. *Procedia-Social and Behavioral Sciences, 116*, 4114-4118.

Darrow, A. A. (2017). Teaching Tolerance in the Music Classroom. *General Music Today, 30*(3), 18-21.

Judiani, S. (2010). Implementasi pendidikan karakter di sekolah dasar melalui penguatan pelaksanaan kurikulum. *Jurnal pendidikan dan kebudayaan, 16*(9), 280-289. [Implementation of Character Education in Primary Schools Through Strengthening the Implementation of the Curriculum. *Journal of Education and Culture*, Vol. 16, Special Edition II.].

Juvonen, A., Ruismäki, H., & Lehtonen, K. (2012). Music education facing new challenges. *Procedia-Social and Behavioral Sciences, 45*, 197-205.

Kuyumcu, F. N. (2012). The importance of "Art Education" courses in the education of prospective teachers. *Procedia-Social and Behavioral Sciences, 51*, 474-479.

Liturgi, T. P. M. (2013). Menjadi dirigen III. Yogyakarta.

Soeharto, M. (1992). *Kamus musik*. Gramedia Widiasarana Indonesia (Grasindo).

Soehardjo, A. J. (2005). *Pendidikan seni dari konsep sampai program*. [Art Education; from Concept to Program]. Malang: Balai Kajian Seni dan Desain.

Topoğlu, O. (2014). Critical thinking and music education. *Procedia-Social and Behavioral Sciences, 116*, 2252-2256.

Wahana, P., & Hum, M. (2004). *Pustaka Filsafat Nilai Etika Aksiologis Max Scheler*. [Value of Axiological Ethics Max Scheler]. Yogyakarta: Kanisius.

Yaumi, M. (2014). Pendidikan Karakter: Landasan, Pilar dan Implementasi. [*Character Education]. Jakarta: Prenada Media Group.*

21st Century Innovation in Music Education – Astuti et al. (eds)
© 2020 Taylor & Francis Group, London, ISBN 978-0-367-11122-9

The educational values in the art of *Randai Sutan Budiman* in Kambang Lengayang

J. Yuliandri & A.N. Machfauzia
Universitas Negeri Yogyakarta, Yogyakarta, Daerah Istimewa Yogyakarta, Indonesia

ABSTRACT: This study aims to identify, express, interpret and find educational values contained in the art of *Randai Sutan Budiman* in Kambang Lengayang sub-district, regency of the south coast of West Sumatra. The research conducted on the art of *Randai Sutan Budiman* uses a qualitative approach. This research was conducted using ethnographic methods and was supported by a hermeneutic approach. Data sources in the study were obtained based on key informants or artists, documents and a literature review. Data collection techniques involved observation, interviews and documentation. The results of the study showed that, in the art of *Randai Sutan Budiman*, there are several educational values such as, religious values, tolerance values, hard work values, creative values, independence values, curiosity values, values of responsibility and friendly/communicative values.

1 INTRODUCTION

Indonesia is a country with many local wisdom and traditions from Sabang to Merauke. This is a strength for the people of Indonesia to unite according to the meaning of the slogan Unity in Diversity. The diversity is contained in its language, customs, morals, art, and science, among others. Cultural diversity throughout the country is a treasure of the nation's culture, such as *Minangkabau*, which is part of the corpus of culture in Indonesia that has its own uniqueness and is an asset of the West Sumatera province.

The *Minangkabau* tribe has fundamental values such as a philosophy of life, a foundation of thinking, systems of kinship, values in the norms of association and laws that must be obeyed. As Syarifuddin (1982, p. 1) suggests, customs according to the *Minangkabau* people are the rules of everyday life of every individual and society in socialising to be a good individuals and communities. In principle, the community which complies with the regulations and norms are a good indication of prosperous country.

The '*Minangkabau Tribe*' is the term of the *Minang* community for their culture, which covers all aspects of *Minangkabau* culture. As expressed by Dewantara (1994, p. 72), culture means everything that is related to 'culture', while culture is derived from the word 'gratitude', which could be interpreted as a human soul that has matured or gratitude. Therefore, a culture or custom is a rule of life which is widespread, similar to *Minangkabau* customs. *Minangkabau* custom is a concept of life developed by the ancestor people of *Minang*, which aims to achieve a happy life in the world and also be happy in the afterlife, as well as in the South Pesisir areas that are also part of the *Minangkabau*.

Southern Pesisir is an area that lies along the coast of the southern part of the province of West Sumatra and is directly confronted with the ocean of Indonesia. Based on that situation, it was very possible for the merchants of ancient kingdoms to anchor in one of the transport centres popular in that period. These activities allowed the South Pesisir region to experience changes either in terms of trade, business or cultural exchange and it is also possible that many new diverse arts came to the area. There is some art that still exists in Southern Pesisir such as *Randai, Gandang Sarunai, Liolo Pupuik* and *Rabab Pasisia*. Of the few that still exist, there is one that will be used as the object of this research, namely the traditional art '*Randai*', which is commonly known as *Barandai*.

Randai, commonly known as *Barandai*, is an artistic tradition that consists of art music, art motion or dance, self-defence arts and drama. This *Randai* art tradition has grown and developed in the *Minangkabau* culture. *Randai* is spread in several parts of the territory and has a specific type and characteristics. The story, the theme and the specification of *Randai* in each region in the *Minangkabau* are different, which can be seen from the story of a play and the shape of a movement. The different stories, drama or movements due to different regional circumstances or location of *Randai* art mean that every region in *Minangkabau* has many local stories, tales and legends. The south coast, as part of the *Minangkabau* culture, is classified as *Rantau Pasisia* according to geo-historians, with a wide coverage area. It is in this region that *Randai* art developed, particularly in Kambang.

Randai, for Kambang society in Lengayang, South Pesisir region, is a traditional art that is very well known by society and the art has frequently been shown from the first *Minangkabau* Kingdom until today. In the past, *Randai* was usually performed at official event of the kingdom. However, as the kingdom system no longer exists, *Randai* is often performed at events of the *Minangkabau* tribe, such as at the crowning of the *Datuak* or village chief, and at events outside the tribe such as a welcome parties for visitors, officials of countries or areas that come to *Minangkabau*, weddings and as entertainment for the youth and communities to maintain relations.

Furthermore, as mentioned, *Randai* art is a place to build friendships, that is, the relationships between young and old generations, parents and children, even a government with society. *Randai* can be defined as a drama musical because *Randai* consists of a *kaba* or life story of someone, which is revived through art activities. The story presented in *Randai* art performances contains the meaning implied in the form of messages, advice, criticism and suggestions, which will be submitted to the community.

However, in the global era, the *Minangkabau* culture seems to be dying out, not only among the *Minang* society living in other areas, but even in the regions of West Sumatra in which the *Minang* tribe are in the majority. They seem embarrassed by the culture that has been passed down by their ancestors. The younger generation is more concerned and proud of the foreign culture started with free signs anywhere, including in the area of the *Minangkabau*. This affects morale, ethical guidelines, style and the way the younger generation behaves. One of the *Minangkabau* cultural arts that have ethical Eastern values is the *Randai* art. It is sad when *Randai*, in its own region, is ignored by the younger generation, and many generations no longer understand the values and messages contained in *Randai*. Therefore, there are many phenomena of moral degradation in Indonesia, especially in West Sumatera, due to the influence of globalisation.

Randai has a lot of educational value for the younger generation, which can be applied in daily life in society, religion and state. *Randai* is a local wisdom that can take a role in addressing the thoughtfulness of local identity so that existential and multicultural crisis can be avoided. Then, as the action to conserve and develop *Randai*, the society and local government have to work together to develop art studios as a place for artists and the younger generation to work and study and know more about *Randai*. The meaning in this art, *Randai Sutan Budiman*, that teaches about educational values is very useful in forming the younger generation so that they always behave politely and courteously and are always positive within society.

2 LITERATURE REVIEW

The study discusses the values, structures and elements contained in the story of *Gedunai* and *Niniak Jiruhun Dang*, which consists of the theme, characters and characterisations, plots, foreground and motifs. The research that has been done by Asnudi (2007) has the similarity with this research in term of the observation of *Randai* and the value in it. However, this research conducted in the origin area *Minangkabau* while Asnudi does not reside in the area of *Minangkabau*.

3 MATERIAL AND METHODOLOGY

Primary data are data obtained through a series of activities such as observations of artists/ practitioners of the art of *Randai* in society, Kambang and the related officials in Lengayang sub-district Kambang, South Pesisir, West Sumatera. While secondary data are data obtained through the collection of data in the form of documentation, references (literature, writing, books, etc.) associated with the arts and cultural customs of *Randai* and culture in *Minangkabau* in Kambang, for use as support of this research. To ensure the validity of the data in this study, the researcher conducted triangulation and the extension of the observation. Further data analysis techniques used in this research are to observe objects, collect data from the observed objects, describe and then analyse. In this study, a flow analysis, as described by Miles and Huberman (1994, pp. 31–33), which involves a condensation of the data, the presentation of data and the drawing of the conclusion is used.

This research was conducted in Kambang sub-district Lengayang, South Pesisir, West Sumatera Province, the majority of which is *Minangkabau* society, such as artists or perpetrators of art and culture who became the informants in this study. This research is using the qualitative methodology. Research methods supporting this study include ethnographic methods to describe the *Minangkabau* culture and understand a view of the life of the indigenous population.

Spradley (1997, p. 3) states that ethnography is the job description of a culture. The main purpose of this activity is to understand a view of life from the perspective of the natives. In terms of other ethnographic approaches, it is a form of research that focuses on the meaning of sociology through closed field observation of the sociocultural phenomenon. Usually ethnographic research can be approached from the perspective of art and culture and is more descriptive than an attempt at an analytical effort. However ethnographic research focusing on analytical aspects and social sciences are also a branch of cultural anthropology (Emzir, 2013, pp. 143–144). To find out more about the art of *Randai Sutan Budiman*, this research uses a hermeneutic approach. Hermeneutics are used to find the hidden values contained in the art of *Randai Sutan Budiman*. Susilo (2017, p. 13), argues that hermeneutics is a science to interpret and understand something abstract and dark to be brighter and able to explain problems that were originally abstract. Therefore, to find value and understand the delivery of messages and news contained in a work of art it is necessary for us to understand the meaning or understanding of a text because the understanding of the text becomes very important in supporting the assessment of its meaning (Ricoeur, 2014, p. 193).

4 RESULTS AND DISCUSSION

Randai is a *Minangkabau* traditional art that has existed since the *Minang* age until today, although *Randai* is experiencing up and down periods due to the impact of globalisation. However, this art still growing, thriving and surviving in *Minangkabau* society because in this *Randai* performance there are stories or *kaba* which contain advice, education and values. Based on the analysis results, eight educational values in *Randai Sutan Budiman* were found.

4.1 *The values of religious education*

The values of religious education are values that concern the relationship of man with God. Educational values of believing in God are based on the story of *Randai Sutan Budiman* in Kambang, such as:

> *Balayia pincalang so'a, balabuah tantang Indogiri, kakak bajalan kami bacokan do'a, mintak salamaik juo pulang jo pai...*
>
> Sailing boat of so'a, anchored in Indogiri, the elder brother went abroad we read of prayers, ask/to survive while home and away...

This sentence shows the belief in the existence of God in every human activity on earth; God oversees every human activity. The evidence of belief in the existence of God is to worship and asking Him for everything so that we get safety and help in the world and in the hereafter.

4.2 *The educational value of tolerance*

The attitude of this tolerance is an attitude which appreciates everything the difference between themselves and the society consciously and openly, such as:

> *Ampun baribu kali ampun, ampun kami kasadonyo, mano sagalo niniak mamak, uarang nan cadiak cado kio. . . .*
>
> Sorry, really sorry, forgive us, to all Ninik mamak/older people, smart and clever people...

This sentence shows the importance of making an apology before doing the activity. This apology must be done so that the program gets bail and permission from the elders, because the position of *Ninik mamak* is very important in *Minangkabau.*

> *Ondeh. . .diak kanduang gadih Rasanah, lah panek denai babansi, rabab tasanda di pamatang, lah penek denai dek mananti, mangapo kok adiak talambek datang. . . .*
>
> O..... Rasanah, I have tired playing the Bansi, Rebab is tied up in a bund, and I have tried to wait, why you are coming late...
>
> *Tuan kanduang si Sutan Rancak, bukannyo padi rambak di pamatang, rantiang rang ambiak ka layu-layu, bukannyo denai talambek datang, jalan manguliliang tampek lalu. . . .*
>
> Sir/Brother Sutan Rancak, doesn't the rice dense in the bund, the people twig takes to wilted, I am not supposed to come late but the way is slue...(The literal translation might be confusing because the purpose of the original sentence arrangements is to make rhymes).

From this sentence we can understand that Sutan Rancak did not dispute the deceased Rasanah who arrived late from the promised time after explaining the reason.

4.3 *The value of hard work*

The value of hard work can be seen in the art of *Randai Sutan Budiman*, that is:

> *Manolah Pandeka nan baduo, denai lah banamo Sutan Budiman, Asa nan dari Batusangka, andak marantau kakampuang urang, mancarikan paruik nan tak barisi, sarato pungguang na tak basaok, pulang ma'alum ke bakeh Pandeka. . .*
>
> The swordsman, my name is Sutan Budiman, comes from Batusagkar, I will go to wander to other village, find some food for the stomach that haven't fulfilled yet. . .

This sentence shows that Sutan Budiman was to go abroad to find a job with the aim to improve the economy. This is in accordance with the philosophy of indigenous *Minang*:

> *Karatau madang di hulu Babuah babungo balun,*
> *Marantau bujang dahulu, di rumah paguno balun.*
>
> Karatau trees grow on the upper, but not yet flowering, fruiting (rhymes)
> Just go outside boy, in the village, you are not functioning well

This proverb describes the attitude of the *Minang* people who have a tradition of wandering with a variety of purposes such as trade, learning and others, with the intent to return to the hometown in order to contribute to advancing the state of the hometown.

4.4 *Creative values*

The value contained in the creative art *Randai Sutan Budiman* can be seen in:

> *Ondeh...diak kaduang Gadih Rasanah, kalau baitu katonyo adiak, itu kato nan sabananyo, putiah kape dapek di liek, putiahnyo hati bakaadaan, ikolah tando putiahnyo hati, cincin ameh cincin timbago, nan bacampua jo Loyang mato, anta katuju diak anta tido, pakai di adiak salamonyo...*
>
> O ... Rasanah, then you said, that the actual words, white cotton can be seen, this is a sign of the white heart, gold ring copper ring, that mixed with brass eye, either you like it or not please wear it forever...
>
> *Tuan kanduang si Sutan Rancak, adatnyo hiduik bale mambale, adat syari'at malu bamalu, iko sapu tangan nan sa'alai, pengipeh jikoknyo palak, pa apuih paluah nan buruak, anta katuju anta indak, pakai dek tuan salamonyo...*
>
> Sir Sutan Rancak, the custom of live is to give to each other, syari'ah custom, this is a handkerchief, you can use it as a fan, a dirty sweat removal, either you like it or not please put it forever...

From this sentence, it can be understood that there has always been a tradition of giving reminders of promises to lovers who have a relationship, even in a modest way. In addition, creative values can also be seen from the uniqueness of customs, rules of conduct, and rhymes in the *Minangkabau* language.

4.5 *The values of independence and confidence*

Independence is the attitude of not being easily tied to others in finishing tasks or activities. In terms of having strong confidence, the independence value is reflected in *Randai Sutan Budiman*:

> *Manolah dek pandeka na baduo, kok kandak indak ka buliah, pintak nan indak balaku, denai nan anak laki-laki sakali malangkah bapantang suruik, alah ka jaleh itu dek pandeka...*
>
> Which one is better, hi swordsman, if the desire cannot be fulfilled, I am a boy, once I go, I won't back to you...
>
> *Asak-asak bajak, den asak ateh pamatang, satapak indak ka den anjak, anta nyawa pandeka nan ka malayang...*
>
> Just move my plough equipment, I will move it to the bund, a footstep I wouldn't go, either your soul will be gone...

We can get the meaning from the text that describes the attitude of independence and strong self-esteem embedded within *Sutan Budiman* in dealing with any issues that arise when travelling abroad. He remains tough and confident in confronting the problems that arise without the help of others.

4.6 *The value of curiosity*

Every human being must have curiosity, which is satisfied when the person finds out what they have been looking for. This curiosity is the attitude and action which is always striving to know more deeply something to learn, see and hear. This attitude is reflected in the art of *Randai Sutan Budiman*, namely:

> *Tabatiak kaba eei,,,ondeh jo barito, Kabakeh diri,,,lah diri Ulak Cumano, Yo Ulak Cumano, Baso Rasanah eeii,,,lah batunangan, Jo diri bujang,,,ndeh mamak eii si Sutan Rancak,, Si Sutan Rancak, Rumah sikolah eii,,,yo nan di Burigo, Buatan anak,,,ndeh aia*

haji, Yo aia haji, Denai nyo ambiak eii,,yo kaharago, Jo Sutan Rancak,,,Rasanah bajua bali, Yo bajua bali, Sinan bajalan eii,,,Ulak Cumano, Yo nak mancari,,,mancari Gadih Rasanah, Yo Gadih Rasanah...

The news of eii is sounded by someone, yes Ulak Cumano, yes Ulak Cumano, that Rasanah eii was engaged, with the bachelor, yes mama eii Sutan Rancak, yes that was in Burigo, yes Hajj Watno will look for, look for Rasanah, yes Rasanah...

The sentence above explains about Ulak Cumano who was shocked knowing that his girl, Rasanah, having been engaged to Sutan Rancak without telling him before. Because Ulak Cumano assumed that Rasanah was his fiancée, Ulak Cumano went to Rasanah to ask the facts of the news. Is it true or not that Rasanah is engaged to Sutan Rancak? It is explained that Ulak Cumano tries to fulfil his curiosity by looking for the truth.

4.7 *The value of responsibility*

The value of responsibility can be seen from the efforts of artists in mastering their roles in Randai performance art. For example, someone who acts as the main character in the Randai story will be responsible for the movement and reciting the dialogue of the characters.

4.8 *The value of friendliness/communicativeness*

This communicative/friendly attitude can be seen from the *Randai* art game itself, such as giving each other a signal in each exchange of movements and can also be seen from the art stories of *Randai Sutan Budiman* in Gadih Rasanah's conversation with Sutan Rancak:

Tuan kanduang si Sutan Rancak, Adatnyo hiduik bale mambale, Adat syariat malu bamalu, Ikolah sapu tangan nan salai, Pangipeh jikoknyo palak, Pa apuih paluah nan buruak, Anta katuju anta indak, Pakai dek tuan salamonyo.

Sir Sutan Rancak, in life, you should help each other, syari'ah rules are about politeness, this is a handkerchief, you can use it as a fan or a dirty sweat removal, whether you like it or not, please put it forever.

5 CONCLUSION

Based on the deliberations and results of this research, it can be inferred that *Randai Sutan Budiman* is a tradition of generations that evolved in the region of *Minangkabau*, in particular in the area of Kambang. *Randai Sutan Budiman* is an art tradition that combines the art of music, drama, motion/dance/martial arts. *Randai* artistry can be classified as music because in the art of *Randai* there are storylines that are played by several people. The story played in the performance art *Randai* contains values, education, moral messages and advice for humans to be individuals who are good, ethical and cultured in order to create a peaceful life for the community.

The recommendations for further research are the study of material objects and formal objects in depth and thoroughly. The weakness in this research is the lack of in-depth study of objects. However, the impact of this study are:

1. This research suggests implementing the character values contained in *Randai* to the community, both through formal and non-formal education..
2. It is hoped this research can be a study to understand the definition of value in the context of local wisdom to understand the analysis of *Minangkabau* society and identity.
3. This research is expected to be a reference base of the development of local wisdom-based education for self-development of students' learning, both formal and informal.

Furthermore, it is expected that this research can be the reference for further research. Very limited research results are expected to be used as a reference for other researchers with regard to arts-based local wisdom traditions (local wisdom), one of them is the art of *Randai Sutan Budiman*. It is hoped that the next regeneration can understand the values contained in works of art so that no art fades away due to the modernization and globalization.

REFERENCES

Asnudi, Y. (2007). *Telaah Struktur, Nilai Budaya dan Konteks Cerita Rakyat dalam Tradisi Lisan Randai pada Masyarakat Rantau Kuantan Singingi Provinsi Riau: Studi Deskriptif-Analitis terhadap Cerita Rakyat Rantau Kuantan Singingi dalam Tradisi Lisan Randai sebagai Bahan Ajar Muatan Lokal untuk Sekolah Menengah Pertama di Kabupaten Kuantan Singingi Provinsi Riau* [Review of structure, cultural value and context of folk stories in the Lisan Randai tradition in the Kuantan Rantau singing community of Riau Province] (Doctoral dissertation, Universitas Pendidikan Indonesia).

Dewantara, K.H. (1994). *Ki Hajar Dewantara* [Ki Hajar Dewantara]. Yogyakarta, Indonesia: Percetakan Offset Taman Siswa.

Emzir, M. (2013). *Metodologi Penelitian Pendidikan: Kuantitatif dan Kualitatif* [Quantitative and qualitative education research methodology]. Jakarta, Indonesia: Rajawali Pers.

Geertz, C. (1973). *Tafsir Kebudayaan* [The interpretation of cultures]. Jakarta, Indonesia: Basic Books.

Koentjaraningrat, K. (1990). *Pengantar ilmu antropologi* [Introduction to anthropology]. Jakarta, Indonesia: PT Rineka Cipta.

Miles, M. B. & Huberman, M. (1994). *Qualitative data analysis: An expanded sourcebook*. New York: Sage.

Pradoko, A.M.S. (2017). *Paradigma-Paradigma Kualitatif untuk Pengkajian Seni, Humaniora dan Budaya* [Qualitative paradigms for art, humanities and culture research]. Yogyakarta, Indonesia: Charissa.

Ricoeur, P. (2014). *Teori Interpretasi Membelah Makna dalam Anatomi Teks* [Interpretation theory: Splitting meanings in the anatomy of the text]. Yogyakarta, Indonesia: IRCISOD.

Spradley, J.P. (1997). *Metode Etnografi* [Ethnographic method]. Yogyakarta, Indonesia: PT Tiara Wacana.

Syarifuddin, A. (1982). *Pelaksanaan hukum kewarisan Islam dalam lingkungan adat Minangkabau* (Doctoral dissertation, Institut Agama Islam Negeri Syarif Hidayatullah, Jakarta, Indonesia).

21st Century Innovation in Music Education – Astuti et al. (eds)
© 2020 Taylor & Francis Group, London, ISBN 978-0-367-11122-9

Meanings in the *Orang Rimba* ritual honey-gathering song in Jambi, Indonesia

D.N. Anggraini & S. Pradoko
Yogyakarta State University, Yogyakarta, Indonesia

ABSTRACT: This study was conducted in order to describe the meanings in the ritual song sung while gathering honey by the *Orang Rimba* people in Bukit Duabelas National Park, Jambi, Indonesia. This study uses ethnographic methods and interpretation analysis via the work of Roland Barthes in semiotics. The results of the study are as follows. The song of the *make madu* ritual performed by the Orang Rimba has seven meanings that consist of: (1) making the body of the tree-climber lighter; (2) asking *Setan Kayu* to make it easy to get to the branch; (3) not being stung by bees and ensuring that humans and bees do not interfere with one another; (4) gathering a lot of sweet honey; (5) unloading the honey that the tree-climber gathered; (6) making opponents feel ashamed, insecure and unwilling to compete; (7) avoiding a honeycomb that has no honey.

1 INTRODUCTION

The Orang Rimba is a group of people who live in the forest. As forests are their home, the Orang Rimba value them. The Orang Rimba is one of the ethnic tribes in Jambi Province, Indonesia with a large presence in Bukit Duabelas, Sarolangun Regency. A land trip from Jambi city to their domain takes approximately seven hours. Orang Rimba is a term for the remote indigenous people who live in the forests of Jambi Province and the province of South Sumatra. According to the Orang Rimba, these two designations are interpretations of their lives since their ancestors depended on the forest and its fruits. The Orang Rimba very much depends on forest products, including plants and animals. Thus, they really appreciate the forest and do not want to damage it because it is their home and source of livelihood.

The Orang Rimba truly trust the gods. They believe that there are good deities and evil gods, and that their god is in the forest. So, whatever the Orang Rimba does with the forest must relate to God, including his art. One of the arts that relates to God is the practice of taking or gathering honey. While collecting honey, the Orang Rimba sing, asking for help and permission from the trees and the bees and to be protected. For example, when Orang Rimba wants to climb a forest tree, they sing to ask for protection. When the tree-climber is close to a *repah* or honeycomb, they sing to the jungle asking for permission. While taking honey, the forest people sing to get lots of sweet honey and when they are finished taking the honey, they sing about the honey as the honey taker climbs down the tree. This practice shows that the arts are interrelated with ritual activities. Therefore, this study aims to find out the meanings in the Orang Rimba ritual song for collecting honey in Jambi Province, Indonesia.

2 RELATED WORKS

Ethnography describes culture and its aspects as knowledge gained by humans and used to interpret and discover the behaviours the subjects under study (Spradley, 1980, p. 3). Dobbert (1982) said that, "the main aim of ethnographic research is to discover and describe the culture of people or organisations" (p. 39). Semiotic etymology is a term derived from the Greek *semeion*, which is a sign (Sudjiman & Van Zoest, 1996, p. 7) or *seme* which means "translator of signs" (Cobley & Jansz, 1999, in Sobur, 2001, p. 16). Thus, semiotics is defined as the study of signs and the ways that signs work. One of the key theorists of semiotics, Barthes (1967), developed Saussure's ideas and tried to apply the study of signs more broadly. Through a productive and exciting career in many cultural phases, included fashion, photography, literature, magazines and music among his many interests. One of Barthes' main preoccupations is "how the meaning enters the image/ image". And that is the key to semiotics: how the creator of an image makes it mean something and how we, as readers, get its meaning (Stokes, 2006, p. 76). Roland Barthes' semiotics can clearly illustrate that the same sign object can be interpreted differently, undergoing changes from the meaning of denotation developing into connotation and metalanguage meanings or synonyms (Pradoko, 2017, p. 236).

The study related to Panjaitan's (2018) work that found that, first, the jungle people are very dependent on the forest; second, that building culture is very disturbing to the lives of the Orang Rimba; and third, that the application of local wisdom values in the Orang Rimba culture includes the perseverance and consistency of the Orang Rimba in maintaining and preserving the forest.

This studi is also related with Islamuddin (2016) research which found that: 1) the Talang Mamak tribe has a culture that includes birth, marriage, death and support. Civic culture is part of people's culture as evidenced by Talang Mamak. Civil values include mutual trust, responsibility, solidarity, consideration, togetherness and mutual cooperation. 2) Local wisdom found in this study that the Talang Mamak tribe is part of civic culture. This is proven by the local wisdom of the Talang Mamak tribe as having civic and cultural values that include patriotism, equality, caring, responsibility, independence and kindness. 3) Cultural preservation carried out by the Talang Mamak provides knowledge about traditional culture to the next generation in competitions and festivals about culture. 4) There are various constraints in the Talang Mamak tribal community, including economic, educational, trust and transportation factors.

3 MATERIAL & METHODOLOGY

3.1 *Data*

The data collection began with a search for data about honey harvesting by the Orang Rimba. This came out of an ethnographic approach in order to learn about the ritual of taking honey. In the ritual of taking honey, there are songs about climbing trees, asking permission from the gods, seducing bees or *sialang* so that they do not sting the tree-climber, singing to get lots of sweet honey, then giving thanks for getting honey and asking for the protection of the gods.

3.2 *Method*

This study used qualitative methods with a semiotic analysis based on the work of Roland Barthes. Data collection techniques were carried out through observation, interviews, literature studies and documentation. Observations and interviews were conducted to obtain information about the lives of forest people and songs to take honey in the Bukit Duabelas National Park in the village of Bukit Suban, Air Hitam District, Sarolangun District, Jambi Province. The literature study was conducted to obtain references related to this research and the documentation was carried out to obtain data in the form of photography and video recording.

In this study the ethnographic method was based on theories by James P. Spradley (1980). The steps taken were: 1) determine who the informant is; in the research the informant was an Orang Rimba headperson and others who were deemed necessary to obtain information; 2) interview informants with a general interview and use the interviews as an introduction by the interviewee to honey taking; 3) make a geographical record during or after an interview; 4) ask descriptive questions about the object of research in more detail; 5) conduct an analysis of the descriptive responses; 6) conduct domain analysis. This kind of analysis looks for semantic relationships between the rituals of taking honey and singing to identify words that have educational meanings and value.

This study will be analysed using the semiotics analysis of Roland Barthes. His semiotic elements focus on two orders of signification, in the form of denotations (markers) and connotations (markers) in each of the lyrics contained in the song about collecting honey by the Orang Rimba. The analysis was carried out by sorting through the text of the song. The entire analysis will be presented in the form of descriptions.

4 RESULTS AND DISCUSSION

This research was conducted in Bukit Suban Village, Air Hitam Subdistrict, Sarolangun District Jambi Province, Indonesia. The precise location was in the Dua Belas National Park. Through ethnography, it was found that the purpose of the ritual of taking honey was to ask permission to take honey, appeal for security, ask permission and protection of the gods, and ask for lots of sweet honey.

Table 1. The song of the ritual of taking honey.

1.Melantak Ooooooooo...... Linjang bagi linjang kundang kecik Ooooooooo.... Kundang kecik baruk andun Oooooo.... Baruk kundak andun kebelas panjang Belas tegantung di awang-awang Betiang tunggal bekeliling	1.Climb a tree Ooooooo..... Excessive girlfriend New little girlfriend walks Just want to go to the long branch I want to take honey on a long soil, is it okay? repah depends on the branch One around already hanging above
2. Near the branch Salammikum dahan jerambang Ooooooo..... Dahan jerambang bagiku lalu Ooooooo...... Kundak lalu ke belas panjang Belas tegantung di awang-awang Lamo lupo lamo tinggal	2. Near the branch Assalamualaikum branch *Jerambang* Jerambang's branches give me a way I want to go to the end of the branch Branch that depends on the air Long forgotten the length of stay
3. To get rid of bees Dedanguy dedangoy alangku elok Budi kundak elok Kunang permai banyak Keola kudang kuini budi baik baso baik Nanti janji semanyo kito Janji jangan bermungkira	3. To get rid of bees Listen, my bee is good Good love stands to many bees I want to cultivate honey with good language Later the promise of the dark moon Later we promise our bond Our promise must not be denied
4. Take honey Mandi dimano ida ingin mandi Diula lasung batu Ati siapo tidak ingin susu nula dalam baju	4. Take honey Bathing where you don't want to take a shower Rubbed with stone Whose heart does not want mother's milk in clothes
5. Lower the honey/get off the tree Alu-alu tutu temalang Tutu tecepa daado diulang Beringat nian bujang pengala Pujang penyambut tembalang Awak bejalan seorang	5. Lower the honey/get off the tree Please help not misfortune To be fast not repeated Very light, single-handed Singles welcome the one who walks alone
6. Singing if there are opponents Main sepat main seluang Main berendam buku balo	6. Singing if there are opponents Playing sepat playing seluang Playing a balo book bath

(*Continued*)

Table 1. (*Continued*)

Main berempat alang kepalang	Playing four is okay
Main sepuloh elok jugo	Play ten better
7. If there is no honey	7. If there is no honey
Itik-itik manggang seluang	Ducks roast seluang
Budak menobo ulu laut	People push the sea
Setitik idak menaruh anak	A drop does not put the child
Aik dalam lintang tanjung	Water in the latitude of cape

a. Song lyrics for taking Orang Rimba honey

b. Based on the song lyrics above, the lyrics can be interpreted through semiotics of Roland Barthes, namely:

4.1 Melantak

In the first stanza, the taking honey ritual song tells the story that when you climb a tree, you are given protection and the body becomes lighter. The singing must be performed so that the gods still remember the tree-climber who takes the honey and then makes the body of the tree-climber light to rise above the branch. In the first verse, "take honey" has the meaning of a denotative primary system which is the initial meaning. Between the initial meanings of the primary system (E1 and C1) there is a relationship so that the process of relations (R) becomes developed which can form a sign. In this study, the denotative meaning of *Melantak* (E1) whose meaning has content (C1) in the primary system is to climb trees. Between (E1) and (C1) there is a relation (R) so that a sign that has developed can become more developed. This development in the E-R-C process in the primary system becomes (C2) in the secondary system, which is connotative meaning. The meaning of the denotative primary system (C1) develops into (C2) which is to relieve the climber's body and to express a wish that the climber does not fall down. In the mythology of the Orang Rimba, if a tree-climber takes honey from a tree height of around 30–50 metres and does not use any safety equipment, the climber believes that a spirit named Dewa Beyuto who is in the tree trunk can protect the climber's body and help prevent them from falling, and they sing to this spirit.

Table 2. First stanza scheme

E1 R1 *Melantak*	C1 climbing trees	Primary system denotative
⇩	Sign	
E2 R2	Lighten the body to climb onto the branch C2 Don't fall down	Secondary system connotative

Explanation:

E1 = Primary system expression | E2 = Secondary system expression
R1 = Primary system relations | R2 = Secondary system relations
C1 = Primary system content | C2 = Secondary system content

4.2 Near the branch, meaning excuse me, "Setan kayu"

Table 3. Second stanza scheme.

E1 Salamikum	R1 C1 Ask permission of "Setan Kayu"	Primary system denotative
Sign		
E2	R2 C2 "Setan Kayu" makes it easy to get to the branch	Secondary system connotative

Explanation:

E1 = Primary system expression	E2 = Secondary system expression
R1 = Primary system relations	R2 = Secondary system relations
C1 = Primary system content	C2 = Secondary system content

The second verse of the taking honey ritual song is about asking permission from the *Setan Kayu* to take honey. This is expressed by greetings when the tree-climber wants to be close to the honey. In the second verse of the ritual song, "take honey" has the meaning of a denotative primary system which is the initial meaning. The initial meanings of the primary system (E1 and C1) have a relationship so that the process of relations (R) becomes developed which can form a sign. In this study, the denotative meaning of excuse (E1) and having content (C1) in the primary system means to ask permission from the "wood demon". (E1) and (C1) have a relation (R) so that a sign that has developed can become more developed. This development in the denotative primary system process E-R-C becomes (C2) in the secondary system, the connotative meaning. The denotative primary system (C1) develops into (C2) which facilitates getting to the branch. In the myth, the climber believes that if he excuses the "wood demon", the wood demon will give the climber permission to get to the branch and also make the climber's steps easier.

4.3 Get rid of bees

Table 4. Third verse scheme.

E1 Janji Semanyo	R1 C1 Get rid of bees	Primary system denotative
Sign		
E2	Avoid being stung by a bee R2 C2 Don't interfere with each other	Secondary system connotative

Explanation:

E1 = Primary system expression	E2 = Secondary system expression
R1 = Primary system relations	R2 = Secondary system relations
C1 = Primary system content	C2 = Secondary system content

The third verse of the ritual song is about seducing the mother of the bees, so that the tree-climber will not be stung while taking honey. Because the Orang Rimba does not use any security equipment, the song is performed to ask for protection so that the bees do not sting the climber. "Take honey" has the meaning of a denotative primary system which is the initial meaning. Between the initial meanings of the primary system (E1 and C1) there is a relationship so that the process of relations (R) becomes developed which can form a sign. In this study, the denotative meaning of seducing (E1) whose meaning is having content (C1) in the primary system is to distract the mother bee. Between (E1) and (C1) there is a relation (R) so that a sign that has developed can become more developed. These developments in the denotative E-R-C system process become (C2) in the secondary system, the connotative meaning. The meaning of the denotative primary system (C1) develops into (C2), that the climber will not be stung by the bees and that the climber and the bees do not interfere with one another. In the myth, the climber believes that by singing to seduce the mother of the bees, the mother and the bees will not sting the climber. At the same time, the climber invites the bees to mutually promise not to disturb one another. This sentiment is sung because the climber does not use any safety equipment.

4.4 *Get lots of sweet honey*

The fourth stanza in the ritual song is in the form of a thank you for gathering lots of sweet honey as well as a promise to take honey again in the future. In the fourth verse in the ritual song, taking honey has the meaning of a denotative primary system which is the initial meaning. Between the initial meanings of the primary system (E1 and C1) there is a relationship, so that the process of relations (R) becomes developed which can form a sign. In this study, the denotative meaning of seducing (E1) whose meaning is having content (C1) in the primary system is for soothing the bees. Between (E1) and (C1) there is a relation (R) so that a sign that has developed can become more developed. These developments in the denotative E-R-C system process become (C2) in the secondary system, the connotative meaning. The meaning in the denotative primary system (C1) develops into (C2), that the climber sings to seduce the mother of the bee in order to be told which honey is sweet and to avoid honeycombs where there is no honey. The belief of the climber is that if with the song the climber is given instructions by the mother of the bee as to which one has sweet honey and many of the *repah* that are marked are empty then there is no mother bee.

Table 5. Fourth stanza scheme.

E1 Susu Nula	R1 C1 *Repah* there is no mother bee	Primary system denotative
	Sign	
E2	R2 C2 Get lots of sweet honey	Secondary system Connotative

Explanation:

E1 = Primary system expression	E2 = Secondary system expression
R1 = Primary system relations	R2 = Secondary system relations
C1 = Primary system content	C2 = Secondary system content

4.5 *Lower the honey/get off the tree*

Table 6. The fifth stanza scheme.

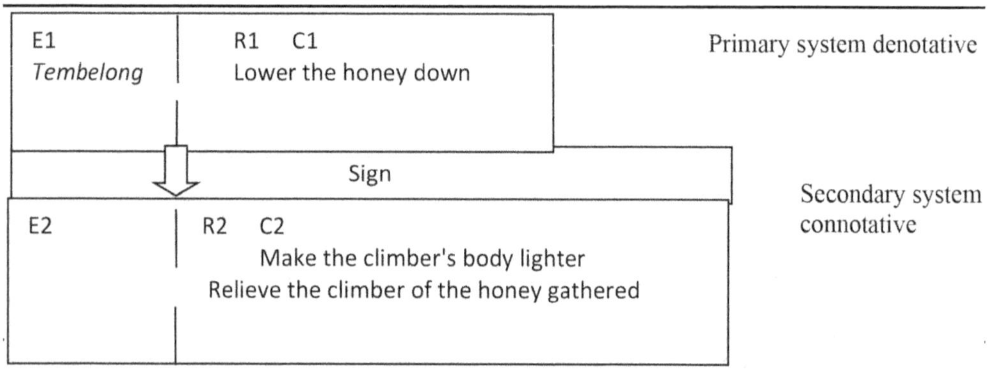

Explanation:

E1 = Primary system expression	E2 = Secondary system expression
R1 = Primary system relations	R2 = Secondary system relations
C1 = Primary system content	C2 = Secondary system content

The fifth stanza in the ritual song of taking honey is in the form of singing chants to make the body lighter and to relieve the climber from the burden of gathering honey, so that the climber does not fall. In the fifth verse, take honey has the meaning of a denotative primary system which is the initial meaning. Between the initial meanings of the primary system (E1 and C1) there is a relationship so that the process of relations (R) becomes developed which can form a sign. In this study, the denotative meaning of Tembelong (E1) whose meaning has content (C1) in the primary system is "to lower honey down". Between (E1) and (C1) there is a relation (R) so that a sign that has developed can become more developed. These developments in the denotative E-R-C system process become (C2) in the secondary system, the connotative meaning. The meaning of the denotative primary system (C1) develops into (C2), that the climber asks for help from the "wood demon" and *Dewa Beyuto* to make the climber's body lighter and to unload the honey brought down by the climber. In the myth, the tree-climber believes that the songs *Setan Kayu* and *Dewa Beyuto* will help them to descend the tree safely. If this song is not performed then the climber will object to bringing the honey he carries and might immediately fall with the heavy load of honey.

4.6 *Singing if there are opponents to compete against*

In the sixth stanza, taking honey has the denotative primary system meaning which is the initial meaning. Between the initial meaning of the primary system (E1 and C1) there is a relationship with the process of relations (R) to develop into a sign. In this study, the denotative meaning of competing (E1) whose meaning has content (C1) in the primary system is that there is an opponent. Between (E1) and (C1) there is a relationship (R) and the sign can become more developed. This development in the denotative system process of E-R-C becomes (C2) in the secondary system, the connotative meaning. The meaning of the denotative primary system (C1) develops into (C2), that the climber challenges an opponent to climb a tree and feels confident to compete against the opponent so that the opponent feels embarrassed, insecure and will move away from the tree. In the myth, the climber believes that the song will make his opponent move away from the tree and will not disturb the tree-climber who wants to take the honey in the *sialang* tree.

Table 7. Sixth stanza scheme.

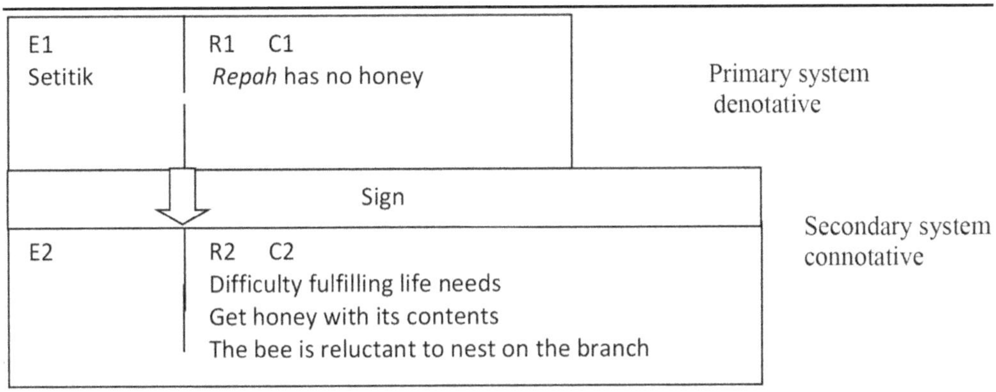

Explanation:
E1 = Primary system expression E2 = Secondary system expression
R1 = Primary system relations R2 = Secondary system relations
C1 = Primary system content C2 = Secondary system content

4.7 *If there is no honey*

In the seventh stanza, take honey has the meaning of a denotative primary system which is the initial meaning. Between the initial meanings of the primary system (E1 and C1) there is a relationship so that the process of relations (R) becomes developed which can form a sign. In this study the denotative meaning of seducing (E1) whose meaning has content (C1) in the primary system is that the *repah* has no honey. Between (E1) and (C1) there is a relation (R) so that a sign can become more developed. These developments in the denotative E-R-C system process become (C2) in the secondary system, the connotative meaning. The denotative primary system (C1) develops into (C2), that the climber will have difficulty meeting the needs of his family's life, that the bee is reluctant to nest on the branch, and that the mother bee will show the climber where to find sweet honey and its contents. In the myth, the climber believes that by singing to the mother of the bee, she will show the climber a honeycomb where there is so much sweet honey that the climber will be able to meet the needs of their family.

Table 8. Seventh stanza scheme.

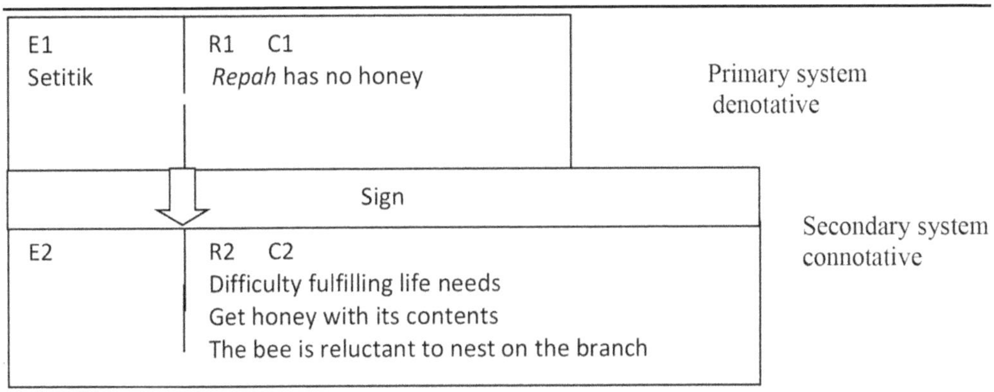

Explanation:
E1 = Primary system expression E2 = Secondary system expression
R1 = Primary system relations R2 = Secondary system relations
C1 = Primary system content C2 = Secondary system content

5 CONCLUSION

In the Orang Rimba ritual song about taking honey, there are seven meanings, which are: 1) making the body of the tree-climber lighter; 2) the "wood demon" makes it easy to get to the branch; 3) the bee does not sting the climber and the climber and the bees do not interfere with one another; 4) gather a lot of sweet honey; 5) relieve the climber of the honey that they brought down; 6) make any opponents feel ashamed and insecure and persuade them to move away from the tree so that there is no competition; 7) avoiding a honeycomb that has no honey.

REFERENCES

Barthes, R. (1967). Elements ofsemiology. London: Cape.

Dobbert, M.L. (1982). *Ethnographic research: Theory and application for modern schools and societies.* California, United States: Praeger Publishers.

Islamuddin, I. (2016). Pengembangan Budaya Suku Talang Mamak Sebagai Nilai-Nilai Kearifan Lokal Dalam Bagian Civic culture [Talang Mamak Tribal Culture Development As Local Wisdom Values In The Civic Culture]. *Jurnal Pendidikan Ilmu Sosial, 23*(2), 55–67.

Panjaitan, N.P. (2018). *Rasionalitas dan Transformasi Sosial Suku Anak Dalam* [The Rationality and Social Transformation of the Anak Dalam Tribe]. *Thesis.* Medan, Indonesia: University of Northen Sumatra.

Pradoko, A.M.S. (2017). *Paradigma-Paradigma Kualitatif untuk Pengkajian Seni, Humaniora dan Budaya* [Qualitative paradigms for art, humanities and culture research]. Yogyakarta, Indonesia: Charissa.

Sobur, A. (2001). *Analisis teks media: suatu pengantar untuk analisis wacana, analisis semiotik dan analisis framing* [An introduction to discourse analysis, semiotic analysis, and framing analysis]. Bandung, Indonesia: Remaja Rosdakarya.

Spradley, J.P. (1980). *Participant observation.* Illinois, United States: Waveland Press.

Stokes, J. (2006). *How to do media and cultural studies.* Yogyakarta, Indonesia: Bentang Pustaka.

Sudjiman, P. H. M., & Van Zoest, A. J. A. (1996). *Serba serbi semiotika* [Everything about Semiotics]. Jakarta, Indonesia: Gramedia Pustaka.

21st Century Innovation in Music Education – Astuti et al. (eds)
© 2020 Taylor & Francis Group, London, ISBN 978-0-367-11122-9

An analysis of *Rokan hilir* hymn song as a learning material for music subject in school

A. Husna, A.M.S. Pradoko, W.O Wiko & G.K.W. Anggraheni
Department of Art Education, Postgraduate Program, Universitas Negeri Yogyakarta, Yogyakarta, Daerah Istimewa Yogyakarta, Indonesia

ABSTRACT: The *Rokan Hilir* hymn song is a compulsory song known by members of Rokan Hilir society in Riau Province, Sumatra, Indonesia. This song has become an identity marker for Rokan Hilir society because the lyrics contain glorious hopes and stories about how the Rokan Hilir Regency was established. This research is a descriptive qualitative research. The analysis of the song in this research has presented several data results: *Rokan Hilir* hymns ong consists of two parts with a sentence order of A and B (A A' B A"). Sentence A serves as a the first theme and sentence B as the second theme, while A' and A" sentences are repetitions of the first theme with an insignificant change of tones.. The *Rokan Hilir* hymnsong generates emotional effects for the society through understanding, memorizing justifying and implementing every lyric in the *Rokan Hilir* hymnsong.

1 INTRODUCTION

Culture, a product of creativity and human intention, is a precious wealth that is owned and continually maintained until today. Every society has a culture that is different from other societies. A diversity of cultures in Indonesia shows that this nation is rich in culture. Many experts interpret that culture is most likely very influenced by a theoretical view that culture will develop from a simple to a more complex stage. Koentjaraningrat (2008) definest culture as, "the whole system of ideas, belongs to human beings, and obtained by learning" (p. 28). In the field of cultural anthropology, culture is the whole system of ideas and tastes, actions, and works produced by humans in social life. The culture that is produced belongs to members of a society and is learnt by society.

In its development, art has become part of a life principle that cannot be separated from the society. In Indonesia, art has been recognized as one of the fields that has had its own place, both traditional and modern or contemporary art. According to Bastomi (1992) art can be traced from the beginning of the word "art" itself (p. 10). In Indonesian language, art is interpreted as subtle/smooth works, inner skills, something extraordinary, the skill of making good quality works and the ability to create something valuable. The word art or 'seni' is also similar with the sanskrit word *sani* which means offering. In old Javanese language, the word *sanidya* means concentration of mind.

Bumi Nusantara is the archipelago that comprises thousands of islands and hundreds of ethnic groups with different cultural backgrounds. The many differences between groups adds to the richness of the country's culture especially folk songs. Folk song is one of the musical works that belongs to a certain region and is created based on the condition of an area, using its simplicity and regional characteristics. The Riau province is one of the island groups in this archipelago with various kinds of cultures living proximate to one another and that consist of music, dance, drama and visual arts.

Syafiq (2003) states that music is, "the art of expressing ideas through sound, whose basic elements are melody, rhythm, and harmony, with supporting elements in the form of ideas, characters, and colors of sound. In its performance, it is still combined with other elements, such as language, motion, or color" (p. 203). Djohan (2003) states that music is a "complex and universal social behavior" (p. 7). Every individual in society can "consume" music, listen, dance and develop/improve music. Thus, there is also an impression that even silence is a music in a certain capacity, to comprehend.

Mack (2001) states, "music has the most powerful elements in influencing mankind, so that music plays the most important role in the context of religion, politics, and social functions" (p. 7). It is said that music is art when the work is in the form of organised sound, for example playing music, songs (singing) and choirs. Songs are part of music art with lyrics that have meanings including the composer's meanings. Songs continue to develop over time in accordance with changing times. Songs live and develop in line with the development of human beings.

Songs are verbal discourses when viewed as based on their particular media, but is also included as written discourse via the text of the song. Song as discourse can be categorised as poetry discourses in terms of gender literature and that is re creative. As a universal and effective media, songs can express the ideas, messages and expressions of composers to listeners through lyrics, musical composition, selection of musical instruments and the way that a song is performed. The ideas in a song can be in the form of expressing love, protesting against something, feeling anger and anxiety and so on. All of the ideas in a song are packaged and strung together with beautiful and poetic words that are not always straightforward.

According to Banoe (2007) a song is, "basic melody; music art to be sung or performed with certain patterns and forms" (p. 233). There are several types of songs found in music. One of these song types is hymn song. Syafiq (2003) interpreted the hymn as "a song to praise the greatness of God" (p. 141). It is also a form of song to pray with, give a great impression or show gratitude through its lyrics.

The Rokan Hilir Regency is part of the Riau Province and its society still holds some traditional rules and customs in social life. The rules are in the form of laws from the local government that are implemented. It is proven that every school learns and studies the *Rokan Hilir* hymnsong. The *Rokan Hilir* hymnsong is sung at the annual anniversary (*HUT*) of the founding of the Rokan Hilir Regency along with other government sponsored events.

The *Rokan Hilir* hymnsong is documented in choir videos, other kinds of recordings and via notations. The *Rokan Hilir* hymnsong has currently been socialised into government schools and other institutions. In every government event, especially at the anniversary celebration of the Rokan Hilir Regency on 4 October, the *Rokan Hilir* hymnsong is always sung. Video and other recordings well as the hymngsong notation that has been standardized, have helped this study's researchers to conduct the research.

The composer of the *Rokan Hilir* hymnsong, H. Misran Rais, who resides in the Rokan Hilir Regency stated that,

> "The process of creating *Rokan Hilir* took approximately for 1 ½ months in 1999. The purpose of creating this song was to be made as the compulsory song of Rokan Hilir Regency and it was accepted at the Great Conference of the establishment of Rokan Hilir Regency. Rokan Hilir was created with a half-month period by conducting a direct survey of each sub-district in Rokan Hilir Regency. At that time, the sub-districts in Rokan Hilir were only Kubu District, Bangko District, Tanah Putih District, Rimba Melintang District, and Bagan Sinembah District. *Rokan Hilir* is a song about showing gratitude for the establishment of Rokan Hilir Regency. This song also invites all residents of Rokan Hilir to make the society prosperous. With the treasure from its oceans and land, Rokan Hilir society can be victorious. This meaning is found in the lyrics of *Rokan Hilir*. (personal communication, 24 September 2015)

The *Rokan Hilir* hymnsong was created by H. Misran Rais on 28 August 1928. H. Misran Rais is a renowned music artist in Rokan Hilir Regency. His artistic soul cannot be separated from the influence of his father. He not only created the *Rokan Hilir* hymnsong, H. Misran Rais also created many other songs including folk songs, hymns, marchs and other music. The songs that Rais created were "Ayam Putih Pungguk", "Mars of Rokan Hilir", "STTI Sri Gemilang Hymn", "Bagan Sinembah", "Intan Kocik" and many more. Because he successfully created hymnsongs and the mars of Rokan Hilir, he received an award from the Rokan Hilir Regent. Furthermore, Rais also received a gift from the BPP Gapensi Pusat Jakarta for the hymnsong and the Mars of *Gapensi*, and he has also received many other awards.

Before analysing the song, it is observed that a song generally consists of elements of music that comprise the song. Certain elements in a song can be observed specifically. Observations of the song forms were conducted by listening to the song repeatedly and by looking at the notation. Thus, the researchers were interested in conducting and analysing in detail, the forms of *Rokan Hilir* hymnsong. The factors that motivated the researchers to conduct this research are: the researchers care about the progress of songs in the Rokan Hilir Regency and the researchers were interested in the lyrics and poetry found in the song that express gratitude for the establishment of the Rokan Hilir Regency and its sub-districts. This expression of gratitude be seen in the lyrics, "puji syukur pada Ilahi terwujudnya cita-cita murni Negeri tercinta telah berjaya menjadi Kabupaten Rokan Hilir Kubu Bangko Tanah Putih Negeri asal di Rokan Hilir pendukung mitra untuk tercipta Negeri Rokan Hilir yang megah" which translated into English reads as, "Praise be to God for the realization of the pure ideals of the beloved country which has triumphed into the Regency of Rokan Hilir Kubu Bangko *Tanah Putih* Land, the home country of Rokan Hilir, the support system to create the magnificent Land of Rokan Hilir".

The researchers' analysis reveals that the composer of the *Rokan Hilir* hymnsong at that time he created the song, was willing to invite Rokan Hilir society to be always united and to live prosperously and happily in order to prosper the nation. The composer's intention is reflected in the song lyrics; "Marilah kita semua warga masyarakat Rokan Hilir satukan tekat membangun Negeri hidup makmur bahagia mari berbakti sepenuh jiwa mencapai kesejahtraan bangsa" or "Let us all the citizens of Rokan Hilir altogether build a prosperous country happily, let's serve our whole soul to achieve the prosperity of the nation".

The composer of the song, H. Misran Rais, is also proud of the Rokan Hilir Regency and its sub-district, Bagansiapiapi, that in the past was the number two producer of fish in the world. Rais inserted that high fish producing status in the *Rokan Hilir* hymnsong lyrics: "Rokan Hilir Bagansiapiapi, a city of fishing, a city full of memory with the plenitude treasure of the sea, living peacefully, forever, that's where I dedicate myself to build the glorious Rokan Hilir Regency".

In conducting this research, capacious insight and knowledge are necessary. In addition, references related to the song forms also help the researchers to analyse the song. This research was conducted because there has rarely been a study analysing these song forms. This research is the first to be conducted on the *Rokan Hilir* hymnsong. Thus the researchers describe, analyse and document the findings into the form of scientific writing with the object of the research entitled as, "The form of *Rokan Hilir* Hymnsong in Rokan Hilir Regency, Riau Province".

2 METHODOLOGY

This research used a qualitative method describing what is seen and heard using an analysis of the song form. This step provides an objective approach for studying the song forms through musical analysis theories.

The instrument of this research was the researchers themselves. The research data was collected by observing and analysing the phenomena occuring in the field. The target of this research was the *Rokan Hilir* hymnsong. The location was in the Rokan Hilir Regency, Riau Province, Indonesia. The interviewee was Mr. H. Misran Rais, the composer of the *Rokan Hilir* hymnsong. The data in this research was collected by using collection techniques such as interviewing, observing and documenting. The observation process was conducted by attending and observing events and other phenomena in the field. Documentation was conducted by analyzing the *Rokan Hilir* hymnsong score along with listening to and looking at existing videos and other recordings. The researchers also conducted an interview with H. Misran Rais, the composer of *Rokan Hilir* hymnsong, to support the data.

3 DISCUSSION

3.1 *The form of rokan hilir hymnsong*

A song form is a composer's idea organised through the composition of all musical elements such as melody, rhythm, harmony and dynamics. The composer's idea unites the parts of the composition that are sounded one by one comprising the main skeleton in a song before completion as a song. A particular song form can be used by a composer in the process of making the music.

A song form that has been arranged becomes a unified composition by creating what is called a sentence. A song lyric can be arranged by using various kinds of sentences. According to Prier (2004) song forms are classified into three types based on the number of sentences: (1) one part song form with one sentence only; and (2) two-parts song form with two different sentences, (3) three-part song form with three different sentences (p. 5).

According to Prier's theory that song forms are based on the number of sentences, a sentence in the *Rokan Hilir* hymnsong is shown as follows:

3.1.1 *Sentence A*

Sentence A in the *Rokan Hilir* hymnsong is in bars of 1 to 8 and is arranged based on several tones so that it becomes a song sentence where each sentence is one completed unit. The entirety of a song sentence can be seen at the end of the sentence, that there is a sign/ impression that the song has finished because the chord progression is included in one tonic chord (one chord). However, there are some songs that are modulated to other chords, for example to the dominant chord and that stops there. In addition, the closing tone of the sentence generally falls at its ending. The chord progression of sentence A is Dm Gm C Dm, Gm Dm Bes A. Chord A gives the impression of a comma at the end of the song so that the Sentence A is repeated in the next sentence but with a slight difference, represented with the symbol A'

Figure 1. Sentence A.

3.1.2 *Sentence A'*

Sentence A' is found in bars 9 to 16. Sentence A' is still the same as sentence A. The difference is the last tone is not so significant. The chord progression in sentence A' is the same as in sentence A. What makes sentence A' different is that D makes the last tone in this section thus becoming a song sentence.

Figure 2. Sentence A'.

3.1.3 *Sentence B*

Sentence B is in bars 17 to 24. Sentence B is a continuation of the sentence A and sentence A'. In sentence B there are repetitions in bars 17 to 20. The chord progression in sentence B is Dm C Gm Dm Gm Dm Bes A.

Figure 3. Sentence B.

3.1.4 *Sentence A"*

Sentence A" is in bars 25 to 34. In sentence A" there is a repetition in bars 29 to 32. Sentence A" is a repetition of sentence A and sentence A' with a change that is not really significant.

Based on the previous analysis of the sentences in the compostion, the *Rokan Hilir* hymnsong is categorized as a two-part song form. The two parts are A and B with a sentence order of A A' B A" meaning that sentence A' is the same as sentence A with the only difference being that the tone is placed at the end of the song. Then when the song moves to sentence B, this is repeated again in sentence A which becomes sentence A". In the *Rokan Hilir* hymnsong, there are several changes that occur with an insignificant tone.

Figure 4. Sentence A".

3.2 *Phrases*

In a song, a sentence consists of two phrases: a question phrase which is in the beginning of the sentence and the answer phrase which is in the end part of the sentence. The question and answer phrases soccur because the chord progression and the tone stops with a floating tone or chord, and the tone or chord stops with a stop. The question phrase usually stops with the dominant chord. The impression for the dominant chord has not been completed and continues as the answer phrase to answer the question phrase and that then stops with a tonic chord or stop.

Based on the chord progressions, the phrases in the *Rokan Hilir* hymnsong are explained as follows:

Bars 1 to 8 is a question phrase. The last stop of the chord is A or called the dominant chord. Thus, it is said to stop with a comma.

Figure 5. Bars 1 to 8.

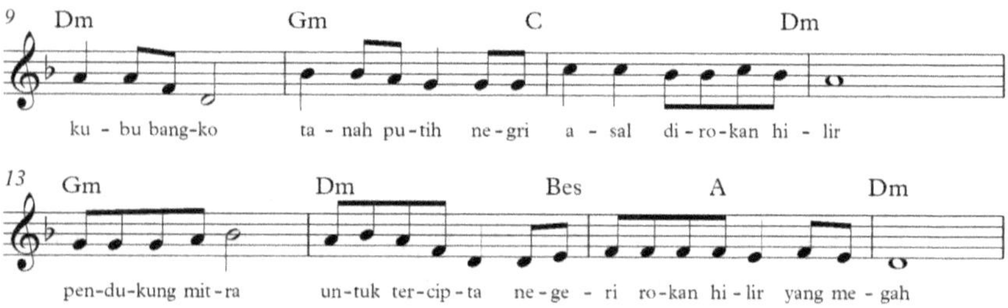

Figure 6. Bars 9 to 16.

Bars 9 to 16 is an answer phrase continuing the question phrase in the previous bars. The answer phrase ends with a tonic chord that is a Dm chord.

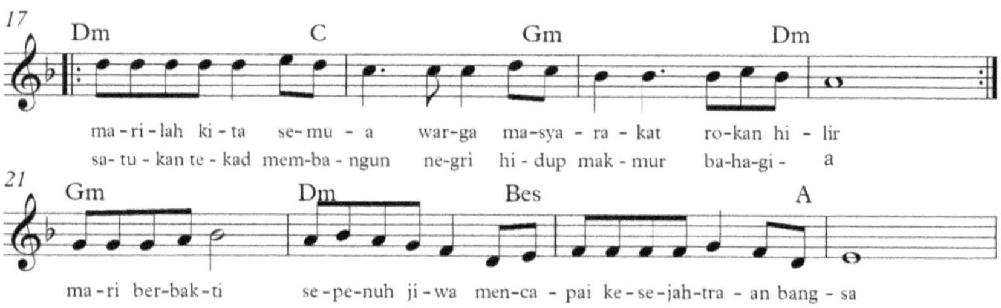

Figure 7. Bars 17 to 24.

The next bars17 to 24 is a question phrase because it stops at the dominant chord, the A chord.

The answer phrase is bars 25 to 34. The answer phrase at the end of the *Rokan Hilir* hymnsong stops at the Dm chord—a tonic chord or one chord.

3.3 *Coda*

In a song composition, there are prefixes and suffixes. The beginning is an opening of a composition and the end is a closing, indicating that a composition work is completed. The ending or concluding passage called coda, is a short addition at the end of a song that concludes the song or concludes a larger composition work (Prier, 2004, p. 23)

In the *Rokan Hilir* hymnsong, the coda is placed in bars 33 and 34.

Figure 8. Bars 25 to 34.

Figure 9. Bars 33 and 34.

3.4 *Motif*

The smallest element in music is tone, but one single tone itself is not yet a musical composition. Thus, tone is arranged into one tone unit with other tones so that it becomes a motif. Motif is a part of song consisting of a group of tones that are one unit and have meaning in each tone and motif itself (Prier, 2004, p. 26). Motif can be seen and heard in melody, rhythm, harmony, dynamics and color of sound, so that motif can be observed in all elements of music. A motif consists of at least two tones and two bars at the most.

There are several forms of motif in the *Rokan Hilir* hymnsong illustrated in the following image:

Figure 10. *Rokan Hilir* hymnsong motif m.

The bar motif at the beginning of the *Rokan Hilir* hymnsong with the symbol m functions as the first motif. This motif consists of 3 tones namely A, F and D.

Figure 11. *Rokan Hilir* hymnsong motif n.

Then the second motif is placed in the second bar. The tones in this motif are B and G with the symbol n.

Figure 12. *Rokan Hilir* hymnsong motif m1.

The motif in the tenth bar is the same as the first motif in the first bar. The number of fourth notes is subtracted from the number of the first motif tones that are from a half note to quarter note and one-eighth that is called a diminuation of the value. The number of the first, second, and third tone values is the same as the first motif. This motif is the development of the first motif with the notes B, A and G.

Figure 13. *Rokan Hilir* hymnsong motif o.

The motif with the symbol o is the third motif in the *Rokan Hilir* hymnsong. This motif consists of notes C and B and is found in bar 11.

Figure 14. *Rokan Hilir* hymnsong motif n1.

In bar 18, there is a development of a motif from the second motif given the symbol n1. This motif includes augmentation of the value. A melody with the rhythm motif process is changed. The value of the first and second tones are doubled to 1½ while the tempo is slow as the beat remains the same.

Figure 15. *Rokan Hilir* hymnsong motif p.

The third motif with the symbol p is found in bar 19. This motif consists of notes B and C.

Figure 16. *Rokan Hilir* hymnsong motif m2.

The symbol m2 is a melodic development from the first motif. This motif includes the substraction of the notes from half note to one-eighth note and one-eighth notes. The changes of the value of the notes are found in the third and fourth tones namely D and A.

Based on the motif analysis of the *Rokan Hilir* hymnsong, there are three motif types with the development of motif m1, m2 and n1. The motifs in the bars that are not mentioned, are repetitions of motifs in motif one and motif two.

3.5 *Learning the Rokan Hilir hymnsong in school*

Learning is a process made possible by an teacher working with students to educate students in a field of study/subject. In a formal education setting such as schools, learning is a responsibility assigned to teachers. As an educator, a teacher acts as a driver of learning activities for the mental and emotional processes of students in achieving progress.

According to Deni Darmawan & Permasih (2011), learning is a process that basically involves an essential effort in forming human personality with various links in his life. Philosophically, learning means reminding people of the process of imitating, understanding, observing, feeling and believing certain truths so that everything can be achieved easily and properly (Deni Darmawan & Permasih, 2011, p. 130). Learning is an accumulation of teaching and learning concepts. The emphasis is on growing via subject activities delivered by the educator. This concept is observed as a learning system comprised of students, educators, materials, objectives, facilities, procedures and various kinds of media that must be prepared.

The *Rokan Hilir* hymnsong is a compulsory song for the society of Rokan Hilir Regency. This song has become learning material that is mandatorily assigned in every school in the Rokan Hilir Regency. The process for leaning the *Rokan Hilir* hymnsong is included in the school subject called, Local Music Arts. The *Rokan Hilir* hymnsong is also studied outside formal learning classrooms, namely extracurricular activities such as choirs, music ensembles and drumband/marching bands.

Based on the work of Darmawan and Permasih (2011), the philosophy underlying learning the *Rokan Hilir* hymnsong is a process of remembering, imitating, understanding, observing, feeling and believing in a truth that is in the form of the *Rokan Hilir* hymnsong. The analysis of the *Rokan Hilir* hymnsong form reveals that are several parts: song sentences, theme, phrases, climax, coda, melody and rhythm. Based on the analysis of the song form, the teacher can deliver the *Rokan Hilir* hymnsong subject by using or implementing the results of this research.

The process of learning the *Rokan Hilir* hymnsong begins with the introduction of the song to students. The introduction includes the song's history, recordings or videos that are played for students, stating aims and benefits to help students understand the song. After completing the introduction stage, the teacher then teaches the *Rokan Hilir* hymnsong to the students in the next stage. The teacher guides the students to sing each sentence of the song followed by the correct tempo and phrasing, with questions and answers about the accuracy of each song sentence and its suitability according to the results of this study's analysis. The climax of the song is taught with the correct/proper technique in order to bring out emotional expressions while performing the song. After students are able to sing the *Rokan Hilir* hymnsong with the proper tempo and phrasing, in the next stage the teacher explains the meaning of lyrics. In addition, song lyrics function to interpret benefits of learning the *Rokan Hilir* hymnsong.

By learning the *Rokan Hilir* hymnsong, it is anticipated that the young generation will be more appreciative, caring, loving and proud of the land they live in. Thus, the sense of nationalism in developing their land will flow through the blood in their body. Thus, the history of the establishment and the glory of the Rokan Hilir Regency will always be remembered, sustained and held by members of the local society.

4 CONCLUSION

The *Rokan Hilir* hymnsong is a local song of the Rokan Hilir Regency created in 1999. The song was initially created to be performed at the Great Deliberation (*MUBES*) of the establishment of the *Rokan Hilir* Regency and was expected to be accepted by members of the local society.

The song form of the *Rokan Hilir* hymnsong is comprised of two parts, A and B, with the sentence order of A A' B A": Sentence A is repeated as sentence A' then immediately arrives as sentence B, and is repeated again as sentence A" from sentence A and sentence A'. Sentence A is the first theme and sentence B is the second theme of the song. Meanwhile, A' and A" are repetitions of sentence A with an insignificant difference of tones.

The *Rokan Hilir* hymnsong consists of three motifs. The *Rokan Hilir* hymnsong tones are made with full notes, half notes, quarter notes, one-eighth notes and one-sixteenth notes. The *Rokan Hilir* hymnsong's distance of the melody is not very large, only using 1 and $1^{1/2}$ distances. The distance between the tones does not jump. The intervals that are used are only prime, seconde and third.

The *Rokan Hilir* hymnsong was assigned to be taught in the local music arts subject and in extracurricular activities such as choir, ensemble music and drumbands/marching band. The learning process applies the results of this analysis and practical learning models in this research. The purpose of learning the *Rokan Hilir* Hymnsong is to know, interpret, understand, comprehend and justify the history of the Rokan Hilir Regency and encourages people who are willing to do so, to become and remain prosperous.

REFERENCES

Banoe, Pono. (2007). *Kamus Musik*. Yogyakarta: KANISIUS.

Bastomi, Suwaji. (1992). *Wawasan Seni. Semarang*: IKIP Semarang Press.

Badan Perencanaan Pembangunan. (2015). *Profil Kabupaten Rokan Hilir:* Pemerintah Kabupaten Rokan Hilir.

Darmawan, D. Permasih. (2011). *Kurikulum dan Pembelajaran.*

Darmawi, G.P Ade. (2001). *Anugrah seni Dewan Kesenian Riau 2001*. Pekanbaru: Riau Pos Media Group.

Departemen Pendidikan Nasional. (2008). *Kamus Besar Bahasa Indonesia*. Jakarta: PT. Gramedia.

Djohan. (2003). *Psikologi Musik*. Yogyakarta: Buku Baik.

Prier, SJ, Karl, Edmind (2004). *Ilmu Bentuk Musik*. Yogyakarta: Pusat Musik Liturgi.

Koentjaraningrat. (2008) *Pengantar Antropologi I: Rineka Cipta.*

Mack, Dieter. (1995). *Ilmu Melodi*. Yogyakarta: Pusat Musik Liturgi.

Sumarsan. (2003). *Interaksi Budaya dan perkembangan Musik di Jawa*. Yogyakarta: Pustaka Pelajar.

Syafiq, Muhammad. (2003. *Ensiklopedia Musik klasik*. Yogyakarta: Adicita Karya Nusa.

21st Century Innovation in Music Education – Astuti et al. (eds)
© 2020 Taylor & Francis Group, London, ISBN 978-0-367-11122-9

The relationship between folk song and people's moral education

A. Yunanto & H. Pamadhi
Universitas Negeri Yogyakarta, Yogyakarta, Indonesia

ABSTRACT: The aim of the study is to describe the relationship between local songs and moral values or moral education. This research is a descriptive qualitative investigation that reveals the interpretation and meaning of songs as a whole. The researcher choose five songs that represented moral values from various region of part of Indonesia. At least five values cultivate by the songs, 1) in addressing a life problem should consider more spiritually and other social aspects; 2) knowing the message from the Guardians that one day they will die and will meet the Creator to account for all deeds 3) advise to look for the treasure of true happiness, do not people obey their own desires or greed.; 4) remind human should not be greedy And 5) to know willingly and sincerely the soul of his body sacrifices in order to defend the country.

1 INTRODUCTION

According to Asamani (2009), in children's education, the use of music is an important part of educational material. Music includes playing musical instruments and singing songs. Music and songs in children's education have two fundamental roles. Firstly, they are used to support the optimization of children's physical, cognitive, and psychomotor development. Secondly, music and songs are used in children's education to stimulate the development of children's musical intelligence (Asamani, 2009: 153, 159, 163-168).

This view is consistent with the fact that children's songs are commonly used as part of children's learning process, in formal, non-formal and informal education. For children's development, music has two main functions – as a means of nemesis and catharsis (soul purification). Music as a media nemesis is shown in 16th century opera performances, where music can be an emotional reflection. The function of music as a means of catharsis is believed to be a medium for expressing oneself.

Music contributes form, regulate, and delivers emotions, and the media train inspiration. In a cultural context, music has more complex functions. Music is present in every aspect of people's lives, such as marriage, death, birth, religious ceremonies, rituals, traditional activities, and dances (Djohan, 2009: 87-88). Regional songs are songs that have distinctive characteristics between one region and another. The difference characteristic lies in the pitch. To find out the origin of a regional song there are several ways, such as whether the language used is the local language, the use of local musical instruments and performing songs *cengkok* according to the local area (Rochani, 2012: 13).

Morality is something that manifests in child behaviour. There are some moral values that we know in our daily lives that relate to ourselves such as honesty, work ethic, firmness, patience, resilience, cheerfulness,, openness, visionary, independence, strength, courage, reflection, responsible answer and discipline. There are other values that relate to other people or creatures such as helpfulness, tolerance, smiling, generosity, cooperation, communication, amar ma'ruf (call for kindness), nahi munkar (preventing bad deeds), caring (human, nature) and fairness. Lastly there are the values associated with divinities that are sincerity, sincerity, faith and piety. Moral teaching is advice and mandate regarding whether or not human attitudes in community life (Nurgiyantoro, 2012: 320).

Furthermore, teaching is useful for humans who live in a society built on good behaviour and attitude (Hadiatmaja and Endah, 2008: 56). The moral word always refers to the good and bad of human behaviour. Morality is a benchmark for determining the fault of human behaviour and attitudes in terms of bad or good. Moral values are based on attitudes, behaviorus that can be seen through actions. Actions that are perceived as commendable and good outwardly will be judged to have good moral values. (Suseno, 1993: 19).

A song is a series of words made with tones and accompanied by musical instruments. Song is very influential on a child's behaviour. With songs, children can learn more by listening to them. Songs can be used as a good medium to educate children to behave well. With songs that are Islamic in nature, children can be influenced by the lyrics of the song and lead to the right path according to the rules of religion.

According to Endraswara in Kusumawati, children's songs are songs that are cheerful and reflect noble ethics (2013: 4). Children's songs are usually sung by children. They contain simple things that are usually conducted by children. Children's song poetry usually tells the story of love for others, God, father-mother, brothers and sisters, the beauty of nature, and the greatness of God. They are written in simple language in accordance with the natural thinking of children. One part of how to examine moral teachings can be seen from the types of moral teachings. The types of moral teachings can be included in events that are free. These types of moral teachings can enter into life and life events, where things have a relationship with human dignity and worth (Nurgiyantoro, 2007: 323).

In cultural terms folk songs can teach children to be disciplined, and maintain harmony with nature, fellow human beings and parents. Hidajat (2008) states that music has three functions in choreography; 1) music as a motion accompaniment; 2) music as an affirmation of motion; 3) music as an illustration. Teaching songs is an antidote to modernization, which generally keeps children from having good morals (Triyono, 2000: 1-12).

From the explanation above, it can be identified that the types of moral teachings can be divided into four categories: moral teachings of human relations with God, moral teachings of human relations with humans, moral teachings of human relations with themselves, and moral teachings of human relations with nature.

2 METHOD

Descriptive qualitative methods are used to explain the objects of study. The analysis used in this research is based on the study of relevant theories. Qualitative research data is collected using observation, interviews, documentation, and combinations (triangulation) techniques. If the researcher uses a questionnaire or interview in collecting data, then the data source is called the respondent, which is the person who responds or answers the researcher's questions, both written and oral questions (Arikunto, 2013: 172). Documentation studies are also part of the research technique for viewing archipelago song documents.

3 DISCUSSION

In this section will analyse the five folk songs including *Gundul Pacul, Cublak Suweng, Lir ilir, Tokecang* and *Yamko Rambe Yamko*. These songs are very popular in folk music for children. This is because they are often sung by an elementary school teacher as a main or supporting material in learning.

3.1 *Gundul Pacul*

Figure 1. Beams Note Gundul Pacul.
Source: Lagu Nusantara, 2017.

Gundul-gundul Pacul cul gembelengan
Nyunggi-nyunggi wakul kul, Petentengan

wakul gelimpang segane dadi sak latar
wakul gelimpang segane dadi sak latar

Gundul are people who put things on their bald heads. This lyric means that if one has knowledge based only in the mind, usually such people are arrogant and haughty. This is because the knowledge of Shari'a is not based on knowledge of the heart/morality is hard-hearted, and such people fall into the category of laymen. People with this trait usually lack gentleness (without the science of essence) and they find it easy to treat others without compassion, and to them it feels justified.

Enjoy the temple's end time. And people like that/bald-headed hoe, if they get (*nyunggi*) knowledge or gift (*wakul*), then they will be arrogant (*petentengan*). They are the people of the bald bark, if they reveal knowledge or something (*wakul gelimpang*) then knowledge (*segane*) they tend to exaggerate it (*sak latar*).

This *Gundul Gundul Pacul* song is written or created with a purpose that is related to Islamic preaching. This song tells us that if you acquire (*nyunggi*) knowledge or gift (*wakul*) then you will become arrogant (*petentengan*). People who lack caring traits will easily justify mistreating others out of self-righteousness. (without the science of essence) The message we get from this song is to be humble and that we should consider responding to life's problems more spiritually or mentally and with other social aspects.

3.2 *Cublak Suweng (Song by Sunan Giri)*

Cublak-cublaksuweng, suwenge teng gelenter,
mambu ketundhung gudel, pak empo lera-lere,

sopo ngguyu ndhelikake, Sir-sir pong dele kopong,
Sir-sir pong dele kopong, sir-sir pong dele kopong.

Cublak Suweng means Suweng's box. *Suweng* is Javanese women's earrings. *Cublak-cublak suweng*, means there is a box of valuable treasure, called *Suweng* (empty, lonely, true) or True Treasure. *Suwenge teng gelenter, suwenge teng gelenter* refers to scattered jewelry. The true treasure in the form of true happiness is already scattered around humans. *Mambu ketundhung gudel, mambu* (smell) *ketundhung* (intended) *gudel* (buffalo child). This means many people try to find that true treasure.

Figure 2. Beams Note Cublak Suweng.
Source: Lagu Nusantara, 2017.

Even fools (likened to *Gudel*) look for the treasure with lust of ego, corruption and greed in their pursuit of true happiness. *Pak Empo Lera-lere, Pak Empo* (toothless father) *Lera-lere* (looking left and right). These words refer to fools, likened to toothless, confused old people.

Even though his wealth is abundant, it turns out that it is a false treasure, not a true treasure or true happiness. They are confused because they are mastered by the lust of their own greed. *Sopo ngguyu ndhelikake, sopo ngguyu* (who laughs), *ndhelikake* (he/she who hides). This shows that whoever is wise will find the Place of True Treasure or true happiness. He/she is a person who smiles all the way through life, even around greedy people. *Sir-sir pong dele kopong, sir* (conscience) *pong dele kopong* (empty soybeans without content). This means an empty conscience. It means that to arrive at the meeting of the True Treasure (*Cublak Suweng*) or true happiness, people must escape from their attachment to worldly possessions, empty themselves, smile happily, be humble, do not demean others, and always use their taste and sharpen his/her sir or conscience.

Cublak Suweng's song tells of a person who is confused because he/she is overpowered by the lust of his own greed. This song conveys the meaning of the true treasure in the form of true happiness which is already scattered around humans. The message of this song is that when looking for the treasure of true happiness, do not let men submit to their own desires or greed, but all should use their conscience, so that the wealth of happiness can overflow and become a blessing for anyone.

3.3 *Lir Ilir*

Song by Sunan Kalijaga

Figure 3. Beams Note Lir ilir.
Source: Lagu Nusantara, 2017.

Lir ilir lir ilir tandure wong sumilir
Tak ijo royo royo
Tak sengguh penganten anyar
Bocah angon bocah angon penekno blimbing kuwi

Lunyu lunyu penekno kanggo mbasuh dodotiro
Dodotiro dodotiro kumintir bedah ing pinggir

Dondomono jrumatono kanggo seba mengko sore
Mumpung padang rembulane

Mumpung jembar kalangane
Yo surak'o surak hiyo

As Muslims we are asked to wake up (become conscious). Wake up from adversity, and wake up from laziness to further strengthen our faith that has been given by Allah. In ourselves, this is symbolized by plants that begin to blossom and turn green. It's up to us, whether to remain asleep and let our plants of faith die or wake up and struggle to grow the plants to become larger and find happiness like a happy new bride. The song refers to a shepherd child (*cah angon*), rather than a king, governor, general or president,*Angon* means that someone is able to persuade the congregation, a person who is able to "graze" his congregation in the right way due to God, and that we have also been given something to graze, which is the heart. Can we graze our hearts away from such strong passions of lust?

Shepherd children are asked to climb starfruit trees (the green color symbolizes Islamic characteristics). Starfruit here symbolizes the five pillars of Islam due to its five segments. Even though it is slippery and difficult we have to keep climbing the starfruit tree, and thusadhere to the pillars of Islam, whatever the obstacles and risks. The point is to wash our clothes that are piety clothing.

Ordinary human beings have a gap of sin. We are asked to always improve and address our potential for sin so that we will be ready when called forth to face the presence of Allah SWT. We are expected to do so when we are still healthy (symbolized by the brightness of the moon) and still have plenty of free time. and if there is one remind then answer with ‘yes’. This song tells us that as Muslims we are asked to wake up. We must build from adversity, wake up from being lazy in order to be more committed to God. The shepherd boy is featured because by Allah, we have been given something to graze, which is the heart. The message conveyed in the lyrics is that as an ordinary person who makes mistakes, we are asked to always improve and fix them so that someday we are ready when called to meet the presence of Allah SWT.

3.4 *Tokecang*

Song by R.CHardjosubroto

Figure 4. Beams Note Tokecang.
Source: Nusantara, 2017.

Tokecang tokecang bala gendir tosblong
Angeun kacang sapependil kosong

Aya listrik di masigit meuni caang katingalna
Aya istri jangkung alit karangan dina pipina

Tokecang tokecang bala gendir tosblong
Angeun kacang angeun kacang sapependil kosong

The meaning of Song Song Lyrics

Tokecang, a bunch of loose slings
Vegetable, peanut vegetable, one empty pot

There is electricity biting until it seems glare
Got a tall, thin wife with a mole on her cheek

Tokecang, a bunch of loose slings
Vegetable, peanut vegetable, one empty pot.

This song explores a belief in Sundanese societies that geckos eat beans, which has a double meaning referring to the greed of humanity. The song teaches us not to eat all the beans, or not to be greedy, because human beings are social beings.

3.5 *Yamko Rambe Yamko*

Figure 5. Beams Note Yamko Rambe Yamko.
Source: Nusantara, 2017.

Hee yamko rambe yamko aronawa kombe.
Hee yamko rambe yamko aronawa kombe.

Temino kibe kubano ko bombe ko.
Yuma no bungo awe ade.
Temino kibe kubano ko bombe ko.
Yuma no bungo awe ade.

Hongke hongke, hongke riro.
Hongke jombe, jombe riro.

Hongke hongke, hongke riro.
Hongke jombe, jombe riro.

The meaning of the song Yamke Rambe Yamko

Meaning:
Hi the way you are looking for dear agreement. Hi the way you are looking for dear agreement. Really homicide in the country. Really homicide - as the nation's interest. Nation flower, nation flower, nation flower - flower scattered - in the hero garden. Nation's interest, nation's flower,

nation's flower - flower grows - in the hero garden. Nation flower, nation flower, flower scattered. National interest, national interest, flowers grow.

The song *Yamko Rambe Yamko* is a theme song about war. Even though the tempo of the song is fast and seems cheerful, this song is a sad song because it tells the story about the area experiencing a war between tribes. The message implied by of this song is the spirit of struggle to become the flower of the nation — the willing hero and sincere soul who will make a sacrifice to defend the country.

4 CONCLUSION

The songs chosen by the researcher are usually folk songs sung by children. The lyrics contain moral values where people should do good to every human being, to always begrateful, humble, and to always worship the Most high power so that when children become adults, their knowledge of religion is stronger. In the first song *Gundul Pacul* teaches us to take a lesson that is, in addressing a life problem we should consider more spiritually or mentally and other social aspects. Second song *Lir ilir* knowing the message from the Guardians that one day they will die and will meet the Creator to account for all deeds. *Cublak Suweng's* third song advises to look for the treasure of true happiness, do not people obey their own desires or greed, but all return to the conscience. The fourth song *Tokecang* reminds us as humans we should not be greedy or eat it all because literally humans are social beings. The fifth song *Yamko Rambe Yamko* to know willingly and sincerely the soul of his body sacrifices in order to defend the country.

REFERENCES

Asmani, J. M. (2009). *Manajemen Strategis Pendidikan Anak Usia Dini*[Strategic management of early childhood]. Yogyakarta: Diva Press. 153, 159, 163–168.

Endraswara, S. (2013). *Metodologi Penelitian Sastra*[Literature research method]. Yogyakarta: CAPS. 4.

Hadiatmaja, l. E. (2008). *Filsafat Jawa, Yogyakarta*[Javanese philosophy of Yogyakarta]. Kanwa Publisher. 56.

Nurgiyantoro, B. (2007). *Teori Pengkajian Fiksi*[Theory of fiction study]. Yogyakarta, Gajah Mada University Press. 323.

Rochani, S. (2012). *Lagu Daerah*[folk song]. Jakarta: Balai Pustaka.13.

Sugiyono. (2014). *Memahami Penelitian Kualitatif*[Understanding qualitative research]. Bandung: Alfabeta. 62–63.

Suseno, F. M. (1993). *Etika Dasar Masalah-masalah Pokok Filsafat Moral*[Basic ethics of main problems in philosophy]. Yogyakarta: Kanisius. 19.

Triyono, B. (2000). *Lagu Dolanan Anak*[Children dolanan song]. Yogyakarta, Tarawang Press.12.

21st Century Innovation in Music Education – Astuti et al. (eds)
© 2020 Taylor & Francis Group, London, ISBN 978-0-367-11122-9

Kuadai dance music and symbolic meaning as a medium for early childhood learning

Hartono & R. lanjari
Semarang State University, Semarang, Indonesia

A.H. Saputri
Lampung University, Lampung, Indonesia

B. Susetyo
Semarang State University, Semarang, Indonesia

ABSTRACT: *Kuadai* dance represents the collective expression of the role of women as a source of life in the Semende community. This paper examines the symbolic meanings in *Kuadai* performances and their musical accompaniment as a medium for early childhood learning. The study adopted an interdisciplinary approach and employed qualitative methodology to collect observational, interview, and documentary data. The results showed that *Kuadai* dance embodies values of gentleness, patience and circumspection that can be effectively instilled in young children.

1 INTRODUCTION

Kuadai dance is part of the cultural repertoire of the Semende community (*Jeme* Semende) of Lampung Province, Sumatra, Indonesia. *Kuadai* is a traditional dance representing the daily activities of Semende women and involves the use of plates held by the dancers. It is performed as a form of respect to important guests. This dance is intended to portray the inner strength in the body of female dancers. As well as a vehicle of communication, it is also a vehicle for the collective expression of the Semende community, especially that of women, who are seen as a source of life and fertility (Karnate, pers. com., 2018). According to Hartono (2016: 36), dance essentially functions as a medium of self-expression that derives from ideas, experiences, surroundings and feelings. Various aspects of *Kuadai* dance performances, such as movements, props, accompaniment and floor patterns, symbolize the rhythm of life of Semende women. The aim of this study was to analyse the symbolic meanings embedded in the performance of *Kuadai* dance and its musical accompaniment in the Semende Lampung community as a learning medium for early childhood.

2 LITERATURE REVIEW

Kuadai dance is one of various symbolic forms created by the Semende community. It can be interpreted as a cultural product that is closely related to the social life of the community. Dance is a representation of the beliefs and practices of daily life (Indrayuda, 2015). *Kuadai* dance, it has been suggested, can be seen to fulfill the basic human need for symbolic expression (Pujiyanti, 2013). *Kuadai* dance is a sociocultural expression of the traditions of an agrarian society. It is performed to the accompaniment of slow tempo music and singing. The music contains symbolic values that children learn from an early age (Suryana, pers. com.).

According to Blanariu (2013), the cultural system of conventions in dance can be interpreted by encoding choreographic signs. These codes represent "natural" spontaneity of repeated movements and feelings in the expression. Similarly, Kusumastuti (2009) proposed that the form and content of symbols contain meaning. Art develops as a result of the expression and creativity of the people who own it (Ardin et al., 2017). Symbols can be viewed as signs that reflect values and meanings (Langer (2006). The meanings of symbols are established through social agreement or historical tradition (Danesi (2010). The socially symbolic elements of *Kuadai* dance are expressed through the aesthetics of the female body.

2.1 *Kuadai* dance

Dance can be seen as a repertoire or presentation of performances that result from creative processes as a means of communicating experience (Jazuli, 2016). *Kuadai* is a solo dance performed by a Semende woman as an expression of respect and gratitude to the community's king or leader. It is characterised by two forms of movement: *sembah adat* (customary worship), which includes a variety of traditional actions of worship (*sejurai, bai tangan,* and *jain-jit*); and movements of the hand that supports a plate with the thumb. The latter is the hallmark of *Kuadai* dance, and can be classified as a gesture that conveys meaningful symbols (Arisyanto et al, 2017). The dance involves the use of 14 plates and lasts seven minutes. It is accompanied by tambourines and poetry chanted in the Semende language. The *Kuadai* costume is typical of the clothing worn by Semende women and is characteristic of ethnic Palembang Malay culture. Currently, *Kuadai* is performed as entertainment at welcoming events.

2.2 *Symbolic meaning of Kuadai*

Art in its various forms can be used to develop character in early childhood. Wahyu (2013), for instance, has shown how *Jarog* dance is used in this way to increase children's self-confidence. Learning to dance can help to develop several different forms of intelligence, including kinesthetic, linguistic, intrapersonal, natural, interpersonal, and spatial intelligence. *Kuadai* dance performances present symbolic meanings (Kusumastuti, 2009) related to the life of the Semende community that can be instilled in children from an early age. The symbolic aspects of *Kuadai* dance are apparent in various elements, including movement, props, accompaniment, stage performance, composition, floor patterns, makeup and clothing (Hadi, 2005). The objects in the performance represent specific aspects of the everyday life of the Semende community (Putri et al., 2015).

The traditional life of this farming community revolves around waiting for *tubang* (pinned). The reliance on agriculture for its livelihood is reflected in the emphasis on fertility symbols, which takes the form of female figures. In Semende culture, the plate represents women, who must be treated gently, carefully, lovingly and reverently. This form of dance imitates the tradition of waiting for *tubang* that is visualized through the strength, patience and familial devotion of Semende women.

Figure 1. *Kuadai* performance.
(Source: Amelia 2018)

Semende society has creatively combined these two concepts into a performance art that symbolizes respect, which is a condition for beauty. As Ardin et al. (2017) have observed, art grows and develops as a result of the expression and creativity of the people who own it. The matrilineal culture of Semende society influences the attitudes and social behavior of Semende women. *Kuadai* dance in this mountain community emphasises both content and context. This is reflected in its static movements, emotional stability and care and its focus on the feelings and strength that come from within the dancer's heart.Semende society is simultaneously human and symbolic. People are free to act, think and make their own decisions.

A cultural system contains four symbolic devices, each of which has its own function for the human being who interacts with them: constitutive symbols, which are manifested as beliefs and are usually the core of religion; cognitive symbols, which represent scientific thinking; symbols of moral judgment in the form of values and rules; and expressive symbols (Fretisari, 2016). The expression of these symbolic devices in *Kuadai* dance will be discussed in relation to the performance elements of motion, props, floor patterns..

2.3 *Symbolic meaning of Kuadai dance movements*

The movements in *Kuadai* dance have meanings that are understood and embodied in Semende community life. *Kuadai* dance comprises two main types of movements—those that convey meaning, often called meaningful motion, and those that only emphasize aesthetic values, or so-called pure motion (Tyas, 2018). This analysis focuses on those movements that contain symbols and meanings. The structure of the *Kuadai* dance movement pattern has two parts: *sembah adat* (customary prayer) and *diatakak*. In the customary prayer section, the dancer begins in a kneeling position, moves with both knees and continues with a Balinese-style motion. On the part of the crepe begins with the movement of a triangle, then the dancer moves her hand while slowly rising to stand and performing a *jainjit* motion.

2.4 *Sembah adat movement*

The traditional prayer movement is performed in a sitting position with the palms held together in front of the chest and fingertips facing up. The seated posture symbolizes respect and humility. The two hands placed together symbolize unity of heart and mind. The downward gaze is a symbol of obedience and submission. Overall, the movement conveys two meanings: first, it is a sign of respect for the king, the traditional leader, older people and the audience; second, it means that women in Semende society have responsibility for protecting the family, supporting the family unit and resolving differences among family members.

2.5 *Bale'k*

Bale'k is performed in a kneeling position with both palms supporting a plate. The various movements convey two key symbolic meanings. First, it symbolises heaven and earth, which must remain in balance and still perform as one unit. This means that Semende society must have harmony and balance between inner feelings and outward behaviour. Second, it symbolises the identity of the Semende people who strongly adhere to the customs and culture of their Malay tribe. Semende people are known as tribes of nomads, but they still maintain and use

2.6 *Sejurai movement*

Sejurai involves the movement in which the two thumbs support the plates with the hands held in front of the chest, then the plates are brought together to generate a teng sound. *Sejurai* is a symbol of combining the right and left sides into one unit. It means that the Semende community traditionally has a single fulcrum for good and evil. It's was known as the *Bemeraje Anak Belai* custom. *Bemeraje Anak Belai* was the most important institution in

Semende social structure, focusing on family support for *Tunggu Tubang* and *Meraje*. Its deeper meaning is that all family members, wherever they are, will return to the home of the *Tunggu Tubang* and *Merajelah.* which is used as the support of the extended family for all the decisions and good things that have happened.

2.7 *Bai tangan movement*

Bai tangan is the movement of the two thumbs supporting the plates with the hand beside the shoulder. The hand movement represents the female figure as a symbol of strength and fertility (*Tunggu Tubang*). The symbol of strength is interpreted as a female figure who is responsible for feeding her family, caring for their physical needs and maintaining their property. Women are identified as the source of life and fertility that gives birth to the next generation. The role of *Tunggu Tubang* is to pass on Semende cultural values to the next generation and to teach the roles they are expected to perform.

2.8 *Jainjit movement*

The *jainjit* movement involves stepping on tiptoe between one plate and another. This movement symbolizes the nomadic nature of Semende society. Historically, the Semende community farmed an area of agricultural land as long as it was fertile, usually about 10 years, then moved on to another area (Interview with Suandi, 2018). Hence in Semende society the *Jainjit* movement means life balance and good decision-making. This custom of migration required *jeme* Semende to be able to be environmentally and socially adaptable, maintaining a balance between rights and obligations in life in order to keep the society grounded in goodness and truth.

3 METHOD

The study was conducted in Sekipi Village, Abung Tinggi District, North Lampung Regency. An interdisciplinary qualitative methodology was employed. Observational, interview, and documentary data were collected. Observations were made of *Kuadai* dance performances, including the role of the dancer and teacher, and of the daily activities of the Semende community. Interviews were conducted with performers, instructors, traditional leaders, arts coordinators, and village residents. A range of documents was also collected, including photos and video recordings of *Kuadai* dance performances, scholarly and news reports on *Kuadai* dance, accounts of the research location and other resources. Validity was established based on the credibility of the data and the use of triangulation. The data were analysed using the techniques developed by Gertrude Prokosch Kurath in her research, which focuses on the symbols and meanings contained in dance.

4 FINDING ANG DISCUSSION

4.1 *Dance props as a medium for early chilhood learning*

The plate is an important medium of expression in *Kudai* dance, where it is used both as a dance prop and a stage prop (Jazuli (2016). As a dance prop, it is moved in both hands. As a stage prop, it is placed in front of the dancer, who steps on it. For Semende women, plates symbolise their responsibility for providing and presenting food to their families. Plates are used as a medium for early childhood learning using a play approach, in which young children learn that a plate is a tool used for daily meals. In dancing, plates are a symbol of the Semende's farming livelihood. Children are introduced to the idea that the plate represents food as a basic need that is close to the source of life; in this case, women are the source of life because they have a uterus. In Semende culture, the plate thus symbolises the mother who must be respected and treated gently, carefully and lovingly.

4.2 *Floor pattern as a medium for early childhood learning*

The *kuadai* dance floor pattern involves circular shapes and straight lines. This pattern symbolises the environment of this agrarian society, which is characterised by slopes and hills. These symbols mean that humans must be submissive to or in harmony with nature. The interaction between humans and nature will go well if there is a balance between them. This floor pattern can be taught in early childhood to develop a positive attitude towards movement in order to ensure optimal physical growth.

4.3 *Accompaniment of Kudai dance as a medium for early childhood education*

Based on Lontoh et al. (2016) research, the poem Syaiful Anam is a functional aesthetic need for the supporting community. The lyrics of the musical accompaniment (*Dundai*) to the *Kuadai* dance also convey meaning. *Kuadai* dance is accompanied by poetry in the Semende language, or *dundai*, which means 'lively'. *Dundai* is both an accompaniment to the performance and a special prayer. The lyrics (shown in Table 1) could be effectively used as a learning medium a in early childhood.

Dundai tells of a mother who wants her child's personality to shine like a diamond. She would be ashamed if her child was not an asset for his or her parents, family, religion and society. There are various ways of obtaining such a "diamond". These include praying and teaching the child to avoid heresy and become close to God. Such expectations, directed towards children, are visualized in the special figure of the *Kuadai* dancer.

The form of the *Dundai* poem makes it suitable for use as a learning medium in early childhood, since the music and words are lively. The poem can teach young children to always be gentle and confident. The accompaniment to a *Kuadai* performance creates an atmosphere by

Table 1. Lyrics of Dundai.

Lirik *Dundai*	English
Ulan Nuli Tembage	*Hope of copper*
Sayang Sekirab Layu.	*Do not let it wither*
Dindanglah Dindang	*Resounding echoed*
Ulan Nuli Tembage	*Expectations of copper*
Sayang Intan Dik Dapat	*It's a pity diamond if it's not obtained*
Badan Dik Malu	*Your body will be embarrassed*
Dindanglah Dindang	*Resounding echoed*
Bunga Pandan	*Pandanus flowers*
Bersuling Dani	*Soft flute*
Asap Kemenyan	*Incense smoke*
Memanggil Dewe (2x)	*Call God*
Sayang Kemane	*Where are you going?*
Nyandarkan Badan	*Leaning body*
Nyandarkan Badan	*Leaning body*
Kepade Dewe	*To God*
Kapal Api Menuju Rembang	*The fire ship headed for Rembang*
Sampai Dirembang Menukar Uang	*Arriving at Rembang exchanged money*
Sangkan Hati Selalu Ribang	*The heart is always happy*
Sanggulan Ading Berisi Kembang	*Sister hair bun contains flowers*
Aik Rawang Bergenai Rawang	*The water is stagnant*
Gadis Sekini La Turun Mandi	*A girl will go down in the shower*
Sikur Hilang	*One is gone*
Sikur La Melayang	*The others floated*
Sikur Mbak Kumbang	*One hope is as big as a beetle*
Lah Putus Tali	*Don't get disconnected too*

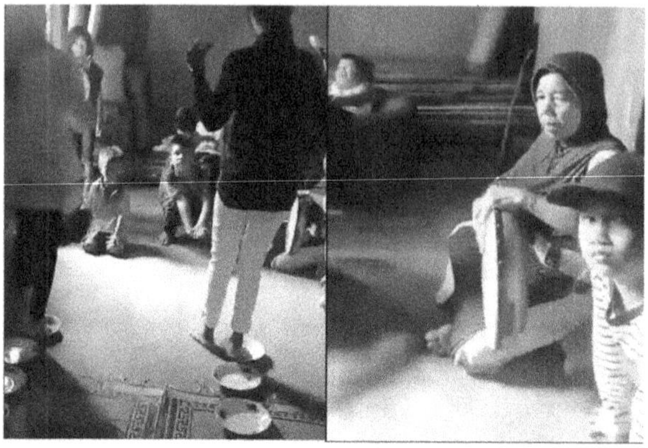

Figure 2. *Kuadai* is taught to preschoolers.
(Source: Amelia, 2018)

emphasising shifts and patterns of movement. A *Kuadai* dance lasts seven minutes and is accompanied by a tambourine-like musical instrument and a poem in Semende. The movements of the dancers are in harmony with the slow tempo of the musical accompaniment. These qualities can instill in young children the values of gentleness, patience, kindness and circumspection in one's behavior.

The symbolic meanings of *Kuadai* dance are conveyed in movement, poetry, floor patterns, and music. Through habituation, these meanings can be taught in early childhood. The movements teach children to be strong, responsible, patient and circumspect. The poetry teaches them to be gentle, patient, circumspect and good. The tambourine music creates enthusiasm and a caring attitude. The floor pattern teaches children to be physically active for optimal growth.

5 CONCLUSION

Kuadai dance is a means of early childhood education that utilises movement, music (*Dundai*), props (plates) and floor patterns to guide children to be gentle, polite, responsible, patient and circumspect. It also plays a role in optimizing children's growth and development and encouraging them to be actively engaged and enthusiastic in their daily activities. These qualities are instilled from an early age through habituation.

Aspects of the form (text) of *Kuadai* dance performances can be seen to have educational value. Analysis of the symbolic elements of the performance discussed indicates that the embedded meanings can be taught in early childhood. Furthermore, the implication for early chilhood, First, constructive symbols are evident in the customary prayer movement as a means of connecting a child to God. Children learn to uphold noble traditions of respect for their elders and other people, which are core Islamic values.

Second, the symbols that form the science of *Kuadai* dance are found in movement, poetry, floor patterns, and music. The meanings contained in these elements can be taught in early childhood through a process of habituation. The movements teach children to be strong, responsible, patient and circumspect in their behavior. The poetry teaches children to be gentle, patient circumspect and good. The tambourine music creates enthusiasm and a caring attitude. The floor pattern teaches children to be physically active for optimal growth.

Third, symbols of morality are evident in the choreography, teaching children from an early age to respect others, uphold traditional customs and culture, maintain a balanced life by caring for their family and fellow citizens, and to have patience and perseverance.

Fourth, as a symbol of expressive disclosure, *Kudai* teaches that each child is unique and should continue to learn to improve their particular abilities.

REFERENCES

Ardin, A., Cahyono, A., & Hartono, H. (2017). The Symbolic Meaning of Linda's Performance in the Karia Ritual Ceremony in West Muna Regency, Southeast Sulawesi. *Catharsis*, 6(1), 57–64.

Arisyanto, P., Cahyono, A., & Hartono, H. (2017). Wong's Wayang Kulit Plays a Role in Upholding the Sky Kissing the Earth: the Study of the Performing Text. *Catharsis*, 6(1), 74–81.

Blanariu, N. P. (2013). Towards a Framework of a Semiotics of Dance. *CLC Web: Comparative Literature and Culture*, 15(1). *Catharsis*, 2(1).

Danesi, Marcel. (2012). *Message, Sign, and Meaning*. Yogyakarta: Jalasutra.

Fretisari, I. (2017). The Meaning of the Nimang Padi Dance Symbol in Traditional Ceremony Naik Dango of the Kanayanti Dayak Community. *Ritme*, 2(1), 68–77.

Hadi, Sumandiyo. (2005). *Sociology of Dance*. Yogyakarta: Penerbit Pustaka.

Hartono. (2016). Yogyakarta Style Klana Raja Dance. *Jurnal Efektor*. 28.

Hartono. (2012). Development of Plural Intelligence in Elephant Melin Dance Learning Activities at the Kendal Community Development District Kindergarten. *Mudra*. 27 (2).

Indrayuda. (2015). "Continuity of Tradition Dance: Acedemicians' Intervention on Artists and Performing Arts Groups". *Harmonia: Journal of Arts Research and Education*. 15 (2).

Jazuli, M. (2016). *Art Education Paradigm Edition 2*. Sukoharjo: CV Farishma Indonesia.

Kusumastuti, E. (2009). Aesthetic Expressions and Symbolic Meanings of Laesan Art. *Harmonia: Journal of Arts Research and Education*, 9(1).

Langer K, Susanne. Terj. FX Widaryanto. (2006). *Problems with Performing Arts*. Bandung: Sunan Ambu Press.

Lontoh, W., Wadiyo, W., & Utomo, U. (2016). Syarofal Anam: Structural Functionalism at An-Najjam Studio in Palembang City. *Catharsis*, 5(2), 84–90.

Madrie. (1992). *Series of Knowing Semendo's Ethnicity: Lampik IV Merdike II Adat Institution, Jurai Besemah and Semende*. Bandar Lampung: Fisipol Universitas Lampung.

Pujiyanti, N. (2013). The Existence of Ireng Mask Dance as Meeting the Aesthetic Needs of the Pandanggungari Parakan Community in Temanggung. *Catharsis*, 2(1).

Putri, R. P., Lestari, W., & Iswidayati, S. (2015). Bedaya Suryasumirat Dance Relevance as a Symbolic Expression of Javanese Women. *Catharsis*, 4(1).

Tyas, G. P., & Kuswarsantyo. (2018). Value of Character Education in Srimpi Pandelori Dance Variety. *Mudra, 33* (Mei), 182–190.

Wahyu, E. (2013). Jarog Dance For Children With Special Needs: Case Study in The Celebration of The International Dance Day in Surakarta. *Harmonia: Journal of Arts Research and Education, 13*(2).

21st Century Innovation in Music Education – Astuti et al. (eds)

Educational values through *Senjang* songs in Sekayu, Musi Banyuasin Regency

A.R. Purnamasari & S. Pradoko
Yogyakarta State University, Yogyakarta, Indonesia

ABSTRACT: *Senjang* is an artistic medium that employs *pantun bersahutan*, a poetic form in which two people or pairs of people answer back the lyrics which are sung in reply. *Senjang* can contain advice or criticism, express happiness or describe a particular event. The purpose of this research was to explore the educational values that are conveyed through the art of *Senjang* in Sekayu city, Musi Banyuasin Regency, South Sumatra Province, Indonesia. The data were collected through observation, interviews and document studies. The results showed that the lyrics conveyed the educational values of cooperation, discipline and responsibility. They reflect wider social values about the importance of education, knowledge, religion, community, and culture that can help to shape and improve the character of the nation.

1 INTRODUCTION

Music in South Sumatera can be played on a single instrument, individually or in ensemble form. Musi Banyuasin (Muba) Regency has its own artistic tradition, among which is the form known as *Senjang*, which is performed in the capital of Sekayu district.

Senjang has been sung in Sekayu from ancient times until the present. It is usually performed on family occasions, such as weddings and the construction of new homes. *Senjang* provides a strong connection between older and younger generations, parents and children, the government and the people. The lyrics of *Senjang Petuah Lame* convey advice, criticism and aspirational values through folk songs. In order to understand the meaning and values contained in the songs of *Senjang*, we used hermeneutics, which is the theory and methodology of interpretation of a text or a set of signs that are considered as a text.

Educational values should include people in a relationship with God, others, the environment and the nation. The educational elements in *Senjang* songs can help children in Musi Banyuasin Regency to understand and apply these values, contributing to shaping their characters.

Values indicate a criterion or standard for assessing or evaluating something, such as industrialisation as a pathway to prosperity. Values can be individual, social, cultural or religious. Education refers to the process of developing knowledge, skills and character, especially through formal schooling and other forms of training. In this paper, the term 'educational values' refers to the wider outcomes that are expected from the experience of interacting in educational institutions and society. In the context of music education, for instance, the goal is not to train people to become musicians but to enhance their mental development. Early education in music has an important influence on the growth and development of children. They do not just learn to sing, but to express the values contained in the song (Zhou, 2015, p. 4)

The educational process is expected to generate positive human values. Because values are abstract, they can only be understood by analysing their manifestation in particular processes. This study analysed the lyrics of *Senjang* songs performed in *Budanyo Senjang* Gallery to identify their educational values.

2 LITERATURE REVIEW

Sukma (2014) discusses the form and function of *Senjang* in its sociocultural context (i.e. Musi Banyuasin Regency of South Sumatera Province). The present study differs in that its focus is on the educational values contained in the *Senjang* songs. The researcher also used the findings of Leo (2016), who examined the form of happy folk songs in Musi Banyuasin society, in this research. Again, the present study differs in its focus on educational values contained in *Senjang* songs.

2.1 *Senjang songs*

Senjang is an artistic form that uses *pantun* as its medium of expression. It can be performed solo or in the form of a dialogue exchange between two people or pairs. It is called *Senjang* because its lyrics and music never coincide; that is, when the lyrics are sung, the music stops and vice versa. Thus, the meaning of *Senjang* in the language of Musi can be interpreted as 'gaps' or conditions that are not aligned (interview with Irawan, 12 July 2017). This is what makes *Senjang* unique (interview with Pahri Azhari, 12 July 2016) in the *pantun* composed. It involves 14 players and eight types of musical instruments – two trumpets, a *jidor* (a type of drum), a stringed drum, two clarinets, two tenor saxophones, two alto saxophones, a counter bass and three alto horns.

Senjang has become a medium through which rural, district, provincial and central governments convey information and advice to children, teenagers and parents. Therefore, *Senjang* is often performed at family gatherings, such as weddings, where all generations are gathered together. The artists convey these moral messages by singing *Senjang* as they perform simple dance moves in an entertaining way.

2.2 *Hermeneutics*

The word hermeneutics is derived from the mythical Greek deity Hermes, who was the messenger of the gods. As a discipline, hermeneutics is a theory and methodology for the interpretation of written texts and other forms of communication. It applies scientific principles of analysis to generate deep understanding of a phenomenon or experience. According to Pradoko (2017, p. 13), the Indonesian form of *hermeneutika* is derived from the Greek *hermeneuo*, which means expressing people's thoughts in words.

Drawing on theoretical and empirical studies, Pradoko (2017, pp. 67–68) developed a conceptual framework that can be used to analyse the values contained in the lyrics of *Senjang* songs. *Senjang* is thus conceptualised as the actualisation of character-building values through verbal messages (see Figure 1).

Ricouer's approach to textual interpretation can be applied to the study of social phenomena, including the material objects of culture (Manen, 2016). Here, the 'context' is the author or artist who creates the work. The environment is the society within which the material object of the art is created; in this case, the local ethnic community and geographical setting. Other relevant texts include theoretical works about the phenomenon under observation. In this study, the reader is the researcher herself, as well as the community of artists who experience the artistic object and artists who live in the local area. Art is a component of culture that develops among a group of people, in this case, those living in Sekayu city. One such art is *Senjang*, which reflects the local culture, including customs and values of the Sekayu community. Over time, the popularity of *Senjang* songs in the community of Musi Banyuasin Regency has grown, and their educative values continue to be expressed to learners.

3 MATERIAL AND METHODOLOGY

This study employed a qualitative method and a hermeneutical approach. Hermeneutics is the theory of understanding in interpreting texts (Pradoko, 2017, p. 101). The hermeneutic technique is used to analyse and interpret the hidden meanings in a song or folk song. Ricouer's

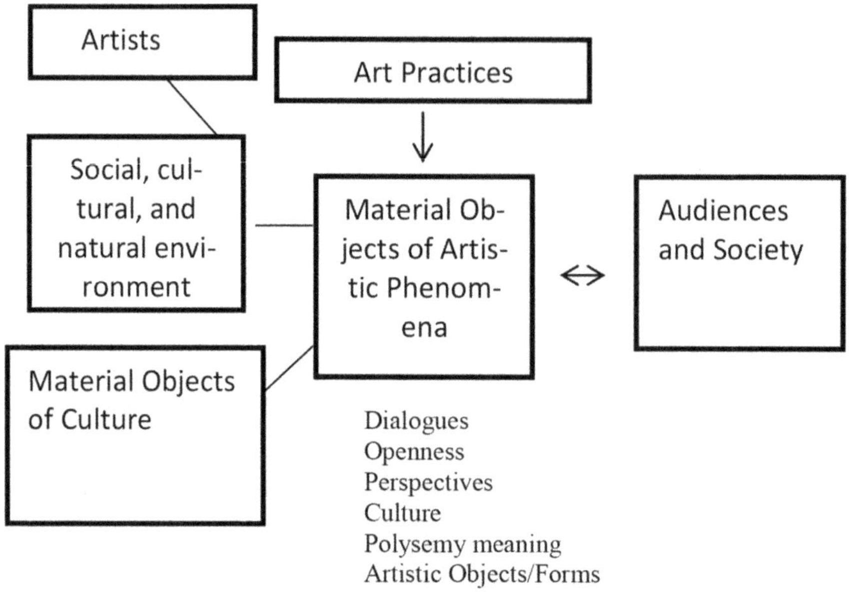

Figure 1. Polysemic text chart (Source: Pradoko, 2017, p. 68).

method of textual interpretation was applied to examine this art as a sociocultural phenomenon (Ricouer, 1981).

The data used in this study were collected from interviews with informants, documentary analysis and a literature review. The data were obtained by a *Budayawan* (cultural observer) in South Sumatera.

The validity of the data was assessed by using triangulation of different data sources (Sugiyono, 2010). The data were analysed according to Miles and Huberman's (Rusmawati, 2017, pp. 31–33) steps of data condensation, data presentation and conclusion.

4 RESULTS AND DISCUSSION

The results showed that the lyrics of *Senjang Petuoh Lame* songs in Musi Banyuasin Regency expressed the values of good manners, hard work, responsibility, helping each other, unity, discipline and avoiding drugs. The songs were sung and the implementation was applied in *Senjang Budayo* Gallery in Musi Banyuasin Regency, South Sumatra Province. The educational values contained in the lyrics are evident in the following:

Dak ke kade masuk lubang
Meski kereto pelake bengkang
Kalu bakereto ati-ati
Ati-ati bagoyo bae
Kade dak sampai walaupun lame

Dak ke kade sare panjang
Kade dak betemu idup senang
kalu banyak babuat kasih
ibarat kate peribahase
sape bebuat pasti nerime

English translation:
there is no way to enter the hole
even though the wheel of the train is tilted
be careful when you train
be careful, just slow down
it's impossible not to arrive, even though it will be long

it may not always be difficult
it is impossible not to meet a happy life
if you do a lot of goodlikened to proverbs
those who do will surely receive

These lyrics convey the message that it is one's responsibility to do and to be good. Taking responsibility is a natural part of human life and each individual must be responsible for his or her behaviour. If a person refuses to accept responsibility, others will take it upon themselves to ensure that he or she does. The lyrics of this *Senjang* reflect the attitudes and behaviours that people need to display and explain that they must fulfil their duties and obligations to themselves, society, the natural, social and cultural environment, the state and religion. Through this *Senjang* song, students learn how to live better lives and are taught to be kind, sincere and selfless according to religious ideals.

amon bekareto dai sukarami
nak ke toman liwat rantau kasih
dusun rantau kasih di pinggir musi
syngai musi tempatnye mandi
penduduknye bacari rante batani

amon lah pacak babuat kasih
babuat kasih ngen senang ati
kasih tanpa ngarep pamrih
dak ke kade ngurang rejeki
rejeki banyan tuhan yang ngenti

English translation:
if you train from Sukarami
to Toman, you must pass through the village of Rantau Kasih
along the Musi river,
Musi river is not only for bathing place
but also a place where people make money

if you can do good
do it with pleasure
kindness without expecting anything
it is not possible to reduce your fortune
fortune is all to gods

These lyrics express the value of cooperation between individuals and groups to achieve common goals. Cooperation is vital because human beings cannot live alone – they need others. Cooperation is possible when individuals have common interests and are aware of the necessity of working together to achieve their interests. We should always cooperate, work hard, help others and give affection without expecting anything in return, since our reward comes from God.

Selesai cerita sepeda Bangkok
Sepeda Bangkok sekarang sudah patah

Bannya kemps karena pecah
Hanya bias dijual jadi barang rongsokan
Dapat seribu saja lumayan

Selesai cerita kami bersenjang
Tinggal pesan sama kaum remaja
bersihkan diri dari narkoba
karna narkoba membuat susah
geleng-geleng seperti orang gile

English translation:
Finish the Bangkok bicycle story
Bangkok's bike is broken now
The tire was flat because it broke
It can only be sold as junk
Getting a thousand is not bad

Our story is complete
Just advice to the teenagers
free yourself from drugs
because drugs make it difficult
shake your head like a crazy person

These lyrics convey a message to young people that they should avoid drugs, which have terrible consequences. They emphasise the importance of discipline, consistency, orderly behaviour and adherence to rules and regulations. Discipline must also be observed in arriving on time for practice and following the rules that govern singing the songs. Drugs are illegal and are strictly forbidden by the government as they endanger the health of individuals, families and the environment.

Senjang songs, with their educational values, are taught to students, young people and the wider Sekayu community in lessons either in school or in the local gallery. Interviews with arts teachers, community members and local artists of Sekayu Musi Banyuasin district revealed a belief that *Senjang* should be taught to students to enhance their good behaviour and creativity.

5 CONCLUSION

The results show that the lyrics of *Senjang Petuoh Lame* songs convey a number of educational values. The process of teaching this art form is comprised of three stages: the songs are sung in *unisono* to begin with; learners practise the song together; and they analyse the meaning of the songs. Because the *Senjang* takes the form of advice from someone with life experience to the younger generation, its educational messages are not perceived negatively.

The process of participating in *Senjang* requires cooperation, discipline and responsibility as well as sensitivity and emotion. Its educational values can help to build the character of the nation. *Senjang*, then, is a medium for inspiring and motivating the younger generation of Sekayu Musi Banyuasin to apply these qualities in their lives.

Future research could explore other values that are conveyed in *Senjang Petuoh Lame* songs and how they are performed and integrated.

This study has made several contributions to knowledge in the field. First, it provides a baseline of information for further analysis of the lyrics of *Senjang* and the educational values they convey. Second, the significance of this cultural art form can be more widely understood and appreciated. Finally, other artists may be inspired to create artworks that are rich in educational values.

REFERENCES

Azhari, P. (2016, July 12). Personal interview.

Irawan, K. (2017, July 12). Personal interview.

Leo, A. (2016). *The form of folk songs in Senjang artistry in Musi Banyuasin Regency*. Palembang: Sriwijaya University.

Manen, M. (2016). Phenomenology of practice: Meaning-giving methods in phenomenological research and writing. London: Routledge.

Pradoko, A.M.S. (2017). *Qualitative paradigms for art, humanities and culture research*. Yogyakarta, Indonesia: Charissa.

Ricoeur, P. (1981). Hermeneutics and the human sciences: Essays on language, action and interpretation. Cambridge: Cambridge university press.

Rusmawati, D. (2017). Strategi Promosi Perpustakaan Di Balai Layanan Perpustakaan Badan Perpustakaan Dan Arsip Daerah DIY [Strategy to promote library]. *Hanata Widya*, *6*(7), 37–46.

Sugiyono, P. (2010). *Metode Penelitian Kuantitatif, Kualitatif, dan R&D* [Quantitative, qualitative, and R&D research methods]. Bandung, Indonesia: CV Alfabeta.

Sukma, I. (2014). *The presence of Senjang in the community of Musi Banyuasin Regency, South Sumatra Province* (Postgraduate thesis, ISI Surakarta, Indonesia).

Zhou, J. (2015). The value of music in children's enlightenment education. *Open Journal of Social Sciences*, *3*, 200–206. doi:10.4236/jss.2015.312023

21st Century Innovation in Music Education – Astuti et al. (eds)
© 2020 Taylor & Francis Group, London, ISBN 978-0-367-11122-9

Efforts to enhance research methodology achievements through classroom interactions

H.S. Mudjilah
Universitas Negeri Yogyakarta, Yogyakarta, Indonesia

1 BACKGROUND

The world of education is currently becoming a concern for those with vested interests, namely, the central government, local governments, teachers, students and stakeholders. The rise of news about education invites a variety of opposing responses. The community has begun prioritizing quality over quantity when it comes to education. The quality of a teacher can be assessed by how competent they are both in their teaching and in their research. Current research competencies for a teacher that are very much needed relate to teacher certification which requires competency in the field of research *Lembaga Perguruan Tinggi Keguruan* (LPTK), or the Institution of the Teacher Training College, produces prospective teachers who need to better master the materials of each subject given. Mastery of materials or competency for a prospective teacher is also determined by how well they have mastered the course, especially in the field of research. One of the subjects that provides the ability to conduct research is the Research Methodology course. Mastery of materials in research methodology courses influences the ability of students to compile their research proposal, which is one of the requirements for achieving their education degree. Lecturers need to facilitate the quality of learning in research methodology courses through innovations to improve the quality of their teaching. The quality of learning expected is one that can provide students with the knowledge and experience necessary for a good level of attainment. The Department of Music Education, FBS-UNY (Languages and Arts Faculty of Yogyakarta State University), is one of the higher education institutions included in the LPTK. The Music Education Department also conducts research methodology courses in Semester 6, and Praktik Pengalaman Lapangan (PPL) or teaching practicum in the special semester. It is expected that after following the Research Methodology course, students can identify problems in the field to be examined as a final assignment to fulfil the criteria for an undergraduate degree. Courses held in the Department of Music Education are divided into two groups: music and non-music courses. Research Methodology of Music Education is one of the non-music courses that support the completion of the students' final project. Based on several semesters of observations in the Department of Music Education, students' understanding and mastery in this course were felt to have not yet reached the expected levels. Students expressed that when it is time to write a thesis or submit their research proposal as a final project, they still feel confused and do not know what to do. The learning process that occurs is often still conventional and does not involve the active participation of students; so, the students are not ready to produce a proposal for their research later. One of the lectures supports completing the final student assignments.

Based on the teaching experience of several years, it is felt that there is a need for innovation in the learning process of the Music Education Research Methodology course. Dimyati (2002) said that learning is essentially determined by several factors, including methods, materials, teachers/lecturers, students, employees, as well as facilities and infrastructure of learning activities that support each other in implementing the learning process.Learning innovations need to be carried out in several areas. The method in the learning process varies greatly. A teacher/lecturer has the right to carry out updates, so that the quality of learners/students can be improved (Muhfida, 2010). The problem that often arises among students is how to begin writing their research proposals. When they have to decide the topic to write on, they are still questioning what they should do: 'what do I want to write?', or 'I want to research X'. Most of the students face difficulties when they arrive at the stage of data collection, because it turns out that the problems they are looking into do not exist. The number of problems and failures is what drives this present study, to anticipate the confusion and unpreparedness of students in writing research proposals. This research identifies that many students are too confused to begin writing research proposals. To overcome this problem, innovation needs to be undertaken in learning research methodology, so that students can better understand and master the problems in writing research proposals. The classroom interaction learning model as part of the learning method largely determines success in learning. It is an interesting and appropriate learning model to be applied in the Research Methodology course. Classroom interaction needs to be built into learning by encouraging interactions between fellow students, between students and lecturers, and between groups. The application of this classroom interaction learning model is expected to be a way of mastering the subject matter of the Music Education Research Methodology course. Motivation and encouragement from lecturers is very meaningful for students who are embarrassed to express their opinions or to ask questions.

2 LITERATURE REVIEW

Social interaction is based on gestalt learning theory, that is, learning in its entirety. The gestalt theory focuses on harmonious relationships between individuals and society (learning to live together) (Dariyantoi, 2014). Social interaction or classroom interaction can make the atmosphere of learning in the classroom come alive, because students play an active role. Bjorklund and Parente (in Kumara & Harsono, 2005, p. 2) state that the quality of classroom interaction affects the quality of student achievement and satisfaction in gaining experience and knowledge of the materials being studied. Furthermore, their research results show that classroom interaction significantly improves professional skills, with constructive feedback delivered by lecturers, and fosters awareness, for both lecturers and students, that the learning objectives will be achieved as expected. Kumara and Harsono (2005, p. 3) also state that the application of gestalt theory in learning includes: 1) insight, 2) meaningful learning, 3) purposeful behaviour, and 4) the principle of life space (Danasasmita, 2008).

Several factors play a role in classroom interaction, including lecturers, students and faculty policies. Lecturers as facilitators in this activity must be able to provide good feedback to be responded to by students as learning participants. Lecturers are required to be able and willing to take the time for discussions and guidance beyond scheduled agenda. Lecturers design lecture activities in such a way that students can interact with other fellow students, as well as with lecturers and between groups. Students are also required to be responsible, to understand their direction and interest in learning, and to be able to manage their time between academic and non-academic activities. This will not work well if it is not supported by faculty policy in an institution that accommodates lecture activities. Furthermore, Danasasmita (2008) states that the model of social interaction includes six learning strategies: 1) group work, 2) class meetings, 3) problem solving, 4) laboratory models, 5) role playing and 6) social simulations.

The action hypothesis in this study is through classroom interaction learning models that can improve student achievement and activity in the Music Education Research Methodology course.

3 MATERIALS AND METHODOLOGY

3.1 *Data*

This study used a classroom action research approach. Classroom action research was first developed by Kurt Lewis in 1946, relying on a teacher to conduct a study that did not interfere with their routine duties. Action research is the way a group or individual creates a condition where they can learn and make their experience accessible to others (Sukardi, 2013, p. 3).

3.2 *Research setting*

This research was carried out for eight months, and was conducted at the Department of Music Education, FBS-UNY, in the Research Methodology course, Class C. The 42 research subjects were all students who were enrolled in Research Methodology course, Class C. This class was divided into two classes. This research was carried out in the classroom that the researchers supported, assisted by two students as members of the research team.

3.3 *Research procedure*

This research consisted of two cycles, with four stages in each cycle: planning, implementation of action, observation and monitoring, and reflection (Kemmis & McTaggart, 2002).

1) Planning: In this stage, researchers identified problems and planned ways to overcome the problems. Researchers made alternative action plans to develop and improve student achievement in preparing a research proposal. They also made plans for implementing activities in accordance with alternative actions chosen to overcome the existing problems.
2) Implementation of action: In this stage, the lecturer divided the class into several small groups consisting of three to four students in each group. The lecturer provided motivation and encouragement to students to interact with each other to solve problems given by the lecturer. Students interacted with each other within a group, with the lecturers, or between groups, thus facilitating positive interactions.
3) Observation and monitoring: Observations were made by noting interactions between students, between students and lecturers, and between groups of students. Monitoring was carried out by lecturers, assisted by students as members of the research team. The students participating in monitoring went through training in advance about how to monitor during the learning process in the classroom in the current cycle.
4) Analysis and reflection: In the fourth stage of the study, a reflection on the results of the monitoring was conducted. Then, the researchers discussed the action given to the first cycle of the research objectives and whether the results of the reflection reveal that the implementation of the action has not yet reached the objectives of the study; they then planned the next cycle.

3.4 *Data collection technique*

Data collection techniques used in this study were observation, questionnaires and achievement tests (Purwanto, 2012). Data collection through observations was carried out when the action was implemented and interactions between students, between students and lecturers, and between groups of students was observed. Questionnaires were distributed to all students and were about classroom interaction learning models: what is the response of students to the learning model? What did students think about the advantages and disadvantages of the learning model?

3.5 *Data analysis technique*

The data analysis technique in this classroom action research used qualitative descriptive analysis, which is supported by student learning achievement data (Sugiyono, 2006). Student achievement data were used to conclude whether there had been a positive or negative change after taking action or after treatment (i.e. application of the learning model).

3.6 *Criteria for successful action*

The success criteria in this study were that the final test results achieved after the last cycle showed students achieving a minimum graduation rate of 75%. The validity was determined as the validity of the results agreed in this study in order to decide on the success of the process, where success was regarded as the stages carried out and what a researcher learned from this study. Then, democratic validity was determined as the agreement between researchers, catalytic validity as the encouragement of participation in orienting, focusing and opening up to the vision and enthusiasm to accept the transformation in their work, and, finally, validity of dialogue as that between researchers and research team members in making their reports so that they can be agreed upon by fellow researchers.

4 RESULTS AND DISCUSSION

This research was conducted in the even Semester IV academic year. The implementation of this study was divided into two cycles as discussed in the following sections.

4.1 *Cycle 1*

4.1.1 *Planning*

Before the implementation of this first cycle, a pre-cycle value was taken from the students to see whether this study was successful or not. The students were given the opportunity to first write their research proposal as is before studying the materials. Lecturers prepared learning materials related to the writing of a research proposal (introduction, problems, and research objectives) and arranged the room as a place to carry out this research. In the planning stage, the research background and planning as an action in Cycle 1 was explained to the students, which was about how to write down all the problems that exist. Matters relating to research methodologies (see Section 3) led to each study. Some chose *Penelitian Tindakan Kelas* (PTK) (classroom action research), some Research and Development (R&D), some experimental research and some chose survey research.

4.1.2 *Implementation of action*

The lecturer gave the students materials for their first chapter: how to write about the background of the problem, explain the problem and the purpose of the research. The lecturer provided an understanding of how to find problems, as well as explained why these problems occur. After everything was explained, the students were instructed to write down the boundaries, the formulation of the problem and the purpose of the research. At first, the students had difficulty writing all of these things, but after being guided they understood what the problem was. Please note that the problem is something neutral, they are not always negative, and can also be positive. After arriving at the formulation of the problem, the next step was to write down the research objectives. The formulation of the problem is taken from the background of the problems described. Then, students proceeded with the research objectives in accordance with the formulation of the problem.

The third chapter concerns research methodologies used in accordance with the research objectives. Thus, the students conducted research using, among others, classroom action research, Research and Development (R&D), experimental research and survey research (Creswell, 1994).

4.1.3 *Evaluation and monitoring*

Lecturers monitor students and their support assists students to write their introductory chapters. In this stage, the lecturers provided careful monitoring, showing students how to write the introduction section correctly.

4.1.4 *Analysis and reflection*

After monitoring, the lecturers discussed with the research team members how this research works. It was found that the first cycle was carried out well enough but still had not achieved the research objectives, as evidenced by the test values obtained displayed in Table 1 (Table 1).

4.2 *Cycle 2*

4.2.1 *Planning*

The lecturer planned the second cycle, continuing with the implementation of the research from Cycle 1. In the research methodology section, several methodologies widely used in research were applied, such as PTK, R&D, experimental research and survey research.

4.2.2 *Implementation of action*

In this stage, the existing problems were further explained and there was more focus on the problems to be studied. The problem was then limited to what the students want to study: students were instructed to write down the formula and purpose of their research.

The lecturer planned to explore music education research methods related to the research methods chosen. PTK (classroom action research) is a research method that is mostly used by teachers and lecturers. This study aims to find out whether there is a positive influence of this research method on student learning outcomes and to improve classroom achievements. Usually, students collaborate with local teachers as collaborators. Research and Development (R&D) is a research method that requires the existence of a product. This research method is rarely chosen by students because the process is difficult. Experimental research refers to research conducted by 'giving treatment' and 'not giving treatment' to the subjects, where the

Table 1. Values obtained in Cycle 1.

No.	Respondent	Pre-cycle	Cycle 1	No.	Respondent	Pre-cycle	Cycle 1
01	EI	65	65	20	NMP	65	70
02	AD	65	67	21	SA	65	72
03	ILVT	50	60	22	PDIR	60	65
04	ARM	70	78	23	LPP	60	65
05	TKI	68	70	24	AAA	60	65
06	DAR	60	60	25	YSPK	70	73
07	YDA	65	72	26	ASMN	60	63
08	ZA	65	73	27	W	56	67
09	AAP	60	67	28	EWN	60	63
10	SP	62	68	29	OP	60	67
11	DTW	60	66	30	BPWN	60	65
12	NJL	65	72	31	FKHA	65	65
13	RFP	60	60	32	RH	65	67
14	LJS	60	60	33	PMDMP	60	65
15	IPPP	60	68	34	DIA	58	67
16	MSL	56	60	35	FC	56	67
17	FXIPW	60	68	36	LAS	60	67
18	DABS	60	65	37	SHSR	60	67
19	FEIR	58	69	38	BGB	60	65
					Average	61.289	66.553

As seen in Table 1, the pre-cycle average test value of X1 = 61.289 changed to a Cycle 1 average test value of X2 = 66.553, an increase of 5.264 points. This indicates that the expected results (75% graduation rate) have not yet been achieved. The next step is to continue research in Cycle 2.

subjects are divided into the experimental group and the control group. Survey research is qualitative research that monitors the events that are experienced, where things related to the research may also be photographed. These last two methodologies are more commonly used by students because they are relatively easy and practical.

4.2.3 *Observation and monitoring*

Observation and monitoring were carried out in the implementation of this research, to assess whether or not it can run well. Students' understanding of the introduction ('Chapter 1') and of the research methodology ('Chapter 3') was well monitored, and most of the students were able to understand how to write down the problems faced and how to choose the appropriate research methodology.

4.2.4 *Evaluation and reflection*

After students understood the problems to be investigated, they then determined the appropriate methodology of their research. Such cases indicate students have really understood the learning materials and done what they should. They are able to precisely write the background of the problem, problem boundaries, problem formulation and research objectives. Also, they are able to choose the appropriate methodology to conduct their research.

The results of this study are recorded as test values obtained from each student participant in the Class C Music Methodology Research Methodology course (Table 2).

As seen in Table 2, the Cycle 1 average value increased from X2 = 66.553 to X3 = 76.816 (Cycle 2 average value), an increase of 10.263 points. This indicates that the research objective of a 75% graduation rate has been achieved. To conclude, this study found an increase in the test results of students. This is evidenced by the test values shown in Tables 1 and 2. It was found that classroom interaction is an indispensable method for the implementation of learning, as many benefits are obtained through classroom interaction in the

Table 2. Values obtained in Cycle 2.

No.	Respondent	Cycle 1	Cycle 2	No.	Respondent	Cycle 1	Cycle 2
01	EI	65	78	20	NMP	70	80
02	AD	67	79	21	SA	72	72
03	ILVT	60	70	22	PDIR	65	76
04	ARM	78	87	23	LPP	65	73
05	TKI	70	82	24	AAA	65	78
06	DAR	60	76	25	YSPK	73	79
07	YDA	72	79	26	ASMN	63	73
08	ZA	73	77	27	W	67	72
09	AAP	67	77	28	EWN	63	77
10	SP	68	79	29	OP	67	79
11	DTW	60	78	30	BPWN	65	72
12	NJL	72	79	31	FKHA	65	73
13	RFP	60	77	32	RH	67	83
14	LJS	60	76	33	PMDMP	65	78
15	IPPP	68	78	34	DIA	67	78
16	MSL	60	77	35	FC	67	72
17	FXIPW	68	77	36	LAS	67	78
18	DABS	65	78	37	SHSR	67	78
19	FEIR	69	72	38	BGB	65	72
					Average	66.553	76.816

course of this research method. This is because the course requires agreement between students and instructors.

5 CONCLUSION

From the results of this study the following conclusions can be drawn:

1) The students are able to understand what must be done in writing their research proposal.
2) The students have written the background of the problem that should be written.
3) The students are able to understand the methodology used in their research.
4) The students have successfully written their research proposals.

6 FOLLOW-UP PLAN

Based on this research, all students wrote down their research proposal thoroughly, and submitted the proposal as the final condition of the Music Education Research Methodology course. For the next step, students will be asked to continue their research proposal as a condition to be able to take *Tugas Akhir Skripsi* (TAS) or the 'undergraduate thesis. Furthermore, the students submitted a request for the final assignment which was considered by *Pembimbing Akademik* (PA) (academic counsellors) for further requirements. From these results, students will be asked to immediately submit a proposal to the Music Education Department to proceed to the next level, that is, writing a research proposal.

REFERENCES

Creswell, J.W. (1994). *Research design: Qualitative and quantitative approach.* London, UK: Sage Publications.

Danasasmita, W. (2008). *Model-Model Pembelajaran Alternatif* [Learning model and approach]. Bandung, Indonesia: UPI.

Dariyantoi, F.N. (2014). *Learning theory gestalt.* Retrieved from https://ferdonan.wordpress.com/teori-belajar-gestalt

Dimyati, M. (2002). *Belajar dan pembelajaran* [Teaching and learning]. Jakarta, Indonesia: Rineka Cipta.

Kemmis, S. & McTaggart, R. (2002). *The action research planner.* Victoria, Australia: Deakin University Press.

Kumara, A. & Harsono. (2005). *Interaksi Kelas* [Classroom interaction]. Yogyakarta, Indonesia: Universitas Gajahmada.

Muhfida. (2010). *Conventional teaching.* Retrieved from http://muhfida.compembelajaran-konvensional/

Purwanto, M.N. (2012). *Principles and techniques of teaching evaluation.* Bandung, Indonesia: PT Remaja Rosdakarya.

Sugiyono. (2006). *Quantitative, qualitative and R&D research methods.* Bandung, Indonesia: Penerbit Alfabeta.

Sukardi, H.M. (2013). *Classroom education action research methods.* Jakarta, Indonesia: Bumi Aksara.

Walgito, B. (2002). *Social psychology (An introduction).* Yogyakarta, Indonesia: ANDI.

21st Century Innovation in Music Education – Astuti et al. (eds)
 ISBN 978-0-367-11122-9

The effects of peer group behavior on attention to the gamelan among teenagers

Y. Delvianita
Universitas Negeri Yogyakarta, Yogyakarta, Daerah Istimewa Yogyakarta, Indonesia

Djohan
Arts Institute of Yogyakarta, Yogyakarta, Daerah Istimewa Yogyakarta, Indonesia

ABSTRACT: This study aimed to identify the effects of peer group behavior on interest in learning gamelan and other traditional arts among teenagers in Yogyakarta City. A survey was conducted at three senior high schools in the Special Territory of Yogyakarta. A Likert scale questionnaire was used to collect data from 117 students who were selected using the area random sampling technique. The Manova procedure was employed to analyze the data. The results showed that there is a positive and significant effect of peer group behavior on interest in learning gamelan, but no peer group effect on attention to the traditional art of gamelan. This indicates that the closer the participants felt towards their peers, the higher was their interest in learning the traditional art of gamelan.

1 INTRODUCTION

Indonesia is home to a wide variety of tribes, languages, religions, faiths, and traditions. Traditional art forms are often viewed as expressions of culture and identity based on the unique characteristics of a local community. In the post-modern era, however, traditional arts are often regarded as out of date (Irianto, 2015:66), and societal interest in Indonesian culture has waned as a result of apathy and unfiltered interaction with foreign cultures. Responsibility for developing and preserving cultural heritage lies not just with government, but also with society, especially artists, art lovers, arts workers, and art critics. Young people today are constantly exposed to western culture and the influences of modernization. As a result, local art and culture are ignored and cast aside. This is reflected in the declining attendance by the young generation, especially in Java, at Javanese artistic performances (Kompas, 2017).

It is not the case that young people are totally disinterested in traditional arts such as gamelan but, rather, they find it boring because of the ways in which it is presented. For example, traditional puppet shows are performed in the Old Javanese Kawi language, which few young people understand. Therefore they are reluctant to watch a performance because they cannot follow the story (Oktadus, 2018). In addition, there is a perception that learning traditional music is old-fashioned and 'uncool'. The impact of such a view is particularly salient if it is held by one's peer group, that is, a group that forms around similarities in age, class, interests or hobbies, or objectives (Chaplin, 2014:357).

The influence of peer groups varies according to the strength of the individual's relationship with the group. According to Slameto (2013: 8), teenagers' behavior is affected by the social environment, including school, family, and community. The significance of peer group relationships in everyday life varies in different environments. The influence of these relationships can be felt in relation, for instance, to interest in acquiring knowledge and skills such as gamelan and other traditional arts. If teenagers' friends are interested in modern or traditional arts, this is likely to have an indirect effect on their own interests.

2 RESEARCH METHOD

2.1 *Research design*

This research is quantitative in nature and employed a survey design. The independent variable was peer group behavior, while the dependent variables were interest in learning gamelan and attention to the traditional art of gamelan.

2.2 *Population and sample*

The study population was all students of State Senior High School 3, State Senior High School 8, State Senior High School 9, and State Senior High School 11 in Yogyakarta. The sample comprised year X and year XI students from the four schools who participated in an extra-curricular program of gamelan traditional art. The area random sampling technique was used to select a sample of 30 year X and year XI students from each of the four schools (N=120). Due to three absentees on the day of data collection, the final sample was 117.

2.3 *Validity & reliability*

The validity of the instrument was measured using the product moment correlation. Meanwhile the reliability of the instrument was measured using Cronbach's alpha

2.4 *Data collection*

Data were collected using a Likert scale questionnaire to measure opinions, attitudes, and perceptions. In this method, the variable to be measured is described in detail, including the variable indicator, which then becomes the starting point for constructing the instrument items, which can be statements or questions (Sugiyono, 2010).

2.5 *Data analyses*

This study involved more than two variables were involved, multivariate statistics were used to analyze the data. Since there was more than one dependent variable, the MANOVA procedure was employed.

3 RESULTS

3.1 *Text and indenting*

Table 1 shows the results of data analysis using SPSS version 21 for Windows to test the null hypothesis that the error variance of the dependent variable is equal across groups. The results of applying Levene's test show that no variable had a significance value of < 0.05 and therefore it can be assumed that there is no error in this variant equation (Ghozali, 2011).

Table 2 shows the results of multivariate tests. The distribution meets the equality assumption, and thus the interpretation of the analysis result used the value of Wilks' Lambda (Pallant, 2005). If the significance level is lower than 0.05, it can be concluded that there is a difference among the variables. The obtained value of Wilks 'Lambda is 0.726, with the

Table 1. Levene's test of equality of error variances[a].

	F	df1	df2	Sig.
Sum of interest	1.216	2	27	.312
Sum of attention	.726	2	27	.493

[a]Design: Intercept + Peer group

Table 2. Multivariate tests[a].

Effect		Value	F	Hypothesis df	Error df	Sig.
Intercept	Pillai's Trace	.950	246.413[b]	2.000	26.000	.000
	Wilks' Lambda	.050	246.413[b]	2.000	26.000	.000
	Hotelling's Trace	18.955	246.413[b]	2.000	26.000	.000
	Roy's Largest Root	18.955	246.413[b]	2.000	26.000	.000
Peer_group	Pillai's Trace	.294	2.326	4.000	54.000	.068
	Wilks' Lambda	.726	2.260[b]	4.000	52.000	.075
	Hotelling's Trace	.351	2.192	4.000	50.000	.083
	Roy's Largest Root	.236	3.183[c]	2.000	27.000	.057

[a]Design: Intercept + Peer_group
[b]Exact statistic

Table 3. Tests of between-subjects effects.

Source	Dependent Variable	Type III Sum of Squares	df	Mean Square	F	Sig.
Corrected Model	Sum of interest	125.840[a]	2	62.920	2.524	.099
	Sum of attention	39.467[b]	2	19.733	1.708	.200
Intercept	sum of interest	10984.900	1	10984.900	440.728	.000
	sum of attention	2000.496	1	2000.496	173.120	.000
Peer_group	sum of interest	125.840	2	62.920	2.524	.099
	sum of attention	39.467	2	19.733	1.708	.200
Error	sum of interest	672.960	27	24.924		
	sum of attention	312.000	27	11.556		
Total	sum of interest	51722.000	30			
	sum of attention	8552.000	30			
Corrected Total	sum of interest	798.800	29			
	sum of attention	351.467	29			

[a]R Squared = .158 (Adjusted R Squared = .095)
[b]R Squared = .112 (Adjusted R Squared = .047)

significance value of 0.075, which is lower than 0.05; therefore it can be said that there is a significant difference between learning interest and attention as a result of behavioral differences among the peer groups.

Table 3 shows the results of tests of between-subjects effects. The results of the MANOVA test at the significance level of 0.05 ($\alpha = 0.05$) shows that the value of *peer group* on the average score of learning interest is 0.099. Since the obtained significance value is lower than the determined significance value, it can be concluded that there is an effect of *peer group* on learning interest. Meanwhile, the significance value of peer group on attention is ≥ 0.05, which is 0.200. Since the obtained significance value is higher than the determined significance level, it can be concluded that there is no peer group effect on attention.

4 DISCUSSION

The results of this study show that there is a positive and significant effect of peer group behavior on interest in learning gamelan and other traditional arts among teenagers in Yogyakarta City. This means that the closer the relationships are between teenagers and their peer group, the higher will be their interest in learning gamelan and other traditional arts. The results also show that the peer group has no effect on attention.

Peer group behavior was found to increase interest in learning and attention to the traditional of gamelan among a sample of teenagers in Yogyakarta City. Teenagers who participate in extracurricular activities involving gamelan and other traditional arts derive enjoyment if their peer group also participates in the activities. Peer group behavior is likely to influence students' interest in such activities and encourage them to participate. In turn, high interest will enhance motivation for learning.

A possible explanation for the role of the peer group in promoting interest in learning to play gamelan instruments is that they have developed their friendship from an early age, have similar characteristics, and are in regular close contact at school and in holiday activities outside school. This ongoing interaction and communication creates common bonds, including shared interests in learning a particular form of the traditional arts, such as gamelan.

Such learning in turn becomes a means of preserving local culture, including the history and important exponents of gamelan music, and developing pride in their cultural heritage as well as fulfilling individual potential.

5 CONCLUSION

The participation of close peer group members in extracurricular activities encourages teenagers to participate in activities involving gamelan and other traditional arts. Peer group behavior can enhance interest in learning such skills, thereby increasing motivation, enjoyment and application to learning and ensuring the successful implementation of such extracurricular programs.

REFERENCES

Chaplin, J. P. (2014). *Kamus Lengkap Psikologi[Dictionary of Psychology]*. Jakarta: Rajawali Pers.

Ghozali, Imam, (2011). *Analisis Multivariate dengan Program IBM SPSS 19[Multivariat Analysis by IBM of SPSS]*. Semarang: Badan Penerbit Undip.

Irianto, A. M. (2015). Mengemas Kesenian Tradisional dalam Bentuk Industri Kreatif: Studi Kasus Kesenian Jathilan[Making traditional art on creative industry: Case study of Jathilan]. *Humanika, 22*(2), 66. doi:10.14710/humanika.22.2.66-77

Kompas, E. (2017, April 25). *Anak Muda Ogah Melirik Seni Tradisional[Youth not interested in traditional arts]*. Retrieved from KOMPAS.com: http://nasional.kompas.com/read/2008/09/14/02422737/anak.muda.ogah.melirik.seni.tradisional.

Oktadus, H. Y. (2018). Hibriditas musikal: strategi budaya?[Musical hybrid: culture strategy]. *Jurnal Warna, 2*(2), 1-14.

Pallant, J. (2005). *SPSS survival guide*. Crow's Nest, NSW: Allen & Unwin.Slameto. (2013). *Belajar dan faktor-faktor yang mempengaruhi[Learning and the the factor of effect learning]*. Jakarta: Rineka Cipta.

Sugiyono, P. Dr. 2010. *Metode penelitian kuantitatif, kualitatif, dan R&D[Method of qualitative Research and R&D]*. Bandung: CV Alfabeta.

21st Century Innovation in Music Education – Astuti et al. (eds)
© 2020 Taylor & Francis Group, London, ISBN 978-0-367-11122-9

Children's learning of *Dolanan* songs in the digital era

E. Nurhayati & K.S. Astuti
Yogyakarta State University, Yogyakarta, Indonesia

ABSTRACT: This study aims to describe the use of children's songs as a learning medium in the digital era. This study is necessary because today, children's behaviors have changed and they tend to not care about the environment. *Dolanan* is a term for Javanese folk songs. These songs are for children that long existed before the 20th century and are still sung today. In Javanese children's songs there are very noble educational values that are useful for the development of cognitive, psychomotor, and affective development. This literature study seeks to uncover the values contained in *Dolanan* children's songs and their relevance in the current digital era. The results of this study conclude that children's songs are very important and must be maintained and developed in accordance with the progress of the times in order to be used as a media.

1 INTRODUCTION

Childhood is a period of play. Parents and childhood experts from the past to the present admit this phenomenon. In Indonesia, the theory that childhood is a period of play has developed since the beginning of the 20th century. Dewantara (1962) has applied this theory to learning in the school that he founded, called Taman Siswa, on beginning of 19th centuryby using *dolanan* songs (games songs) as a classroom subject matter.

Playing means that not everything is taken seriously. Playing a game means that an activity can be performed in a relaxed manner, not always referring to the overt meaning of the game, and does not have to be done perfectly. Likewise, the period of childhood should be a rewarding time for children because they have not been burdened with serious matters. Thus, every activity can be fun.Childhood is also an opportune time for various competencies to develop making it possible for a child to be exposed to the widest opportunities possible to develop his or her talents. Childhood is a golden age, a period in which all competencies can develop as well as possible. Mutiah (2010) states that various studies show that this period is a golden age during which the growth of the child's young brain is very fast because of the creation of billions of brain cells.

Older people understood this growth period so that parents of the past provided stimuli for their children to stimulate the development of various competencies in the cognitive, psychomotor, and affective domains. In Indonesia, nearly all tribes have songs to put the child to sleep (lullaby) with contents that can be in the form of advice, prayer, or life values. Most of the regions in Indonesia have songs that accompany games.

In Java, especially in the Special Region of Yogyakarta, Central Java and East Java there are children's songs. These songs can be used to accompany a game or be used in school to teach noble values that must be developed so that children can become successful people later in life.

In the era of globalization, the advancement of information technology has taken place so rapidly and has changed the culture of children's play. Gadgets are inherent in children's lives and better known than physically active games that develop the body and that encourage children to interact with their social environment. In this era, children are more familiar with their gadgets than their peers. This kind of familiarity makes the behaviors of children today different from children in previous generations. Today's children are more individualistic, selfish, and ignorant.

In the past, Javanese children were more concerned and responsive to their surroundings compared to Javanese children today. In the past, Javanese children wereeager to help when someone worked or did other activities. Today's children are not eager to help and tend to only watch someone who is working. Furthermore, the values and behaviors of children who care less about the environment must be changed and improved. One of the ways to improve appreciation of the environment is by reintroducing games in children's songs that involve physical movement so that children care about the environment. This caring attitude is also very important for developing an attitude of friendship in order to minimize confrontation. Reintroducing *dolanan* songs to children is one of the efforts that can foster positive attitudes in children. The verses that are in the children's songs not only provide entertainment but also give advice so that children and grandchildren behave as expected as stated in the contents of the song. This paper seeks to describe the meaning of children's *dolanan* songs and the values contained therein, and to examine the strategies for teaching these songs as a medium for the formation of children's character in the digital era.

2 LITERATURE REVIEW

2.1 *The children's dolanan song*

The children's *dolanan* song or *tembang dolanan* in Javanese, is one of the Javanese vocal traditions (Endraswara, 2005) classified as a folk song (Endraswara, 2009). As the name suggests, children's songs are usually sung while playing. For example, the *Cublak-Cublak Suweng* song is used by children as a medium of play to hide an object that is held in the hand, then another child has to guess which hand the object is hidden inside. Children's *dolanan* songs in the past were also used as a subject matter in the schools where teachers often used children's songs to start classroom lessons. An example is the song *Siji Loro Telu* (1, 2, 3), a *dolanan* song that is used to signal that the learning will begin. When the song was sung by the teacher, the students immediately sang the song with the teacher and the students had to sit neatly with their arms crossed to listen to the teacher so that when students were asked questions by the teacher, students could answer.

Hans Overbeck has identified that the number of children's songs in Java is actually more than 690. However, according to Overbeck (1930) there were only around 400 songs (Astuti, 2018). Therefore, it can be suggested that from the time of Overbeck until now there are probably fewer than 400 known songs. For this reason, efforts to revitalize this song tradition are necessary.The need for the revitalization of children's songs has been understood by academics such that current studies on children's toys are widely available. Winarti (2014: 51) reveals that *dolanan* songs are included in the Javanese song classification because they have a linguistic order in which they must be sung or preferred.*Dolanan* means "game", so in the song there is an element of simplicity. Winarti (2014) explains that children's songs can be in the form of one stanza such as *Pak Tejo wis teko* (*Pak Pelangi wis teko*). This song is sung when children see a rainbow while playing. Generally, the form of the song is more than one verse. On average there are 4 stanzas while the next stanza is in the form of development. An example is the song, *Nami Kulo Kamit*.

Nami kulo Kamit (My name is Kamit)
Kulo dados dokter (I'm a doctor)*Jampeni sing sakit* (heals a sick person)
Kulo sampun pinter (I'm already capable)

Radi keleresan (very coincidental)
Kulo sakit wudun (I have had an ulcer)
Sampun tigang wulan (for three months)
Mboten mantun-mantun (But have not recovered)

The song is typically sung by two groups of children. The first verse is a statement sent by the first group of children, while the second verse is sung by the second group of children. Furthermore, Winarti (2014: 55) explains that most of the *dolanan* songs are known as songs for children (Javanese children's games) while other songs do not have to be accompanied by certain activities.

2.2 *The meaning behind the creation of children's songs*

As previously mentioned, *dolanan* songs were created with the intention of developing noble values and attitudes through children singing the songs. The creators of children's *dolanan* songs convey advice through song lyrics with a positive message. Additionally, when these songs are more closely examined, positive values are not only contained in the lyrics of the song. Positive values are also contained in the game through physical movements that support good health or a particular motor skill.

Wang (2015) revealed that the *Jamuran* song is a song that can develop language skills. The song is like a quiz that asks children to explain different types of fungi and therefore can enrich children's vocabulary.

In another example, the song *Cublak-Cublak Suweng* is used to accompany a game where children guess which child is hiding objects in his or her hand and is an exercise for children to sense the feelings of others. For example, by observing each other's faces to discover which child is holding a hidden object, children can learn the meanings of facial expressions. Thus, this game can develop children's empathy and as previously stated children's songs are also created for the learning process in formal schools in the past. One of these songs is the following called, *Siji Loro Telu.*

SIJI LORO TELU

The song was formerly used by the teacher to start the class lesson. The lyrics "*siji loro telu tangane sedheku*" means "one two three, arms crossed". The meaning of the lyrics is that students must be ready to take lessons by sitting cross-legged because the lesson will begin soon. The next sentence, "*mirengake bu guru*", means "listen to the teacher". The sentence continues with "*menowo didangu*" where the teacher asks students to answer well.The poem in the second verse "*papat nuli limo*" means "four then five", followed by the lyrics "*lenggahe sing ta ta*" which means "to sit regularly". The next sentence, "*ojo podho sembrono*", means "do not do carelessly". The song ends with the phrase, "*mundhak ora*", which means "later it cannot".*The dolanan* song has important meanings with very deep values. Habituation to acceptable attitudes and discipline is the main foundation for children's success in the future. Getting used to a habitualized life of discipline will make it easier for students to achieve goals or targets that support them. With a positive attitude everything can be achieved.The song also teaches children to pay close attention to important subject matters delivered by the teacher. Sitting in the correct position can help students concentrate. This teaches students to focus on important things that must be learned so that goals will be easier and faster to achieve.At the end of the song, the poem "*Ojo Podho sembrono mundhak ora bias*" means "do not do carelessly and then it can't". In the poem it is implied that at the end of the lesson there will be an evaluation so that students do not need to work on the assignment when the evaluation is positive.It is favorable to invite students to start learning by singing simple songs because singing makes the learning environment serious but not tense. Students are happy to take lessons because of the singing. In addition, the song is very simple and easy so that students do not fear being wrong while singing.

2.3 *Characteristics of children's songs*

Children's *dolanan* songs are intended for children. Therefore, children's songs are adjusted to the characteristics of the child. If the soul of the child is happy, the song is happy or in Javanese, *sigrak*.

Fun in the song can be viewed from the element of tempo. Happy songs generally use a medium or fast tempo or in western musical expression, tempo andante (like people walking) or allegro (fast). The song "*siji loro telu*" will be more appropriate if the song is sung with the tempo andante and it is easy to follow because the tempo corresponds to the pulsing of the human heartbeat. Another characteristic is that the song's lyrics and melody must be simple. The lyrics used in children's songs are easily understood by children with verses that are not long so that children can easily remember the words. Likewise, the melody is adjusted to the child's vocal register which is between c1 and d2.

The children's *dolanan* song brings forth a happy character so generally, major scales are used as in the song "*siji loro telu*". However, there is also the possibility of using

KEMBANG MELATHI

minor scales in children's songs. An example is the following song "*Tak pethik-pethik Kembang Melathi*".

The song seems simple, but it is able to generate feelings so that the child can feel its beauty. In general, songs with minor scales will easily evoke feelings and make a deep impression on the child. Awareness of these beautiful feelings will stimulate children to be able to feel and like other beautiful phenomena that may be more complex. Beautiful feelings will make children happy and happiness will bring children to always think positively and be able to be thankful for everything they receive.

Another interesting aspect is that in the song *Kembang Melathi,* there is an activity where children bring jasmine flowers and then the flowers are sprinkled. This moment is a very pleasant moment for children especially girls.

3 DISCUSSION

The digital era grew exponentially with the rapid development of computer information technology bringing consequences that must be faced. Digital technology makes it easier for people to receive information faster and different communities can enjoy digital facilities freely. Today, all levels of society including children can access information and communicate via various kinds of gadgets. But Setiawan (2017: 3) argues that the era of digital technology holds a variety of potentially negative impacts that can harm humans. The ease of all jobs with various applications and technology actually makes a person move less so that physical activity decreases and laziness arises.

Thus, efforts need to be taken to better activate children's physical activities and skills. This is necessary because with gadgets only motor and nerves at the fingertips will be developed while a greater range of motions such as running, throwing, and catching are not trained.

The use of children's songs in daily games can be used as a solution to develop children's movements. Not only will physical skills increase but the cognitive and affective aspects of the child will also develop more fully.

It can be said that games have become part of children's lives today. For children to not feel alienated from the surrounding world does not mean that digital games must be abandoned. Children can enjoy application games but at the same time they can also develop cognitive, psychomotor and affective balance.

With digital technology, children's songs can be accessed and children can recognize and imitate songs through social media. But that does not mean that the children should be left alone to sing children's songs on social media. Parents and schools must facilitate this process. Parents can show and accompany children playing with their friends singing and playing children's toys. This kind of activity stimulates very high-quality communication. Alia (2018: 73) explains that quality communication in early childhood will help children to be able to recognize and distinguish right from wrong, make it easier to know the root of problems and provides for the best interests of the child.

The teacher can assign students the task of finding children's songs through their gadgets and invite them to play with children's songs. The teacher can also invite their students to find traditional games that the children can use to play. If possible, children with parents and teachers can together plan a play project based on children's songs. This is in accordance with children's needs as studied in the field of child development. As stated by Hermoyo (Alia, 2018: 68), children communicate using their body words and gestures. Additionally, children must always be encouraged to express their desires and establish relationships with other people.

Singing children's songs directly guided by parents or teachers brings children closer to parents and teachers. Direct communication between children and their parents and teachers will lead to stronger bonds so that children trust parents and teachers. Under these conditions, the child will obey adults making it easier for the child to be guided and directed.

In addition, singing by imitating and direct contact with parents, teachers, and instructors can not only transfer singing skills but can also develop a child's connection to his or her emotions. This kind of development will not happen if the child just imitates videos on YouTube or uses gadgets only. The presence of a "living" model is needed to develop children's competencies as a whole person.

However, considering that children can no longer avoid digital technology, digital use in the learning process must not be discouraged as the use of gadgets is still needed. Sukirman (2017) reveals that in many studies it was found that games have the great potential to be used as learning media. This is very possible because in video games there are rules and tasks that must be completed to achieve the goal. Through games, teachers and parents can help children at the unconscious level by including pedagogical elements.

4 CONCLUSION

Planting noble values to shape character must be done early so that a child will become a complete human being and can achieve a good life. Javanese people in the past taught noble values to children in various ways including using children's songs. By using children's songs, there are direct and indirect ways to instill desired values and develop cognitive, psychomotor, and affective domains that are very important for the formation of a child's character. In this digital era, planting values through children's songs can still be done. Parents and teachers can combine digital technology with children's songs as part of the process of exploring songs and traditional games while maintaining communication through direct interactions with children. In this way, children can develop cognitive, psychomotor, and affective skills comprehensively without having to avoid digital influences.

REFERENCES.

Alia, T. (2018). Pendampingan Orang Tua pada Anak Usia Dini dalam Penggunaan Teknologi Digital [Parent's Guiddance in Early Children by using digital technology]. *A Journal of Language, Literature, Culture, and Education. POLYGLOT, 14*(1), 65-78.

Astuti, K. S., & Widyantoro, A. (2018, September). The teaching of morality through musical elements based on learning models of the Netherlands and Indonesia. In *Character Education for 21st Century Global Citizens: Proceedings of the 2nd International Conference on Teacher Education and Professional Development (INCOTEPD 2017), October 21-22, 2017, Yogyakarta, Indonesia* (p. 389). Routledge.

Dewantara, H. (1962). *Karja Ki Hadjar Dewantara: Bagian 1[Work of Ki Hadjar Dewantara 1st]. Pendidikan.* Madjelis Luhur Persatuan Taman Siswa.

Endraswara, S. (2005). *Tradisi lisan Jawa: warisan abadi budaya leluhu[Javanese oral tradition: the eternal heritage of ancestral culture].* Narasi.

Endraswara, S. (2009). *Metodologi penelitian folklor[Research Method for folklore].* Media Pressindo.

Mutiah, D. (2010). Psikologi bermain anak usia dini[Games Psychology of Early Children].

Setiawan, W. (2017). Era digital dan tantangannya.

Sukirman, S. (2017). Peningkatan Atmosfer Belajar Siswa Generasi Digital Native Melalui Pembelajaran Berbasis Game. Prosiding Seminar Nasional Pendidikan Berkemajuan dan Menggembirakan [Atmospheric Enhancement of Native Digital Generation Student Learning Through Game-Based Learning. Proceedings of the National Seminar on Advancing and Joyful Education](The Progressive & Fun Education Seminar) 2nd.

Wang, J. C. (2015). Games Unplugged! Dolanan Anak, Traditional Javanese Children's Singing Games in the 21st-Century General Music Classroom. *General Music Today, 28*(2), 5-12.

21st Century Innovation in Music Education – Astuti et al. (eds)

The revivals of Haydn's oratorio *The Creation*

T. Bramantyo
Indonesia Institute of the Arts of Yogyakarta, Yogyakarta, Daerah Isitmewa Yogyakarta, Indonesia

ABSTRACT: This study has conducted an analysis of Haydn's oratorio *The Creation.* It found that each phase of the six days of Creation that made up Part One and Part Two of the oratorio began similarly with a passage of biblical narrative sung by one of the three archangels, Gabriel, Uriel and Raphael. The passages also contained materials of a lyrical or descriptive nature, amplifying the objective narration. These sections contained some of Haydn's most colourful music. Most importantly, there was something inside of this magnificent oratorio of Haydn, despite its musical structural standpoint. Something of 'most dramatic music', which could be understood when it is studied toward in a more satisfactory level of study, that was to say, through its aesthetics observation, because *The Creation* has so many symbolisms inside itself by which its meanings could be appreciated only through the study of its structural plans.

1 INTRODUCTION

In the event of Haydn's oratorio *The Creation's* 217-year celebration this year, I would like to contribute to the story of this oratorio from its early history to the present, which I refer to as its revivals. In an introduction to Haydn's musical career written by Greenberg (2000), it was described as follows:

> No composer did as much as Haydn to create and develop the Classical symphony and quartet and, to a lesser extent, other Classical genres, such as the trio and piano sonata. 'Papa' Haydn earned his nickname because of his good nature, the care he took of the musicians who worked in the orchestras he directed and his sense of humour, which often found its way into his music.

The author has excerpted the long story behind the success of Franz Joseph Haydn, who was born in 1732 in the small town of Rohrau in Eastern Austria near the border with Hungary. In 1758, Haydn obtained his first permanent musical post, to Count Ferdinand Maximilian von Morzin. Early in 1761, Count Morzin was forced to disband his orchestra through financial difficulties so that Haydn had to look for another post. Prince Paul Anton Esterhàzy heard that Haydn was unemployed and at once offered Haydn the appointment of assistant conductor of his orchestra at Einstadt, his palace near Vienna (Butterworth, 1977).

The story progresses to Haydn's musical career in Vienna, then as a *kappelmeister* (chapel master) and composer. An opportunity was given to him to visit London in January 1791 and he was much impressed by the vastness of the city. In May 1791, a Handel festival was held in Westminster Abbey on a huge scale, with over a thousand singers and players. Haydn attended these concerts and the experience had an enormous effect upon him. After one performance of *The Messiah*, and of the 'Hallelujah Chorus' in particular, he said of Handel, "he is the master of us all". It was as a direct result of this festival that Haydn began to consider the possibility of composing an oratorio himself.

The dream came true when in April 1798, the first two performances of *The Creation* before a selected audience in the Palais Schwarzenberg in Vienna were announced. The '*Gesellschaft der Associerten*' (Society of Associates), a group of music-loving noblemen, organised these two concerts. Haydn himself conducted the orchestra and the chorus with Antonio Salieri on the keyboard. Its success surpassed all expectations.

The success of *The Creation* was told by the critics of the *Allgemeine Musikalische Zeitung* who reported that, "it is difficult to imagine the silence and attention with which the whole oratorio was listened to, only broken by quiet exclamations in the most dramatic places, and received with enthusiastic applause at the end of each piece and each section" (Kluge, 1992).

The reasons why I restudied and revived this oratorio were, on the one hand, I saw the decline of Haydn's reputation in the concert repertoires, while on the other hand I saw the potentialities of Haydn as someone who got the credit for the invention of the symphony, string quartet and sonata form. Some of the credits came from Tovey, d'Indy, Heinrich Schenker, Arnold Schoenberg and Arturo Toscanini, who gratitude to Haydn for his works have made each of their life's have been wonderful. *The Creation* certainly was the example that might have provided the initial spark of the revival just by maintaining a nominal presence in the history (Proksch, 2015).

2 THE LIBRETTO

An attractive account concerning the search for the subject of Haydn's *The Creation* tells how during his stay in London his friend Salomon, an impresario, gave him a novel written by Milton entitled *Paradise Lost*. The story of this novel was based on the Bible and Haydn, with the help of Baron Gottfried van Swieten, the prefect of the Imperial of Library in Vienna, had started to write a libretto for his first oratorio. (Olleson, 1968).

Meanwhile, in another account, Olleson impressively wrote that the earliest of the reports is found in a letter by van Swieten himself, which was printed in the *Allgemeine Musikalische Zeitung* in January 1799, and is the only one to come directly from someone (van Swieten) who had been immediately involved in the libretto (Olleson, 1987).

Another account was written by G.A. Griesinger who had come to know Haydn well in his capacity as intermediary between Haydn and the Leipzig publishing house of Breitkopf & Härtel. According to Griesinger, the first idea of the oratorio *The Creation* belongs to an Englishman by the name of Linley, and Haydn was to compose Linley's text for Salomon (Griesinger, 1965).

The third account, still according to Olleson (1987), was a conversation between Salomon and Haydn concerning the plan for writing *The Creation*. Salomon suggested that since he had been successful in so many musical enterprises to date (a success to which Haydn had contributed in no small measure), he became ever more venturesome for new undertakings.

Hence, Haydn only remembered the English text when Baron van Swieten said to him, shortly after his arrival in Vienna; "Haydn, we are still waiting to hear another oratorio from you!" He told the Baron how things stood and showed him the English text. Van Swieten offered to prepare with a free translation of it in German and, once this was finished, it was wrong to think of that the oratorio was of Haydn's work that it had been written for London.

There are structural parts of the libretto that can be divided as follows: in the first two parts of the oratorio one can see a consistent plan for the six days of the Creation, and there are signs that the plan may have been still more consistent in the original libretto. Each act of *The Creation* was first stated in the words of the English-language Bible (the first chapter of Genesis, with one verse from the second account, Genesis 2:7) and then provided with a free paraphrase and commentary. In the case of the third, fifth and sixth days, this procedure was followed in biblical style but not directly from the Bible, introducing a song of general phrase by the heavenly host.

The writer confirmed that the first two parts were clearly articulated into the six days of the Creation. Each day had its own internal organisation, whereby it opened with a biblical narration, had a middle section of descriptive or lyrical commentary and closed with a song of praise. The first day ended up with a descriptive chorus, rather than one of praise; a particularly demanding day, such as the sixth, which had to produce the entire animal kingdom, with freer, more lyrical passages. As befits the seventh day, the oratorio relaxed from the rigours of *The Creation* on Parts Three; here there was no biblical narrative, and the mood was entirely lyrical and idyllic.

The writer also found that the libretto of *The Creation* literarily could be divided into parts (*teile*) and scenes (*auftritten*). The indication for each part is shown in Table 1.

The arias including *terzet* (trio) of the three archangels and duet of Eve and Adam with their text sources can be described as shown in Table 2.

3 MUSICAL SOURCES

Landon wrote that, with the exception of Handel's *Messiah*, perhaps no oratorio captured the imagination of the musical public as completely as Haydn's *The Creation*. Yet, whereas Handel's various revisions to accommodate the principle principal singers had been carefully documented by modern scholars, up to now it had been assumed that *The Creation*, as edited by Mandyczewski in the early 1920s for the old Breitkopf & Härtel collected edition (published in 1803), provided the definitive text (that is, it contains van Swieten's word-books). H.C. Robbins Landon had clearly described that Breitkopf & Härtel issued the score in two versions, one with German and French texts, and another one with German and English ones (Landon, 1977).

However, the writer found the fact that, as mentioned by an authentic conductor named Brown (1986), Mandyczewski's edition (1920s), whose only source was the first edition

Table 1. Indication of each part in *The Creation*.

Indication	Translation	Movement	Days of Creation
Erster Teil	1st part	1–13	
Erster Auftritt	1st scene	1–2	1st day
Zweiter Auftritt	2nd scene	3–4	2nd day
Dritter Auftritt	3rd scene	5–10	3rd day
Vierte Auftritt	4th scene	11–13	4th day
Zweiter Teil	2nd part	14–28	
Erster Auftritt	1st scene	14–18	5th day
Zweiter Auftritt	2nd scene	19–28	6th day
Dritte Teil	3rd part	28–34	7th day

Table 2. Texts sources of the arias including trio (*terzet*) and duet.

Movement	Voice	Texts sources
2	Uriel	*Paradise Lost* VII: 236–242
6	Raphael	*Paradise Lost* VII: 285–287, 298-300, and 304–306
8	Gabriel	*Paradise Lost* VII: 315
15	Gabriel	*Paradise Lost* VII: 421
18	Terzet: Tree Archangles	*Paradise Lost* VII: 399–400 and 412
22	Raphael	*Paradise Lost* VII: 499–505
24	Uriel	*Paradise Lost* VII: 288–305
27	Terzet: Three Archangels	Psalm 104:27–30
30	Duet: Eve and Adam	Source unknown
32	Duet Eve and Adam	*Paradise Lost* IV: 610–-630

Table 3. Texts sources of the recitatives sung without accompaniment (*secco*) or with accompaniment (*accompagnato*).

Movement	*Secco/Accompagnato*	Source
3	Combined	Genesis 1:7
5	*Secco*	Genesis 1:9–10
7	*Secco*	Genesis 1:11
9	*Secco*	Source unknown
11	*Secco*	Genesis I:14 and 16b
12	*Accompagnato*	Genesis I:16–19
14	*Accompagnato*	Genesis I:20
16	Combined	Genesis I:21–22
17	*Secco*	Source unknown
20	*Secco*	Genesis I:24
21	*Accompagnato*	Genesis I:25
23	*Secco*	Genesis I:27
25	*Secco*	Genesis I:31a
29	*Accompagnato*	Source unknown
31	*Secco*	*Paradise Lost* V: 17–31
33	*Secco*	*Paradise Lost* IV: 774–5

published by Haydn (through Breitkopf & Härtel in 1803), had subsequently been the basis for nearly every modern performance. Ironically, it seemed that Haydn never used this edition (Breitkopf & Härtel, 1803) for any of the performances with which he was directly involved. Indeed, like Handel, Haydn was constantly tinkering with the music, so that we could not speak, wrote Brown, of a definitive version.

The author also studied *The Creation's* basic plans, which were generally as one recitative followed by an aria, with the characteristic of instable tonal areas in the recitative, followed by a stable tonal centre in the aria. The orchestration plan of *The Creation* was one of sparse instrumentation in the recitative followed by full instrumentation in the aria, with the two linked by common vocal forces. Not unsurprisingly, the text matches this plan. The recitative texts were generally from Genesis and were narrative. The aria texts were generally from *Paradise Lost* and were more reflective.

More importantly, the author also found that the division between overture, *arias*, *recitativo*, *secco*, *accompagnato* and chorus was very delicately balanced. The only duet was between Adam and Eve in Part Three, but the soloists often sang with the chorus (for example, Duet and Chorus No. 30); also Part Three, the beautiful *Terzetto* No. 27, which was really of a larger structure, and of course the famous conclusion to Part One, the Chorus No. 13 'The Heavens are Telling'.

As for the general pictures of the tonalities used in almost all the parts of this oratorio, it was certainly in the progressive tonality. The reasons for this elaboration of these tonalities seem to have been the concern of the composer, done as early as the introduction of Adam and Eve into the world of angels, of God and Satan, and of the fallen angel s Part One) .

4 DISCUSSION

4.1 *Part One*

4.1.1 *The overture*

Here the author begins to study the overture of *The Creation* that is opened up by the orchestra and then concludes with the recitative of Raphael at bar 60, '*Im Anfange schuf Gott*' ('In the beginning God Created'), Chorus '*Und eine neue welt*' ('A New Created World'). The overture is entitled '*Die Vorstellung des Chaos*' ('Representation of Chaos').

Looking into this overture or *Einleitung* closely, it can be noted that it is an entire work of symbolism without words; it is, in a sense, up to the listener to make up his or her own *programme* when listening to what is officially called '*Die Vorstellung des Chaos*'. The clue to appreciating this symbolism might start with the question of how the chaos can be represented without words.

In answering the question, however, let the author refer to Charles Rosen, who, in his book entitled *The Classical Style*, wrote almost exactly the same enquiries, "By what is chaos represented, and how can Haydn's musical language express this and still remain language?" The answer from Rosen (1972) was, "Simply by the absence of clear articulation in large phrase-groups, which merge and blend with each other and by the clear at definite cadences".

Meanwhile, Sir Francis Donald Tovey, in his essays in 'Musical Analysis', argued that it is the isolation of the musical event that creates the chaos. "In other words, they are admirably chaotic: they are not nonsensical, for their resolution is quite orthodox, though they occur as shocks for which their antecedents did not prepare us" (Tovey, 1972).

Different views should also be added here. According to Landon, *The Creation* text is full of symbolism; it is also full of descriptions that automatically suggest *Tonmalerei* (tone painting), or realisation of these descriptions. Throughout *The Creation*, Haydn explored symbolism in two forms; first, tonality, and second, the use of instruments to express a concept (Landon, 1985).

It seems that in this overture, in order to describe this symbolism, it can be seen that when Raphael enters as the first human voice after the description of 'chaos', Haydn describes the beginning of the world as we know it from Genesis I:1–2: "In the beginning God created the heaven and the earth. And the earth was without form, and void: and darkness was upon the face of deep."

4.1.2 *No.2 aria (Uriel) with chorus*

Now when the author looks into the whole set of arias in *The Creation*, generally, all of the movements are unified by their metric flows and musical contents. Take, for example, movement No. 2; the differences between sections are delineated by their non-idiomatically related tonalities, contrasting texts, moods, texture and orchestration-approach to the extreme of the accompanied recitatives.

Technically, No. 2 is an aria with a chorus in a complex and symbolic key relationship of A major - C minor - A major. This symmetric concept of structure is a typical form of the eighteenth-century concept of symbolism and it was worked well in Haydn's *The Creation*. The content of the text according to Landon is 'The shadows of night flee to the deep of abyss, to endless night', from A major down to C minor, and when A major reappears 'A New Created World springs up at God's command' (Landon, 1970).

4.1.3 *No. 3 recitative (Raphael)*

This is the first of the famous descriptive recitatives in which the music describes some aspect of the universe; these sections are so constructed that the music comes first, the words afterwards. Here we have: (1) 'outrageous storm'; (2) 'as chaff by the winds are impelled the clouds'; (3) 'lightning'; (4) 'thunder'; (5) 'showers of rain'; (6) 'wasteful hail'; (7) 'the light and flaky snow' – all portrayed with relish and a great sense of instrumental colour.

4.1.4 *No. 4 chorus with soprano solo (Gabriel)*

For the writer, it is very impressive to find that the Archangel Gabriel appears for the first time in this solo in a stunning allegro with an oboe *obbligato*. The chorus comes in at measure 16 (m. 16). The violin's rhythm is doubled by the timpani (m. 16) and by an oboe *arpeggio*. Structurally, the layout seems to be like one huge *ritornello*: solo oboe (mm. 1–4), solo soprano (mm. 4–16), then comes the chorus in Handelian manner (mm. 16–30), and then the chorus and solo soprano (mm. 31–46) recalling the first theme, but now progressing to a stupendous passage where the soprano soars above the chorus and orchestra to a high C (m. 44) with only the first flute in the same range. No. 4 rounds out the first C major cycle of the oratorio; the key (C Major) that will not return until the end of Part One.

4.1.5 *No. 5 recitative (Raphael)*

What can be described about this recitative is that this is a *secco* recitative that leads to the first aria. The text is Genesis I:9–10. The melodic content of the beginning and ending of this recitative is very similar to that of the other recitatives. Each of the melodic cadences that conclude the two sentences of Movement 5 ends in a descending perfect fourth. This is similar to the final cadences in all of the *secco* recitatives except Movements 9, 11 and 20. It should be noted here that while this may not have been a conscious linking on Haydn's part, it, along with the overall syllabic declamation, harmonic language, tonal instability and Spartan orchestration, does add unity to the *secco* recitatives in *The Creation* (Lucas, 1977).

4.1.6 *No. 7 recitative (Gabriel) and No. 8 aria (Gabriel)*

The writer should specifically note that this beautiful aria entitled '*Nun bout die Flur das frische Grün*' ('With verdure clad the fields appear') in B flat, is probably the most famous single aria in *The Creation.* In its simplicity of melody, its elegant off-the-beat phrasing and its limpid orchestration (no oboes: only flutes, solo clarinet, bassoons and B flat alto horns), it has become a touchstone for the Viennesse Classical Style in vocal writing. There is just enough *coloratura* to enable the soprano soloist to display her agility, but never too much.

4.1.7 *No. 9 recitative (Uriel) and No. 10 chorus*

In order to study this recitative and chorus, the writer should also mention that according to Landon's opinion that if there is any part of the oratorio that suggests a direct comparison with Handel, this chorus is surely it; partly because it is in Handel's characteristic trumpet key of D, partly because we have fugue in the middle. But furthermore, Landon wrote, we must recall that this is modernised Handel, with the orchestration of the Mozartised *Messiah* and with the trumpets restricted (as they were in Mozart's much criticised adaption). One wonders if Haydn would not have re-orchestrated Handel more fittingly than Mozart, to whom this style was basically foreign: we mean not, of course, the Baroque, but rather the oratorio. Mozart was certainly the most universal genius in the history of music, but oratorio writing did not suit him at all (Landon, 1985).

4.1.8 *No. 11 recitative (Uriel) and No. 12 recitative accompagnato (Uriel)*

A *secco* leads into the justly celebrated first sunrise, the first moonrise and the sons of God, gathering together and shouting for joy, and leading without pause or even cadential resolution into No. 13, the chorus with the Three Archangels Soli (Landon, 1977).

However, Brown described that the sunrise of the No. 12 also has precedent in French pictorialism, as does the high flute colour of No. 27, which suggests both pastoral bliss and the celestial world on earth (Brown, 1986). He also added that Haydn's most densely pictorial music is concentrated in three extended accompanied recitatives for solo voice (Nos. 4, 13 and 21), which must have delighted every *liebhaber* (lover).

4.1.9 *No. 13 chorus (with the three archangels)*

It was so impressive to look over this chorus that, according to Landon, is the most famous chorus in *The Creation*, and for many years Haydn's most played choral piece altogether, 'The heavens are telling' sums up the first part in a glorious outpouring of C major (Landon, 1985).

4.2 *Part Two*

4.2.1 *No. 14 recitative and No. 15 aria (Gabriel)*

As the sketches show, Haydn always intended this to be a *bravura* (florid passage requiring great skill and spirit) piece. The author should mention that Landon refers to Tovey's statement that though the accompanied recitative starts on the six chord of C major, as if to continue a narrative that has been punctuated rather than interrupted, the *ritornello* of the aria is so long (34 bars) that it was obviously made to serve as the introduction to Part Two. As in the previous soprano aria, the oboes are dropped and the clarinets have a prominent part (Tovey, 1972).

4.2.2 *No. 16 recitative (Raphael)*
This is a very impressive recitative and a good example of the great description of life multiplying with the divided lower strings. Here the divided violas and divided cellos with the double bass in its lowest register, together in the key of D minor, create a mysterious, veiled sound – as mysterious as the secret of life itself. The whole passage remains piano and the complexities of the divine spark of life are mirrored in the intricate, almost Bachian lines of the stringed instruments.

Brown (1989) impressed us with his statement; "next to the appearance of light, perhaps the most impressive of these accompanied recitatives is Raphael's 'Be fruitful all, and multiply!'" (No. 16). Again, it can be admired by the amateur for the sensuous orchestration of the lower strings and its declamatory majesty, and by the connoisseur for its impressive part writing.

4.2.3 *No. 17 recitative (Raphael) and No. 18 Terzetto, leading to No. 19 chorus with soli*
A very short *secco*, wrote Landon (1977), leads to the A major *Terzetto*, scored with the characteristic delicacy that Haydn always reserves for this key (flutes, oboes, bassoons, horns, strings). The words are first given to Gabriel, who sings of the 'sloping hills' and the 'crystal drops' of the fountains. Uriel then describes "the cheerful host of birds with their wings shining in the sun", and Raphael follows with "the fish in the ocean and the immense leviathan (who) sports on the foaming wave".

4.2.4 *No. 20 recitative (Raphael) and No. 21 recitative (Raphael)*
Here we find another short *secco* that leads to what is probably the most famous of the descriptive accompanied recitatives. The tawny lion is portrayed, 'cheerful roaring' with *ff* trombones and double bassoon, followed by the 'flexible tiger' and the 'nimble stag'. For the stag's music, we move to the hunting metronome of six-eight but, more surprising, the music is a quotation from the recapitulation of the 'Surprise' Symphony's first movement, bars 207 *ff*. There then follows the horse with 'flying mane and fiery look' and the bucolic scene (pastoral poem) with 'fleecy, meek and bleating flock', and finally the famous imitation of the worm creeping 'in long dimension'. The final bass note of the recitative, D, is usually dropped down an octave here, a liberty of which Haydn would certainly have approved.

4.2.5 *No. 22 aria (Raphael)*
This is another famous aria of this oratorio. Here Raphael sings of heaven's glory '*Nun scheint in pollen Glanze*' ('Now heaven is fullest glory shone') with strident horn, trumpet and timpani fanfares, the basic rhythm of which colours much of this music. Beneath the bass solo's words '*den Boden drückt der Tiere Last*' ('By heavy beasts the ground is trod'), the famous bassoon and double bassoon passage suddenly growls out.

4.2.6 *No. 23 recitative (Uriel) and No. 24 aria (Uriel)*
Genesis I:27 described that "God created a man in his own image … He breathed into his nostrils the breath of life, and man became a living soul". This is the text of this *secco* recitative, which leads to the aria in C major '*Mit Würd und Hoheit an getan*' ('In native worth and honours clad').

About this magnificent aria, Landon gave us his impression that if one had to choose the greatest single aria in *The Creation*, it would probably fall to this majestic, poetic and original piece. Its freedom of form, he wrote, is extraordinary and has been much admired. Landon quoted Tovey's comment, which described that "we have not only the quintessence of Haydn but the perfection of *bel canto*" (Tovey, 1972).

The formal shape of the text of this number is formed into two distinct sections. The trumpets and the timpani leave the orchestra (m. 54) as soon as Haydn described man, and a beautiful *obbligato* violoncello part emerges (m. 55) and dominates the texture to the very end. To illustrate 'the breath and image', Haydn introduces a fantastic modulation to the flattened submediant Ab (mm. 48–51). The second part is a song to 'love and joy and bliss', which dies away in pianissimo (mm. 55–10l).

Brown (1990) described how this aria moves from activity to relaxation. It becomes martial, with dotted rhythms as man is declared 'the lord and king of nature all', but the music for the concluding three lines of text provides tonal stability, rhythmic relaxation, regularity of phrasing and a new counter-melody in the cellos that anticipates the later duet of Adam and Eve.

4.2.7 *No. 25 recitative (Raphael), No. 26 chorus, No. 27 Terzetto and No. 28 chorus*
In analysing these numbers, it is found that after an eight-bar *secco*, we begin the mighty 'Finale' with which Haydn concludes the second part of *The Creation*. No. 26 is a shortened version, with different *fugato* to No. 28, but the final *ritornello* of No. 26 (mm. 33–37) cleverly introduces material from which the great double fugue in No. 28 (mm. 10–68) will be formed.

In No. 27, which has a striking projection of words to music, Haydn uses several methods to treat the words symbolically: (a) he breaks the flow of the music by suddenly switching to triplets; (b) he leads the bass solo (Raphael) to the bottom register; (c) he takes us to Gb major (home key: Eb). In the English version, the low Gb of the Archangel coincides with the word 'dust'; in German it is equally effective, coming earlier but on the note Cb. Notice too, wrote Landon (1970), how Haydn prepares the words '*Du wendest*'. Up to this point the orchestra has consisted of wind instruments only (flute, oboe, clarinets, bassoons and horns) with Gabriel and Uriel. They finish their cadence at bars 33 *ff*, then the music 'turns away': (a) from the wind band; (b) from the major; (c) from the soprano and tenor *soli*.

In the final part of this solemnly beautiful trio, Haydn depicts 'now force and now delights' with a motif that rises up an octave in all three voices, starting with the Archangel Raphael (mm. 85–89).

4.3 *Part Three*

4.3.1 *No. 29 recitative (Uriel)*
Looking closer at this recitative, there are three flutes used here, symbolising 'the morning young and fair' and 'still innocent Adam and Eve'. In the first section, or until the recapitulation of the opening material, only flutes, horns and strings are used. Measures 29–32 present us with the repetition of the beginning but in its original state, that is, just the three flutes unaccompanied. These unaccompanied flutes were to open No. 29, but Haydn later added the *pizzicato* strings. After the flutes have repeated the opening material, they are silent. Similarly, the horns have a short solo section (mm. 39–40) and are the removed, making way for bassoons and oboes. By the time we have modulated from E to G, we can see No. 29 for that which it really is, namely 'a fast slow introduction' to No. 30.

4.3.2 *No. 30 duet and chorus (Adam and Eve)*
This movement has a very slow build-up. First, a fine oboe singly emerges from the strings triplets, then Adam and Eve begin their song of thanksgiving. It is not until measure 24 that Haydn adds more woodwind (flutes, oboes, bassoons). At the end of that measure, the chorus enters, almost chanting. At measure 31, the extraordinary timpani solo begins and continues throughout this adagio. This entire No. 30 is unique in late-period Haydn. The ease with which the key structure is planned is as admirable as the invention and inspiration with which the whole piece is informed.

4.3.3 *No. 31 recitative (Adam–Eve) and No. 33 Duetto (Eve–Adam)*
We need the widest possibilities of contrast after nearly four hundred measures of No. 30, and Baron van Swieten provided Haydn with the simplest method, that is, a long *secco* recitative for Adam and Eve. It can be noted that the Bb chorus ending Part Two could also be regarded as the dominant key of the forthcoming Eb duet. To underline this relationship, Haydn has recourse to a clever device.

5 CONCLUSION

From the analysis above it is clear that Haydn has put into his *Creation* immensely as a work of composition that brought experience of his lifetime: the technique of thematic work came from the symphony (he wrote as many as 108 symphonies according to the 1980 listing of Grove's Dictionary), the detailed knowledge of character and the sound effect of individual instruments betrays a composer of chamber music, the melodic diversity and seemingly natural relationship between words and music, the writer of operas and songs. In the powerful fugues and double fugues of the choruses, the organist and master of counterpoint demonstrates his skills. But in the tone paintings of storm and rain, birds and fish, sea and mountains, it is an imaginatively serene artist. The end result of this collaboration is an oratorio that denotes both a temporary climax in, and an end to, the development of the style.

Finally, it would be wise to cite Hogwood's (1983) impression, which has quite impressed us, arguing that '*The Creation* is more than a work for musical amateurs and connoisseurs, conservatives and progressives, and more than a variety of national tastes: it is as two centuries have verified, an oratorio of all time'.

REFERENCES

Brown, P.A. (1986). Haydn's chaos: Genesis and genre. *The Musical Quarterly*, 73(1), 18–59.

Brown, P.A. (1989). *Performing Haydn's The Creation, reconstructing the earliest renditions*. Bloomington, IN: Indiana University Press.

Brown, P.A. (1990). *The Creation*: An oratorio for all tastes and times, a program note on the CD of The Academy of Ancient Music Orchestra & Chorus, Conducted by Christopher Hogwood, Edition de L'Oiseau-Lyre, The Decca Record Company Ltd, London.

Butterworth, N. (1977). *Haydn*, in the series of *The illustrated lives of the great composers*. London, UK: Omnibus Press.

Greenberg, R. (2000). *Haydn: His life and music*. Los Angeles, CA: The Teaching Company.

Griessinger, G.A. (1965). Correspondence with Breitkopf & Härtel, In *Haydn Yearbook No. 3*.

Kluge, A. (1992), Choral works in the spirit of the Enlightenment, Haydn's Creation, a program note to a Sony Classical CD, The Royal Edition, No. 36OF 100.

Landon, H.C.R. (1970). Haydn's Creation, in *Essays on The Viennese Classical Style*. London, UK: Creset Press.

Landon, H.C.R. (1977). *Haydn, chronicle and works, Vol. IV: Haydn, The Years of "The Creations"* 1796–1800. Bloomington, IN: Indiana University Press.

Landon, H.C.R. (1985)."Foreword" in *The Creation and The Seasons, complete authentic sources for the word-books*. Cardiff, UK: University College Cardiff Press.

Lucas, J.A. (1977). *A conductor's analysis of 'The Creation' by Joseph Haydn* (Master's dissertation, University of Iowa, USA).

Olleson, E. (1968). The origin and libretto of Haydn's Creation. In *Haydn Yearbook, No. 4*. London, UK: Universal Edition.

Olleson, E. (1987). Joseph Haydn: *The Creation*, a program note on the CD of the Chor und Symphonie-Orchester des Bayerischen Rundfunks, Live recording from München Hercules-Saal, 6/1986, Conducted by Leonard Berstein, Polydor International GmbH, Hamburg, Germany.

Proksch, B. (2015). *Reviving Haydn: New appreciations in the twenty-first century*. Rochester, MN: University of Rochester Press.

Rosen, C. (1972). *The classical style*. New York, NY: W.W. Norton.

Sadie, S. (Ed.), (1980). Joseph Haydn, in *The New Grove's dictionary of music and musicians*, Book 8. London, UK: Macmillan.

Temperley, N. (1983). New light on the libretto of 'The Creation'. In C. Hogwood & R. Luckett (Eds.), *Music in eighteenth-century England: Essays in memory of Charles Cudworth*. New York, NY: Cambridge University Press.

Tovey, D.F. (1972). *Essays in musical analysis*, New York, NY: Oxford University Press.

21st Century Innovation in Music Education – Astuti et al. (eds)

The perspective of Western music in *wayang kulit* gamelan

Y.E. Susilo
Institute of the Arts of Yogyakarta, Yogyakarta, Indonesia

ABSTRACT: The Javanese culture has facilitated the introduction of diatonic music to the island of Java thanks to its trait of recognising and accepting other cultures. There are two pathways in the development of *wayang kulit* (shadow puppetry). The first is that of the *Kraton* (palace) (up to this time, the original honourable principle of shadow puppetry has been maintained), while the second pathway of development happened outside the palace and has been influenced by other cultures. One of these was the combination of Western diatonic music with shadow puppet gamelan accompaniment in either the *limbukan* or *goro-goro* scenes of the play. Although there are differences in pitch between gamelan *slendro* and Western diatonic music, both instrument types (gamelan and Western diatonic) can be unified successfully while presenting *gendhing* (songs) in the *limbukan* or *goro-goro.* The differences in pitch and interval are clearly described in this study. The method used in this study was a qualitative one with musicological and historical approaches.

1 INTRODUCTION

The essence of music is sound. Sound can come from the natural voice. Humans are part of nature. Nowadays, the presence of Western diatonic music in traditional Javanese music is regarded as a commonplace thing, in the music of wayang, *campursari* (crossover), gamelan, and so on. In the gamelan accompaniment of *wayang kulit* (shadow puppetry), diatonic music usually appears during the *limbukan* and *goro-goro* scenes. Because the tuning systems of diatonic Western music and gamelan pentatonic music are different, this represents an interesting aspect worthy of study.

The tuning system in Western music is based on equal temperament, just intonation, meantone temperament and wide tuning. However, all Western musical instrument tuning is based on the frequency standardisation of the A tone, which is about 440 hertz. Sumarsam (1995) said that the key to understanding a musical culture lies in its historical view. One of the main characteristics of Javanese history is its acceptance and recognition of external culture and ideas. In essence, the development of Javanese culture is stimulated by its relationship with foreign cultures. Furthermore, Sumarsam said that it would be inappropriate if, from the 14th century until now, we talked simply about Javanese culture as an independent culture, and not about plural culture itself. After all, in any historical period, the characteristics of this plurality of culture are always realised because there are, for example, always differences in racial groups (Javanese, European, Eurasian, Chinese, Arabic and Malay), class systems, inhabitants, and sexuality.

Once, when Raffles served as a ruler in Indonesia and lived in Java, he specifically ordered a gamelan with a Western diatonic scale. This gamelan is now stored at Claydon House, London (Sumarsam, 2013). Merriam (1964), an observer of musical culture, argued that a lot of things can be learned from the importance of the function of music in a society. He mentioned three functions: (1) explaining the cohesiveness of social norms, (2) validating and religious rituals, and (3) sustaining the continuity of cultural stability. He also said that "every society has events with music indication that can evoke togetherness of its citizens and remind their unity".

In terms of this notion, Sumarsam (1995) responded that "I do not question the validity of this functional idea of music. However, we [are] faced with other problems when we encounter musical events that [are] quoted above. Inevitably, the combination of two types of gamelan and [a] European string orchestra that plays the Dutch national anthem will produce a chaotic sound of music, especially since the three ensembles play different compositions with completely contrasting tone systems. This tone poisoning will occur regardless of the time or the context in which the events are displayed." From here, the question is indirectly focused at music as a symbol of social unity (how can a combination of sounds as described in the previous description illustrate a unity?) but also as an allegory of Javanese social and cultural historical dynamics. The combination of musical sounds explained previously happened in the 19th and early 20th centuries of Javanese culture; therefore, other cultural aspects that affected the Javanese culture in that period need to be explained.

2 LITERATURE REVIEW

Garrahan (1957) explained the utilities of the historical method and the main things that have to be traced. These are periodisation of time (determination of time), explanation of each time, and making connections holistically and chronologically to determine the golden thread running from the beginning of the specified time until now. However, there are no explanations of how to utilise this method in musical research.

Meanwhile, Kasidi (2001) described the changes in Javanese puppetry settings and the infiltration of diatonic musical material into shadow puppetry, especially in the *limbukan* and *goro-goro* sections. Changes also occur due to the alteration of broadcast times and the utilisation of social media as another medium for shadow puppetry performance. However, this study did not discuss the differences in intervals and pitch between pentatonic and diatonic forms.

Sumarsam (2003) discusses the history of gamelan, the infiltration of diatonic music in Indonesia, the influences of foreigners and foreign traders who brought diatonic musical culture to Indonesia, and the response of Javanese natives. Nevertheless, he does not explain the diatonic and pentatonic differences, nor the combinations between the two.

3 MATERIAL & METHODOLOGY

The data were collected from various journals and books related to the topic. The data were also obtained from a discography of musical recordings, video CDs (VCDs) available on the market, and webliography. This study employed a qualitative method with a historical musicological approach.

4 DISCUSSION

4.1 *Karawitan as accompaniment music in shadow puppet shows*

Western music in Java can be traced from the music brought by sailors of the large ships that stopped at the Javanese coast at the end of the 16th century (Sumarsam, 2003). The voyage of Francis Drake landed on the southern coast of Java. In his travel report, he mentioned that ship musicians played music for a king, or perhaps local dignitaries. After the performance was over, the ruler responded with his own musical performance, although there is no information about this local music, whether *gamelan* or another type of music. The ship musicians consisted of a trumpet player and four string music players. Likewise, Cornelis de Houtman and Jacob van Neck, who landed in Tuban and Ternate at the end of the 16th century, brought similar musicians. According to Sumarsam, the introduction of Western music did not affect Javanese music in the first instance. The second introduction of Western music occurred through Portuguese merchants (Kornhauser, 1978). The Portuguese base was in the Maluku and Malacca areas; the Portuguese themselves never settled in Java. Their music was brought and performed in Java by their slaves, comprising Indians, Africans and Southeast Asians. In the 17th century, the music of the slave orchestra was commonly heard in Dutch merchant houses. One report of a Dutch woman in Batavia said that, in 1689, she had a slave orchestra (consisting of harp, violin and bassoon), which performed various musics when their masters had a feast together (Sumarsam, 2003).

During the 1970s to 1980s, the performance of shadow puppet works emphasised the development of non-physical aspects. In the 1990s, massive transformation occurred in the addition of physical structural elements, almost becoming uncontrollable. There are many factors that helped this transformation to grow rapidly, such as the development of technology, the mindset of society including that of shadow puppetry practitioners, and market demands. As a result, the puppeteers competed with each other in order to find innovations to create a fashionable and innovative *pakeliran* (show), without considering artistic limitations (Heriwati, 2000).

In addition, Kasidi (Heriwati, 2000) also explained that violations of artistic limitation also occurred when the puppeteers began the *ndhodog kothak* (the act of hitting a wooden box) in which the *kayon* (the image of Javanese mountain as represented in a traditional puppet show) was revoked, the accompaniment was specially arranged, and it also did not begin with a scene of a palace of a king who is sitting on a throne with the nuances of a king who was *gung binanthara* (a supreme ruler). The original starting scene involving a king who asked what state and situation the country was in and *bagebinage* (greetings) were completely missing. Instead, the first scene was altered to the sudden appearance of *Semar* (one of the shadow puppet characters), and starts a free dialogue without giving any attention to Javanese culture and manners because Indonesian or other languages specific to other environments were introduced, with the intention to be more communicative. Musical instruments such as keyboards, trumpets, cymbals, drums and electric guitars were also combined with traditional ones in order to change the performances and make them more modern, even though the resulting combination of tones seemed to be discordant (less harmonious). However, the exceptional audience responses indicated that they appreciated this.

A more extreme example happened in a village in Japran, Sragen, when the village held an event that brought a famous puppeteer from Central Java, nicknamed "*Dalang Edan*" (crazy puppeteer). It turned out that his performance resembled his nickname, involving a massive shift from the noble shadow puppetry closed-art concept to an open concept through the combination of tradition and modernity. This was supported by the presence of *dangdut* singers, a *pesindhen* attitude, which was singing while standing with makeup similar to the regular singers, in combination with other *dagelan* figures (comedians) who dressed differently from puppetry characters. When a *neneko* (entourage) group suddenly appeared in the *pakeliran* (a screen for puppet show), *Dalang Edan* turned in a suggestive hip-shaking performance, while on the right and left sides of the stage, the *waranggana* (choir), *niyaga* (gamelan musicians) and puppeteer sat crossly and watched the *neneko* art.

Technological developments also made a contribution to the *pakeliran*. The 1999 birthday celebration of the TVRI Yogyakarta television station included a performance of a puppet show with a young puppeteer from Yogyakarta. In the *limbukan* and *goro-goro* scenes, the audience around the stage can usually request the desired *gendhing* (song) but, in this case, the television audience at home was able to directly order their desired *gendhing*. To smooth the event, there was a presenter who acted as interpreter between the puppeteer and audiences at home. In relation to this, Kasidi said that, subsequently, every puppeteer performing on stage had not only to arrange his Western musical instruments but also to prepare a computer with an internet connection.

In a VCD entitled 'goro-goro', along with *Dagelan Gareng Gayeng* volume two, a puppeteer named Ir. Warseno, M. Si., known as Warseno Sleng, exhibits many *campursari* genre songs that have Western diatonic musical scales. One of these is the song entitled 'Sido Opo Ora'. In the VCD recording, it can be seen that gamelan instruments try to follow the Western diatonic tunes played by keyboard instruments supported by percussion instruments (drumkit). Meanwhile, the *kendang* musical instrument follows the song with the technique known as the *kendang patutan* (Martopangrawit, 1972). Waridi (2006) stated that musical works that live and develop in today's society are quite diverse. In addition to the legacy works of *Karawitan* music masters, there are also *Karawitan* musical works of various types and colours appearing in society, such as the *dangdut* style, rock style, jazz style, *langgam* style, and mixtures called contemporary composition. The types include independent compositions for musical concerts, musical compositions for dance performance purposes, *wayang purwa*, *wayang wong*, *ketoprak* (humorous and serious), and *ludruk*. These are used for a variety of purposes including folk art performances, films and soap opera accompaniments.

In the VCD entitled 'goro-goro', recorded on the 1st of *Suro* month (19 January 2007), the performance of *bimo kurdo* by the shadow puppeteers Ki Anom Suroto and Bayu emerged with Western diatonic musical instruments (keyboard, drumkit) and a pentatonic gamelan. The guest stars of this event were Yati Pesek, Rabies and Cak Dikin. One of the performances was of Gesang's *langgam* '*Caping Gunung*', which is usually played as *keroncong*-style music with Western diatonic music as its basis. The song was beautifully sung by Yati Pesek and other available *pesindhen* with the Western diatonic scale as the original. The keyboard is not heard as the accompaniment of this song, but the gamelan musical accompaniment tried to accompany it by playing *patutan* notes. In the middle of the song, the puppeteer asked the accompanists to change the song's rhythm to a *dangdut* rhythm.

Karawitan has a strong potential to adapt through eras. History has provided sufficient evidence for this assertion. One concrete example can be seen in *Karawitan*'s location in Central Java, West Java, East Java, Bali, West Sumatra, North Sumatra, Banyuwangi and other regions. On the other hand, Javanese musicians have given considerable tolerance to the addition of *birama bermatra 3 (*triple meter) and the addition of choirs in *gendhing* works (Sumarsam, 2010).

The culture of Western diatonic music has become a part of traditional art itself and now goes unnoticed by society. For example, violins have become an inseparable part of *Karawitan* "*Gandrung Banyuwangi*" and *rebab pesisir* ensembles in West Sumatra. Likewise, the trumpet, flute, percussion and other musical instruments have been integrated into the traditional Yogyakarta *Kraton* music.

James R. Brandon, in Soerdasono (2002), claimed that the number of Indonesian performing art forms that had been developed up until the beginning of the 1960s exceeded half of the number of performing art forms in Southeast Asia. Furthermore, Soerdasono said that the number would certainly increase in the era of globalisation. In addition, it is also enriched by contact with outside cultures. The contact between local and foreign performing arts adds considerably to the local and externa; cultural activities. When cultural contact happens, other human activities also occur. According to Koentjaraningrat (1974), with this adjustment activity, the condition can be described as modernisation.

In the early 20th century, *keroncong* music has spread rapidly with the evidence of the *concours* appearance in the night markets. In addition, this type of music is seen as a cultural heritage. The next development was that *keroncong* music dominated Javanese cultural regions, especially Yogyakarta and Surakarta. *Keroncong* music can be categorised as folk music, defined as music that comes from people, is created by the people and is carried by the people. It seems that *keroncong* music can also be referred to as popular music, which means that the music is created for the public by professional musicians, and delivered to the public by these professionals (Kornhauser, 1978).

4.2 *Keroncong, campursari, dangdut, and other genres of music in shadow puppet shows*

Generally, *keroncong* musicians assume that the problem with *campursari* music is that its musical style is influenced by *Karawitan*, while *Karawitan* players assume the opposite. Meanwhile, there once were conflicts in the world of puppetry between the older puppeteers and the younger ones who supported the contemporary style concerning the future of the world of puppetry; it even became a national issue. The agreement reached was that developments in this era demand changes in various aspects, including in the world of puppetry and puppeteer, leading to development through the eras and allowing processes to attain their own maturities. The reality in society is that both traditional and contemporary have their respective supporters and should not need to cause any concern. The development of *campursari* music has revived the gamelan or *Karawitan* music at the heart of the art community and become an alternative outlet for *Karawitan* artists, as well as traditional music appreciators. To enter the world of the arts industry, traditional arts must be able to deal with the concepts of modern art created for urban and industrial societies.

The combination of Western diatonic instruments and gamelan can be seen in the shadow puppetry performance VCD containing the play of "*Sugriwo-Subali*" with Ki Enthus Susmono as the puppeteer, and Thukul Arwana (comedian) and Cucu Cahyati (singer) as guest stars. Western diatonic musical instruments and guest stars appeared in the *goro-goro* section. The genres of the songs performed are *dakwah*, Western pop, *dangdut*, *campursari*, and so on. Although the songs use the Western diatonic scale, gamelan instruments support the diatonic music with a *patutan* accompaniment. This VCD was recorded from one of the shadow puppetry performances that had been aired by the Indosiar television channel. Soerdasono (2002) wrote that to observe the development of Indonesian performing arts from the past to the era of globalisation, it is necessary to trace its history from the pre-historical era to the present. Soedarsono added that this does not mean looking back without considering the future; rather, it is the opposite because by being familiar with the performing arts of the past, the fluctuations that happened in the favoured arts can be understood. Commercial or professional *ketoprak* performance art, which is also an historical practice, represents one of the Indonesian idioms called "*Hidup tak mau mati tak hendak*" (life does not want to die). The extinction of an art form generally occurs because the values of life that surround it, in terms of ideology, politics, economics, socioculture, defence and security, and so on, have shifted. Meanwhile, Raymond Williams (1981) said that to examine the sociology of a culture there are three essential components to consider: (1) cultural institutions, (2) cultural material, and (3) cultural effects or norms contained in the culture. As a cultural institution, you have to understand who is producing the culture, who controls it and how the control is implemented. The meaning of the generated symbols must be understood from cultural perspective. It is stated that on the cultural side, it must be understood what the meanings of the generated symbols are.

The 1970s to 1980s was a golden era for "*gagrak enggal*" *gendhing* performance, which featured "*gendhing-gendhing dolanan*" (traditional game songs) by Ki Nartasabda, and from the end of 1993 until now can be considered as a golden era for *campursari*. Thus, the "*Campursari Mantous*" album sold about 4.5 million copies, which was an enormous number for a music album. Ki Nartasabda became the frontman of the puppeteers who were constructing the *goro-goro* scene at that time. Right now, *campursari* music tends to have more *langgam* rhythms while *dangdut* dominates the Limbuk/Cangik and *goro-goro* scenes. In various tape and disc stores in Yogyakarta, *campursari* music dominates rather than any Javanese *gendhing* tapes. Apart from that, *gendhing-gendhing dolanan* compositions by Ki Nartasabda were sunk by the explosion in *campursari* music (Sumaryono, 2002).

Although *campursari* has nuances of tradition, as a pop nuance, *campursari* music is very competitive with other types of music. Although the development of creativity is slow, *campursari* music still shows a clear and solid form and will differentiate itself from pop music, which is intended more for urban society and commercial interests (Kayam, 1981).

Traditional art in general prevents external influence, so that a typical art is preserved in its original form and prevents the developments that will provide opportunities for the influence of external elements. It can be seen that people are concerned about the contamination of traditional art or the displacement of original art, fearing that there will be loss of elements that form the basis of the original cultural personality. However, with that closed attitude, people are not aware of human experience throughout the ages that the acceptance of cultural contact between cultures and nations throughout history has developed in each of them. Even within that development, the establishment of cultural identity is more likely to be developed. With cultural contact with other nations, each party raises its awareness and is more convinced of the advantages of its own culture. So, in intercultural relations, there are mutual giving and receiving, filtered according to the suitability of each type of culture on both sides (Soebadio, 1991).

The variety of arts all over the world has developed fast, which generally begins in nations from outside the region. On the other hand, isolated tribes that live in remote areas have their own culture and art that remains the same and difficult to develop. Therefore, there should be no need to worry about the tendency for the development of art through cultural contact between nations. This kind of cultural contact should be used as a development of the local art itself. It can be seen that each culture maintains its original personality because the cultural roots are firm. Through such cultural contacts, local cultures adjust and enrich their own original cultural traits.

Slendro is a unique scale system (from a musicology perspective) that is considered one of the key characteristics of Javanese culture. However, its fame has decreased significantly and has been misinterpreted by some people who set out to develop *slendro*. This disaster also leads to different perceptions that affect Javanese music creation. *Laras slendro* is a five-tone scale system. The intervals between tones have been divided equally but this does not mean that the intervals are exactly the same. In *Karawitan* terms, even though *slendro* has equal intervals, their cent intervals are not exactly the same. This is termed *bedantara.* This differentiator requires familiarity with Javanese culture itself, such that the steps to preserve and develop the understanding of *slendro* are not lost. One of the unfamiliar actions in introducing *slendro* is comparing the diatonic musical scale that consists of do–re–mi–fa–sol–la–do' (movable doh system) with *slendro* that consists of 6 - 1 - 2 - 3 - 4 - 5 - 6 '.

For those who are familiar with *slendro*, it is not the case that the diatonic musical scale do–re–mi–fa–sol–la–do' represents *slendro*. However, people who are unfamiliar with *slendro* consider the diatonic musical scale to be similar. If the do–re–mi–fa–sol–la–do' and *slendro* scales are written out, as shown in Table 1, the difference can be seen.

For do–re–mi–sol–la–do' intervals, the cent interval pattern is: 200–200–300–200–300. In the Java *slendro* scale, these intervals are much more even; although each of them differs slightly, it is not by as much as the 100 cents here, so it is a different scale. Unfortunately, certain people cannot differentiate these two scales and are unaware of this. Even worse, right now, our environment is filled with diatonic-scale music, so the original intention to introduce *slendro* in terms of the diatonic scale has become misunderstood, with a diatonic scale now thought of by society as *slendro* itself. Nowadays, most middle-aged people, young people and children cannot use the proper *laras* (scale) in singing *slendro*. Their singing tones are: do–re–mi–sol–la–do', which are different from the *slendro* scale of Javanese gamelan (Sri Hastanto, 1991).

Indications of globalisation in music have been recorded since the 17th century in Europe, which started with the tuning concept of Johann Sebastian Bach known as equal temperament and creation of the diatonic scale system. The system involving the major and minor scales has survived to the present and has been used by a variety of cultures and nations. Bach certainly did not anticipate that his discovery would result in the establishment of a global musical culture that could unify humanity from various cultural traditions into one. There was also the paramount invention of Western musical instruments, including a sophisticated mechanism for keyboards created by Christofori, a technician from

Table 1. Cent intervals in Western diatonic and *slendro* pentatonic scales.

Diatonic :	C – D – E – F – G – A – B – C'
Cent intervals :	200 200 100 200 200 200 100
Diatonic :	C – D – E – G – A – C'
Pentatonic diatonic :	do – re – mi – sol – la – do'
Cent intervals :	200 200 300 200 300

Florence. His discovery is known as the '*pianoforte*', derived from the Italian meaning 'soft-hard' (the point being that the instrument can play music either with soft or loud dynamics) But the name carried by the musical instrument does not reflect the more crucial essence, which is the substance of equal temperament that can be absorbed by musicians. These events are hugely significant, given that pianos became globally known. Christofori might not have thought that his discovery would last into the advanced technological era of today. In fact, it has been able to create a history in the form of a well-known musical instrument that is not only accepted by various cultural traditions but has also helped to create a new musical culture in this modern era.

Muslim and Parwata (1918) wrote that, twenty years ago, when looking at traditional arts, WS Rendra had said that traditional arts were like "*kasur tua*" (an old mattress). The *kasur tua* reference implies that it is a pity to throw it away but it will become useless if it is preserved. Rendra's jokes, of course, are not directed at the traditional arts as a whole. At that time, it was said that some of the traditional arts still had springs that would overcome thick foam mattresses. Of course, this flexibility is determined by the supporting artists. Furthermore, Rendra said that traditional arts consist of three groups: 1) traditional arts that are sluggish but merit their existence, 2) fierce but transparent traditional arts, and 3) traditional arts that are full of vitality but isolated. All of them have their own uniqueness and require special attention and discernment.

The first group relates to traditional society and *Kraton* traditional arts. *Kraton* art is often promoted as an *adiluhung* (honourable) art or classical art. Meanwhile, popular art is often called folk art. Both arts involve products and cultural wealth that deserve their existence and continuation.

The problem faced by folk art is that it has become culturally unfunctional. At first, folk art was beneficial to the surrounding community (mainly in the countryside) as a vehicle for fostering kinship and other common interests, when it was still homogeneous and traditionally agrarian. Now, due to technological advances, countryside societies are becoming identical to those of urban communities, such that the function of folk art has become unnecessary.

One important development in ethnomusicological history was the invention of the cent musical interval system by Alexander J. Ellis, a mathematician and philologist from England, and his assistant A. J. Hipkins. A variety of musical tunes originating from various regions of the world have been characterised. Thus, *laras* music, in principle, is different from Western music. In other words, Ellis has opened the eyes of the world to the fact that the scales of world music are different in principle. This variation in scales was captured in the article included in the *Journal of the Society of Arts* published in March 1885 entitled "On Musical Scales of Various Nations". The significant thing is that the research includes a scale sample originating from one of the Javanese gamelan that has harmonisation consisting of *slendro* and *pelog*, proving that the musical scales of Indonesia have received serious attention from ethnomusicologists for a long time (I Wayan Rai S., 2001).

Yulius Dwijo Purnomo, a priest of the Betani Church located in Dukuh Padan, Kraguman village, Jogonalan subdistrict, Klaten, wanted to make *Sabda*-type puppetry from the performance of *Warta* puppets presented by R. Sumijan at a Christmas event at a Pentecostal church. In 1987, the *Sabda* puppet gamelan accompaniment was changed from a diatonic to a pentatonic form due to the assumption that Javanese people felt that it was not appropriate for a puppet accompaniment to use a diatonic scale (Marsono, 2003).

To understand traditional arts within the Indonesian national cultural context, it is only natural that there are groupings based on time, place, type, and so on. Holt (2000) said that Indonesian arts were arranged and divided into three groups that overlapped and were arranged chronologically according to the history of Indonesia.

The first group is called "inheritance", which includes the preserved historic-times works. The works are made from materials with excellent durability and are still preserved in both domestic and foreign museums. The second group, called "living traditions", is the captured work that has traditional conceptions of form and content although often applied in a new medium. The third group includes contemporary "modern art", which is a developed urban phenomenon.

In an imaginary "present period", if we then try to touch the historical depths of art in the different groups, it will be found that the development of modern art has accumulated its impetus over the last three decades. In fact, the living traditional arts must have assumed their latter forms for the past 200 or 300 years after experiencing a slow evolution in earlier times.

According to Koentjaraningrat. (1987) in *Repelita* II, part III, chapter 24, Indonesian national culture contains plans for art development programmes (ancient heritage, historical heritage, as well as regional heritage), development of national and regional languages, and science (not in terms of the research aspect but about scientific book and magazine publishing). If the Indonesian national culture is based on a specific trait and high-quality characteristic conception, then the case of the relationship between regional culture and national culture that has been the subject of debate from the national movement period until now becomes unimportant. Apart from regional issues, each Indonesian work from an ethnic group is essentially unique and is of high quality, so that Indonesians have their own cultural identification and pride in their works. This is the 'Indonesian national culture'.

It has to be appreciated that the development of globalisation has a tendency to give and receive. It has become a reality that similiar types of art exists in lots of places in the world. Perhaps in Indonesian traditional art there are still special characteristics, such that it can be termed absolute. In other words, it must be possible to accept and promote an appreciation of quality art produced by Indonesian artists. Without pursuing its uniqueness in the context of a globalised future, it may be difficult to maintain. It cannot be denied that the significance of art assesment lies in its quality (Soebadio, 1991).

In the post *Orde-baru* era, shadow puppetry performance involves the development of the shadow puppetry that has gone before. In the past, shadow puppet culture was accepting of various external changes. In the next development, shadow puppet shows became an entertainment medium maintained as a popular art performance and as a government tool to give messages to its citizens. This approximate six-hour show was dominated by performances other than *pakeliran*, such as *limbukan* or *goro-goro* scenes, that consisted of *campursari* or *keroncong* songs accompanied by gamelan, dance and comedy.

Sometimes *campursari* or *keroncong* music is performed in the *limbukan* scene. It usually starts with the appearance of a singer on a gamelan stage in a standing position, who occasionally sings together with a comedian. The emergence of the singer and comedian tends to substitute for the Limbuk and Cangik characters in the *limbukan* scene. Smier (2003, p. 81) writes that in the era of globalisation, artwork is considered as a form of communication that offers people the possibility to obtain and learn information. Humans can freely express their feelings at a deeper level. This situation is contrasted with the bestowal by UNESCO, on 7 November 2003 in Paris, France, on Indonesian *wayang* (puppet theatre) as a piece of Oral and Intangible Heritage of Humanity. Since the *Orde-baru* period, Indonesian puppetry has become a popular art. Nevertheless, there are some puppeteers who are not affected by popular art. Furthermore, the people's interest in *wayang* is diminishing. As a result, puppeteers have tried to find ways to increase society's demand for puppetry. Ki Nartasabda became the first puppeteer who dared to deviate from the traditional shadow puppets, in 1969. On the other hand, the changes that he made have led to views both for and against. As a result, the puppet show is not as solemn as before, when it only focused on the "*Mahabaratha*" or "*Ramayana*" stories and could only be enjoyed by certain groups. The positive thing that happened is that the puppet show has risen in popularity again due to the other parts in shadow puppetry performances, called the *limbukan* and *goro-goro*.

Up to this time, shadow puppetry performance, with its accompanying arts, continues. The *limbukan* part raises both pros and cons, due to its long duration and its deviation from the original shadow puppet stories. The *limbukan* scene, played by the characters Limbuk and Cangik, is widely used by officialdom or shadow puppet performers to convey messages in the social, political, economic, and other domains. But in its application, it is no longer played by puppet characters but by celebrities or comedians.

Ki Nartasabda started as a *kendang* player in the *wayang wong* (a puppet performance which uses a man/woman instead of a puppet) performance group named "*Ngesti Pandawa*", owned by Pujasumarta, a puppeteer who had compiled Javanese *gendhing* since 1952. He created many new works that developed the Javanese *gendhing*. Ki Nartasabda is known as "*dalang lucu*" (the funny puppeteer) because of his courage in differentiating his shadow puppet performances and performing shadow puppetry with new perspectives. The humorous part of his performance began when Ki Nartasabda literally showed the ugliness of some *Kurawa* figures such as Raden Dursasana and Prabu Duryudana. This was considered taboo in shadow puppet performance but Ki Nartasabda intentionally violated it because he wanted to attract the attention of shadow puppet audiences.

As a result of his efforts, he was invited by Radio Republik Indonesia (RRI) in Jakarta to air his shadow puppet show. It was a desirable situation for Ki Nartasabda himself, because radio was the fastest mass media that could transmit information to all regions at that time. With the creativity of Ki Nartasabda, new updated works of Javanese *Karawitan* were made and welcomed by society. *Gendhing Karawitan* (Karawitan songs), which was known as a soft, lonely, and undynamic music has became more dynamic, entertaining, and fresh in Ki Nartasabda works. His works are loved by society because his music is not monotonous and has more power. As Ki Nartasabda became a famous puppeteer in 1969, his new creations of Javanese *gendhing* began to be included in his shows. There are lots of *dolanan* songs, *campursari, dangdut* and even Javanese *langgam* performed in the *limbukan* or *goro-goro* sections.

Due to his creativity, Ki Nartasabda managed to become a famous puppeteer, as demonstrated by the many shadow puppet live shows and audio recordings offered by various studios (Lokananta, Ira Records, Kusuma Records, Fajar Records, and Dahlia Records). Incidentally, in the early 1970s, the availability of radio-cassettes gained popularity due to their affordable prices, allowing a lot of people in society with a range of wealth to buy the tapes. The success of several recording studios in selling their own tapes made Ki Nartasabda a public figure in society. He is regarded as the one who brought a reformation to the shadow puppet world and is considered controversial. In 1985, the last year of his life, Ki Nartasabda still accepted recording offers from Dahlia Records on five occasions. These events indicate that Ki Nartasabda works shifted from popular art to *seni massa* (people's art). The achievements of Ki Nartasabda were envied by the other puppeteers of his era. The other puppeteers tended to imitate Ki Nartasabda's *pakeliran* style and it was true that the puppeteers who imitated the style of Ki Nartasabda gained popularity in society, while the puppeteers who were still following the old rules began to disappear. Thus, it appears that the puppet community had begun to abandon its original role as palace art (*seni istana*) and turned to popular art, which is more in demand in a market society.

The changes in Ki Nartasabda's *pakeliran* style were also influenced by the policies of government officials as they used puppeteers as communicators of their infrastructure programme; told through the *limbukan* section, the infrastructure programme became less offensive. The change that occurred in the *Karawitan* aspect of the shows involved its *gendhing*. The *pakeliran* forms in the post *Orde-baru* period are not only used as entertainment but also as one medium by which to convey messages about the government's infrastructure programme. The scene structure is dominated by *limbukan* or *goro-goro*. This show takes more time than before and renders the shadow puppet show itself less sacred. It has another impact on the audiences because they can no longer comprehend the moral messages and social critiques provided by the puppeteer.

The presence of *limbukan* in shadow puppet performance brings a new atmosphere to society. This does not happen only in Yogyakarta but also in various cities in Central Java. In this case, the *limbukan* scene is favoured by the community, especially in the Surakarta area, until it reaches Yogyakarta. However, *limbukan* became completely different from the *pakeliran* storyline, plus Limbuk and Cangik converse in their own language and give little attention to Javanese manners. There are also instruments consisting of diatonic keyboards, drums, trumpets, cymbals, and electric guitars that participated in order to smooth the *limbukan* scene. So, even though the *pakeliran* has switched to popular music and tried to combine it with the gamelan in order to maintain the standard, which feels out of tune, the audiences remain happy and are still keen to watch the *limbukan* scene (Kasidi, 2009, p. 75).

In fact, the *limbukan* scene was a trial by Ki Nartasabda as a substitute for the "tense" scene of *goro-goro*. However, the public was very enthusiastic to watch Ki Nartasabda's performance so that such puppet performances were maintained by other puppeteers, even beyond the end of Ki Nartasabda's life. The style of performance continued and grew. However, Ki Nartasabda's creativity transformed the *limbukan* scene, causing him to be threatened by other puppeteers. Ki Nartasabda was considered to have changed the values of the *pakeliran*, not only in relation to the *panakawan* (clown servants) who can joke around, but also other figures, including the king himself, who joke such that the palace scenes are similar to those of the *kelurahan* (village). Hence, Ki Nartasabda's nickname as "*Dalang Mbanyol*" (funny puppeteer) (Suparno, 2007, p. 74).

After the Yogyakarta earthquake catastrophe (27 May 2006), Bantul held an event called *Recovery* in October 2006, which took place at the Gabusan Art House. The event was enlivened by the presentation of shadow puppets with the *lakon* (roleplay) of "Dewa Ruci" and Ki Seno Nugroho as the puppeteers. The existing gamelan instruments at the time consisted of a set of puppet accompaniment gamelan and a set of *campursari* gamelan, which had adapted diatonic tuning. The songs performed in the *goro-goro* event were '*Caping Gunung*' and *Langgam* '*Bengawan Solo*'. When the songs started, both diatonic and traditional musical instruments accompanied it together. Traditional instruments were playing *patutan* tones (Junaidi, interview April 20, 2007).

In April 2007, Ki Dalang Warseno Sleng participated in a *Gunung Kidul* shadow puppet performance entitled "*Semar Mbangun Kayangan*". During the performance there were gamelan instruments that consisted of the ordinary gamelan usually used as an accompaniment in shadow puppetry and a gamelan with diatonic tones for accompanying diatonic songs. Meanwhile, on 7 May 2007, the Yogya TV channel featured a shadow puppet show with a puppeteer known as Ki Joko Edan. This show also employed two special gamelan instruments usually utilised by Ki Seno Nugroho and Ki Warseno Sleng. In addition, Junaidi also provided information about a gamelan maker named Wagimin from Klaten, who frequently received orders for making gamelans with diatonic scales.

4.3 *Three main parts of shadow puppetry shows*

There are various scenes in shadow puppetry shows. Traditionally, the show lasts for about eight hours, running from around 9 p.m. to 5 a.m. It is divided into three main parts that are adapted from the Javanese philosophy that consists of *pathet nem*, *pathet sanga*, and *pathet manyura*, and each of them has their own scenes. *Pathet nem* tells of the youth situation, *pathet sanga* narrates the adult life, and *pathet manyura* recites an old life perspective. Before the *pathet nem* section starts, there is a section called *Patalon*, which is a series of *gendhing* and is played for approximately 30 minutes in order to create the atmosphere leading to a *lakon* that will be performed by the puppeteer, while waiting for the audience to arrive (Suparno, 2007, p. 135).

The *pathet nem* section usually runs from 9 p.m. to 12 midnight. In this section, the audience is first introduced to the time and location of the story, the characters, and the relationships among the characters. Puppet performance always starts with a good, calm and peaceful situation. After that, it discusses the main problems according to the corresponding actors.

Then, the *pathet sanga* section begins around midnight and runs until 3 a.m. The scenes in this section are called *goro-goro*, which are about unpleasant manifestations in nature. In this scene, the *panakawan* group, which consists of Semar, Gareng, Petruk and Bagong, appear to entertain their master, the main character. The *pathet sanga* section usually uses the wild forest or the hermitage as its background. The *panakawan* try to entertain their master and also the audience with their jokes. The *pathet sanga* has tones consisting of *seleh* 5 (*mo*), 1 (*ji*), and 2 (*ro*). Generally, the Javanese *gendhing* played in the *pathet sanga* have a relatively higher pitch than in the *pathet nem*, with the exception of the *pathet sanga* that is used to accompany particular scenes where it is necessary to create another nuance.

The last section is *pathet manyura*. This section runs from around 3 a.m. until 5 a.m. This part starts with long characters' scenes, followed by war between the characters who act as enemies and then continuing to additional war scenes. The battle ends with a triumphant dance by a particular character, followed by a *tancap kayon* scene that depicts the character's gratitude. The *pathet manyura* tones consist of 6 (*nem*), 2 (*ro*) and 3 (*lu*) (Suparno, 2007, pp. 107–122).

Generally, *slendro* gamelan music uses the same three types of *pathet*, called *pathet nem*, *pathet sanga*, and *pathet manyura*. Furthermore, use in each *pathet* is divided into two, *slendro* and *pelog*. Earlier, many *Karawitan* experts tried to define *pathet* by referring to pieces of gamelan melody, especially in the *balungan* section. In fact, no theory has been found to define *pathet* until now. Udith Becker, in R. Anderson Sutton (1991, p. 289), suggests three basic *pathet* related to each other, which are:

a. melody pattern, formula, or contour
b. pitch from the mentioned pattern
c. pattern position in the structure of the pieces.

To define the *pathet* concept, Javanese society often equates *pathet* to a scale in Western music but they are different. The difference can be seen in the pitch interval calculations. Western diatonic music uses a hertz-based frequency counting scale (the number of vibrations per second for each tone), while Javanese music in gamelan uses a tone interval counting scale (cents). Both scales have their own calculations and cannot be translated from one to the other.

Usually, *limbukan* is a scene showing an empress waiting for the arrival of a king who had left the palace for a long time. After such a long wait, the king returned and was greeted very happily by the empress who wanted to meet the king immediately. They poured out their happiness in the *limbukan* scene that contained songs and jokes from the Limbuk and Cangik characters. The background used in *limbukan* is usually the same as in *goro-goro*. The theory of *kempyung* is known as the theory to determine *pathet* in Javanese music. In diatonic music, *kempyung* means a perfect quint tone interval. On the other hand, to find *pathet* in the Javanese gamelan in the *laras* (scale) of *slendro*, one has to start with the *Dhodo* tone (3 / *lu*). Based on the calculation of *kempyung*, the obtained notes start from the *Dhodo* tone (3 / *lu*) as the lower *kempyung*, so that the *kempyung* slendro tones are 3 (*lu*) – 1 (*ji*) – 5 (*mo*) – 2 (*ro*) – 6 (*nem*). With this kind of theory, *pathet sanga*, *pathet nem* and *pathet manyura* intervals can be derived. Sri Hastanto (1991, pp. 21–22) states that the *pathet sanga seleh* tone has to be 5 (*mo*). It is a simple thing to find another *seleh* tone in *pathet nem* and *pathet manyura*, which are only increasing 1-quint intervals from tone *mo* because of the 5 (mo) *pathet sanga seleh* scene. With this approach, it can be found that the *pathet sanga* consists of 5th to 12th intervals.

Laras slendro is a structure of notes that contains five notes within one *gembyang* (octave) with almost the same interval between each. These are the notes contained in *slendro*:

1. *Barang*. Symbolised as "1" and read as *siji* or *ji*.
2. *Gulu*, or *Jangga*. Symbolised as "2" and read as *loro* or *ro*.
3. *Dhodo*, or *Jaja* or *Tengah*. Symbolised as "3" and read as *telu* or *lu*.
4. *Lima*. Symbolised as "5" and read as *limo* or *mo*.
5. *Nem*. Symbolised as "6" and read as *enem* or *nem*.

With this development, when *laras slendro* is played by one of the gamelan instruments named *saron penerus*, the tone consists of 6 - 1 - 2 - 3 - 5 - 6 - 1′ (*nem–ji–ro–lu–mo–nem–ji′*).

The identification of the tones above is based on a gamelan located at the Puppetry Department of the Indonesian Institute of the Arts, Yogyakarta. The research process began with a *saron penerus* sound recording. The resulting recordings were tuned using a smartphone app named "Soundcorset". The result of the tuning is shown in Figure 1. The tone adds the – or + symbol because the tone is not exactly the same as Western diatonic tone so that the symbols – and + are added to indicate whether the frequency is lower or higher than the Western diatonic tones.

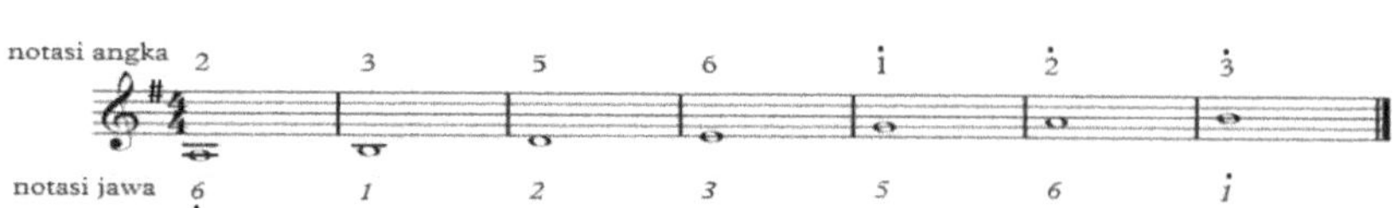

Figure 1. Javanese (*jawa*) and numerical (*angka*) notation for G major.

Nevertheless, the tones mentioned are not definite in frequency (hertz) because the tones in the gamelan are different in measure compared to the tones of Western musical instruments. The corresponding B flat is 1 (*ji*) in *laras slendro* and considered the same as the B flat tone on the Western piano. In the first phrase of the song '*Caping Gunung*', the Javanese notation is *ji–ro–ji–ro–mo–ji–nem*, while its diatonic notation is mi–sol–mi–sol–do–mi–re (movable doh). Based on the melody of the first phrase in '*Caping Gunung*', it can be concluded that the tone of mi (3) is equivalent to *ji* (1), sol (5) to *ro* (2), do (1) to *mo* (5), and re (2) to *nem* (6). It can be concluded that the gamelan plays in G major.

As shown in Figure 1, the notation is made using the C major scale because this is the natural scale that has not been touched by any accidental (sharp or flat) sign. The C major scale is used as a benchmark tool in the process of calculating the pitch of *laras slendro*. It is different from the *slendro* tone (*saron* voice) which is successfully recorded and sought for its similarity to the piano instrument. This is an example of *slendro* tone notation (Javanese notation) that has been compared with the closest diatonic tones (numerical notation) after it is played by piano and written in notation with G major as its scale and A flat as the supertonic.

Table 2 explains that 6 (*nem*) is the same as A flat or re (2), while 1 (*ji*) is the same as B flat or mi (3). The 2 (*ro*) is the same as C or sol (5), while 3 (*lu*) is equal to the tone of D1 or la (6). Lastly, 5 (*mo*) is equal to G1 or do (1). From the two images above, it can be concluded that if the tones in the *slendro* and diatonic tunings are aligned, it will appear as in Table 3.

Table 2. Comparison between Javanese and diatonic notations.

Javanese notation	6 (*nem*)	1 (*ji*)	2 (*ro*)	3 (*lu*)	5 (*mo*)	6 (*nem*)	$\dot{1}$ (*ji*)
Numerical notation	2 (re)	3 (mi)	5 (sol)	6 (la)	1 (do)	$\dot{2}$ (re)	$\dot{3}$ (mi)
Diatonic notation	A♭–15	B♭–14	C+12	D+36	G+35	A♭–17	B♭–22

Table 3. Cent differences between Javanese and diatonic intervals.

Cent interval in Javanese notation (Sukotjo)		240	240	240	240	240	240	240	240	240
Javanese notation	5< (*mo*)	6< (*nem*)	1 (*ji*)	2 (*ro*)	3 (*lu*)	5 (*mo*)	6 (*nem*)	1> (*ji*)	2> (*ro*)	3> (*lu*)
Numerical/diatonic notation	1/G	2/A♭	3/B♭	5/C	6/D	1/G	2/A♭	3/B♭	5/C	6/D
Cent interval in numerical notation	100		200	200	200	500	100	200	200	200

To strengthen this study, laboratory observation was conducted by calculating the cent intervals for the *slendro* pentatonic tones and the diatonic notes referenced by the *slendro* pentatonic scale in Table 3. The pentatonic tones were measured using the cent unit based on the UPT ISI Library (Yogyakarta) online journal created by Sukotjo.

On this basis, it can be concluded that there are differences in the values between tones in the Western and cent notations. The calculations show that the order of *slendro* tones in gamelan cannot be equated with the diatonic musical scale. Nevertheless, it seems that the difference will not become material in either the diatonic or pentatonic music communities. Furthermore, although the application of cent unit intervals provides further proof that the gamelan instrument sounds are different from those of diatonic instruments, not everyone can sense the difference.

Laras slendro saron sounds were measured using the Soundcorset smartphone application, and it was discovered that A flat is approximately the same as 6 (*nem*). Similarly, B flat approximates to 1(*ji*), C to 2 (*ro*), D to 3 (*lu*), and G to 5 (*mo*). However, if these tones are compared in frequencies, the precise results are as follows:

a. 6/*nem* = 466 Hz (= A flat–15)
b. 1/*ji* = 523 Hz (= B flat–14)
c. 2/*ro* = 622 Hz (= C+12)
d. 3/*lu* = 698 Hz (= D+36)
e. 5/*mo* = 783 Hz (= G+35)

The data above shows the results of *saron* sounds with *slendro* scale (*laras*) tuning frequency. If the results are compared with the diatonic scale tones, for example, 6 (*nem*) (= A flat–15), it can be seen that the result is 466 Hz. However, if A flat is being precisely tuned, there will be a difference in frequency, as shown below:

1. A flat = 415 Hz
2. B flat = 466 Hz
3. C = 261 Hz
4. D = 293 Hz
5. G = 392 Hz

It can be concluded that although *Karawitan* academicians claimed that 1 (*ji*) in gamelan tones is equivalent to B flat, it is different from the diatonic music perspective. The difference in frequency between diatonic and pentatonic music scales can be seen in Table 4.

The notation system applied by the Javanese people has its own system, so it should not be difficult for Javanese to learn a song or *gendhing*. This approach will become more difficult if Javanese song notation is altered to Western music notation because of the difficulties in exactly measuring the gamelan tones. However, this problem is not an obstacle to analysing Javanese *gendhing* because the author has conducted laboratory observation and interviews with a subject. There are some suggestions that have to be noted when converting Javanese notation to Western diatonic notation, which are (Malm, 1996, p. 46):

1. recognise the scales used in the music
2. determine the key
3. look for the biggest intervals within the instrument
4. measure the intervals
5. cadence pattern
6. melody formula
7. contour.

If these elements are carefully observed, it can be established whether or not the diatonic and pentatonic can exist in harmony. However, the most important thing is that if the elements are discovered and the transposition from Javanese notation to Western notation succeeds, it will feel different when played. Nevertheless, universally, it will become a connector by which to learn existing Javanese music.

Table 4. Frequency intervals between diatonic and pentatonic.

Diatonic (Hz)	Pentatonic (Hz)	Interval (Hz)
A flat = 415	6/*nem* (A flat – 15) = 415	51
B flat = 466	1/*ji* (B flat – 14) = 523	57
C = 261	2/*ro* (C + 12) = 261	361
D = 293	3/*lu* (D + 36) = 698	405
G = 392	5/*mo* (G + 35) = 783	391

5 CONCLUSIONS

From the explanations above, it can be concluded that the shadow puppet performing arts have undergone significant changes since their original appearance. At first, shadow puppetry was employed as a royal art that could only be enjoyed by the royal family but, in the present conditions, all kinds of societies from various places can enjoy the shadow puppet entertainment.

Following a variety of observations, it can be concluded that the songs presented in the shadow puppet performance are the innovation of pentatonic artists, especially puppeteers and gamelan musicians. Although gamelan vocalists are performing songs that were originally based on pentatonic songs, they still tend to follow the melody with a diatonic tuning system, although sometimes the vocalists use the gamelan vocal inflection to create timbre in their vocals. Meanwhile, gamelan players play the accompaniment section with *patutan* tones that are different from the traditional gamelan standards, and diatonic music players are still playing Western diatonic tones together with existing vocalists. Although it has been proven that gamelan instruments use a pentatonic music system that is different from diatonic music, these types of music are still performed and have a lot of enthusiasts.

Despite the differences in pitch, there is no difficulty occurring when gamelan tones and diatonic music tones are played simultaneously. Even though the combination of Western diatonic tones and pentatonic Javanese gamelan produces discordant sounds according to musicology, this condition is considered to be a new generation of the shadow puppet performance, which is no longer considered as an exclusively classical (*adiluhung*) art.

REFERENCES

Garrahan, G.J. (1957). *A guide to historical method*. New York, NY: East Fordham University Press.

Hadiprayitno, K. (2009). *Suluk Wayang Kulit Purwa Gaya Yogyakarta* [Suluk's Purwa puppet shadow of Yogyakarta style]. Yogyakarta, Indonesia: Bagaskara.

Hastanto, S. (1991). The Ramayana epic and traditional Indonesian music. *SPAFA Journal (Old series 1991–2013)*, *1*(2), 30–33.

Holt, C. (2000). *Track the art development in Indonesia* (R.M. Soerdasono, Trans.). Bandung, Indonesia: Masyarakat Seni Pertunjukan Indonesia.

Heriwati, S.H. (2016). *Tindak Tutur Direktif dalam Pertunjukan Wayang Relevansinya dengan Pembentukan Karakter Bangsa* [Directive approach in puppet shadow show and the relevance to cultivate character of nation]. *Prosiding Prasasti*, 204–212. Retrieved from https://jurnal.uns.ac.id/prosidingprasasti/article/view/494/0.

Kayam, U. (1981). *Seni, tradisi, masyarakat* [Arts, tradition, community]. Jakarta: Penerbit Sinar Harapan.

Koentjaraningrat. (1974). *Development of mentality and culture*. Yogyakarta, Indonesia: Gramedia.

Koentjaraningrat. (1987). *Sejarah Teori Antropologi I* [History of anthropological theory I]. Jakarta, Indonesia: Universitas Indonesia Press.

Kornhauser, B. (1978). In defence of kroncong. In M.J. Kartomi (Ed.), *Studies in Indonesian music* (pp. 137–139). Melbourne, Australia: Centre of Southeast Asian Studies, Monash University.

Malm, W.P. (1996). *Music cultures of the Pacific, the Near East, and Asia* (3rd ed.). Upper Saddle River, NJ: Prentice Hall.

Martopangrawit. (1972). *Titilaras Kendanga* [Musical Scale in Kendangan] Surakarta, Indonesia: Badan Research Konservatori Karawitan Indonesia Departemen P & K di Surakarta.

Merriam, A.P. & Merriam, V. (1964). *The anthropology of music*. Evanston, IL: Northwestern University Press.

Muslim, A.B. & Parwata, P. (2018). Dari seniman ke birokrat: biografi enthus susmono 1984–2014 [From artist to bureaucrat: The biography of Enthus Susmono 1984–2014]. *Publika Budaya*, *6*(2), 101–109.

Rai, I.W. (2001). Metode Penelitian Laras (Tuning System) Musik Tradisi Nusantara: Sebuah Model [Tuning system in traditional Indonesian music: A model]. *Selonding Jurnal Etnomusikologi Indonesia*, *1*(1).

Smiers, J. (2003). *Arts under pressure: Protecting cultural diversity in the age of globalisation.* London, UK: Zed Books.

Soebadio, H. (1991). "Penelitian Naskah Lama Indonesia" dalam Buletin Yapena [The Indonesian old manuscript research in the Yapena Bulletin]. *Tahun VII, Juni.*

Soerdasono, R.M. (2002). *Seni Pertunjukan Indonesia Di Era Globalisasi* [Indonesian performing arts in the age of globalization]. Yogyakarta, Indonesia: Gadjah Mada University Press.

Sumarsam. (1995). *Gamelan: Cultural interaction and musical development in central Java.* Chicago, IL: University of Chicago Press.

Sumarsam. (2010). Gamelan de Solo—Le jeu des Sentiments (A Garland of Moods) (review). *Asian Music*, 41(1), 173–175.

Sumarsam. (2013). *Javanese gamelan and the West.* Rochester, NY: Boydell & Brewer Group.

Suparno, T.S. (2007). *Seni pedalangan gagrak Surakarta: butir-butir kearifan lokal sebagai solusi problimatik mutakhir* [The art of puppetry gagrak Surakarta: The points of local wisdom as the latest problematic solution]. Surakarta: ISI Press.

Sutton, R.A. (1991). *Traditions of gamelan music in Java: Musical pluralism and regional identity.* Cambridge, UK: Cambridge University Press.

Waridi, P. (2006). *Seni Tradisi dan Apresiasi Masyarakat Terhadap Pembelajaran Seni*[Traditional art and the appreciation community toward art lea . Surakarta, Indonesia: Pusat Studi Budaya dan Perubahan Sosial, Universitas Muhammadiyah Surakarta.

21st Century Innovation in Music Education – Astuti et al. (eds)

Opening melody for *patet* sound space on shadow puppet theatre

H.B. Prasetya, I.K.N Putra & T.W. Widodo
Indonesia Institute of the Arts, Yogyakarta, Daaerah Istimewa Yogykaarta, Indonesia

ABSTRACT: This study aims to understand the various types of melodies that play a role as opening *patets* (gamelan mode) in *wayang* (shadow-puppet theatre). The data for this study was obtained by observing the *Kakrasana Wanengpinta* shadow-puppet performance. Observations was focused on the parts that have changes in *patet*. Based on the research it can be concluded that there are two types of melodies used in the transition of *patet* changes, namely *thinthingan* and *grambyangan,* additionally, the vocals of the puppet master (*dalang*) who mimics *thinthingan*. *Thinthingan* and *grambyangan* are carried out by the *gender*, one of methallophone gamelan instruments. *Thinthingan* melodies are needed for light transitions, while *grambyangan* melodies are used for heavy transitions.

1 INTRODUCTION

Gamelan music plays an important role in *wayang* or shadow-puppet theatre. The resulting musical sense plays a role in strengthening the atmosphere needed in the scene and gives a certain sense of character to the characters played in the shadow-puppet theatre (Lysloff, 1993). In other words, gamelan music functions to support the atmosphere in the scenes so that it supports the essence of the play that is performed. Therefore, without gamelan music, shadow-puppet theatre cannot take place.

The importance of gamelan music in Yogyakarta-style shadow-puppet theatre is reflected in the *jêjêran* (main scene) in the performance. Traditionally, shadow-puppet performances are divided into three segments determined by the *patet*, namely *patet nem*, *patet sanga*, and *patet manyura* (Weiss, 2006). During the overnight performance of *wayang*, *patet nem* takes place between 9:00 p.m. to 12:30 a.m, *patet sanga* between 12:30 p.m. to 03:00 a.m. and *patet manyura* between 3:00 a.m. to 04:00 a.m. In these large segments most of the musical events are associated with *patet*. In shows with shorter duration, the *patet* division adjusts the duration of the show. Generally, *patet nem* and *patet sanga* are longer than *patet manyura*.

Each *patet* consists of *jêjêran* and each *jêjêran* often consists of scenes *(Prasetya, 2013)*. *Patet nem* comprises *jêjêr* I or *jêjêr kawitan*, *jêjêr* II or *jêjêr sabrangan* and *jêjêr* III or *jêjêr bondhèt*. *Patet sanga* comprises *jêjêr* IV or *jêjêr pandhita* and *jêjêr* V or *jêjêr uluk-uluk*. *Patet manyura* comprises *jêjêr* VI or *jêjêr sumirat* and *jêjêr* VII or *jêjêr rina-rina* (see Table 1). The terms *bondhèt*, *uluk-uluk, Ssmira*t and *rina-rina* refer to *gending* or the gamelan piece name that is usually used in that scene. Although other *gending* is sometimes used in *jêjêran*, the *jêjêr* stays the same.

The distribution of *patets* in shadow-puppet performance is related to the type of *gending* or gamelan piece used. In the *patet nêm* section, the compositions are *nêm*-bound compositions. In the *patet sanga* section, the compositions used are *sanga*-themed compositions. Likewise, in *patet manyura*, the *gending* used is the *manyura*-style compositions. However, in some parts, the *patet* changes. For example, the *budhalan* scene is in the *patet nêm* segment, but often uses a number of *sanga*-themed compositions, and likewise in the *goro-goro* scene. This scene is actually in the *patet sanga* segment, but often uses *manyura*-style compositions (see Table 3).

Table 1. The relation of *Patet* with scenes and *Jêjêran* in the play of *Kakasrana Wanengpinta.*

Patet	*Jêjêr* Name	Scene/Act Name	Character
Nem	*I. kawitan*	Mandura country	Prabu Basudewa, Haryaprabu, Patih Saragupito, Tumenggung, Narayana.
		Kundur kedaton	Prabu Basudewa
		Paseban Jawi	Haryaprabu, Narayana, Patih Saragupita, soldiers.
		Budhalan	Haryaprabu, Patih Saragupita, soldiers.
		Ampyak war	Soldier's journey of Mandura country
	II. sabrangan	Girigathok country	Patih Nagarangsang, Bedongbau, Bramangkara
		Simpangan war	The battle of Mandura soldiers against Girigathok soldiers
	III. bondhèt	Jongringsaloka heaven	Batara Guru, Narada, Brama, Endra
		The Repatkepanasan	Narada, Brama, Endra, Sambu, Yamadipati
		Gagal war	The battle of Patih Nagakilat, Gedongbau, and Bramangkara against Brama and Endra
Sanga	*IV. pandhita*	*Goro-goro*	Semar, Gareng, Petruk, Bagong
		Ringin Gedabya Hermitage	Begawan Wanengpinta, Margana, Semar, Gareng, Petruk, Bagong
		Begal war	The battle of Margana against Cakil and soldiers of Girigathok
	V. uluk-uluk	Surateleng heaven	Nagaraja, Dewi Pertiwi, Setija, Narayana
		Tanggung war	-
Manyura	*VI. sumirat*	Girigathok country	Prabu Nagakilat catching up Nagarangsang
		Tandang war	The lost of Margana in the war against Nagakilat. The war between Wanengpinta against Nagakilat and Nagarangsang. The lost of Nagakilat, then he turns into a weapon called Nanggala while Nagarangsang turns into Alugara.
	VII. rina-rina	Mandura country	Prabu Basudewa, Kakrasana, Narayana, Permadi, Haryaprabu, Semar, Gareng, Petruk, and Bagong
		Brubuh war	The batle of Werkudara against soldiers of Girigathok.

Source: Hanggar and Nuryanto (2018).

Before the 1970s, shadow-puppet shows were only accompanied by the gamelan in *slendro* scale. Along with the development of gamelan music and the shifting of the function of shadow-puppet as part of ritual ceremonies into performing arts, shadow-puppet performances today are often accompanied by gamelan in *pelog* scale. When there is a transition, either from *slendro* to *pelog* or between *patets* with one another, there is a melody that allows a pleasant transition to be felt both by the *pengrawit* (music players) and the listener. This article explores various transitional melodies found in shadow-puppet shows. The transitional melodies are known as *thinthingan* and *grambyangan. Thinthingan* and *grambyangan* are played by the *gender,* an ensemble of methalophone gamelan instruments, and in some cases *thinthingan* is sounded by the *dalang*'s vocal mimicry of the *gender*.

2 THE SOUND SPACE IN THE SHADOW-PUPPET SHOW

Prasetya's research (2012) ascertained that *patet* was a sound space. A person who plays gamelan can be analogous to being in a sound space. In general, there are three sound spaces, namely *patet nêm* room, *patet sanga* room and *patet manyura* room. One who originally plays *gamelan* in the *patet nem* room could not suddenly play *gamelan* in the *patet sanga* or *manyura* room. In order to be able to play on different *patets*, one must open the door first (Prasetya, 2012). Three *patet* doors, starting from the simplest one are *thinthingan*, *senggrengan* and *grambyangan*. Prasetya's research did not explain why *thinthingan*, *senggrengan* and *grambyangan* were able to open the door of the *patet* room.

Shadow-puppet shows, both in the form of overnight shows (five to seven hours), concise (two to four hours) or solid (less than one hour) can be seen as sound spaces. Based on observations during shadow-puppet shows, the music is continuous. Gamelan music continues to be played with different intensities to complement each other and strengthen the feeling of shadow-puppet shows. When puppet characters deliver a dialogue or when a puppet master or *dalang* gives a narrative, the music sounds with a low intensity. Only one or several music instruments are played. Conversely, when puppets are moved, arranged on the *kelir* (screen), walking, dancing or making war, the music sounds with high intensity. Nearly all of the music instruments are played (see Table 2).

Table 2 shows that during the performance music always accompanies the actions. To understand this, I observed one of the overnight *wayang* performance plays called *Kakrasana Wanengpinta,* which was performed by Ki Cerma Suteja, one of the *dalangs* from Yogyakarta.

Table 2. Relationship among performance aspects with the music is played.

No	Performance aspect	Music played	Instrument
1.	Puppets are in dialogue	*Grimingan*	*gender*
2.	Puppets are moving (walk, fight)	*Gending* (gamelan piece)	all gamelan instruments
3.	Puppets in a certain mood (sad, angry, happy, etc.)	*Sulukan* (mood song)	*gender, rebab, gambang, seruling, cempala, keprak*, gong
4.	The *dalang* is giving a narrative	*Grimingan*	*gender, keprak*

Source: Hanggar and Nuryanta, 2018

Table 3. Sound space in shadow puppet performance.

Patet	Jejer/Scene	[1]Gending (Name of Gamelan piece)
Nem	Jejer I. The begining. Mandura Country. Haryaprabu, Patih (Prime minister) Saragupita, Narayana meets Prabu Basudewa. Basudewa would crown Kakrasana to replace him as the king, but Kakrasana leaves the kingdom without saying a word.	Ayak-ayak Slendro Patet Nem, Gending Karawitan Slendro Patet Nem, Ladrang Krawitan Slendro Patet Nem, Suluk Lagon Patet Nem Wetah, Suluk Kawin Girisa Patet Nem, Grimingan Patet Nem, Suluk Ada-ada Patet Nem Wetah, Grimingan Patet Nem, *Playon Patet Nem*
	Kundur Kedhaton; Prabu Basudewa returns to the palace	*Ladrang Sri Katon Patet Manyura, Ayak-ayak Lasem Patet Nem*
	Paseban Jawi: Patih Saragupita and Tumenggung meet Haryaprabu.	Suluk Ada-ada Patet Nem Jugag, Grimingan Patet Nem, Playon Patet Nem, Grimingan Patet Nem, *Suluk Ada-ada Patet Nem Wetah*
	Budhalan: Haryaprabu and the soldiers join in a campaign to find Kakrasana.	*Suluk Ada-ada Kawin Sekar Asmaradana Patet Manyura, Suluk Kawin Sekar Pangkur Patet Sanga, Lancaran Gagak Setra Patet Sanga*
	Jejer II. Sabrangan. Girigathok Country. Bedongbau and Bramangkara meet Patih Nagarangsang. They leave for Kayangan	*Playon Lasem Patet Nem*, Suluk Ada-ada Patet Nem Wetah, Grimingan Patet Nem, Playon Lasem Patet Nem

(*Continued*)

Table 3. (*Continued*)

Patet	Jejer/Scene	[1]Gending (Name of Gamelan piece)
	Kandhawaru to propose to Gagarmayang and Parijatha	
	Midway scene. Patih Nagarangsang are accompanied by Togog and Bilung.	Playon Lasem Patet Nem, Grimingan Patet Nem
	War Scene: Patih Saragupita fights against Gedongbau. Gedongbau loses, and then he continues the journey	Playon Lasem Patet Nem, *Suluk Ada ada Patet Nem, Grimingan Patet Nem*
	Jejer III. Jongringsaloka Heaven: Batara Guru order Narada, Brama, Indra and Yamadipati to send Patih Nagarangsang back.	*Ladrang Mega Mendhung Pelog Patet Nem,* Suluk Plencung Jugag Pelog Patet Nem, Ada-ada Pelog Patet Nem wetah, Grimingan Pelog Patet Nem
	Repat Kepanasan Scene. Brama, Indra, Sambu, and Mahadewa meet Batara Narada	Playon Lasem Pelog Patet Nem, Grimingan Pelog Patet Nem
	Midway scene. Patih Nagarangsang is accompanied by Togog and Bilung.	Playon Lasem Pelog Patet Nem, *Suluk Ada-ada Pelog Patet Nem, Grimingan Pelog Patet Nem*
	The scene of Brama meeting Patih Nagakilat.	*Suluk Ada-ada Slendro Patet Nem*
	War scene: Brama fights Nagarangsang, Brama loses.	Playon Lasem Patet Nem, Grimingan Patet Nem
	Brama meets Narada and reports his loss.	Playon Lasem Patet Nem, Grimingan Patet Nem
	Brama release a fog so that the Nagarangsang cannot see.	Playon Lasem Patet Nem, Grimingan Patet Nem
	Narada tells Indra to look for a champion who can defeat Nagarangsang.	*Playon Lasem Patet Nem*
Sanga	Gara-gara (consists of humor and messages)	*Suluk Lagon Patet Sanga Wetah,* Ilir-ilir song with Patet Sanga, Suluk Ada-ada Gara-gara, Ayak ayak Patet Sanga, Playon Patet Sanga, Sampak Patet Sanga, Playon Patet Sanga
	Petruk sings a Patet Sanga song.	Lagu Slendhang Biru Patet Sanga, Playon Patet Sanga, Sampak Patet Sanga
	Semar sings songs of Patet Sanga and Patet Manyura.	Suluk Jingking Wetah Patet Sanga, Jineman Patet Sanga, *Grimingan Patet Sanga Suluk Sekar Pucung Patet Manyura*
	Jejer IV. Pandita: Ringin Gedabya. Margana, Semar, Gareng, Petruk, and Bagong meet Begawan Wanengpinta. They are visited by Indra. Indra asks for help to expel Patih Nagarangsang.	*Ladrang Sri Kaloka Patet Sanga,* Suluk Lagon Patet Sanga Wetah, Grimingan Patet Sanga, Suluk Ada-ada Patet Sanga
	War scene. Margana fights against Cakil.	Suluk Ada-ada Patet Sanga, Grimingan Patet Sanga, Playon Patet Sanga,
	War scene. Margarana fights against Patih Nagarangsang. Patih Nagarangsang is thrown away.	Playon Patet Sanga, Grimingan Patet Sanga
	Jejer V. Surateleng Heaven. Pertiwi and Setija meet Nagaraja. Setija wants to meet his father Narayana. Nagaraja summons Narayana and hands him the Wijayakusuma Flower to revive the person who died prematurely.	Playon Patet Sanga, Suluk Lagon Patet Sanga Wetah, Grimingan Patet Sanga, Playon Patet Sanga, Suluk Lagon Patet Sanga jugag, Grimingan Patet Sanga, *Playon Patet Sanga*
Manyura	Jejer VI. Girigathok country. Prabu Nagakilat will follow Patih Nagarangsang who had not gone home.	*Suluk Lagon Patet Manyura Wantah, Grimingan Patet Manyura, Suluk Ada-ada Pelog Barang.*
	Kandhawaru scene: Nagakilat meets Nagarangsang.	Playon Pelog Barang, Grimingan Pelog Barang

(*Continued*)

Table 3. (*Continued*)

Patet	Jejer/Scene	Gending (Name of Gamelan piece)[1]
	War scene: Nagarangsang meets Margana and they fight. Margana dies from a neck bite.	Playon Pelog Barang, Grimingan Pelog Barang, *Suluk Ada-ada Pelog Patet Barang*
	Begawan Wanengpinta meditate to bring Narayana. Narayana arrives and revives Margana using the Wijayakusuma flower. Margana transforms into Permadi.	*Playon Patet Manyura*, Grimingan Patet Manyura
	Begawan Wanengpinta face Nagakilat. Wanengpinta is bitten and turns into Kakrasana. Nagakilat is held by Kakasrana using the power of *lebur pamor* in his hand. The Nagakilat tongue is pulled into a *Nanggala* weapon. Nagarangsang pulls his tongue and turns into *Alugara* blade.	Ada-ada Patet Manyura Jugag, Grimingan Patet Manyura, Playon Patet Manyura
Galong	Jejer VII. Kakrasana, Narayana, Permadi, Bima, Punakawan meet Narada. Narada gives Kakrasana two elephants, Puspadenta and Yudaketi, to accompany him later when he becomes king.	*Suluk Galong Wetah*, Grimingan Galong, Playon Manyura
	Brubuh war: Bima drive all the guards from Girigathok.	*Sampak Galong, Ayak-ayak Manyura*, Gangsaran Patet Manyura, Bubaran Wasana Patet Manyura

[1]Italic letters to indicate a patet transition.

This show tells the story about the two sons of the king of Mandura, Narayana and Kakrasana, who receive a gift from a god so that later it would be revealed who is worthy to become a king. Narayana managed to get the Wijaya Kusuma flower to revive the dead whose time had not come yet, while Kakrasana obtained *Nanggala* and *Alugara* weapons and two elephants, Gajah Puspadenta and Yudakethi. This show lasts for 6 hours.

3 TRANSITIONAL MELODY OF MUSIC FOR *WAYANG*

To facilitate the discussion, the following transitional melodies are included in each *patet* section.

3.1 *Patet Nem*

In the *Patet Nem* segment, the *patet* changes occur six times, namely when the show has started, the Kundur Kedhaton scene, the Paseban Jawi scene, Jejer II, Jejer III and the scene of Brama meeting with Patih Nagarangsang.

3.1.1 *The Beginning of the show*

When the puppet will begin, a transition takes place between *patet sanga* and *patet nem*. In the transition from *patet sanga* to *patet nem*, there is a transition melody in the form of *grambyangan* of *patet nem*. Before the puppet show begins, gamelan music has been sounded by playing *patalon* pieces. *Patalon* pieces consists of compositions or pieces with *patet sanga*, namely: Gambir Sawit, Ladrang Pangkur, Ketawang Langen Gita, Ayak-Ayak Sanga, Playon Sanga, and Sampak Sanga. The melody of *grambyangan Patet Nem* is as follows

After the *grambyangan*, the *dalang* gives the cues or comand in the form of *dhodhogan* (tapping the wooden puppet box with a *cempala* or a mallet) five times, followed by Ayak-ayak Patet Nem.

652. 3

...6 .

53..   21..   ...3   ....

..53   ..21   ..23   .123   .332

.653   ...2   ...5   ...6

....   .532   ..35   .356

...5   .3.3   ...2

53565   .356   ...2

.612   1616

....2

Figure 1. Grambyangan patet nem.

3.1.2 *The scene of Kundur Kedhaton*

This scene is accompanied by Sri Katon Patet Manyura Ladrang compositions. Thus, in this section there is a transition from *patet nem* to *patet manyura*. The transition melody used is in the form of Manyura *grambyangan* as follows.

```
3216   53..   .61.

....   ....   ....

....   ....   .321

.561   .111   ...1

.232   ..16   ..53   ..21   ....

...2   ....   16..   53..   21..

..23   ....   ....   .161

....   .123   .333   ...3
```

Figure 2. Grambyangan of manyura.

The above melody is played by *gender* players after the *dalang* gives a code or *sasmita* with the phrase "Katon Tejane" as a sign so that the gamelan players play the Ladrang Srikaton piece. After playing the Ladrang Srikaton Patet Manyura piece, the gamelan players continue the Ayak-ayak Lasem Patet Nem, Srepeg Lasem Patet Nem, and Playon Lasem Patet Nem pieces without the transition melody of *thinthingan* or *grambyangan*. The transition melody used is Ayak-ayak with the tone of 2356. Although patet changes occur, this part is not a problem because the taste between Patet Nem and Manyura is very close.

Ladrang Srikaton Patet Manyura

```
Opening: 3.32   3.32   3.32   3.3(2)
[:5653   2121   3565   3212

  5653   2121   3565   3212

  66.5   3561   3265   3235

  3353   5616   3565   321(2):]
```

Ayak-ayak Lasem Patet Nem 2.3.5.(6)
Srepeg Lasem Patet Nem

2626 2626 1561 2321

5235 2356 1656 5323 1232

Playon Lasem Patet Nem

[: 5653 5653 6526 5235

1232 6523

5353 5235 1653 2132

6632 3565

2321

2132 6356

3263 6532 :]

Closing: 11 321(6)

Figure 3. Ladrang sri katon patet manyura.

Ayak-ayak Lasem Patet Nem

Figure 4. Ayak-ayak lasem patet nem and srepeg lasem patet nem.

In the last row of the Ladrang Srikaton piece following the 3565 tone, the tone does not go to 3212 but to 2356 with a slowing tempo. Although the gong tone is different in which the gong of 2 (3212) is replaced by the gong of 6 (2356), it does not create any problem because tone 6 is the *kempyung* of tone 2. Musically, the *kempyung* tone is close.

3.1.3 *The scene of paseban jawi*

In this section, there was a change in the *patet* from *patet nem* to *patet manyura*, then *patet sanga*. Originally, the gamelan music pieces played are Playon Lasem Patet Nem - Suluk Ada-ada Kawin Sekar Asmaradana Patet Manyura - Song of Suluk Ada-ada Sekar Pangkur -

Lancaran Gagaksetra Patet Sanga. Therefore, the transition is as follows: Playon Lasem – *thinthingan* Manyura (*gender*) – Suluk Ada-ada Sekar Asmaradana – *thinthingan* Sanga (vocal *dalang*) – Lancaran Gagaksetra. The transition from *patet manyura* to *patet sanga* is done by the *dalang* vocally imitating *thinthingan* with the sounds of the *bende*, drum and *terumpet* when there will be a departure of soldiers within songs as in the following:

Mung mung jir Tret tret trit tret tret tret tret tret trit tret drang drang durr
6 6 2 6 6 $\dot{2}$ 6 3 5 3 5 6 2 6 6 $\dot{2}$

Mung mung jir Tret tret trit tret tret tret tret tret trit tret drang drang dur.....mung mung jir
6 6 2 6 6 $\dot{2}$ 6 3 5 3 5 6 2 6 6 $\dot{2}$ 5 5 1

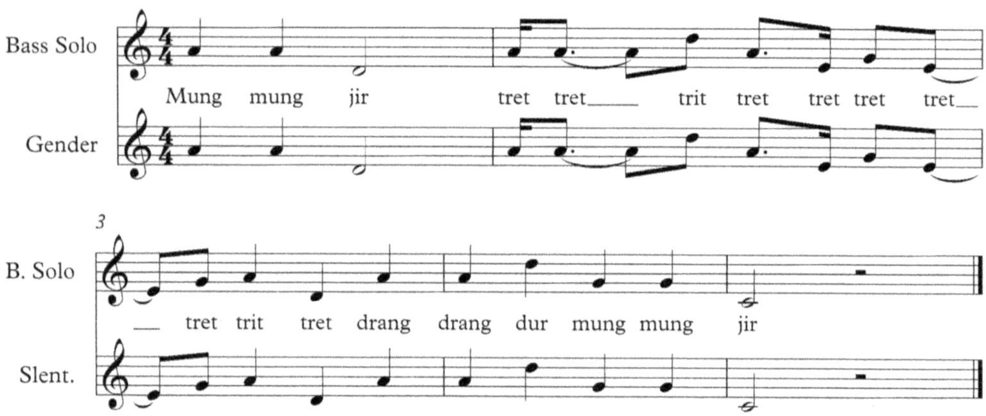

Figure 5. Vocal *dalang* as *thinthingan.*

The song is actually *thinthingan* as the transition from *patet manyura* to *patet sanga*. The initial sounds are tones 6 and 2 which are the *patet manyura* tone areas, while the final tones 5 and 1 are the *patet sanga* tone areas. After the *thinthingan* was continued by a vocal song of the *dalang* singing Kawin Sekar Pangkur Patet Sanga, it was continued by the Lancaran Gagak Setra Patet Sanga piece. This music is used to accompany the departure of soldiers.

3.1.4 *Jejer II*

This section uses the Playon Lasem Patet Nem piece, although in the previous case, the music was Lancaran Gagak Setra Patet Sanga. So here a transition occurs and the transition melody used is the *grambyangan patet nem.*

3.1.5 *Jejer III*

This section is accompanied by the Ladrang Mega Mendung Pelog Patet Nem piece. The previous scene used the Playon Lasem Slendro Patet Nem piece. Here there is a change in the type of gamelan used, namely from the *slendro* gamelan scale to the *pelog* scale. The transition used is the *grambyangan* Pelog Patet Nem melody as in the following:

```
3216   ..61   ....   .111   .321
____   ____   ____   ____   ____
....   53.1   .561   .111   ....

...2   16..   53..   21..   .3..
____   ____   ____   ____   ____
.232   ..16   ..53   ..21   23..

....   ....   .161
____   ____   ____
.123   .333   ...3
```

Grambyangan Pelog Patet Nem

Figure 6. Grambyangan pelog patet nem.

After completing the *grambyangan*, it is followed by gamelan pieces and vocal songs in the *pelog* scale until it ends with the scene of Brama meeting Patih Nagarangsang.

3.1.6 *The scene of Brama meeting Patih Nagarangsang*
In this section there is a change in scale, from *pelog patet nem* to *slendro patet nem*. To begin this scene there is *grambyangan patet nem*. This was done because previously it was gamelan pieces and vocal songs on *pelog* scale. The gamelan pieces with *patet nem* continue until the scene of Gara-gara which is played on *patet sanga*.

3.2 *Patet sanga*

In the *patet sanga* segments, there are four *patet* changes: in the scenes of Gara-gara, Petruk singing, Semar singing and *jejer* IV.

3.2.1 *The Scene of Gara-Gara*
The scene of Gara-gara applies the earliest *patet sanga*. Thus, a process of change from *patet nem* becomes *patet sanga* that occurs in this scene. The *patet* transition uses *patet sanga*'s *grambyangan* as follows.

```
2165   .3.2   .3.5
____   ____   ____
....   ....   ....

....   ....   .165   .212
____   ____   ____   ____
.222   .555   ....   ...5
```

Figure 7. Grambyangan sanga.

After the *grambyangan,* the atmosphere brings a feeling of *patet sanga.* This taste is strengthened by the mood song of Suluk Lagon Patet Sanga.

3.2.2 *The scene of Petruk Singing*

The character of Petruk sings in *patet sanga* because Petruk is supposed to be sung in the atmosphere of the *patet sanga* segment. Since there is dialogue and *grimingan* (*gender* sounds accompanying the *dalang* as he speaks) beforehand, during this performance the *dalang* did not hear well enough. Because the *dalang* lost the atmosphere of the *patet*, through the character of Petruk the *dalang* sang *ma ma ma ji ji ji* vocals followed by *gender* voices of the 5 and 1 tone. After this, the *dalang* then sings the song of Slendang Biru Slendro Patet Sanga with the initial tone of 5.

3.2.3 *The scene of semar*

In this section, there is a transition from *patet sanga* to *patet manyura.* The previous music was Sampak Patet Sanga. When Semar arrived, the gamelan piece stopped (*suwuk*). When Semar sings, the gamelan musicians make *thinthingan* by ringing *ning ning ning nong nong nong* followed by *gender* tones 666 222. This *thinthingan* is able to turn *patet sanga* into *patet manyura.* After this *thinthingan,* Semar sings Suluk Sekar Pucung Patet Manyura.

3.2.4 *Jejer IV*

This section uses the music of the Ladrang Sri Kaloka Patet Sanga piece. Previously, it was a piece with *patet manyura.* Thus, there is a change in *patet* here. The transition melody used was *grambyangan patet manyura.* The Ladrang Sri Kaloka piece is followed by vocals and playing the *patet sanga* piece and lasts until jejer VI.

3.3 *Patet manyura*

Patet manyura starts on *jejer* VI. In the *patet manyura* section there are two transitions, in *jejer* VI and in the scene of the Wanengpinta meditating to bring Narayana.

3.3.1 *Jejer VI*

Jejer VI begins with *grambyangan manyura* to change the atmosphere from *patet sanga* to *patet manyura.* After the *grambyangan* continues, the *dalang* vocalises a mood song in the form of Suluk Lagon Patet Manyura Wetah. However, the atmosphere of *patet manyura* only lasts for a short time only to open the *patet.* After that, there is *grambyangan pelog* to prepare the piece with *pelog* scale. After singing the Suluk Lagon mood song, the *dalang* narrates the next scene. When the *dalang* narrates the scene, the *gender* musicians begin to play *grambyangan pelog barang* to change the atmosphere to *pelog.* The following is the *grambyangan pelog barang.*

2765 ..35 ....

.... 32.. .235 .555

.765 ...6 ...5

.... .356 53565

.3.3 ...2 .672 7676

.356 ...2 2

Grambyangan Pelog Barang

Figure 8. Grambyangan pelog barang.

After *grambyangan pelog barang*, all the mood songs and gamelan pieces play at *pelog barang*, the *pelog barang* atmosphere lasted until the scene of Begawan Wanengpinta meditating.

3.3.2 *The scene of* begawan wanengpinta *meditating*
In this scene, there was a transition from *pelog barang* to *slendro manyura*. This transition was in the form of *grambyangan manyura*. During the fieldwork this atmosphere lasted until the show was finished.

4 PATET AS A FORM OF MUSICAL TASTE

Patet is a kind of music mode, but it has different modes. Every *patet* has a different musical taste. This musical taste is influenced by the intervals of each *patet* tone. *Patet nem* has a pitch of 6-5-3-2, *patet sanga* has intervals of 2 - i - 6 - 5, while *patet manyura* has intervals of 3 - 2 - I - 6. At a glance, *patet manyura* is a note above *patet sanga*, and *patet sanga* is a two tones above *patet nem*. However, it is not as simple as this because the intervals between the tones in the gamelan are not the same. For example, the gamelan at the Puppetry Department of Indonesia Institute of the Arts, Yogyakarta has the following intervals:

	Interval among note n with note n+1 (cent)												
Note	6̣-1̣	1̣-2̣	2̣-3̣	3̣-5̣	5̣-6̣	6̣-1	1-2	2-3	3-5	5-6	6-1̇	1̇-2̇	2̇-3̇
Interval	260	237	254	250	229	250	243	238	243	223	263	236	248
Patet Nem			■	■				■	■				
Patet Sanga					Patet Sanga					Patet Sanga			
Patet Manyura						Patet Manyura					Patet Manyura		

Source: Prasetya, 2013

The intervals between the tones in the gamelan will form a different sense of feeling in each *patet*. If the interval is categorised into three, near, medium and far, then it can be categorised as follows:

223 – 236 = near
236 – 249 = medium
249 – 262 = far

Table 5. Tones in *patet.*

Tone Area		Impression
Patet Nem	6 5 3 2	223 – 243 – 238 near – medium – medium
Patet Sanga	2̇ 1̇ 6 5	236 – 263 – 223 near – far – near
Manyura	3̇ 2̇ 1̇ 6	248 – 236 – 263: medium – near – far

The feeling of *patet* is characterised by a leap in tone. This also applies to the repository of *patet nem*. For example, the piece that is always used in shadow-puppet shows in the *patet nem* section is the *Playon Lasem Patet Nem* as follows.

Opening: kendang: (5)

A
65 6565 2356
1̇656 **2353** 2121
2121 5235 2356
1̇656 **2353** 1232
3263 6532

B
[:5653 5653 6526 5235
1232 6523
5353 5235 1̇653 2132
6632 3565
2321
2132 6356
3263 6532:]

Closing (Suwuk): 11 321(6)

Figure 9. *Playon lasem patet nem* piece.

5 CONCLUSIONS

There are two transitional melodies when there are changes of *patets* in shadow-puppet shows, namely *thinthingan* and *grambyangan*. *Thinthingan* is usually played if there is a melody that has prepared the transition. *Grambyangan* melodies are used when there is a strong transition that was not preceded by a strong musical sense, for example, a change in the scale from *slendro* to *pelog* or vice versa. Likewise, this happens when there is a transition between *patets* but with a strong distance, for example, the transition from *patet sanga* to *manyura* or vice versa. The transition from *patet nem* to *patet manyura* or vice versa, can be played with *thinthingan* because the taste of melody between *patet nem* and *manyura* is not too wide.

REFERENCES

Brinner, Benjamin. (2008). *Music in Central Java*. New York: Oxford University Press.

Lysloff, Rene TA. (1993). A Wrinkle in Time: The Shadow Puppet Theatre of Banyumas. *Asian Theatre Journal. 10(1)*, 49-80.

Prasetya, Hanggar Budi. (2012). *Pathet: Ruang Bunyi dalam Karawitan Gaya Yogyakarta.*[Pathet: Spaceof Sound in Yogyakarta's Style Karawitan]. *Panggung Jurnal Seni dan Budaya. 22 (1)*, 67-82.

Prasetya, Hanggar Budi. (2013). Mlesed dan Nggandhul dalam Karawitan Pedalangan Gaya Yogyakarta. *Disertasion.* [Mlesed and Nggandhul in Yogyakarta's Style of Pedalangan's Karawitan]. Universitas Gadjah Mada Yogyakarta.

Prasetya, Hanggar Budi. (2013). *Fisika Bunyi Gamelan.* [Sound of Gamelan in Physics]. Yogyakarta: BP ISI.

Sutton, Anderson. (1991). *Traditions of Gamelan Music in Java: Musical Pluralism and Regional Identity.* New York: Cambridge University Press.

Weiss, Sarah. (2006). *Listening to an Earlier Java.* The Netherlands: KITLV.

21st Century Innovation in Music Education – Astuti et al. (eds)
 ISBN 978-0-367-11122-9

The meaning of the Koyunbaba song (Suite Für Gitarre Op.19) in Carlo Domeniconi's work

B. Walidaini & S.A. Sayuti
Universitas Negeri Yogyakarta, Yogyakarta, Indonesia

ABSTRACT: The objective of this research is to describe the meaning of particular signs contained in the song lyrics or score of the Koyunbaba (Suite Für Gitarre Op.19) song by Carlo Domeniconi using the semiotic approach of Charles Sanders Peirce. This study is descriptive-qualitative research. The focus of the research looks at particular matters or the uniqueness inserted by the composer, derived by studying the form and structure of the score. The meaning of the song was then identified and analysed by using the typological signs semiotic approach of Charles Sanders Peirce. The data was obtained by observation, interviews, documentation and literature study techniques. The legitimisation of the data was made possible by conducting a validity (peer debriefing) test. The results showed that signs in the Koyunbaba text was a representation of a modern suite with its own particularities.

1 INTRODUCTION

Music is one expression of art that reflects culture in society. In music there are values and norms that are part of the cultural process, considering that the functions possessed by music are very significant for human life. Carlo Domeniconi is one of the modern composers who created great works for guitar musical instruments, one of which was Koyunbaba (Suite Für Gitarre Op. 19). The advantages of Carlo Domeniconi's works are: 1) Long enough duration; 2) Requires high guitar playing techniques; 3) Harmonies that are highly varied in all of Domeniconi's compositions; 4) A dramatic song atmosphere in all of his works; 5) Elements of traditional music; and 6) Use of signs in scores that are unique or different from other works. Over the course of his life, Carlo Domeniconi was not only known as a composer but also as a teacher. This experience as a teacher made it possible for him to move from one country to another. One country that was once a place of residence was Turkey where he became a teacher at the Conservatory of Istanbul, an institution that highly influenced the works he created. Most of his compositions tell about places in Turkey.

One of Carlo Domeniconi's famous works is Koyunbaba (Suite Für Gitarre Op. 19). Koyunbaba is a work written for solo guitar that takes the title from one of the places in the country of Turkey. One of the unique of Koyunbaba is that the composer does not give sukat or bar signs in the song, even though the bar sign is very important in writing music because it determines the beat, accent and judgement on each note that will be played. In addition, in the Koyunbaba composition the composer used a tuning system that is not commonly used in ordinary classical guitars. Generally, the classical guitar string tuning is EBGDAE. However, in this work the composer wanted to give a different nuance so that the tuning of the strings was changed to DGDADF. The composer created this work because he wanted to convey a message from inside of himself that can be formed from phenomena, events and the background of life that has been experienced or learned. The effort to convey this message is called communication and the communication process can be said to be successful if the communication participants can understand the meaning of the message being communicated.

From the background above, the researcher became interested to study the song Koyunbaba (Suite Für Gitarre Op. 19) by Carlo Domeniconi. The text in the Koyunbaba song can be analysed using a semiotic approach so that the meanings and messages communicated through Koyunbaba (Suite für gitarre Op. 19) can be understood by the text reader.

2 IDENTIFICATION OF PROBLEM

Based on the background description several problems can be identified. Koyunbaba (Suite Für Gitarre Op. 19) can be referred to as one of the unique classical guitar song that: 1) uses string tuning arrangements that are different from guitar tuning in general; 2) uses writing notations and signs that are rarely used in classical guitar compositions; 3) does not use signs birama or sukat and; 4) is instrumental music, without the use of language lyrics. Thus, meanings in the music when performed, certainly requires a more in-depth and complex study.

3 DESCRIPTION OF THEORY

3.1 *Suite*

Suite is a form of dance accompaniment that began in Europe in approximately the 16th century. The word *suita* is taken from French which means row. Prier (1991) says that another name for *suita* is *partita*, meaning consisting of parts from the word Italian *partire* or dividing (p. 9). The peak of the development of *suita* was in the 17th century until the middle of the 18th century. At that time there had been a standardisation of *suita* creation namely by assembling four parts in one section which previously only contained two parts in the 16th century. As stated by Prier (1991), over time it becomes a habit or standard that the parts of *suita* are called *Allemande*, *Courante*, *Sarabande* and *Gigue* (p. 10).

Toward the turn of the 19th into the 20th century, entering the modern age, there was a change in European music, including the emergence of the notion of symbolism in music. Mack (1995) explains that musical symbolism at the turn of the 20th century may be more interpretable from the point of view of a composer who is able to give the impression of a secret both in his work and in his own way of life (p. 13). The turn of the century and new styles in creating works indirectly influenced modern composers to abandon the style of music iof previous eras. Mack (1995) suggests that these modern composers were somewhat opposed to the attitude of other conservative composers who were full of romantic styles (p. 23). However, in these early modern times composers still created many forms of music that were popular in the previous era. In the early 20th century era, the existence of *suita* music was still recognised, but only in terms of the style of writing and the series of parts in it while its use deviated from *suita* in the era before the modern period. Mack (1995) also argues that of course, Debussy did not create this music to accompany dance and that this function did not apply to Bach's *suita* during the baroque era (p. 25). This is just a kind of stylisation for processing patterns on the historical model. The above-mentioned quote provides an example, using Debbusy who was a pioneer among modern composers and was followed by many other composers.

3.2 *Music as a sign in aesthetic communication*

There are several opinions from experts regarding signs, including Danesi (2012) who argue that signs are all colours, cues, blinking eyes, objects, mathematical formulas and others that represent something other than themselves (p. 6). According to Martinet (2010), a sign is something that can be captured that shows something other than itself (p. 45). In addition, according to Nurgiyantoro (2005) a sign is something that represents something else that can be in the form of experience, thoughts, feelings, ideas and so on (p. 40). So, the real sign is not only language, but various things that surround on self in life, although it must be admitted that language is the most complete and perfect sign system. These signs can be in the form of limb movements, eye movements, mouth movement, writing forms, colours, flags, shapes and pieces of houses, clothing, works of art such as literature, painting, music, dance and other phenomena around us. From this standpoint, the score or text of the song Koyunbaba can be said to be a sign because a score can give a signal, formula or command that is used by text readers to present what the composer wants in a musical text or score. While a composer does not just create a piece of music, there must be a meaning stored in the sign that was written into the score.

To know the meaning in a musical work one can use the semiotic approach. The term semiotics comes from the Greek word *semien* which means sign. Hoed (2008) says that semiotics is a science or method of analysis used to examine signs. Barthes (2001) expresses Saussure's opinion about signs in the context of human communication by making a connection between signifier (marker) and signified (sign) (p. 30). A signifier can be a meaningful sound or meaningful graffiti, namely what is written or read. The signified is a mental image, namely the thought or concept of language. Saussure (1980) also confirms that in addition to serving as explicit markers, signs also bring a pair of other language signs (p. 208). Wibowo (2012) quotes Umberto Eco, who refers to the sign as a lie, where in the sign there is something hidden behind it and is the sign itself (p. 9). Therefore, it can be concluded that the semiotic method as a search for meaning is an appropriate in this study.

One concept of semiotic thinking comes from Charles Sanders Peirce (1839-1914) who discussed the trichotomy theory. Trichotomy is a theory that discusses signs that are sign, object and interpretant. This means that this theory discusses the relationship of the sign with the sign itself (sign/representment), the sign with the object and the sign with interpreter. Peirce said that in semiotics there is a relationship that cannot be separated from the three different elements. This relationship is called triadic, namely the sign or representment, object and interpreter (Peirce in Wahono and Kustap 2007, p. 53).

3.3 *Three sign trichotomies of peirce*

The first trichotomy is the sign relationship with the sign itself, in which the sign is divided into qualisign, sinsign and legisign. Qualisign is a sign based on its quality, sinsign is a sign of its existence and legisign is a sign of general rules.

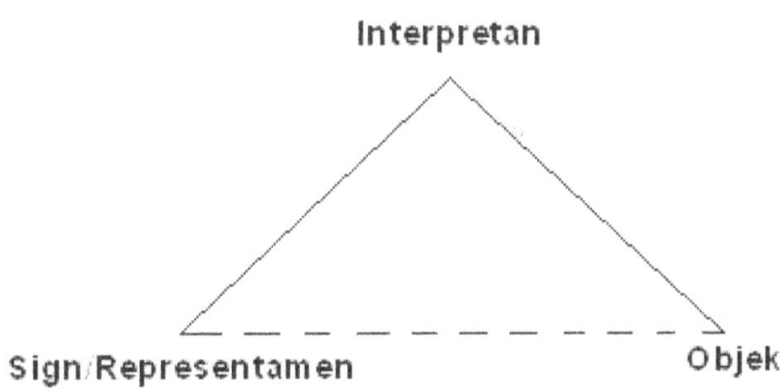

Figure 1. Peirce triangle trichotomy (Wahono and Kustap, 2007, p. 53).

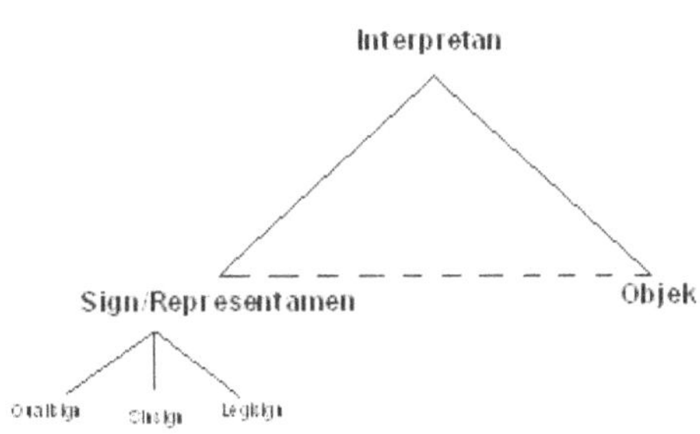

Figure 2. Sign in the relationship in the sign itself (Wahono and Kustap 2007, p. 53).

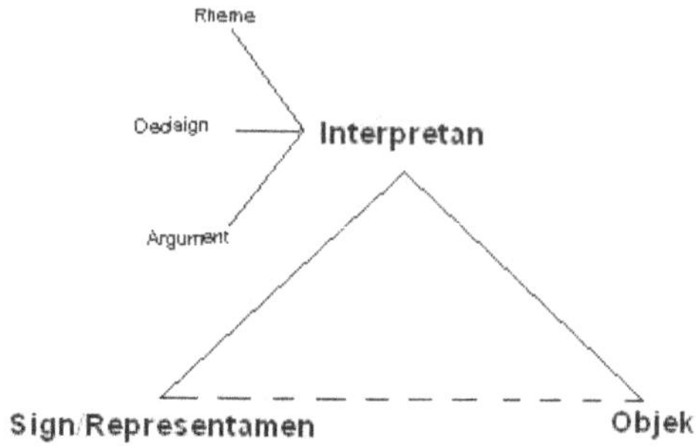

Figure 3. Signs in relationships in signs to objects (Wahono and Kustap, 2007, p. 53).

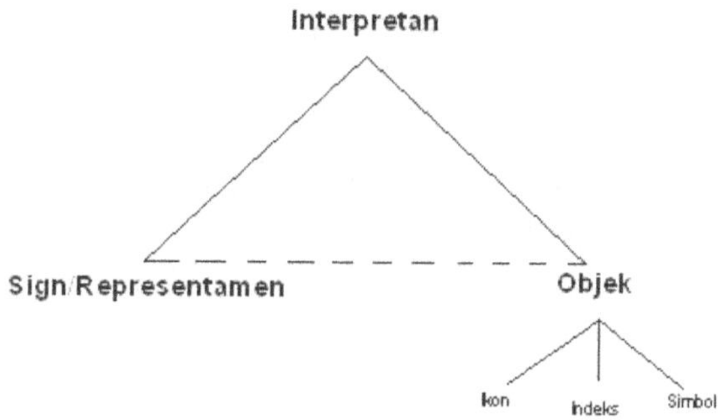

Figure 4. Signs in relationships in signs to objects (Wahono and Kustap, 2007, p. 54).

The second trichotomy is the relationship of the sign to the object. Inside are icons, indices and symbols. The icon is if the sign has a relationship in common with the object. An index is if the sign is a pointer to the object, while the symbol is a conventional sign.

The third trichotomy is the relationship of signs to the interpretan as a sign of qualitative possibilities (rheme), about facts (decisign/dicentsign) and about thinking (argument) (Wahono and Kustap, 2007, p. 54).

The theory of the trichotomy concept from Peirce is considered appropriate for this study because it contains the discussion needed for the initial step in searching for the meanings found in musical signs in the song Koyunbaba (Suite Für Gitarre Op. 19) by Carlo Domeniconi. However, in order to focus research and analysis, this study will only use sign analysis based on the second trichotomy, namely the relationship of the sign with the object. The purpose of this research is to classify and describe the signs in the Koyunbaba text.

3.4 *Charles S Peirce's version of typology*

Peirce distinguishes the types of signs into the following: icons, indices and symbols that are based on the relationship between the representamen and the object.

The icon is a sign that contains a 'likeness' so that the sign is easily recognised by its bearer. Inside the icon, the relationship between the representatmen and their objects manifests as there are similarities in several qualities. For example, most traffic signs are iconic because they describe a form that has similarities to the actual object. The index is a sign that has a phenomenal connection or existence between the representamen and its object. In the index, the relationship between the sign and the object is concrete, actual and usually in a sequential or causal way. The footprint on the ground is an index of someone or animal that has passed there. The door knock is an index of the 'guest' presence in the home. Symbols are a type of sign that is conventional in accordance with an agreement or convention of a number of people or communities. Signs of language are generally symbols. A few of the traffic signs are also symbolic. From Peirce's point of view, the signification process can produce a series of endless relationships, so that in turn an interpretant will become representamen, become an interpretant again, become representamen again and so on. Thus, it can be emphasised that the second trichotomy of Peirce is the most appropriate theory in an effort to analyse the results of the signs found in the text so that the signs found can be identified and reveal their meaning.

4 RESEARCH METHODS

4.1 *Research approach*

This study uses a qualitative descriptive approach, namely research that provides an overview of the situation that occur within the context of the object under study. This is because the collected data are in the form of words or images and not in the form of numbers. As expressed by Bogdan and Biklen (2011) qualitative research is more descriptive, the collected data are in the form of words or images and do not emphasise numbers (via Sugiyono, p. 13). In qualitative research data comes from documentation, supervision, evaluation, preliminary observations and statements from trusted sources. The hypothesis in qualitative research is to find a theory rather than formulate or specify a hypothesis clearly before going into the field. Thus, this research is a study conducted using qualitative methods and is descriptive in nature in order to find a detailed explanation of the sign system phenomena that exist in the Koyunbaba text. This study uses several work steps, namely reading (playing) text, listening, observing and analysing music texts (partitur/music score) in the song Koyunbaba which are supported by data sources such as articles, books, statements from informants and the internet, which were recorded for processing.

4.2 *Research data*

The main data used in this study is the score text of the song Koyunbaba (Suite Für Gitarre Op. 19) by Carlo Domeniconi. This was coupled with supporting or external data in the form of video performances for audio-visual purposes, books, articles and interviews with resource persons for the purposes of analysis and identification. Official external documents according to Moleong (2000) are documents that contain information materials produced by a social institution such as magazines, bulletins, statements and news broadcast to the mass media (p. 121).

4.3 *Research instrument*

According to Moleong (2000), qualitative research deals with the researcher himself or uses someone else's help as the main data collection tool (p. 19). Researchers as research instruments function in taking initiatives related to research. This includes searching data, creating questions for interviews and as data processors. From this understanding, the research instrument in this study is the researcher.

4.4 *Data collection technique*

The researcher observed through participating directly as a presenter and audience member in several musical performances that presented the repertoire of Koyunbaba (Suite Für Gitarre Op. 19) by Carlo Domeniconi. In October 2011, the researcher participated as a presenter, namely a reading and playing of the Koyunbaba (partitur) text in a classical guitar performance at the Karawitan Laboratory of the Language and Arts Faculty of Yogyakarta State University. In May 2012, the researcher participated as an audience member in a classical guitar concert at the Indonesian French Institute, Yogyakarta with Thiboult Cauvin, a classical French guitarist. In addition to observations in musical performances, the researcher also observed the Koyunbaba text (Suite Für Gitarre Op. 19). The researcher conducted an interview with Rahmat Raharjo, S.Sn in October 2013. Raharjo is a classical guitar player from Yogyakarta who first played Koyunbaba (Suite Für Gitarre Op. 19) in Indonesia at the national classical guitar festival in 1992. The researcher also conducted an interview with Bakti Setiaji, S.Pd in October 2013. Setiaji is a composer who dared to explore in his works and was an expert in the analysis of the shape and structure of music. The results of the documentation were obtained from the recording of a statement from Carlo Domenikoni regarding the work Koyunbaba. This statement is the result of an interview with Classical Guitar Alive radio, an online radio channel that broadcasts the development of classical guitar in the world (see http://www.guitaralive.org/).

4.5 *Data analysis technique*

According to Creswell (2010), data analysis is an ongoing process that requires continuous reflection on the data, asking analytical questions and writing brief notes throughout the study (p. 274). The interrelationships of these affect the level of understanding and interpretation. As explained by Creswell (2010), researchers need to prepare data for research, conduct different analyses, deepen understanding of the data, present the data and make interpretations of the broader meaning of the data (p. 274). Then according to Rohidi (1992), the acquisition of the data is organised into one place to be used and interpreted as findings to answer research problems (p. 95).

In this study, data processing techniques were carried out, namely using qualitative description. The steps used in this technique are as follows: First, researchers try to understand and translate what is in the Koyunbaba's text into the guitar playing technique, then present it in the staging of classical guitar. Aside from being a presenter, the researcher was also involved in a musical performance that presented the Koyunbaba repertoire as an audience member. Both of these steps were taken so that the researcher experienced direct involvement in the process of revealing or turning text into a presentation of guitar playing. Both steps were successfully recorded on video.

Next, the researcher looked at the Koyunbaba text in terms of shape and structure with reference to music analysis books. The researcher categorised based on the largest division in the structure of music, namely the period until the division of phrases and motives. This step was done so that a music text became clearer distinguishing between one sentence with another sentence and to find out what was contained in the text. Instrumental music or music without lyric also contain sentences. The sorting of motives aimed to see what patterns are in the Koyunbaba text.

After analysing the shape and structure in the Koyunbaba text, the researcher conducted interviews with two speakers. The first speaker Rahmat Raharjo, S.Sn., a classical guitar player who first performed the work of Koyunbaba in Indonesia in Jakarta in 1992. Apart from being a guitar player, he was also an expert in the form of music. From the results of the interview, information was obtained about the work of Koyunbaba, information on the forms of *suita* music and about Carlo Domeniconi. The second interview was conducted with Bakti Setiaji, S.Pd., a music composer and expert in the analysis of the shape of musical structures. In the interview, the researcher obtained information about the forms of compositions that deviate from the rules.

After getting the results of the analysis from the Koyunbaba text, the text becomes easier to observe because it has been divided into small parts based on type. Next, the researcher identified the signs contained in the text based on the typology of the sign from Charles Sanders Peirce, which included identification and classification of signs. This step produced an identification and classification of signs of the Koyunbaba text into three types, namely icons, indices and symbols based on the Charles S Peirce version of typology. This step is supported by a literature study, and the final step in the data analysis technique of this research checked the validity of the data.

In this study, the technique used to check the validity of the data was to conduct question and answer with fellow researchers in a peer debriefing. This technique is one of the various data validity techniques. According to Creswell (2010), this strategy involves other interpretations besides interpretations of researchers who can add validity to the results of the study (p. 288). Researchers chose a question and answer technique with fellow researchers because the results of the research in this study could be shared with other researchers and the primary researcher can receive responses and reviews from fellow researchers. In study, the researcher conducted a question and answer interview with the Kustap, S.Sn, M.Sn., a lecturer at the Yogyakarta Indonesian Art Institute who has conducted research on semiotics and music. The interview was conducted in December 2013.

5 IDENTIFICATION CLASSIFICATION AND MEANING OF SIGNS IN THE KOYUNBABA TEXT (GITARRE SUITE FUR OP. 19)

Identification, classification and meaning of markings in this research was carried out by adapting the types of signs based on an object's relationship with the sign proposed by Peirce. In the Koyunbaba text after being analysed the structure of the music can be identified and classified into several types of signs in the text structure as the unit of analysis studied.

5.1 *Icon*

The first icon is a scordatura. Based on its relationship with the object, the scordatura has the same form with the original scores so that at a brief glance the Koyunbaba text may seem to be written as the same in two scores. However, after researching scordatura is an imitation of scores that functions similar to a map that is in not scordatura and does not refer to the intended tone but symbolises the tone before the tuning changes occur. The shorter scordatura functions to show the position directly on the fret board on the guitar. The function of scordatura is nothing but to make it easier to read the text itself.

Another icon is the first arpeggio motif in movement I (Moderato). Arpeggio is a series of chords or trellises that are sounded one by one alternately and played in a free and floating way. In this case, the arpeggio in question has similarities to the prefix of the opening in each work which has an introduction. The arpeggio is one type of introduction in the structure of a *suita* as stated by Michels in Prier (2013, p. 141). A type of arpeggio is a sequence chords which are broken into sixteen notes that flow constantly, creating a calm melodic current that is associated with the final rhythm that floats like a heartbeat.

The next sign of the icon is the t movement (Moderatro, Mosso, Cantabile, Presto), which is a division of a work into several movements which is characteristic of a work in a *suita*. Like the meaning of the word *suita*, from the French meaning 'row', every work is divided into several movements.

The fourth icon has a G clef. sign. This key is used for musical instruments whose registers tend to be high, including a guitar that cannot use other keys besides the G key because the guitar sound register ranges from e to b ".

5.2 *Index*

On the number one index there is a sign in the title of Koyunbaba. This sentence is taken from the Turkish language and can separately be interpreted as *koyun* meaning 'sheep' and *baba* meaning 'father'. However, if combined the word *koyunbaba* can be defined as 'shepherd'. This refers to the Turkish culture of herding livestock. The interpretant formed is the image of a shepherd who is grazing his livestock with all his activities as outlined in a musical work. Another interpretation formed by this sign is the composer himself, Carlo Domeniconi, because he was an artist who carried out cultural crossings in music so Carlo Domeniconi could be called a cultural herder. The next interpretant from Koyunbaba is the name of a barren place filled with mystical aura.

At index number two there is a sign of the year of writing (1984). Indirectly, this sign refers to the year when there were innovations from composers who began to abandon the romantic notion in the creation of his works. At this time, composers began to abandon the rules and provide different nuances to composers who were still conservative and began to emerge with the understanding of musical symbolism. Composers who succeeded in innovating and giving particularity in creating musical works at that time included Debussy, De falla, and others.

At index number three there is the name Carlo Domeniconi. The name Carlo Domeniconi refers to the origin of the name and the country is one of the countries in Europe, namely Italy. The interpretant that was formed, the Italian state, was very influential in the development of classical music in the world as evidenced by many composers from Italy including: Antonio Vivaldi, Nicono Paganini and others. In addition to the emergence of reliable composers, the influence of the Italian state in the development of classical music is also due to the fact that music education in the country was very advanced compared to other countries in Europe.

At index number four there is another sign. Interpretants are formed from this index sign, namely a sign that is used to repeat the time referred to, in the sense that a sign can be likened to an arrow that refers to a text reader to turn around. At index number five, there are all types of coda, fine, arm 19, d.s al fine, arm 7, rasgueado, come sopra. These signs are words from the Italian language which function the same as the direction of the arrow but with the word command literally taken. That is, if we are reading the text and finding the signs, then we must follow what is said on the sign like following the arrow that has been provided. At index number six there is a segno sign. The name of this sign also comes from the Italian language with a function that is the same as the direction of the arrow. That is, if we are reading the text and finding the signs then we must follow what is seen on the sign like following the arrow that has been provided. At index number seven there is an emptying of sukat. In passing, the sukat void means there is no sign and there is no meaning because sukat is one of the most important parts in music. However, if examined further, this void is a sign and can be classified into an indiscriminate type such that in an index, this sukat void refers to other signs that are related to the sukat emptiness itself, finding the meaning of sukat emptiness can be extracted from other signs associated with it. Other interpretations formed from this succinct blanking sign show that technically, the koyunbaba text has a complex structure whereby the number of notation values between one and the other is not the same, so that the sign serves to give freedom to the reader because if the composer gives succor, the text will become confusing to read. At index number eight there is a sign DCA-B, DCA-B poi C - D, rip. A - B. These signs often appear in movement IV (Presto), whose function is the same as the direction of the arrow. That is, if the text reader is reading the text and finding the signs then the reader must follow what is said on the sign such as following the arrow that has been provided.

5.3 *Symbol*

The number one symbol denotes the use of ascending sequences and among other considerations, there are at times, 2, 3, 14, 16, 46. The development of motives in this technique is widely used for question sentences because the impression that arises from this processing creates a tense atmosphere. Thus in the Koyunbaba text, most of the ascending sequences are used in the question sentence and if some are used in the answer sentence then there are not many but only a few. The use of this ascending sequence is also a representation of the pattern of saz musical instruments originating from Turkey, because the koyunbaba music works describe one of the places or events in Turkey.

Symbol number two, namely the use of descending sequences, among others, is in the number 26, 28, 48, 50, 77, 78. The development of motifs in this technique is widely used for sentences because the impression that arises from this processing is to reduce tension. As in the Koyunbaba text, most of the down sequences are used in the answer sentence or at the end of the motive. If they are used in the initial sentence there are not many, only a few. The descending sequence is the opposite of the ascending sequence, and answers the question sentence which mostly uses ascending sequences. The use of this down sequence is also a representation of the pattern of saz musical instruments originating from Turkey, because the koyunbaba music works describe one of the places or events in Turkey.

Symbol number three uses sentences, phrases, and motives. The term is an agreed upon way of analysing a musical work. Like language, music also has coupled verses that form musical sentences so that it can be seen how the storyline or path of the music goes.

The symbol number four is the use of kadens. Kadens is a chord progression at the end of a sentence. The answer sentence is usually denoted by progression V - I. However, in the Koyunbaba text, it is rarely found at the end of the answer phrase that uses the last chord progression with a chord I. Most of the answer phrases end with an I minor chord, and the impression that emerges from this minor I kadens is an unresolved atmosphere. In Koyunbaba's text, the chords used are also not varied, due to the use of nonstandard tuning so that it only allows reaching certain chord positions.

Symbol number five is the use of tuning. In general, guitar string tuning consists of E B G D A E. However, in the Koyunbaba text the composer changed the guitar tuning to D A D A D F. This change was used to make it easier to reach a series of melodic tones and accompaniment intended by the composer, because the Koyunbaba text is not possible if a standard guitar tuning was played by default. The weakness in this change in tuning is that it cannot use a variety of chords.

The symbol number six is the use of literal repetition, a technique of repeating a motif or an exact sentence without any change. The impression that arises in this repetition is an affirmation.

The symbol number seven is the use of the title movement. This is a sign that is agreed upon as one of the characteristics of music with a suita form. The next interpretant in this sign is the names of movement- Moderato, Mosso, Cantabile and Presto. The names of these movements are in agreement with the names of the movements in the work of Koyunbaba.

6 DISCUSSION OF ANALYSIS RESULT AND MEANING OF SIGNS

Type of Icon: From the results of the sign analysis on the Koyunbaba text above, there are six iconic signs found. From this sign group, suita music is represented through the existence of deviations written by the composer, where the tuning system in this work is changed and the text is written using scordatura. The existence of this sign deviation turns out to only occur on the use of a single skordatura. However, when looking at the relationship with other signs in the Koyunbaba text, it turns out that meanings can occur in this musical work. If it is associated with the iconic number two, the arpeggio pattern that is attached to the beginning of the dance shows that this work still holds the rule of composition because the arpeggio pattern at the beginning of birama can be viewed as an introduction which is an important part in a suita musical work.

If connected with the number three icon, the division of movements can strengthen meanings that the Koyunbaba still maintains the rules of creating suita music works even though the composer does not give a name that exactly matches the movement names Allemande, Courante, Sarabande, Gigue. The movements in the Koyunbaba text are clearly different, but their meanings and use in music can be said to be similar.

Icon number four is an iconic representation of a guitar instrument, although there are many musical instruments that use the G key. However, in the Koyunbaba text it is associated with a sentence under the title (Suite Für Gittarre) which in German means suita for guitar. The use of the German language in the title is based on the work of Koyunbaba, which was completed in the city of Berlin, Germany.

Type of Index: From the results of the analysis above there are eight index signs. The representation of suita music in this group of signs refers to a place and time as its object. In addition, there are also signs that function as arrows and if all these signs are connected to each other hidden meanings are revealed in accordance with the nature of meanings that will continue to develop if associated with related signs. The researcher derived several interpretions in this discussion.

Koyunbaba is a modern suita music created by an Italian composer where the Italian state is undoubtedly the quality of its composer. Koyunbaba was different from other compositions of Carlo Domeniconi even though many of his works were written because of his admiration for the country of Turkey and even though Koyunbaba was not his first work. The seriousness of Carlo Domeniconi in creating this work gave him notoreity as a composer who broke a conservative style but without leaving the foundational aspect of the music itself. The composer's seriousness can be seen in his courage to change the tuning system on the guitar strings that really gives the impression of traditional Turkish music and with the use of signs that are not commonly used in creating guitar compositions. The emptying of sukat in this work is very interesting to discuss where the readers of the Koyunbaba text are required to be critical in listening to this text. Critical listening is not from only one point of view or from one sign but from various signs that relate to one another so that readers can find the sign behind a sign and the meaning behind a meaning. By avoiding a critical approach, the reader will be trapped in the uniqueness itself and assume that Koyunbaba is a work that cannot be read or cannot be played.

Symbol: From the results of the analysis, seven symbolic signs are found. In the signs of this type the interpretant refers more so to Koyunbaba as a form of suita music that is intact because it has parts and grooves as that of a suita with the presence of signs that are common and agreed upon in music including the use of ascending sequences, descending sequences, cadens, sentences, phrases, motives, literal repetitions and movement. The signs above are general signs and have been agreed upon in the creation of a musical work so that the work can be identified in the type of symbol. Deviant symbols are rare and use tuning systems that are different from the guitar in general and different in movement names. However, if examined more deeply the tuning system deviation can also be identified in the symbol type because the D A D A D F guitar tuning system has been agreed to as Koyunbaba tuning. If played, the series of movements in Koyunbaba (Moderato, Mosso, Cantabile, Presto) does not leave the flow of the suita movement series: the first movement is the introduction which flows with the arpeggio pattern at first, followed by the second movement which is more alive than the first movement and contains the new rhythm, then the third movement is a slowing down with singing and the fourth movement is closed with a fast or very fast tempo so that the deviant tuning and differently named movements can be said to be a symbol in the Koyunbaba song and not a symbol in creating general musical works.

This type of symbol sign can be divided into two categories, namely categories of music symbols that have been agreed upon in general and musical symbols agreed upon in the work of Koyunbaba. Thus the researcher argues that this work of Koyunbaba deviates but does not leave the foundation of the essence of music creation.

7 CONCLUSION

In conclusion, the representation of suita music in the Koyunbaba text contains two meaning triangles in the semiosis chain. The first chain explains that representation occurs in the Koyunbaba text which is divided into four movements, even though movement names are not in accordance with the rules of suita in the sense of dance music but still holds the groove which is the basic benchmark in suita music.

The division of the four movements also represents suita music form through the support of other signs, such as the use of one of the introduction patterns which is indeed an introduction pattern in suita music. Emptying the sign of sukat is actually a new sign that requires the text reader to be more critical in paying attention to the surrounding signs that are very likely to influence the meaning of the sign of emptying sukat itself.

Furthermore, the Koyunbaba text as a form of suita music is present as the second semiotic chain, namely the Koyunbaba text as a suita music refers to modern suita music. In modern times there has been a shift in the meaning of music that was once considered absolute, to becoming flexible music. Modern suita music has lost its exclusive function as dance music but does not eliminate its basic foundation. When there has been a lot of musical symbolism in the writing of musical works, then a shift also occurs in its function. Suita music was created by obeying the rules that apply, while modern music is created to describe other objects outside the music itself with its own freedom and in accordance with the wishes of the composer. The development of the function of suita music and its growth also involves the adoption of elements of traditional music.

The objects from outside the music that can be described in the Koyunbaba text include the following:

The meaning of the word Koyunbaba, which in Turkish means shepherd. A shepherd is a person who grazes his cattle and experiences problems in the midst of his pastoral care. The problem can be seen in one of the sentences in movement IV that sounds like a group of farm animals running. The meaning of the word shepherd can also refer to a composer who can be called a cultural herder and that is related to the life background of Carlo Domeniconi: Domeniconi moved from Italy to Turkey, the displacement made it possible for him to cross cultures between one place and another. Koyunbaba also means the name of a place in the country of Turkey, a place that is very arid and wild filled with frightening myths including epidemics and suicides. The composer summarizes all of these phenomena and represents them in this musical work. Thus, when heard, Koyunbaba's work will create a haunting and frightening interpretation. The traditional music used is from Turkey and is known as the *taqsim* musical pattern.

The seriousness of Carlo Domeniconi in elevating traditional music is seen in the use of the Turkish title in this work, Koyunbaba. Domeniconi also changes the guitar tuning system so that it really resembles the nuances of traditional Turkish music *taqsim*. *Taqsim* music is typically played with traditional saz instruments that resemble guitars.

In the Koyunbaba text, the representation of suita music is not different. When viewed from the context and time of the creation of Koyunbaba, signs that are not commonly used are still within the normal scope of modern times and are still surrounded by the basic identity of suita music. However, more than just musical knowledge is needed to be able to understand and criticically interpret the meanings of signs in the Koyunbaba text.

REFERENCE

Creswell, J. W. (2010). *Research Design Pendekatan Kualitatif, Kuantitatif, dan Mixed* [Research Design Qualitative, Quantitative, and Mixed Approaches] *Yogyakarta: Pustaka Pelajar.*

Danesi, M. (2012). Pesan Tanda dan Makna [Message, Sign, and Meaning] 2004. *Yogyakarta: Jalasutra.*

Jeanne, M. (1973). Clefs pour la sémiologie. Translated by Stephanus Aswar Herwinarko.Yogyakarta: Jalasutra.

Hoed, B. H. (2008). *Semiotik dan dinamika sosial budaya: ferdinand de saussure, roland barthes, julia kristeva, jacques derrida, charles sanders peirce, marcel danesi & paul perron, dll.* Fakultas Ilmu Pengetahuan Budaya, Universitas Indonesia.

Kurniawan. (2001). *Semiology Roland Bhartes* [Semiology of Roland Bhartes] Magelang: Indonesia Tera.

Mack, D. (1995). *Sejarah Musik Jilid III* [Music History Volume III] Yogyakarta: pusat musik liturgi.

Moleong, L. J. (2000). *Metodologi Penelitian Kualitatif [Qualitative Research Methods] Bandung: PT Remaja Rosdakarya.*

Nurgiyantoro, B. (2005). *Teori Pengkajian Fiksi* [Fiction Study Theory] Yogyakarta: Gajah Mada. University Press.

Prier, K. E., Mack, D., & Liturgi, P. M. (1991). *Sejarah musik.* Pusat Musik Liturgi.

Rohidi, T. R. (1992). *Analisis Data Kualitatif* [Qualitative Data Analysis] Jakarta: Universitas Indonesia.

Saussure, F. D. (1988). *Pengantar linguistik umum.* [Course de Linguistic Genetale] Translated by Rahayu S. Hidayat. Yogyakarta: Gadjah Mada University Press.

Sugiyono. (2011). *MetodePenelitian Kuantitatif, Kualitatif, dan R&D [*Quantitative, Qualitative, and R & D Methods] Bandung: Alfabeta.

Wahono and Kustap. (2007). *Adagio Dari Concierto de Aranjuez untuk solo gitar dan Orkestra Karya Joaquin Rodrigo Prespektif Semiotika* [Adagio From Concierto de Aranjuez for guitar solos and Orchestra by Joaquin Rodrigo Semiotics Prespective] "RESITAL". *Performing Arts Scientific Journal Volume 8 no.1.*

Wibowo, A. (2012). *Pendidikan karakter: strategi membangun karakter bangsa berperadaban.* Pustaka Pelajar.

21st Century Innovation in Music Education – Astuti et al. (eds)

A study of *stambul II* singing style in the Muntilan *Keroncong Putra Kasih* orchestra

K.W.I. Cahyani & K.S. Astuti
Universitas Negeri Yogyakarta, Yogyakarta, Indonesia

ABSTRACT: This study aims to describe *stambul II* singing style in the Muntilan *Keroncong Putra Kasih* Orchestra by using proper vocal techniques. The research method for this study is descriptive qualitative method. The data sources were obtained from experts and singers mastering *stambul II* singing style. The data was obtained through the process of observation, interviews and documentation. The data analysis was undertaken in four stages: analysing the data collection, data reduction, data presentation and drawing conclusions or data verification. The results showed that in singing *stambul II* songs in the Muntilan Keroncong Putra Kasih Orchestra, the singers have not yet acquired mastery of the vocal techniques. Those vocal techniques include breathing, intonation and articulation. The styles used in singing *stambul II* were recetative or *rall*, *cengkok*, *luk*, *gregel*, *embat* and *nggandul*.

Keywords: *keroncong*, *Stambul II*, singing, style

1 INTRODUCTION

In this sophisticated era, art is very important in humans' lives. Without art, life would feel barren. A life without art narrows the horizon of knowledge or, generally speaking, life would be colorless. Music is a branch of art that is also very influential in humans' lives. The techniques of vocal production, melody, harmony, rhythm and tempo are meant to express the feelings or emotions felt by the creators. Music does not discriminate against age. Children and elderly people have the right to play music.

Keroncong music is one of the music genres in Indonesia. This music existed in the 17th century when the Portuguese settled in Batavia. Some people think that *keroncong* music is music of the elderly people and the young generation studying this music are considered "old school". *Keroncong* music has three types: original *keroncong*, *stambul I* and *II* and *langgam*. Based on the research observations, *stambul II* songs are optional songs that have a high level of difficulty for singing competitions.

In *keroncong* music competition, singers often face obstacles in performing *keroncong* songs within the *stambul II* style. The competition songs are standard songs that aim to make singers learn the songs earnestly. However, there are many singers who have not noticed that techniques in singing must first be mastered in order to perform the *stambul II* songs. Problems arise with singers who lack mastery of specific vocal techniques for breathing, intonation and articulation. Additionally, what has not been mastered are the singing style and the interpretation of the song performed. The singing style of *stambul II*, inspired by local Javanese songs are *cengkok*, *embat*, *gregel*, *luk* and *nggandul*. In this digital era, communication media does help people and yet new technology can also make people lazier. This can be seen from the fact that there are many *keroncong* singers who only imitate the songs from tapes, CDs and audio videos taken from the Internet.

In a previous study conducted via interviews one of the *keroncong* experts named Mrs. Hj. Sri Hartati mentioned that when performers sing the *stambul II* songs, singers should be singing while also understanding the style of singing *keroncong* songs and the characteristics of performing the *stambul II* songs. However, singers generally lack vocal techniques for breathing, articulation, intonation, phrasing and expression. Other factors that created more problems in singing *stambul II* songs well and properly, is that singers did not rehearse with *keroncong* experts and singers did not want to read existing notations written by composers.

2 LITERATURE REVIEW

Vocal techniques generally refer to ways of producing sounds properly, so that the sounds produced sound clear, beautiful, melodious and sonorous. According to Suharto (1975), vocalizing is a human activity using vocal cords as a source of sounds (p. 1). Vocal art is the art of making sound in all activities using the human voice as a basic tool such as speaking, declaiming and dialogue delivery between role players in theater and film. The focus on vocal technique in this study is how to sing properly. A good singer needs vocal techniques that integrate proper breathing, intonation, articulation, resonance, body posture and improvisation. However, in singing *keroncong* songs, vocal techniques focus on breathing, intonation, articulation and expression. Technique is not a style because singing *keroncong* has characteristics of *cengkok*, *embat*, *gregel*, *luk and nggandul.* Based on the preceding discussion, it can be concluded that singing *keroncong* songs requires a vocal technique and singing style that fits the character of *keroncong* music.

3 MATERIAL & METHODOLOGY

3.1 *Research method*

The method used in this study is a qualitative research method with a qualitative descriptive type. Qualitative descriptive research requires an explanation of a person or phenomenon that occur without losing focus on the object being studied. Qualitative descriptive method is used to factually test a hypothesis related to its variables, phenomena or circumstances (Suharsimi, 1995, p. 310).

3.2 *Time and place of the research*

This research was conducted at Balaputradewa street No. 32, Brojonalan, Borobudur, Magelang Regency, the location where the Muntilan Putra Kasih Keroncong Orchestra conducts rehearsals. Rehearsals occurred every Wednesday starting at 08.00 local time from the beginning of February to March 2017.

3.3 *Data source*

The data sources were experts and singers who have mastered *stambul II* with the proper vocal style and technique.

3.4 *Data collection techniques*

The data was collected through observation, interviews and documentation.

3.5 *Instrument*

The research instrument was the researcher herself as qualitative research employs people as the instruments by which research is conducted.

3.6 *Data analysis technique*

The data analysis technique used in this study was performed in four steps: 1) analysing data collection; 2) data reduction; 3) data presentation; and 4) drawing conclusions, better known as data verification (Suprayogo, 2001, p. 192)

3.7 *Validity of the data*

To check the validity of the data, this research used the credibility criteria of the degree of validity which includes extension of the participation, perseverance, observation and triangulation (Moloeng, 2006, p. 327). Participation means that the researcher also participated as a singer in the Muntilan Keroncong Putra Kasih Orchestra, while triangulation is a technique of validating the data by testing data credibility by examining the data from several sources.

4 RESULTS AND DISCUSSION

4.1 *Results of the research in the muntilan* Keroncong Putra Kasih *orchestra*

The results of this research showed that the *stambul II* singing styles used by the singers were *rall*, *cengkok*, *luk*, *gregel*, *embat* and *nggandul*. The *stambul II* singing styles in this orchestra are common styles used by *keroncong* singers, especially the *stambul II* style. This is related to what was stated by one of the *keroncong* experts, Mrs. Hj Sri Hartati, during the interview conducted on 20February 2017 that, "Singing *the stambul* II songs, as same as singing the original *keroncong* songs and *langgam* styles, uses *cengkok, luk, gregel, embat* and *nggandul* styles". The *stambul II* songs are different from original *keroncong* songs. For example, at the beginning of *stambul II* songs, the singer sings without being accompanied by any musical instruments after the introduction section is played by a solo guitar or violin player. This way of singing is known as *rall*. Based on the reviews, the researcher suggests that to properly sing *stambul II* songs, the singers in the Muntilan Keroncong Putra Kasih Orchestra need to pay attention to the singing styles of the *keroncong* songs and know the characteristic of one of the the *stambul II* styles that is called *rall*. *Rall* is one of the *stambul II* singing styles used in the Muntilan Keroncong Putra Kasih Orchestra. In Western music, *rall* is known as acapella and is also called recitative. The term recitative is an expression by singers who sing and perform songs related to the nuances of song lyrics that are sung without any accompanying musical instruments. This statement is based on the interview with Mr. Hendyatno Hendro Sujongko in which he stated that, "*Rall* is singing a song without being accompanied by any musical instruments and is sung with the heart by comprehending fully on the meaning of the songs" (personal communication, 1 March 2017). Based on those reviews, the singers in the Muntilan Keroncong Putra Kasih Orchestra sang the *stambul II* songs by using their hearts and by comprehending fully on the meaning of the songs without using any musical instruments. One of the *stambul II* songs known by many *keroncong* singers is called "Stambul Tinggal Kenangan". This piece is often sung by singers in the Muntilan Keroncong Putra Kasih Orchestra. The song lyrics describe a situation through sad and sorrowful nuances about a broken-hearted person who thinks about someone who has left, leaving his/her loved one. Therefore, the expression of the singers' voices and their facial expressions should describe sadness and the song should be sung smoothly and sound flexible. The singers need proper breathing vocal technique before they begin to sing in order to complete the full sentence lyric and not cut it off so that the sound will be more intact. However, the singers in the Muntilan Keroncong Putra Kasih Orchestra paid little attention to proper breathing technique so that the singers did not sing with *fermata* nor with proper breath inhalation. The beautiful vibration was not fully reached due to the lack of proper technique.

The singers in the Muntilan Keroncong Putra Kasih Orchestra also delivered a great improvisation while singing the song "Stambul Tinggal Kenangan". The beginning of the song was given a touch of the *rall* style and sounded flexible so that listeners enjoyed the song and knew that singing "Stambul Tinggal Kenangan" requires strong vocal technique and a touch of properly singing the *stambul II* style. In addition to the characteristic of *rall* in singing *stambul II* songs, the style used by the Muntilan Keroncong Putra Kasih Orchestra was the *cengkok* style. In the music there was also the movement of tones such as *cengkok*, tones that are known as ornaments of melodies. The movement of these tones is similar to *gruppetto* technique. The *Muntilan Keroncong Putra Kasih* Orchestra singers gave a touch of *cengkok* style that was slightly different from *cengkok* in general. *Cengkok* is sung quickly but is still flexible and relaxed. In the *stambul II* songs, as in the original *keroncong* songs, not all phrases are given a touch of *cengkok*. Each song is usually only given one or two touches of *cengkok*. For singers in the Muntilan Keroncong Putra Kasih Orchestra, *cengkok* shows the beauty of the songs performed. The beauty occurs when the singer has one note in one beat, but with a touch of *cengkok* singers are able to sing four notes quickly. *Cengkok* will not sound perfect if the singers do not give a touch of *cengkok* quickly and flexibly. If this *cengkok* tone is sung in a way that excedes or is less than a ½ tone, it will sound false and does not match with the instruments played. The strength of a *keroncong* singer is that he/she can carry a proper and flexible *cengkok*. The singers in the Muntilan Keroncong Putra Kasih Orchestra attempted to sing *stambul II* songs with a *cengkok* style. However, they did not pay execute the proper vocal intonation technique, so that it made the song sound false. Additionally, the singers in the orchestra did not express their feelings in such a was so as to blend with the tempo of the songs that they sang. Furthermore, the *stambul* singing style used in the Muntilan Keroncong Putra Kasih Orchestra was instead a touch of *luk* style. *Luk* is a style of singing that can be used to sing any kind of *keroncong* song. In Western music terms, *luk* is better known as a kind of legato or a continuous tone sung by being pulled or *slurred* before ascending to the main tone. Legato is a tone that is sung continuously (Suharto, 1975, p. 84). The following statement comes from the interview with Adi Nurcolis: "*Luk* is the previous tone that is sung again that to come up to its main tone, it is sung by being drawn" (personal communication, 24 February 2017). In the *stambul II* songs, *luk* is widely used because the drawn tone will beautify the *stambul II* songs with sad and rumbling nuances. In the Muntilan Keroncong Putra Kasih Orchestra, *luk* functions to beautify the songs and make it easier for the singers to reach the high interval tone. When the singers give a touch of *luk*, what must be considered is the accuracy of capturing the tone before coming up to the main tone. This is because the singers in the Muntilan Keroncong Putra Kasih Orchestra often gave a touch of *luk* in the tone that was sung so that the tone produced sounds swerving from the music.

Beside *rall, cengkok* and *luk*, the Muntilan Keroncong Putra Kasih Orchestra singers also used *gregel*—another *stambul II* singing style. *Gregel* is a tone ornament that is similar to *cengkok* tone and is sung by lowering or increasing by one tone or a half tone. In Western music terms, *gregel* is a kind of musical ornamentation of a tone or that is similar with the upper mordent. In the Muntilan Keroncong Putra Kasih Orchestra, the style of *gregel* was mostly placed on the tone after the tone was given a touch of *luk*. The same thing was expressed by one of the *keroncong* singers, Hendyatno Hendro Sujongko: "*Gregel* is a word that is sung in three different tones quickly. These three tones are the original tone, the upper tone of the original tone, and the original tone (back to the original tone again)" (personal communication, 1 March 2017). *Gregel* in the Muntilan Keroncong Putra Kasih Orchestra, is generally sung by reflecting the note on the main note quickly but still sounds flexible. *Gregel* is one of the styles that every *keroncong* singer must possess, that should be mastered. By mastering this singing style, the singer will beautify the *stambul II* songs. *Gregel* in the Muntilan Keroncong Putra Kasih Orchestra also has a function to beautify songs.

The next style is *embat*, a way of singing *keroncong* in general. In the Muntilan Keroncong Putra Kasih Orchestra, the *embat* style is part of *stambul II* singing. *Embat* will be fully heard if the singers in the orchestra use proper breathing technique and requires the addition of vibrations after the final tone with a touch of *embat* style. However, in the Muntilan Keroncong Putra Kasih Orchestra, several obstacles were observed in attempts to sing with *embat* style because there was a lack of attention toward proper vocal breathing technique. During an interview, Hj Sri Hartati noted that, "*Embat* is the style of singing that at the end of the song given pressure (being reflected) and a vibration. It requires proper breathing technique, and to make a good breath taking, the singers need to shorten some sentences" (personal communication, 20 February 2017). In addition to vibrations, expressions and dynamics, a touch of *embat* style also beautifies the *stambul II* songs. The addition of this vibration will be beautiful when the singer's voice becomes smoother (decrescendo) and the duration of the tone is long.

Another *stambul II* singing style used in the Muntilan Keroncong Putra Kasih Orchestra is *nggandul. Nggandul* is one of the characteristics of *the keroncong. Nggandul* style can be enjoyed if the singers have mastered and understood the style of singing *stambul II* by using their feeling and appreciation. If the singers have not mastered and understood the style of *nggandul*, then the song will sound weird and as if left behind by the actual tempo. This will lead listeners to conclude that the singers cannot sing the *stambul II* songs properly. *Nggandul* style in this orchestra functions to add the characteristic of *keroncong* itself that is identical to its delay in entering the tone or does not match with the first beat. The style of *nggandul* must be owned by *keroncong* singers in general, as *keroncong* songs will be more flexible if it is given a *nggandul* touch at a certain part in a song. This type of *keroncong* song is unsettling when singers start to sing, as it songs are not sung in the right beat and thus distinguishes *nggandul* from the other types of songs.

4.2 *Discussion*

The results of this research show that the singers used singing styles such as *cengkok, luk, gregel, embat* and *nggandul.* These singing styles are used to sing *keroncong* original songs, *stambul and langgam.* The singers in the Muntilan Keroncong Putra Kasih Orchestra also used these styles while singing the *stambul II* songs. However, it was observed that the singers in the Muntilan Keroncong Putra Kasih Orchestra have not yet mastered vocal techniques of proper breathing, intonation, articulation and expression. Additionally, the singers in this *keroncong* orchestra did not often pay attention to the tempo, feel and meaning of the songs they sang. Thus, while singing the songs, both the singers and the songs were not intertwined, not connected with one another. The *stambul II* songs would certainly sound beautiful and the meanings contained in the songs would be delivered, understood and well received by the listeners if the singers use proper vocal techniques. Proper technique would help singers to fully attain the *stambul* II singing styles of *rall, cengkok, luk, gregel, embat* and *nggandul.*

5 CONCLUSION

Based on the results of this observer-participant study of the *stambul II* singing style in the Muntilan Keroncong Putra Kasih Orchestra, the following conclusions are made: the singers performed the *stambul II* songs in ways that are characterictic of the *stambul* song by using *rall* as well as other styles of singing *keroncong* songs—*cengkok, luk, gregel, embat* and *nggandul.* These singing styles are necessary and must be mastered by *keroncong* singers especially for *stambul II* songs. Proper technique would make the songs *ngroncongi* or sounding like a rich *keroncong* song that is flexible. However, this research found that the singers in the Muntilan Keroncong Putra Kasih Orchestra have not yet paid attention to proper vocal techniques for effective breathing, intonation, articulation and expression. Overall, proper vocal technique training is needed in order to help the singers master the various *stambul II* singing styles.

REFERENCES

Ali, M. (2006).*Senu Musik SMA untuk kelas XII.* [High School Music Art for Class XII]. *Jakarta:ESIS.*

Budiman BJ. (1979). *Mengenal Keroncong dari dekat.* [Get to know Keroncong up Close]. *Jakarta: Perpustakaan Akademi Musik LPKJ.*

Harmunah, S. Mus. 1987. Musik Keroncong. *Sejarah, Gaya dan Perkembangan.[*History, Style and Development of Keroncong Music].
Margono, D., & Sumardi, S. A. (2007). *Apresiasi Seni Seni Rupa & Seni Teater 2*. [Appreciation of Fine Arts and Theater Arts 3]. Yudhistira Ghalia Indonesia.
Mark, D. (1995). *Sejarah Musik Jilid 4 cetakan kesatu.*[Music History Volume 4 1st Printing]. Yogyakarta: Perpustakaan PuSAT Musik Liturgi.
Moenzir, I. A. (2013).*Gesang Mengalir Meluap Sampai Jauh*. [Gesang Overflowing Far Away]. Jakarta. Gramedia Pustaka Utama.
Moleong, L. J. (2006). *Metodologi Penelitian Kualitatif Edisi Revisi*. [Qualitative Research Methodology (revised edition)]. Cetakan keduapuluh dua, Bandung: PT. *Remaja Rosdakarya Offset*.
Nawawi, H. H. (1991). *Metode Penelitian Bidang Sosial*. [Social Research Methodology]. Yogyakarta: Gajah Mada University.
Pranadjaja. (1976). *Seni Menyanyi*. [Art of Singing]. Jakarta: C.V. Baru.
Pramayudha. 2010.*Buku Pintar Olah Vocal*.[Smart Book for Vocal].Yogyakarta: Buku Biru.
Poetra, A. (2006). *1001 Jurus Mudah Menyanyi*. [1001 Easy Tips for Singing]. Dar Mizan.
Simanungkalit, N. (2013). *Teknik Vokal Paduan Suara*. [Choir Vocal Technique]. Pt Gramedia Pustaka Utama.
Soedarsono, (1992). *Pengantar Apresiasi Seni*. [Introduction to Art Appreciation]. *Jakarta:Balai Pustaka*.
Soeharto, A. H., Soenardi, A., & Sunupratomo, S. (1995). *Serba-serbi keroncong*. [All about Keroncong]. Jakarta. Musika.
Soeharto, M. (1975). *Belajar notasi balok*. [Learning Beam Notation]. PT Gramedia Pustaka Utama.
Sukardi. (2003). *Metodologi Penelitian Pendidikan*. [Educational Research Methodology]. Yogyakarta: Bumi Aksara.
Suprayogo, I and Tobroni. (2001). *Metodologi Penelitian Sosial-Agama* [Social Religion Research Methodology]. Bandung:PT. Remaja Rosdakarya. http://www.tjroeng.com/?p=865. Diunduh pada hari Minggu 12 Februari 2017.

21st Century Innovation in Music Education – Astuti et al. (eds)

Malay song forms in Kuansing, Kampar and Rokan Hulu regencies, Riau Province

Nurmalinda
Islamic University of Riau, Indonesia

ABSTRACT: The purpose of this study is to explain Malay song forms in Kuansing, Kampar and Rokan Hulu regencies, Riau Province. This study is an in-depth and detailed research into the form of Malay songs compiled by artists from the province. Five songs were selected as representative of Riau Province: 'Panjek-panjek Tabalusui', 'Menganyam', 'Nosib Pamotong', 'Baghandu', and 'Tuanku Tambusai'. The research shows that the characteristic form of such Malay songs has: 1) repetitive patterns, 2) a variety of motifs, 3) two phrases, 4) a variety of themes, especially about phenomena that occur in the local communit.

1 INTRODUCTION

Regional songs are a form of music that comes from a particular area and is sung both by the people of the region and other regions. Generally, the songwriter is anonymous. Regional songs are similar to the national anthem, but regional songs can only achieve the status of a temporary national anthem. Regional songs generally have lyrics in their regional languages. Usually, local songs are sung or played for certain traditions in each area, for example, as stories for children who are naughty, as children's games, and as community entertainment. Folk songs usually refer to a song that has a rhythm particular to the region. According to Purnomo and Subagyo (2010), in preserving the culture of the Riau region, there is one element of art that has not received attention, which is the regional song in Kuantan Singingi (Kuansing), Kampar and Rokan Hulu. Kuantan Singingi, Kampar and Rokan Hulu regencies are part of Riau Province, where people still abide by the traditional rules of the community. Habits in the Kampar community have been integrated so that they are applied in daily life, such as ceremonies of marriage, inauguration, death, and birth. People who are highly respected in this society are those with the lineage of the King and nobles. Malay songs in Kuansing, Kampar and Rokan Hulu regencies in Riau Province are one form of community culture that is currently maintained.

It should be noted that songs consist of elements which are part of the song form. Certain elements in a song can be observed specifically, such as melody, tempo, and expression (Prier, 1996, p. 2). Another element is that of form. Observations about songs can be done by listening to songs repeatedly, or by looking at the song notation.

Besides this, the lyrics may involve satire of government that commits corruption or takes people's rights. Malay songs are very popular in Kampar and Rokan Hulu regencies. They are generally performed at entertainments, celebrations, and so on. Most of the people of Kampar, both young and old, know the song 'Moncik Badasi', because the song reflects events that are happening.

Based on this phenomenon, this study is an in-depth and detailed research (Sugiyono, 2010) into the form of Malay songs compiled by artists from Kuansing, Kampar and Rokan Hulu. This study came about as a result of the author's interest in the development of folk songs in these regencies. Furthermore, the author also wants to analyse songs in terms of musical theory, which is very useful for students in discussions and lectures at the Art Department of the Islamic University of Riau, Riau, Indonesia.

2 DISCUSSION

2.1 *Forms of songs*

Soeharto (1986) states that the formation of folk songs is as follows: (i) themes and contrasts, where the most common form of song is AABA, with part A being the melody unit that is the theme of the song, while part B is a melody unit that is often called the contrast part, with a melodic style that contradicts the theme, although it still has continuing harmony; (ii) pattern of repeating song units; (iii) climax of songs; (iv) ending of songs. To discuss the analysis of the form of the song 'Panjek-panjek Tabalusui', the theory of Karolyi (1995, p. 2) will be used as a reference, which states that musical form consists of smaller units: (i) song motifs, (ii) song phrases, (iii) song lyrics.

2.2 *The form of Malay song 'Panjek-panjek Tabalusui'*

The theme of the main melody is the main element in a musical composition. One of the songs of the traditional *Randai* performance in Kuantan Singigi regency is 'Panjek-panjek Tabalusui'. The main melody of the song is formed by a repetitive pattern.

Figure 1. Themes of *Panjek-panjek Tabalusui.*

Repeated units in 'Panjek-panjek Tabalusui' appear in part A, with each of the two questions and answers repeated twice with different lyrics. Based on observations of 'Panjek-panjek Tabalusui', the climax of the song is at bar 32, with a 1/2 beat to the end of the song which is in bar 40 on the first beat. The notation (Soeharto, 1975) of this climax is shown in Figure 2.

Figure 2. Climax of 'Panjek-panjek Tabalusui'.

The coda is the end of a song, or its climax, and is usually the conclusion of all compositions or harmony, and is illustrated in Figure 3.

Figure 3. Coda of 'Panjek-panjek Tabalusui'.

2.3 *'Menganyam' song form, compiled by Roslan Roseli*

'Menganyam' is a popular Malay song in this region. Roslan Roseli analysed that in part one there are eight motifs. The first motif is from bars 1 to 3 on the first beat; the second motif is at bar 3 on both beats up to bar 4; the third motif is at bar 5 on both beats until bar 6 on the first beat; the fourth motif is from bars 7 to 8; the fifth motif is at bars 9 to 10 on the first beat; the sixth motif is in bar 10 on both beats to bar 12; the seventh motif is from bar 13 on the first beat to bar 14 on the first beat; the eighth motif is from bar 13 on the second beat to bar 16 on the first beat. In part two there are also eight motifs. The first motif is in bar 18 to bar 19 on the first beat, the second motif is in bar 19 in the fourth beat to bar 20, the third motif is in bars 21 to 23, the fourth motif is at bar 24 until bar 25 on the first beat, the fifth motif is at bar 25 on the fourth beat to bar 27, the sixth motif is in bars 28 to 29 on the first beat, the seventh motif is in bar 29 on the fourth beat up to bar 30 on the first beat, and the eighth motif is in bar 30 on three half beats to bar 31 on the first beat. The motifs of 'Menganyam' are shown in Figure 4.

Figure 4. Motifs of 'Menganyam'.

Based on observations, 'Menganyam' consists of five question phrases and five answer phrases. There are two question phrases and two answer phrases in part one, while in part two there are three question phrases and three answer phrases. For more details, see Figure 5.

Figure 5. Phrases from 'Menganyam'.

In part one of the song 'Menganyam', the first question phrase runs from bar 5 to bar 8, and the second question phrase is at bars 13 to 16, while the second part consists of three question and answer phrases. The first such question phrase runs from bar 18 on the second beat up to bar 20, and the first answer phrase is at bar 21 and ends at bar 23. The second question phrase is in bar 24 to 25, while the second answer phrase is in bar 25 on the second beat to bar 27. The third question phrase is in bar 28 on the second beat until bar 29 on the first beat, and the third answer phrase starts at bar 29 in the fourth beat to bar 31.

The theme included in part one of 'Menganyam' is arranged as one question and answer phrase while the theme of the song in part two is arranged as three question phrases and three answer phrases. For more details, see Figure 6.

Figure 6. Themes of 'Menganyam'.

2.4 *'Nosib Pamotong', a song from Kampar, compiled by Suhardeli*

Based on the results of the analysis, the 'Nosib Pamotong' song has three themes, because the melodic patterns that appear in the subsequent lyrical sections have a different tone from the initial melody. Bars 9 to 16 have theme A because it appears at the beginning of the song.

Figure 7. Themes of *Nosib Pamotong*.

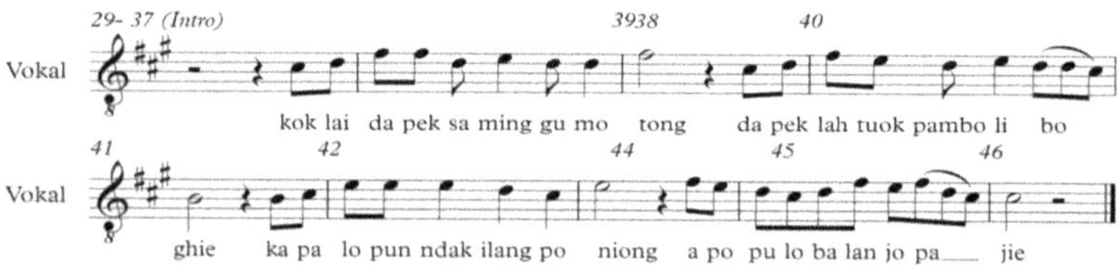

Figure 8. Theme C of 'Nosib Pamotong'.

Then, theme B, which starts in bar 17, continues to bar 24, and theme C runs from bars 38 to 46, which is the refrain of the song.

The refrain units of the 'Nosib Pamotong' song begin in bars 25 to 28, in bar 29 to the intro of bar 37, then continue from bars 38 to 62, and then repeat bars 9 to 16. In the analysis of the song 'Nosib Pamotong', there is no climax or peak in the song. Because it has a refrain within the melody and tone, it does not show any climax. According to Soeharto (1986), a song does not always have a climax or peak and this does not mean that 'Nosib Pamotong' is weak or boring, because the song's strength is not defined by its climax. The coda is the ending or closing part of a song or even its climax., and is the conclusion of all compositions or harmony.

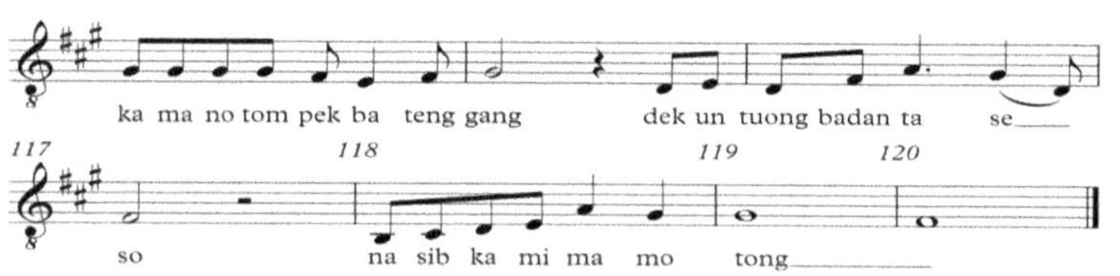

Figure 9. Finale of 'Nosib Pamotong'.

Figure 10. Motifs of 'Baghandu'.

2.5 *Local form of 'Baghandu' song, compiled by Fian Pantogi*

Based on the results of the analysis, 'Baghandu' songs have eight motifs. Figure 10 is a representation of the notation of 'Baghandu'.

Motif 1 starts at bar 1 on the second beat and ends at bar 3 on the first beat. This consists of three 1/8 notes, four ¼ notes and one ½ note. Then, motif 2 starts on the second beat of bar 3 and ends in bar 5 on the first beat. It also consists of six 1/4 notes, and three 1/8 notes. Motif 3 starts on the second beat of bar 5 and ends in bar 7 on the first beat. It consists of three 1/8 notes and six ¼ notes. Motif 4 starts on the second beat of bar 7 and ends in bar 9 on the first beat. It consists of six ¼ notes and three 1/8 notes.

Furthermore, motif 5 starts on the second beat of bar 9 and ends in bar 11 on the first beat. It consists of three 1/8 notes and five ¼ notes. Motif 6 starts on the second beat in bar 11 and ends at bar 13 on the first beat. This consists of four 1/8 notes and four ¼ notes. Motif 7 is set on the second beat in bar 13 and ends at bar 15 on the first beat. This consists of three 1/8 notes and five ¼ notes. Finally, motif 8 starts on the second beat in bar 15 and ends in bar 17 on the fourth beat. This consists of three 1/8 notes, three ¼ notes and one ½ note.

The 'Baghandu' song is composed of four phrases. Further explanation can be seen in Figure 11.

Phrase 1 starts in bar 1 on the second beat and ends in bar 5 on the first beat. Phrase 2 starts on the second beat of bar 5 and ends in bar 9 on the first beat. Phrase 3 starts on the second beat of bar 9 and ends in bar 13 on the first beat. Phrase 4 starts on the second beat of bar 13 and ends in bar 17 on the fourth beat. Based on the results of the analysis, there is no question and answer sequence because the song starts from the base of tone at the end of the sentence.

Theme 1 consists of the first and second phrases, and theme 2 consists of the third and fourth phrases. More details can be seen in Figure 13.

Figure 11. Phrases of 'Baghandu'.

Figure 12. Hukuman 'Baghandu' song.

Figure 13. 'Baghandu' song themes.

Based on the results of the analysis, the 'Baghandu' song only has two themes, because the melody pattern appearing in the second lyrical section has the same tone as the previous pattern. The repetitive unit in 'Baghandu' is in the form of ABAB and appears four times, but each repetition has different lyrics. Based on the results of the analysis, 'Baghandu' does not have a climax or peak, because the melody and tone are repeated without a climax.

The coda is the conclusion of 'Baghandu', in the final bar. 'Baghandu' is in the form of 4/4 within A major, and the tempo used is *moderato*. According to Prier (1996, p. 2), the sentences in a song are a number of bar spaces, which usually have 8 to 16 bars as a single unit. For general sequences, letters A, B, C and so on are used. When a sequence is repeated with a change, the letter is followed by a (prime) indicator, e.g. A'.

2.6 *The form of the regional song 'Tuanku Tambusai', compiled by Husni Thamrin in Rokan Hulu Regency, Riau Province*

Based on the results of the analysis of the song 'Tuanku Tambusai', it was found that bars 1 to bar 8 used the sentence form A because it appeared at the beginning of the song (Figure 14). This sentence is repeated in bar 9. Sentence B, which starts in bar 15, is a song refrain and acts with bars 15 to 19 as sentence questions (Figure 15), with the response running from bars 20 to

Figure 14. Sentence A, bars 1 to 8 of 'Tuanku Tambusai'.

Figure 15. Sentence B, bars 15 to 19, is part of the question in 'Tuanku Tambusai'.

Figure 16. Sentence in bars 20 to 23 forms part of the answer in 'Tuanku Tambusai'.

Figure 17. Refrain of 'Tuanku Tambusai' appears from bars 8 to 15.

23 (Figure 16). Based on the explanation above, the 'Tuanku Tambusai' song was classified as a two-part song of the form A-A'-B.

The refrain of this song appears from bar 8 and continues to bar 15, as shown in Figure 17.

The climax and peak of this song is from bars 15 to 19. The end of this song is from bars 20 to 59. The entire song is illustrated in Figure 18.

B = Do

Composer: Husni Thamrin

Moderato

Dalam nukinan Sejarah jasa Dan baktimu Nyata Sudah Kau beri

Juang menjunjung marwah berbilang negri rantau Dan Lembah Takzim tuanku

Tambusai bergelar Paderi harimau Rokan Pesan amanahmu tak

Terlerai tak berarah tak berbatas Sampai tiada Madah Dan sanjungan tak

Bintang tersemat didada Walau pun bakti mu Setara Hanya kini nisan

Pusara DENGAN syair kukalamkan Engkau pahlawan Patut disebut andalanan

Negeri bumi Bertuah oh Tambusai ‘Tuanku Tambusai’ *Takzim Tuan ku*

Tambusai bergelar Paderi harimau Rokan Pesan amanahmu tak

Terlerai tak berarah tak berbatas Sampai tiada Madah Dan sanjungan tak

Bintang tersemat didada walaupun baktimu Setara Hanya Kini nisan

Pusara DENGAN syair kukalamkan Engkau pahlawan Patut disebut andalan

Negeri bumi Bertuah oh Tambusai ‘Tuanku Tambusai’ *andalan*

Negeri bumi Bertuah oh Tambusai 'Tuanku Tambusai'

According to Blaukopf (1992), the form of the song 'Tuanku Tambusai' can be explained as follows: from bar 1 to bar 8 is the sentence A at the beginning of the song (see Figure 18). This is separated into a question from sentence A from bars 1 to 4, and the answer in sentence A, which runs from bars 5 to 8. Sentence B (see Figure 20) starts in bar 15 as a refrain of the song, with bars 15 to 19 being the sentence question and bars 20 to 23 the sentence answers, which is also the climax or peak of this song. Finally, from bars 24 to 59 is the end of the song. Referring to the analysis above, it can be concluded that the 'Tuanku Tambusai' song is classified as a two-part song form. This means there are two different sentences (A and B) in one song with sentence A repeated with different variations in bar 9 (see Figure 19), thus labelled A', giving a sentence sequence of AA'B.

2.7 *The song form of Borudah poetry in the community of Kepenuhan Barat Mulia Village, Kepunahan Subdistrict, Rokan Hulu Regency*

Based on data obtained in the form of observations, the climax or peak of 'Borudah' is at bars 18 to 35. As a conclusion of the meaning of the song, the ending or closing part of the song runs from bar 73 to bar 81 (see Figure 23).

From the notation above, bars 1 to 9 form sentence A because it is at the beginning of the song. There is a refrain of the sentence A, and sentence B is the climax of 'Borudah', which is from bar 10 to bar 81. The song uses the tempo of *adagio*, which is very slow and full of sentiment. This song consists of 365 bar notations, and there are 234 1/8 notes, 99 ¼ notes and 18 ½ notes. Moreover, it has 81 bars with a single refrain from bars 1 to 9, with tones compiled in an Arabic melodic style.

Figure 18. Sentence A.

Figure 19. Sentence A'.

Figure 20. Refrain of Sentence B.

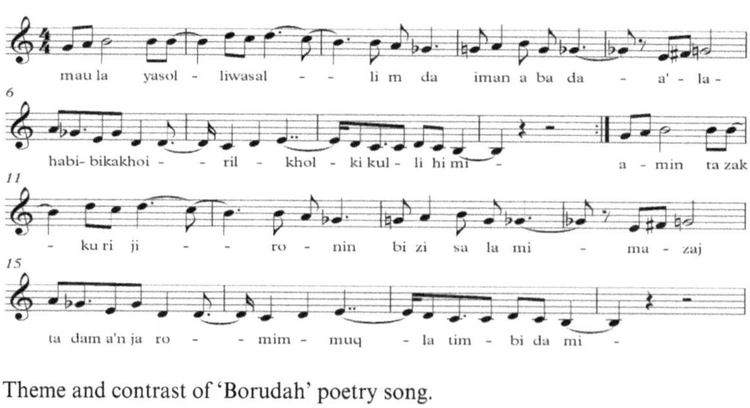

Figure 21. Theme and contrast of 'Borudah' poetry song.

Figure 22. Refrain of 'Borudah' poetry song unit.

Figure 23. Ending of 'Borudah'.

3 CONCLUSIONS

In general, there are many repetitions in Riau Malay songs. Kuansing songs are composed by artists in the area. Malay songs tell about phenomena that occur in the local community, both in the social and cultural life of the community, and their daily activities. Based on the assessment above, local artists are becoming more creative in making original songs from their region, for entertainment of the community and especially the younger generation. This also improves the economic situation for creative artists. If viewed in terms of the existing regional songs, the end of the song is generally sung with expressions of soft and hard excitement and dynamics in the parts of the song. Regional songs always have a good refrain of poetry, music or lyrics.

REFERENCES.

Blaukopf, K. (1992). *Musical life in a changing society: Aspects of music sociology*. Portland, OR: Amadeus Press.

Karolyi, O. (1995). *Introducing modern music*. London, UK: Penguin Books.

Prier, K.E. (1996). *Ilmu Bentuk Musik* [Science of music form]. Yogyakarta, Indonesia: Pusat Musik Liturgi.

Purnomo, W. & Subagyo, F. (2010). Terampil Bermusik [Skillful Musical Performance]. Jakarta, Indonesia: Pusbuk Kemdiknas.

Soeharto, M. (1975). *Belajar notasi balok* [Learning of Western music notation]. Jakarta, Indonesia: Gramedia.

Soeharto, M. (1986). *Belajar MEMBUAT Lagu* [Learning to create songs]. Jakarta, Indonesia: PT Gramedia.

Sugiyono. (2010). *Metode Penelitian Kuantitatif, Kualitatif, dan R&D* [Quantitative, qualitative, and R&D research methods]. Bandung, Indonesia: CV Alfabet.

21st Century Innovation in Music Education – Astuti et al. (eds)
© 2020 Taylor & Francis Group, London, ISBN 978-0-367-11122-9

The process of learning *Tifa* instruments in the Asmat tribe

F.E.I. Rahayaan & K.S. Astuti
Yogyakarta State University, Yogyakarta, Daerah Istimewa Yogyakarta, Indonesia

ABSTRACT: A *Tifa* is a traditional musical instrument, which is mandatory in every celebration of traditional events held by each tribe in the Papuan region. The focus of this study is on how the music is passed on through the learning process. This study was carried out using a qualitative research approach. Data collection involved observation, interviews and documentation in Agats, Asmat Regency, Papua. The study found that learning the *Tifa* in the Asmat tribe refers to traditional rituals that involve a process of initiation (*Emak-Cem*) where children who will learn the *Tifa* are dried in the sun in front of '*Jew*' in the daytime. The types of rhythms taught in the learning process are *sarpi*, *dames* and *damesnab*. Children who are proficient in playing the *Tifa* will be included in the group at an Asmat tribe traditional party.

1 INTRODUCTION

1.1 *Background*

The diversity of ethnic groups in Indonesia is a uniqueness that distinguishes Indonesia from other countries. With regard to ethnic diversity, Papua is one province in Indonesia that has attractive, distinctive and diverse cultures to explore. Papua is the largest island in Indonesia with an area of 317,062 km^2, a large area that is also directly proportional to the diversity of its tribes and cultures (Mentansan, 2013). The number of tribes in Papua, according to data from the Papuan government, amounts to 1,068 (Papua, 2015). The cultures of Papua clearly have characteristics that are different from the cultures in other Indonesian provinces. One of the Papuan cultural aspects that is interesting to explore is the *Tifa* instrument. The *Tifa* is a traditional musical instrument that is a mandatory instrument in every celebration of traditional events held by each tribe in the Papuan region. Usually, the *Tifa* is used as an accompaniment of war dances, guest welcoming dances, arrow dances, social relations dances and other traditional ceremonies (Leirissa & Latuconsina, 1999, p. 118).

The *Tifa*, as a traditional instrument, comes in several types including the *Tifa Jekir*, *Tifa Dasar*, *Tifa Jekir Potong* and *Tifa Bas* (Kristi, 2012, p. 120). The different types of *Tifa* are seen from the size and how they are played; several different types of *Tifa* are ussually played in combination in one performance. The *Tifa* instruments are essential to study, especially to pass the playing techniques to students or the next generation.

The *Tifa* culture, as a musical art that is typical of every tribe in Papua, certainly has a distinctive playing pattern which depends on the background culture of the tribe that preserves it. The Asmat tribe is one of the tribes in Papua which is famous for its typical and unique wood carvings. The population of Asmat is divided into two, those who live in the coastal areas and those in the mountain areas. The differences between the two populations of the Asmat tribe lie in their dialect, way of life, social structure and traditions (Pram, 2013, pp. 75–76).

The Asmat tribe uses the *Tifa* as a cultural instrument to complement traditional events carried out with the accompaniment of traditional music to add values to the original culture. *Tifa*'s music art is an outline of knowledge which can be developed as a reference to delivering the instrument to the public. The *Tifa*, as the Asmat tribe's traditional music, is indeed a culture that needs to be preserved in order to maintain traditional music as an identity of the Asmat tribe's existence.

1.2 *Research focus*

Traditional music is music that is created, developed and inherited periodically by a certain area and has distinctive values (Nakagawa, 2000, p. 27). Departing from the notion of traditional music as a culture that is developed and inherited, the *Tifa* instrument is part of music art developed and inherited by the Asmat people. Through this study, the focus is given to how the music is inherited. In academic terms, this process is called the learning process. The inheritance process of the music accompaniment produced by the *Tifa* or how the Papuan people make efforts to preserve the culture of this instrument in the Asmat tribe are essential to examine.

2 LITERATURE REVIEW

2.1 *Learning*

Learning is not limited to the nature of the technique, formality or requirements for the activity called teaching learning. Activities in which a mother tries to teach her child how to do things correctly in her environment can also be called learning. In other words, learning has a broad nature, not bound by conditions or terms. Before education emerges, traditional learning has already been applied as the methods aimed at providing knowledge to students.

The learning process is categorised into three forms, which are formal, informal and non-formal (Hashim et al., 2004, p. 157). Formal learning is academic learning, where a learning plan arrangement is carried out systematically by an institution. Informal learning is a learning process that is carried out indirectly and unconsciously. Non-formal learning is a learning process that is carried out as a complement or enhancer of knowledge that has been gained from formal learning. Learning the *Tifa* instruments in the Papuan Asmat tribe is not carried out by an institution or the auspices of Papua's educational office. Learning the *Tifa* instruments, as carried out by the Asmat Papuan tribe, is assumed to be more towards the form of informal learning. Hashim et al. (2004, p. 159) state that some of the characteristics of informal learning are explained in several points, which are: 1) Informal learning is not done with written debriefing or modules; 2) Informal learning is not accepted in awareness or, in another sense, not in activities that are bound by a curriculum or educational guidelines from a particular institution; 3) Informal learning is not limited by time or certain schedules; 4) Time management is done by habit, not conditioned by anything; 5) Informal learning is carried out openly, not bound by a place except due to certain conditions; 6) Knowledge and learning carried out in informal learning are obtained from experience; 7) Informal learning is carried out not to establish a systematic learning pattern; but it is direct or spontaneous. Schugerensky as cited in Bakar (2014, p. 7) states that there are forms of informal learning as seen in Figure 1.

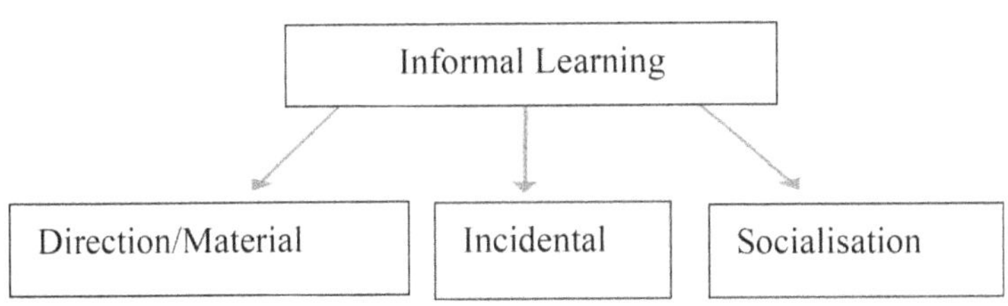

Figure 1. Chart form of informal learning.
(Source: Bakar, 2014)

Figure 1 above shows the interrelationships between several elements that lead to informal learning processes. Direction or material is material that is taught indirectly to students involved in the informal learning process. The next element is incidental which shows that informal learning is done without attachment to a schedule, place and material to be taught. Then, the last element is socialisation in which informal learning processes can take place during the socialisation process.

The informal learning process is a form of learning activity that is carried out freely in the absence of certain standard requirements. The characteristics of informal learning activities that are open allow anyone to become a teacher and a student. Informal learning can be found in the pattern of parents teaching children, traditional learning related to a particular culture and others.

2.2 *The Asmat tribe*

The Asmat tribe is one of the tribes in the province of Papua. They are famous for their unique art and carvings. The tribe is one of the most famous tribes among many tribes in Papua. Most Asmat people can be found in the Agats district, which is located 4–7° S and 137–140° E (Papua, 2015). The Asmat tribe is dense with indigenous Papuan culture which makes it an icon or representative of other tribes in Papua.

The wealth of culture possessed by the Asmat tribe involves not only the carving and craft works but also extends to their way of life. The Asmat tribe is led by a tribal chief. The tribal chief is an 'indigenous elder' who guides the Asmat people in their daily lives based on the customary rules. Dumatubun (2002, p. 1) writes how the Asmat tribe still holds the culture from their ancestors to the present day. For example, in the perspective of the Asmat tribe, a disease that afflicts a person is not caused by medical reasons but is due to interference from paranormal or supernatural activities. The Asmat tribe's traditional house is divided into two; *Jew* and *Tysem* houses (Bawantara & Ekaristi, 2011, p. 69). The *Jew* house is a traditional house specifically used for traditional activities, while a *Tysem* house is a house that is specifically for families as a place to live. The Asmat tribe still very often organise spirit evictions in the *Jew* house, which can help eliminate 'diseases'. The ceremony is led by a tribal leader who then asks the Asmat people to take part in the activity. This culture, which is still held in high esteem, allows the Asmat tribe have a deeply rooted local culture.

The strong culture and ethnicity that still exists in the Asmat tribe in terms of art are also still maintained. Famous for its sculptures and carvings, the Asmat tribe has high artistic power that comes from the personality of their ancestors. The Asmat tribe distinguish two spirits that influence their lives: the life-threatening spirit and the spirit that does not endanger life. Spirits that endanger life, for example, are the spirits of pregnant women, spirits on banyan trees, and spirits carrying diseases and disasters (*Osbopan*). Spirits that do not endanger life are spirits that only scare or interfere with them, usually the spirits of ancestors called *Yi-ow*. In the kinship system, the Asmat tribe recognises three family forms that are monogamous and polygamous family, broad uxorilocal family (married families living in the wife's family's house) and local *Ovunku* family (married families living in the wife's mother's house). The Asmat tribe also has a very distinctive dance full of its customs called the *Kobe* dance. This dance is often used as an accompaniment to traditional ceremonies such as wedding ceremonies and guest receptions. The *Kobe* dance is always accompanied by the *Tifa* instruments. The *Tifa* is an instrument that must be a companion in every custom event carried out by the Asmat tribe. Although the form of *Tifa*'s instrument is simple, the Asmat people believe that the accompaniment of musical instruments by the *Tifa* will bring blessings from the ancestors.

2.3 *The Tifa*

The *Tifa* is one type of traditional music in Indonesia. The *Tifa* is one of the instruments to accompany dances in the Papuan Province. It is often used in traditional events as a complement to traditional events held by local residents. In Papua, the *Tifa* generally has

Figure 2. Types of *Tifa*: (from left to right) *Tifa Jekir*, *Tifa Dasar*, *Tifa Jekir potong* and *Tifa Bas*. (Source: Google Image, accessed on 29 October 2016)

the same use by all tribes living in the Papuan region. More than a thousand tribes in Papua use the *Tifa* as one of the instruments that help to enliven their traditional events. The *Tifa* has several types including *Tifa Jekir*, *Tifa Dasar*, *Tifa Jekir Potong* and *Tifa Bas* (Kristi, 2012, p. 120). One *Tifa* is often used together with other types of *Tifa*. This is because each type of *Tifa* has different characteristics both in terms of size and sound. The *Tifa* is a traditional instrument similar to the *Jimbe*, which is also played by hitting, but the *Tifa* is hit only by one hand. The *Tifa* is made of wood, which is emptied and then decorated by carving on the outside part.

3 METHODS

3.1 *Research design*

This study on the learning process of the *Tifa* instrument of the Asmat people in Papua was carried out using a qualitative research approach. As a qualitative research approach, this study emphasises the research process with an in-depth examination of the data. The qualitative approach is a research approach that aims to understand and explore a symptom or phenomenon (Creswell as cited in Semiawan, 2008, p. 7). The data in this research is qualitative data that was taken in the natural setting (Sugiyono, 2011 p. 55). Researchers analysed the data as they are observed during the pre-field process, field work and post-field work. The data were then analysed to obtain the results that answered the problem statements.

The study was descriptive in that all types of data were explained in the form of sentences describing what had been done. The study highlighted images that explained how the teaching and learning process took place related to the *Tifa* instrument. Data collection was conducted in Agats, Asmat Regency, Papua between 13th January and 18th January 2017. Three individuals were asked to be the research informants: Erick Sarkol, as curator of the Asmat museum; David Jimanipits S.Sn. as a figure of the Asmat tribe; and Aloysius Ari Bainepe, as a *Tifa* player (*Wair Ow*).

The data were in the form of words and images from the results of the interviews and documentation. The researchers took the role as the research instruments. In this role, they played a position as an instrument or research tool (Moleong, 2000, p. 168).

3.2 *Data collection techniques*

This study used observation, interviews and documentation as the data collection techniques. Observations were made on January 13 2017 in Agats, Asmat Regency, Papua. Interviews were conducted with the three speakers at the Asmat Museum and in the *Jew*. The documentation technique collected recordings and documents in the form of photos and videos. The validity of the data in this study was processed through a triangulation technique that tested the correctness of the data obtained during the study.

3.3 *Data analysis technique*

The collected data were analysed descriptively in the light of the existing reality. The purpose was to describe, in complex terms, the learning process of the *Tifa* instrument in the Asmat tribe. The data analysis procedure in this study consisted of data reduction, data presentation and conclusion drawing.

4 FINDINGS AND DISCUSSION

4.1 *City agencies*

The study was conducted in Agats, a city of a new division of Asmat Regency, which is part of the province of Papua. The city is located on the southern coast of the island of Papua, adjacent to the Timika region in the Mimika Regency. Because of its closeness to Timika, it is easier to reach Agats from the city of Timika by using a boat or pioneer aircraft. Agats has its own uniqueness in that it stands on muddy soil and swamps, with a road in the form of a board which, at a glance, resembles a pier. All roads in the city of Agats resemble bridges made of ironwood, but over time, these bridges began to be refined using stronger concrete. Until now, the development of the city of Asmat Regency was carried out on this unique road. All buildings in this city are also adjusted in the form of stage houses. In fact, the main means of transportation in the city of the bridges is a motor that uses electricity. Agats' condition, which is still very limited, makes the local government and the entire population

Figure 3. The city of Agats.

adjust to the situation. The wide wooden road bridge in the whole area of Agats is unable to withstand the heavy motorbike load of an engine. Moreover, the car is certainly not to be a means of transportation. Another limitation Agats has is the lack of clean water supplies. The Agats community has survived with rainwater which is stored in water tubes. The condition of the swampland makes it difficult to provide clean water. Therefore, it is not surprising that taking a shower using this reservoir of water feels more slippery and it is difficult to rinse the soap used. However, people's creativity allows them to continue to survive in these conditions. Figure 3 shows the atmosphere of one of the streets in the city of Agats.

4.2 *Tifa instrument learning process*

The learning process of the *Tifa* instruments by the Asmat people in Papua begins with a selection process carried out by the elders based on the talents or skills that the children have. Although *Tifa* skills are not always hereditary, the parents (*Tifa* musicians) generally want their children to inherit their *Tifa* skills. If no talented children are to be seen, then they will look at their grandchildren to inherit their *Tifa* playing skills. When a child has been chosen by the chairman, the child is then educated in the *Jew*. The *Jew* is an Asmat tribal house; wherein someone can learn a lot of things from the customary leader. Moreover, the *Jew* is the only place that is used as a gathering of children to learn to play the *Tifa*. Furthermore, the children are elected by the customary leader and parents who feel their children have talent can send their children to the *Jew*. This initial selection process is confirmed from the interview with Speaker 3, Mr. David (Jimanipits) S.Sn, the tribal art figure, as shown in the following field note:

> *Para Tua-Tua adat biasa melihat dan memperhatikan anak-anak kecil yang sedang berkembang, kalau mereka merasa anak tersebut mempunyai bakat, lalu dia ditarik dan dididik ke Jew. Orang tua juga berperan, jika dia merasa anaknya mempunyai bakat, orang tua bisa meminta agar anak bisa di ajar di Jew.*
>
> Traditional elders can see and pay attention to the growing children. If they feel the child has talent, then he is taken and educated in the *Jew*. Parents also play a role; if they feel their child has talent, they can request that the child be taught in the *Jew*.

From any group, there will be someone who has a role in the tribe in the future, where electability is not adjusted to the dependence of one of the people who has the previous role. For example, not all children of *Tifa* players will later become a teacher of the *Tifa*; it all depends on the people who have talent or who have been selected by the customary leader. Each role must be resumed from the learning process in the *Jew*. According to the information given by the guest speaker Aloysius (Bainepe), *Tifa* musician from the Bismam family:

> *Anak-anak yang telah terpilih untuk belajar Tifa harus melalui proses inisiasi sebelum memulai pembelajaran Tifa, di mana anak-anak itu dijemur di depan jew pada siang hari.*
>
> Children who have been selected to study the *Tifa* must go through an initiation process before starting *Tifa* learning, where the children are dried in front of the *Jew* in the daytime.

Tifa learning uses the lecture and demonstration methods of learning. At first, the children are taught the Asmat tribe's traditional rules in beating the *Tifa*, introducing songs from the Asmat tribe and showing how to make the *Tifa*. Then, a song that has been taught begins to be sung by a *Wair Ow* while beating the *Tifa*. Because the *Tifa* is a rhythmic instrument, children are only taught about beats, rhythms and how to

Figure 4. Initiation process for the Asmat tribe children.
(Source: Doc. Ricko, 17 January 2017)

Figure 5. *Tifa* learning process in the *Jew*.
(Source: Doc. Ricko, 17 January 2017)

play the *Tifa*. For the initial stages, the children are invited to participate by clapping their hands or hitting their hands on the floor following the *Tifa* rhythm played by the teacher. This phase is called *Banbiem*. After starting to understand, the children will try to beat the *Tifa*. In playing the *Tifa* in the *Jew*, the children must form a circle and have a leader in charge of giving orders and determining what song will be played. The leader of the circle is called a *Wairipit*. Children who are already adept at playing the *Tifa* will be included in a group at a celebration of the Asmat tribe. The *Tifa* beat rhythms are divided into three types of fast rhythm or *sarpi*, moderate rhythm or slow rhythm or *damesnab*. The following are examples of rhythm pattern notations from the types of rhythms of the *Tifa* instrument playing:

Tempo: 145–147

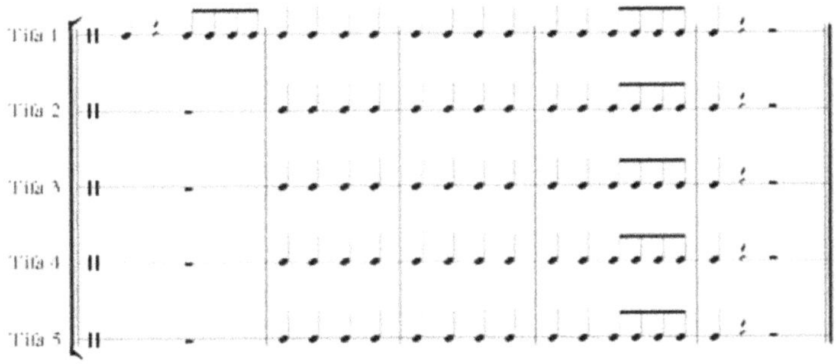

Notation 1 Examples of the rhythm of *Tifa* playing of the fast rhythm or *sarpi*.

Tempo: 117–120

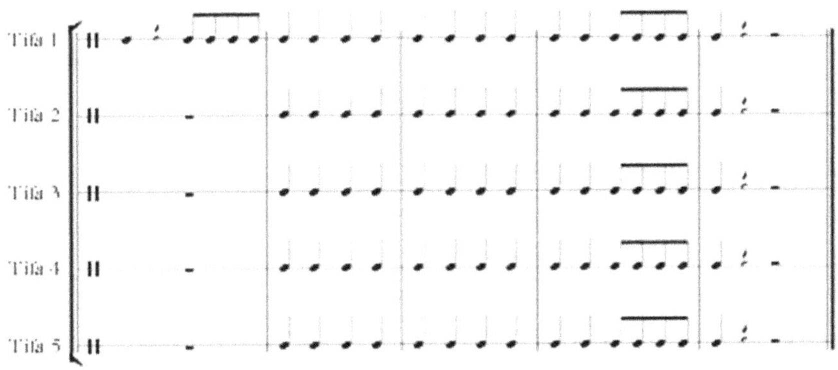

Notation 2 An example of a moderate rhythm or *dames*.

Tempo: 75–78

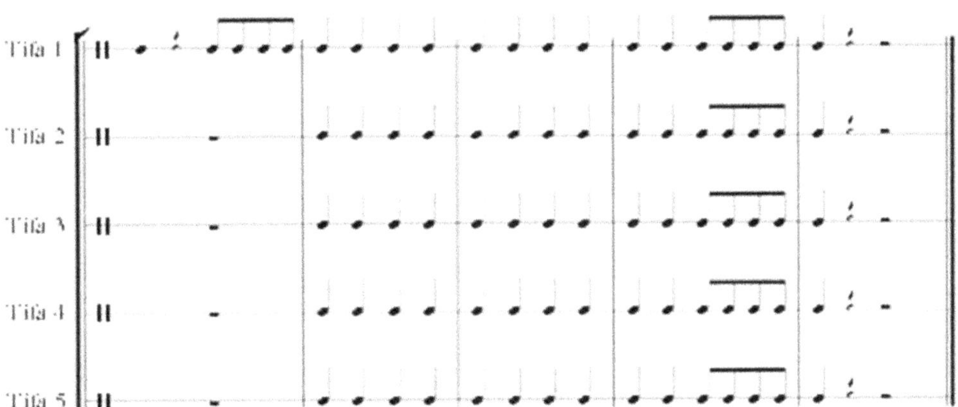

Notation 3 Examples of the rhythm of the slow *Tifa* game rhythm or *damesnab*.

Diaru

Notation 4 One song with *Tifa* accompaniment that is taught to children in the *Jew*.

The rhythmic pattern of the *Tifa* instrument in the Asmat is relatively similar in terms of rhythm; only the intro and outro parts are different. The *Tifa* game is played by five to six people. One person serves as the group leader or circle leader (*Wairipit*) who will play the intro. When the song played with a *sarpi* rhythm, the musicians of the *Tifa* will play the *Tifa* while standing and dancing. However, there are some rhythms called *dames* that must be played while standing only.

4.3 *Identification of the Tifa and the types of Tifa in the Asmat*

The *Tifa* is made from stretched logs. For the Asmat, the *Tifa* is also a tool used to provide motivation and encouragement, which can bring hope in life and become an interpretation of their lives. The shape is round and elongated like a drum. On the surface of the *Tifa,* there are carvings in the form of ancestral patterns or animals that have become symbols of strength for the sculptor such as cuscus, birds, monitor lizards and so on. In addition, carving patterns are also taken from the motifs of shields. Carvings made on the *Tifa* are also abstract based on the contents of the *Tifa* makers' thoughts. At the top, it is wrapped with monitor lizard skin and the skin is tied with rattan. The *Tifa* is usually named according to a person who has died. *Tifas* are usually made and carved by *Tifa* sculptors known as *Wow-ipits*. Beating the *Tifa*, for the Asmat tribe, is still governed by very strong traditional rules, so only men are allowed to play or beat the *Tifa*. The musicians of *Tifa* or *Wair Ow* must have and make their own *Tifas*. Furthermore, they are also obliged to understand the language of Asmat oral literature. Asmat people themselves do not understand written literature; they only recognise the language of oral literature. The *Wair Ow* is obliged to pass on the talent to play the *Tifa* to his descendants. If, for example, his son does not have an interest in *Tifa* playing, he will turn to his grandson. The *Tifa* has four types which, in a play, are usually used simultaneously. In the Asmat tribe, the *Tifa* is not differentiated by the type of function, but it is distinguished by its physical characteristics. As informed by the first resource person, Mr. Erick (*Sarkol*), curator of the Asmat museum, it is shown in the following interview that: 'The physical characteristics of the typical *Tifa* from the Asmat tribe living in the coastal areas are large and long, whereas, the physical characteristics of the typical *Tifa* from the Asmat tribe in the mountain areas are short'.

Figure 6. Numbers 1 and 2 are *Tifas* originating from the Asmat tribe living in the mountain area, while numbers 3 and 4 are *Tifas* from the Asmat tribe living in the coastal area.

(Source: Doc. Ricko, 16 January 2017)

In general, there are only two types of *Tifa* in the Asmat tribe, which are distinguished by their physical and sound forms. The *Tifa*, which has a shorter size, will produce a louder sound, while the *Tifa* with a long size will produce a denser sound. Factors that make a difference in the physical size of *Tifa* between the Asmat tribe in the coastal and mountain areas are the wood used to make them. The mountain Asmat tribe makes the *Tifa* shorter in size to save more material on *Tifa* making; this is because it is difficult to get suitable wood in the mountainous area. Another distinguishing factor in the *Tifa* from the Asmat tribe is the highly artistic carving pattern on the surface of the *Tifa*. The *Tifa* carving of the Asmat tribe is the main attraction for domestic and foreign tourists. The Asmat tribe has five clusters namely *Bismam*, *Betscembub*, *Safan*, *Joerat* and *Unir Sirau*, with each having different carvings and their own distinctive characteristics. There used to be traditional rituals in the *Tifa* engraving process; when the *Wow-ipits* or *Tifa* engravers had to let out their blood to later be used as the colouring in the engraving. The process was also aimed at making the ancestral spirits of the *Wow-ipits* merge with the engravings. However, that tradition has been abandoned because it is considered sacred. Figure 7 is a picture of the types of engravings on the *Tifa* of each family:

Figure 7. Types of carvings from the *safan* family.
(Source: Doc. Ricko, 16 January 2017)

Figure 8. Types of carvings from clumps.
(Source: Doc. Rome, 16 January 2017)

Figure 9. Types of carvings from clumps. (Source: Doc. Ricko, 16 January 2017)

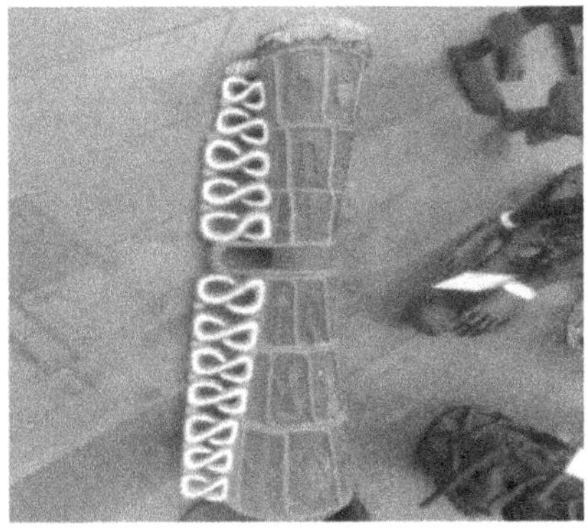

Figure 10. types of carvings from clumps. (Source: Doc. Ricko, 16 January 2017)

4 DISCUSSION

The *Tifa* is a musical instrument originating from Papua. One tribe that preserves the *Tifa* music playing is the Asmat tribe. There are several types of *Tifa*; namely *Tifa Jekir*, *Tifa Dasar*, *Tifa Jekir Potong* and *Tifa Bas*. However, in the Asmat tribe, the *Tifa* is only divided into two physical types based on the location of the residence of the tribe in Asmat. The first is the *Tifa* from the tribe who lives in the coastal area, which has large and long features. The second is the *Tifa* from the mountainous areas, which is shorter in size and has a louder voice.

The *Tifa* learning process, in general, cannot be said to be easy because, for the Asmat tribe, *Tifa* players are those who have been chosen by their traditional leaders. Starting from childhood, the growth and development of a child in the Asmat tribe has been considered; if the child has talent, he will be chosen by the traditional elders and then educated in the *Jew*. Furthermore, if parents feel that their children have talent, they can send them to the *Jew* to take part in the *Tifa* learning. Before starting the *Tifa* learning process in the *Jew*, the selected children must follow the initiation process (*Emak-Cem*). This is a process where the children must be dried in front of the *Jew* in the daytime. After the initiation process is complete, they can begin the process of learning the *Tifa* music. In the learning process, the teacher uses lecture and demonstration methods in which the children are invited to learn the traditional songs and customs of the Asmat tribe. Then, they are shown how to play the *Tifa*. The children can try the *Tifa* rhythm by clapping their hands or beating the floor rhythmically. After that, when they have begun to understand how to play the *Tifa*, they are given the opportunity to play the *Tifa* while singing traditional songs they already know. When they are proficient, they are given the opportunity to join a *Tifa* playing group in an event. They will be given the opportunity to play the *Tifa* in turn for a song that is being played.

Regarding procedures and materials, the learning process of *Tifa* playing does not have a formal module for teaching the instrument. The *Tifa* learning instruments are based on experience systems. This is explained by the benchmark of the learning system using a simple learning method and is continued with routine practice.

The *Tifa* is made from wood, monitor lizard skin and rattan. The wood used in the making of the *Tifa* is the *chi* wood or *waru* wood. This wood is chosen because it has good quality and can last a long time. The monitor lizard skin is used as the top cover of the *Tifa*. The monitor lizard skin commonly used is that of one to two year-old monitor lizards. This skin is thick and lasts a long time. The last ingredient is rattan; it is used as a rope to tie the monitor lizard skin to the *Tifa* so that it remains firmly attached to the *Tifa*. The process of making a *Tifa* takes a long time. The making of a *Tifa* is considered sacred to the Asmat tribe. It requires several rituals that must be completed. The rituals start from looking for wood as the material for making the *Tifa*, someone who makes the *Tifa* must not be close to women, as well as in the manufacturing process which needs the blood of the *Tifa* maker himself. For the Asmat youths, the *Tifa* has been considered as their identity, so that the manufacturing process is full of sacred rituals. Moreover, the *Tifa*'s carvings are adjusted to the power symbols of the owner of the *Tifa*. The Asmat *Tifa* has an important role. Besides being a musical instrument used to accompany dances, the *Tifa* is also used as a musical instrument in every traditional ritual, traditional ceremony or party. However, the *Tifa* is not used at grieving events. Some examples of the presentation of *Tifa* in the Asmat tribe are during the traditional plate party or *An Pakamu*, the feast of making a bachelor house (*Jew*) or *Jew Pakamu*, the sago caterpillar party or *Amos Pakamu* and so on. The *Tifa* is included as a rhythmic instrument because it is only played by being hit, not by being pitched.

5 CONCLUSIONS

The conclusion of this study is presented in the following items. First, the learning process of the *Tifa* instrument in the Asmat tribe includes rhyme and rhythm, because the *Tifa* instrument is a rhythmic instrument. Second, the learning of the *Tifa* instruments in the Asmat tribe refers to traditional rituals that are initialised by the process of initiation (*Emak-Cem*) where children who will learn the *Tifa* are dried in front of the *Jew* in the daytime. Third, in the learning process, the *Tifa* mentor uses lecture and demonstration methods, where the instructor or traditional elders will explain the history of the Asmat customs, how to make the *Tifa* and the customary rules in beating the *Tifa*. Then, the children are shown how to hit the *Tifa* without learning material or modules from the teacher. The process starts from inviting children to know their tribal songs, and then the traditional elders play the song accompanied by *Tifa* instruments and the children follow the rhythm by clapping their hands. The types of rhythms taught in the learning process are *sarpi* (fast rhythm), *dames* (moderate rhythm) and *damesnab* (slow rhythm). Fourth, children who are proficient in playing *Tifa* will be included in a group at an Asmat tribe traditional party such as the traditional plate party (*An Pakamu*), the spirit party (*jipai pakamu*), bachelor house making party (*Jew Pakamu*), boat party (*cimicim pakamu*), stacked sculpture party (*pakamu bus*) and sago caterpillar party (*Amos Pakamu*).

The suggestions put forward are: 1) Teachers should open opportunities for people who want to learn and are interested in learning *Tifa* musical instruments even from outside the lineage or family, to be able to achieve the goal of preserving the *Tifa* instruments; 2) For instructors, it is better to create modules related to learning *Tifa* musical instruments so that they can be learned by a wide audience.

REFERENCES

Bakar, Z.A. (2014). *Psikologi Pendidikan* [Educational psychology: Guidelines for teachers and parents]. Singapore: Partridge Publishing.

Bawantara, A. & Ekaristi, M. (2011). *Khazanah Negeriku; Mengenal 33 Provinsi di Indonesia* [The wealth of my country; Get to know 33 provinces in Indonesia]. Jakarta: Media Kita.

Dumatubun, A.E. (2002). Kebudayaan dan kesehatan orang Papua dalam perspektif antropologi kesehatan [Papuan culture health in the perspective of anthropology]. *Jurnal Antropologi Papua (ISSN: 1693-2099)*, *1*(1), 65-72.

Kristi, N. (2012). *Fakta Menabjubkan Tentang Indonesia: Wisata Sejarah, Budaya, dan Alam di 33 Provinsi* [Wonderful Indonesia: Tour of history, culture, and nature of 33 provinces]. Jakarta: Katalog Dalam Terbitan.

Hashim, S., Mahani, R. & Ramlah, J. (2004). *Psikologi Pendidikan* [Psychology of Education]. Pahang: PTS Publications & Distributor Sdn. Bhd.

Leirissa, R.Z. & Latuconsina, D. (1999). *Sejarah Kebudayaan Maluku* [History and Culture of Maluku]. Jakarta: Directorate General of Culture.
Moleong, L.J. (2000). *Metodologi Penelitian Kualitatif* [Qualitative research methods]. Bandung: PT Remaja Rosdakarya.
Mentansan, G. (2013). Surga terumbu karang Raja Ampat sebagaI objek wisata unggulan: ambisi dan tantangan [Raja Ampat's paradise of coral reefs as a leading tourist attraction: ambition and challenge]. *Tifa Antropologi*, *1*(1), 210–219.
Nakagawa, S. (2000). *Musik dan Kosmos: Sebuah PengantarEtnomusikologi* [Music and cosmos: An introduction of ethnomusicology]. Jakarta: Yayasan Obor Indonesia.
Papua, B.P. (2015). Papua Dalam Angka Tahun 2015 [Papua 2015 in number]. Papua: Central Bureau of Statistics.
Ricko, R. (Photographer). (2017, 17 January). Initiation process for the Asmat tribe children [photograph]. Jakarta, Indonesia: Asmat Museum.
Ricko, R. (Photographer). (2017, 17 January). *Tifa* learning process in the *Jew* [photograph]. Jakarta, Indonesia: Asmat Museum.
Ricko, R. (Photographer). (2017, 16 January). *Tifa* from mountain and coastal area [photograph]. Jakarta, Indonesia: Asmat Museum.
Ricko, R. (Photographer). (2017, 16 January). Types of carvings from clumps. [photograph]. Jakarta, Indonesia: Asmat Museum.
Ricko, R. (Photographer). (2017, 16 January). Types of carvings from the *Safan* family [photograph]. Jakarta, Indonesia: Asmat Museum.
Rome, R. (Photographer). (2017, 16 January). Types of carvings from clumps. [photograph]. Jakarta, Indonesia: Asmat Museum.
Pram, P. (2013). *Suku Bangsa Dunia & Kebudayaannya* [Tribe of world and culture]. Jakarta: Swadaya Group.
Sugiyono. S. (2011). *MetodePenelitian Kuantitatif, Kualitatif, dan R&D* [Quantitative, qualitative, and R & D methods]. Bandung: Alfabeta.
Semiawan, C. (2008). *Metode Penelitian Kualitatif: Jenis, Karakteristik dan Keunggulannya* [Qualitative Research Methods: Types, Characteristics and Strengths]. Jakarta: Grasindo.

21st Century Innovation in Music Education – Astuti et al. (eds)
© 2020 Taylor & Francis Group, London, ISBN 978-0-367-11122-9

Counter rhythmic and counter melodic methods for *angklung* music arrangement's

B.P. Korantalaga & A.M.S. Pradoko
Universitas Negeri Yogyakarta, Yogyakarta, Indonesia

ABSTRACT: The purpose of this study is to identify students' reactions to experimental counter rhythmic and melodic methods for the introduction of music arrangements in *angklung* ensembles at the Binatama Health Vocational Schools in Yogyakarta, Indonesia. The study uses a qualitative research method to describe the students' reactions. The subjects involved a teacher and tenth grade students at Binatama Health Vocational School. It was found that the introduction of counter rhythmic and counter melodic methods could make it easier for the students to understand music arrangements. The theories learnt after using the counter rhythmic and counter melodic methods include understanding the arrangement itself, elements of the arrangement such as the main melody or the main rhythm, the adherent melody or rhythm, dynamics of sound including expression and tempo, and even the meaning of the songs. Students also like to play *angklung* without consciously playing parts of the music arrangement.

1 INTRODUCTION

The *angklung* is a traditional musical instrument from West Java. There are several types of *angklung* that can be found in Indonesia. These types include the melodic *angklung* and *angklung* that produce chords. The *angklung* represents more than one chord at a time, for example, in the *angklung* with a C Major chord, there are C, E and G chords in one *angklung*. The angklung was developed for the first time by Daeng Sutisna and then continued by Mang Udjo. It is not difficult to use this instrument. The most common way to play *angklung* is by shaking the instrument holding it with the left hand while shaking the instrument with the right hand. The musical notes used in the *angklung* are the same as the ones used in piano music but can be changed according to the desired alignment.

The angklung is widely used in schools today. The learning process can be carried out by playing *angklung* in an ensemble or it can also be played solo by using *angklung* sets that are specifically used for solo performance. For learning in schools, the *angklung* that are used is in accordance with the notes on the piano, so that it can be easily integrated with other musical instruments such as keyboard, guitar, violin and so on. The twenty-first century learning trend is a collaborative system. Thus, playing music within the ensemble format is very supportive of learning focused on collaboration. In this ensemble system, all students are given one *angklung* that represents one melody so that when the *angklung* is played there will be a melodic sound.

Creativity is important in learning. To support the process of creativity in learning, *angklung* arrangements can be used. Westrup and Harrison (1959, p. 42) stated that arrangements are adaptations of parts of music in other forms to make them suitable to perform but do not change the essence of the original composition. Westrup and Horison argue that creating an arrangement is about converting existing music into another form without changing the essence of the original song. Students still have difficulty understanding arrangements especially when it is carried by using ensemble music. The process in an ensemble format is different from the process of playing solo. The process in an ensemble format does not only focus attention to individuals but also encourages listening, seeing and paying attention to other individuals in the group. For this reason, a method is needed to make it easier for students to understand the concept of music arrangements.

1.1 *Background of the research*

Learning musical arrangements requires developing a way for students to understand arrangements easily, whether by using supporting media or integration with today's technology. Learning traditional music such as *angklung* also pays attention to effective and sufficient time allocation. The allocated time for one music lesson at Binatama Health Vocational School is three hours per class. If students are given a difficult concept to learn, the allotted time will not be enough, and the lesson will be less effective. The teacher should develop a learning concept that transmits knowledge in a way that is simple and easy to understand.

The *angklung* is a musical instrument that can be applied in an ensemble format. Each student holds one *angklung* with the exception that if there are very many tones some students will hold more than one *angklung*. A lesson where the students only play song melodies will not be enough to explain the main melody, the accompanying melody, dynamics and so on. By leaning to play *angklung* musical instruments via an appropriate method, students can understand the music more easily.

The *angklung* is usually learned by playing melodies in traditional songs, for example, using the *angklung* to play the melody of a song called, "Suwe Ora Jamu" or any traditional songs. As mentioned above, conventional learning is not sufficient to being able to understand a copious amount of material. This study suggests that by using counter rhythmic and counter melodic methods, students can learn many musical terms and also understand the arrangements.

1.2 *Identification of the problem*

This research is important, foregrounding a new method that can meet the limited time allocation provided by the Binatama Health Vocational School, making learning effective and efficient. The research discusses a method of teaching referred to as, "Counter Rhythmic and Counter Melody for *Angklung* Ansambel Arrangement Introduction in Binatama Health Vocational School". The study found that with this method, it is easier for students to play and apply arrangements using a concept that is easy to understand. This method is useful for student development in playing and applying traditional music arrangements in the community.

The conventional learning method needs to be modified because it is less effective and less efficient for the limited time provided. Usually, one lesson is wasted because only the first concept is taught through theory delivered by the teacher. The next lesson hour is used to learn the ways that the instrument is used. In the next hour lesson, the students do not really understand the sounds of a melody written on the classroom board or shown on a projector. Thus, a new method is required to solve these problems.

1.3 *Formulation of the problem*

Base on the identification of the problems, the formulation of this study is 1) how is the delivery system in counter rhythmic and counter melodic methods used for students to learn music arrangements at Binatama Health Vocational Schools?; 2) How do students react to the method?, and 3) What arrangement theories can be delivered with the method?

2 LITERATURE REVIEW

2.1 *Arrangement*

Westrup and Harrison (1959, p. 42) states that arrangements are adaptations of parts of music with other forms to make them suitable to perform but did not change the essence of the original composition. Stainer and Barret (2009, p. 36) states that an arrangement is the selection or adaptation of parts in the composition so that they are suitable to be displayed with sound or other instruments compared to their original composition. According to the fourth edition of the Great Bahasa Indonesia Dictionary (1988, p. 84), the word "arrangement" is defined by adjusting the composition of music to the singer's vocal number or another instrument based on a composition that already exists so that the essence of the music does not change.

2.2 *Musical elements*

Music elements include melody, rhythm, harmony and expression. Miller (1994, p. 37) states that melody is a series of related tones which usually vary in height and length and short tones. Meanwhile Safrina (1998, p. 196), states that melody is the arrangement of a series of tones (sounds with regular sequences) that sound sequential and rhythmic and express an idea of thoughts and feelings.

Regarding melody, Mudjilah (1998, p. 4) states that rhythm is determined by the length of a sound. Bonoe (2003, p. 198) mentions that rhythm is a certain pattern that is stated with names such as: wals, mars, bossanova and so on. Rhythms are generally related to beats, time and tempo. According to Safrina (1998, p. 168), rhythm is the sequence of motion which is a basic element in music and dance. Rhythm in music is formed from a group of sounds and silence with various lengths of time.

Miller (1994, pp. 48-49) said that harmony is a musical element based on the simultaneous incorporation of tones as distinguished from the series of tones of melody. If melody is a horizontal concept, harmony is a vertical concept. Safrina (1998, p. 236) states that harmony or chord is a composite arrangement of two or more different tones, the height of which we hear simultaneously.

Definition of Interpretation and expression suggested by Prabowo et. al (in Roby, 2011, p. 7). He said that interpretation is the understanding and inspiration of the content of songs performed by listening to poetry or paying attention to melodic patterns. The fourth edition of the Great Bahasa Indonesia Dictionary (2008, p. 543) adds that interpretation is the giving of an impression, opinion, or theoretical view of something. Safrina (1998, p. 273) states that expression in music is a manifestation of thoughts and feelings that includes all the nuances of tempo, dynamics, and tone/sound of the basic elements of music embodied by music artists or singers, delivered to their listeners.

2.3 *Music creativity*

Djohan (2009, p. 175), states that as part of intelligence, the concept model of creative thinking in music involve product, ability, expertise, and aesthetic sensitivity. Product is deals with three ways of involving individuals with music which include: a) Composition, namely, concepts and recording of sound forms to be presented; b) Presentation or improvisation, the delivery of sound forms that have been prepared by the player when shown; and analysis, the process of understanding and expressing written or verbal forms of voice.

Ability that allows the thinking process consisting of a) Musical intelligence, namely, the ability of individuals who are influenced by the environment from the beginning of their development period to adulthood, b) Conceptual understanding, namely, cognitive facts as material for understanding music; and c) Expertise, the ability to practice real knowledge in through the action of demonstrating music.

Further more, aesthetic sensitivity is a response to a form of sound. The sound produces a high level of emotional response and is obtained through long practice.

2.4 *Adherent music*

Stainer and Barrett (1876, p. 6) state that adherent music is a separate part, which is used as an addition to solo or playing together. Suharto in Priyadi (1992, p. 2) states that adherent music is the background music that accompanies a single play or song and other forms.

3 METHODOLOGY

The sampling technique in this study involved saturating sample points or making all students as samples. The students who were sampled were members of Class X Pharmacy 1 and Class X Pharmacy 2 with each class totally 20 children. Data collection techniques used in the research are observation, documentation, field notes and questionnaires. The analysis was made using a qualitative method. Cresswell (2010) states that qualitative research is a method for exploring and understanding meanings which/by a number of individuals or groups of people are thought to originate from social or humanitarian problems (p. 4). Tohirin (2013) states that the data analysis is possible through data reduction, categorizing, synthesizing, and making temporary conclusions (p. 148). Data will be presented in the form of sentences, pictures and examples of music scores.

Experimentation is the research method. The data was analysed qualitatively by describing the students' reactions to the experimental method in the form of counter rhythmic and counter melodic methods for the introduction of musical arrangements in *angklung* ensembles in the Binatama Health School. As described above, qualitative research is a method used to explore and understand meanings derived from social and humanitarian problems. Using the old method of explaining the elements of the arrangement one by one is quite time consuming so that a new method is required to disseminate information within the short time frame provided by the school. Thus, a new experimental method for learning was used in this study.

The musical score uses number notation under which rhythmic and counter melodic elements are written and presented by the teacher. The score uses the song "Suwe ora jamu" because the notation is not so complicated and focuses more on the rhythmic and counter-melodic counter.

Melodic counter score are illustrated in Figure 1 below.

Figure 1. Counter melodic.

While the rhythmic counter can be seen in the following figure.

Figure 2. counter rhythmic.

4 RESULTS AND DISCUSSION

4.1 *Results*

Samples of data collection were carried out for all students of Class X 1 Pharmacy and Class X 2 Pharmacy. Each class has 20 students. To find out how students responded to this method, a questionnaire was given.

Based on the responses to the questionnaire, nearly all of the students answered that it was easier to understand the arrangement by using experimental approaches, namely, the counter rhythmic and counter melodic methods. The students also responded that they were happy that only one or two children were confused because they did not get an *angklung* musical instrument. Nearly all of the students who received an *angklung* felt happy. Students also learned several theories through the music arrangements. The theories learned were about melody (the main melody and its accompaniment), rhythm (main rhythm and its accompaniment), the dynamics (expression and tempo) of tempo is the speed of a song and changes in the speed of the song (Safrina, 1998, p. 273), sign or a dynamic term is a sign to state the level of volume, or the softness of the sound and the soft changes of the sound (Safrina, 1998, p. 274), and understanding the arrangement itself. Stainer and Barret (2009) state that the arrangement is the selection or adaptation of parts in the composition so that they are suitable to be displayed with sound or other instruments compared to their original composition (p. 36). After playing the music using counter rhythmic and counter melodic methods, students understood that arrangement involves making changes by adjusting or changing a composition/song into another form but that an arrangement does not change the essence of the song. By adding a rhythmic and counter melodic tone, students can rearrange the song.

From the students' direct responses, students expressed that they were happy with this method because they could directly play *angklung* musical instruments and acquire new information in through this way of learning. At first, students looked awkward because they listened to two kinds of melodies and rhythms that sound similar. Next, the teacher explained the main melody must be clearer than the accompaniment. The teacher explained that in a section of the arrangement, the accompaniment music should be softer than the main melody. The students then learned rhythmically through the numbered notes to understand how to read notes correctly. At first, they played rhythmically in a way that was not as desired. After half an hour of running rhythmic, counter-sounds start. The main melody shows 4/4 time and the accompanying melody shows 3/4 time followed by 5/4 notes as shown in Figure 4 and receives a beat which is sounded one, two, three, one, two, three, four, five. For time, the song is sounded according to the original rhythm while the rhythm of the accompaniment is noted within its own bar. After combining the main rhythm and accompaniment rhythm, there will be a rhythmic counterpoint between the main rhythm and accompaniment. Rhythmic counter is the opposite rhythm between one rhythm and another. For counter-melodies, students played the original melody while another group of students played the melody in the opposite direction, starting with do = C, see Figure 5. After combining the two melodies, the music will sound like counter-melodies against the main melody. When the main melody moves up, the melody goes down and will sound harmonious. After students played for one hour, they began to feel the beauty of harmonization or melody combination.

4.2 *Discussion*

This experimental method is expected to help students understand the music arrangement. The method helped students to respond to questions such as: What is the meaning of the arrangement itself? What are the elements in the arrangement? How can students appreciate arrangements created by other people? The experimental method using counter-rhythmic and counter-melodic contra methods as an introduction to music arrangement is in accordance with the target set, namely, introducing music arrangements for students of Class X SMK. However, musical background also has an influence. Students who can live fully are students who have musical experience and have played works that have been arranged by other composers or that the students themselves arranged. Unfortunately, not all students have that experience. Thus, a simple method is useful in helping students understand concepts of music arrangement.

The present research refers to a previous and related work titled, "Image Streaming Methods in Improving the Creativity of Music Arrangements by Atip Nurharini and Putri Yanuarita Sutikno in 2017". The purposes of this research and the previous research are nearly identical, namely, nurturing the aesthetic soul toward music arrangement. However, the research conducted by Atip Nurharini and Putri Yanuarita is more focused on fostering creativity. For Class X students at the Binatama Health Vocational School, they needed to know in advance about the arrangements such that they are more extensive in developing their musical activities.

In my opinion, the use of rhythmic and counter-melodic counter methods can be applied in various high schools and vocational schools but with the condition that those who have *angklung* musical instruments or similar musical instruments can be divided into one melody and other melodies, or one rhythm and other rhythms. These approaches apply not only to *angklung*, but also to gamelan or even the band format and that can be implemented in the schools.

5 CONCLUSION

Introducing music arrangements by using counter rhythmic and counter melodic methods for learning in Class X Binatama Health Vocational students can make it easier for other students to understand music arrangements. What the students in the study learned after using counter rhythmic and counter melodic methods are: the understanding of the arrangement itself, elements of arrangement of arrays such as the main melody or rhythm, melody or accompanying rhythm, dynamics namely soft sound including expression of tempo and even the meaning of the song listed. Students also like playing *angklung* without consciously playing a part of the arrangement.

For the development of music arrangement learning in the future, the research suggests that it is better to use methods that can disseminate a multiplicity of ideas at once and that is in accordance with a twenty-first century learning approach, namely, collaboration. While the use of rhythmic and counter-melodic counter methods to introduce how to make musical arrangements in *angklung* ensembles has proved successful, these methods would be more effective if accompanied by today's technology to increase student interest in learning.

The benefit of this study is that the data shows that experimental methods can be used to easily explain music arrangements with learning that is stimulated by counter rhythmic and counter melodic methods. These methods can be used for future studies related to music arrangements. These methods can minimize the time required to learn *angklung* particularly if the allotted class time is short. Furthemore, students can use the counter rhythmic and counter melodic theories to arrange songs on their own. These experimental methods are easily accepted by students and students have also unconsciously learned many elements of music arrangement.

REFERENCES

Banoe, Pono (2003) *Kamus Music* [Music Dictionary]. Yogyakarta: Kanisius.

Creswell, John W (2010). *Research Design (Qualitative, Quantitative and Mixed Approach)*. Yogyakarta: Pustaka Pelajar.

Djohan (2009). *Psikology Musik* [Music Psikology]. Yogyakarta: Penerbit Best Publisher.

Miller, M Hugh (1994). *Introduction Music a guide to good listening*. Yogyakarta: Yayasan Pustaka Nusatama.

Mudjilah, Hanna Sri (1998) *Teori Musik Dasar* [Basic Music Theory]. Yogyakarta: Institut Keguruan dan Ilmu Pendidikan Yogyakarta.

Nurharini dan Sutikno (2017). *Image Streaming Method in Improving Music Arrangement Creativity*. Universitas Negeri Semarang.

Priyadi (2014). *Efforts to Improve the Ability to Play Ensemble Using Accompaniment Music Media in MTS Muhammadiyah Tawangsari*. Yogyakarta: Universitas Negeri Yogyakarta.

Safrina, Rien (1998). Pendidikan Seni Musik [Music Arts Education]. Yogyakarta: Departemen Pendidikan dan Kebudayaan.

Stainer, John and Barrett, William (2009). *A Dictionary of Musical Terms*. New York: Cambridge University Press.

Roby, T. (2011). Frasering Technique in Musical Interpretations of Eros Djarot's Matahari Song. Yogyakarta: Institut Seni Indonesia.

Westrup, Sir Jack and Harrison (1959). *Collins Encyclopedia of Music*. Lodon: Chancellor Press Michelin House 81 Fulham Road.

21st Century Innovation in Music Education – Astuti et al. (eds)
© 2020 Taylor & Francis Group, London, ISBN 978-0-367-11122-9

The implementation of canon training to increase harmony sensitivity of choir members

F.A. Pamungkas & K.S. Astuti
Universitas Negeri Yogyakarta, Yogyakarta, Indonesia

ABSTRACT: This study aims to increase the harmony sensitivity of choir members in Senior High School 1 Rembang. This study uses classroom action research methods consisting of two cycles by using Kemmis & Taggart model. The subject of this study was members of the choir. This research uses descriptive qualitative and quantitative techniques. The results showed that by implementing canon training, the harmony sensitivity of choir members in Senior High School 1 Rembang increased. This can be seen from the average results of pre-cycle, the cycle I, and the Cycle II. Harmony sensitivity was measured by harmonising choir members at the last cycle test. The result of the average value of pre-cycle was 61.17 and this increased to 70.45 after the action of Cycle I and to 79.71 after the action of Cycle II. Based on the results of the average value, it can be concluded that the implementation of canon training increased the harmony sensitivity of Senior High School 1 Rembang's choir members.

1 INTRODUCTION

The choir is synonymous with the art of vocal production (Daugherty, 1999). Good vocal processing in a choir has two important elements to consider and comprehend: technical and non-technical. Technical elements include techniques in expressing melody, harmony and rhythm, while non-technical elements emphasise beauty including taste, dynamics, tempo and interpretation.

The progress of music in Indonesia has developed quite rapidly including vocal arts, especially choirs. There is evidence of this by the participation of Indonesians in the classic and national folklore (folk songs) categories at various choir competitions held at regional, national, or international levels.

Previous observations of the Senior High School 1 Rembang's choir revealed that the mastery of materials consisting of intonation, rhythm, and harmony of the choir members in singing songs arranged for soprano, alto, tenor, and bass had not been maximised. The choir members were not able to distinguish between the main melody and the accompaniment melody. There was often a lack of sensitivity of the choir members in animating harmony and stabilising the frequency of the tones when the melody was sung together. It was therefore established that effective training was needed so that the song materials could be understood and sung properly.

This research was conducted by applying canon training to the choir and having a melody sung together with different durations and repetitively performed by the melody leader and melody follower. The implementation of canon training in this research was expected to increase the harmony sensitivity among members of the Senior High School 1 Rembang's choir, so that the inconsistencies in harmony could be improved, or, in other words, so that the choir could produce better harmonies. Thus, the choir would be able to perform a musical work well and properly. The purpose of this research was to increase the harmony sensitivity of Senior High School 1 Rembang's choir members.

2 RELATED WORKS/LITERATURE REVIEW

2.1 *Harmony*

According to Banoe (2003, p. 192), harmony is a form of conformity and beauty. It means that a harmony will be formed in two possibilities called consonance or dissonance; beautiful or not beautiful. Harmony will be formed if there is a melody or tone of more than one sound which is sounded simultaneously. To harmonise a melody requires an effort of processes that will produce a harmony and is known as harmonisation. There are several ways to combine tones that produce a harmony/beauty according to Banoe (2003, p. 192) and they are:

1) Two-voice harmony. This is a very basic form of harmony that uses a combination of two tones. This harmonisation can determine major and minor qualities, for example the tone "do" combined with the tone "mi" becomes major third/m3, while the tone "re" combined with the tone "fa" becomes a minor third/m3.
2) Three-voice harmony is the development of the two-voice harmony. It is also called a harmony of triad because it has a combination of three tones that are sounded simultaneously, for example, the tones "C", "E", "G" when sounded together will produce a C Major triad.
3) Four-voice harmony, the development of triad harmony, adds the triad harmony to the fourth tone from outside the triad tone by doubling one of the triad tones. One example is the combination of the tones "C", "E", "G", "c", which when sounded together will produce a C Major triad.
4) Five-voice harmony is commonly used for musical performances of an instrumental family such as the family of saxophone instruments (full saxophones), or family of string instruments consisting of violin 1, violin 2, viola, cello and contrabass in which each musical instrument has a role as a high note, middle note or low note. However, five-voice harmony might be applied to a variety of other musical instruments.
5) Eight-voice harmony is a doubling of the four-voice harmony. The sound produced will be solid and attractive.

According to Kamien (1988), harmony is a combination of two or more tones that are sounded simultaneously, so that the sound produced is harmonious. According to Simanungkalit (2013), harmony in the narrow definition is "a simultaneous sound which has at least three tones or called chords". According to experts, harmony is various sounds or tones that are sounded or played simultaneously.

2.2 *Choir*

Choral singing in a choir is a simultaneous singing activity that focuses on following the harmonious harmony and providing a good interpretation of the songs sung/performed. A choir is a set or a number of singers or vocal groups and is based around each type of voice. According to Banoe (2003, p. 320), a choir is a group of sounds grouped into several sound lines, namely Soprano, Alto, Tenor and Bass (SATB). A children's choir is not able to produce SATB, but the sound lines may at least be divided into two. It can be concluded that a choir is a number of singers forming a team or vocal group and that these singers are grouped according to their type of voice.

2.2.1 *Types of choirs*

Prier (2003, p. 13) stated that there are four compositions of choirs in Indonesia and that they are: 1) children choirs; 2) teenage choirs; 3) adult choirs; 4) similar choirs. The subject of this research is the adult choir. The definition of an adult choir, Prier (2003, p. 14), is that the choir members are generally over the age of 17 and that the members have voices that already show the character of adults' voices. The number of members in an adult choir is at least 20; there is no maximum number.

The divisions between the choir members are: Soprano = 3, Alto = 2, Tenor = 2, Bass = 3. SATB choirs for adults are considered to have the most rounded and balanced sound because each of the sounds can stand alone, especially if the song is in a polyphonic style. If the members are trained properly, the choir will develop to a professional level and may be recruited for musical performance accompanied by dance and so on. In this research, there are more than 20 members of Senior High School 1 Rembang choir and the members met the range of adult soprano, alto, tenor and bass.

As stated in his journal article titled 'A Prescribed Alternative Mainstream: Popular Music and Canon Formation' (2006), Karja explained that 'The fundamental criterion by which Bohlman separates different folk music canons is 'the extensiveness of the community'. Other decisive factors include transmission and origin of repertories, specialisation of musicians and the relationship to other musical genres. It is worth noting that Bohlman does not deny that there may be other patterns of canon formation which are based on different criteriaMeanwhile, Falck (1972) identifies that there are many types of canon as he said that, canon and voice exchange are commonly regarded as a peculiarly English specialty, and the priority of the Summer Canon is the earliest example of its kind. This has led to the assumption that only in England was this particular form of music known before the 14th century. There is also the discovery of two round canons from the second half of the 12th century, one of which (Leto leta concio) may even be German.

2.2.2 *Vocal technique*

In this research, vocal techniques that become the focus to assess the level of harmony sensitivity are explained as follows:

1) Intonation

According to the PML (2013), intonation is a technique in sounding tones properly. Meanwhile Kamien (1988), intonation is a technique that can determine high-low pitch accuracy.

It can be concluded that intonation is a high and low tone that must be reached with the right frequency. However, not everyone can sing the tone correctly.

The Music Centre Team (*Liturgi*) (2013) argues that there are several reasons why the tones sung are less precise and they are: a) the nuance in singing activities is too strained; b) lack of concentration in singing; c) singers are out of breath when singing; d) tones repeated or retained is tiring; e) singers are less sensitive to harmony when voices are combined; f) singers are not good at producing tones; g) certain tones are difficult to master; h) certain tones are difficult to sing; i) letters with dark and light colours determine the pitch; j) singers tend to follow other tones; k) failed when switching between tones.

Intonation can be interpreted as a reference in knowing the sensitivity level of the harmony of choir members because if one of the choir members doesn't reach the tone correctly, the harmony sung sounds inharmonious.

2) Rhythm

Simanungkalit (2013) stated that rhythm is a sequence of differences in the value of the tone sung. An example is sol tone (5) sounded for four beats that can be written as || 5. . . ||, while rhythm is the length short of the note and the light weight of the pressure or accent on the note. According to Kamien (1988), rhythm is the soul of a song. This means that in determining the perfect scale in interpreting songs, rhythm is needed.

3) Balance

Simanungkalit (2013) stated that there are three things that need to be considered to make a harmonious choir and they are: (1) balance; (2) blending (integration); (3) sonority (loudness of sound). This research puts more emphasis on the balance in determining the level of harmony sensitivity. A good choir must have a sound balance between soprano, alto, tenor and bass in singing a song that has been arranged. This balance needs to be understood to avoid making a group of dominant sound. The balance can include the volume level of sound (the loudness level of sound produced).

3 MATERIAL & METHOD

3.1 *Types of research*

This research was conducted by using Classroom Action Research (CAR). According to Suharsimi (2011), CAR is a study that describes the occurrence of causation from treatment, that observes what happens when treatment is given, and that describes the entire research process from the beginning to the impact of the treatment. Class action research is an action research conducted with the aim of improving the quality of learning practices in a class.

3.2 *Time and place of the research*

This research was conducted for two months from November to December 2016 in Senior High School 1 Rembang, which is located at Gajah Mada Street No. 5, Magersari, Rembang District, Rembang Regency, Central Java. The research was conducted because the members of the school choir were not yet sensitive to singing harmonies simultaneously. Thus, canon training was held as an effort to increase the harmony sensitivity of the choir members.

3.3 *Subjects*

The subjects of this research were 23 members of Senior High School 1 Rembang's choir consisting of 13 male students and 10 female students. The members were 11th and 12th grade students.

3.4 *Data collection techniques*

The data collection techniques used in this research were participatory observation, test and documentation. The research instrument was an assessment sheet with aspects of intonantion, rhythm, chord and balance.

3.5 *Data analysis technique*

The data analysis technique used in this research was the descriptive qualitative method used by describing developments in the cycles that were carried out continuously during the data collection until the indicators of harmony sensitivity and quantitative descriptive were achieved. The observational data were processed by giving a score on each aspect observed as follows: 4 (Very good); 3 (Good); 2 (Sufficient); 1 (Insufficient).

4 RESULTS AND DISCUSSION

4.1 *Pre-cycle*

The research began with pre-cycle activities, which were carried out with the aim of establishing the level of harmony sensitivity among the members of the choir before any action was taken. Pre-cycle activities were held at one meeting on 28 September 2016.. The choir members only sang their own melodies without hearing other melodies.

There were five choir members from the 23 members who met the standard of the successful criteria. The average result of the Senior High School 1 Rembang's choir members' harmony sensitivity at the pre-cycle stage was 61.17. Thus, the average value of the choir group including 18 members of the choir did not yet meet the standard of the criteria in the good category (70-89).

4.2 *The implementation of Cycle I actions*

Cycle I is the beginning of the actions that were implemented in the canon training to choir members. Cycle I activities were carried out for four meetings in November 2016. The actions in Cycle I were carried out as an effort to improve the early conditions (pre-cycle). The implementation of class action research in Cycle I included planning, action, observation and reflection.

Based on the observation results with the co-labourers conducted for four times per meeting, several things were obtained, and they were: 1) At the first meeting, the choir members were still not careful about producing tones and determining the rhythm when performing vocals and trying to sing a melody of canon 1 song entitled 'Magnificat'. The choir members had not yet sung smoothly the canon 'Magnificat' melody. Each choir member was not independent in listening to different voices; 2) At the second meeting, 11 choir members were not present due to illness and other agendas. The 12 choir members who attended were able to follow canon training activities with good enthusiasm and were willing to help the author to encourage the 11 absent members to prioritise canon training activities so that the research process could run smoothly. Due to the absence of 11 choir members, the mastery of canon 2 intonation and rhythm was not successful; 3) At the third meeting, all the choir members attended and followed the training activities properly and enthusiastically. Choir members could sing the canon 1 song 'Magnificat' well, but were not successful with canon 2. The song 'Bagimu Negeri' could be sung with quite good intonation and chords. However, the determination of rhythm and balance was not well mastered because the choir members paid less attention to the one-sixteenth (♪) and triol (♫) of the rhythm contained in partiture and to the vocal volume; 4) At the fourth meeting, the choir members could follow the Cycle I test well. The choir members also wanted more meetings as they realised that they still needed canon training to increase harmony sensitivity.

The results of harmony sensitivity increasing in Cycle I can be seen in Figure 1.

Based on Figure 1, it can be seen that the results of the Cycle I action obtained an average of 70.45 and showed an increase of 15% after the action was held. However, there were still 12 members of the choir who did not meet the criteria standards of the good category (70-89), so Cycle II needed to be carried out to increase the harmony sensitivity for all members of Senior High School 1 Rembang's choir.

4.3 *The implementation of Cycle II actions*

Cycle II is a follow-up action of the Cycle I. The action was carried out as an effort to improve the actions of Cycle I. The implementation of the Cycle II action is generally almost the same as the Cycle I and included planning, action, observation and reflection. The difference was that the steps chosen were adjusted following suggestions from collaborators. Cycle II activity was held over four meetings in December 2016.

Based on the observation results with the collaborators conducted for four times, several things could be concluded: 1) At the first meeting, the choir members were careful in producing tones and determining the rhythm when performing vocals. Then, when trying to sing a song melody entitled 'Are you sleeping?', the choir members were able to sing with proper intonation and

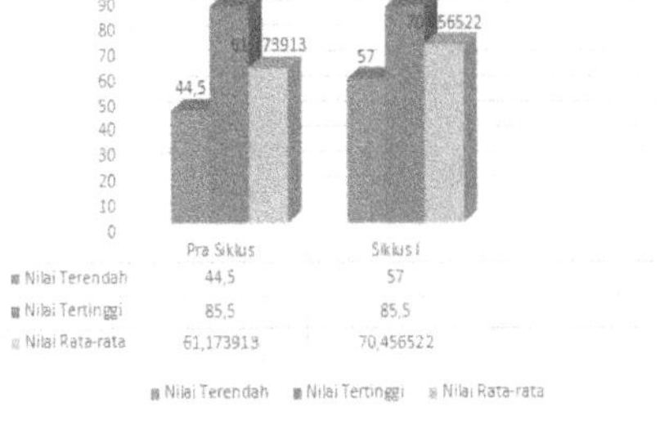

Figure 1. Assessment result graph of Pre-Cycle and Cycle I (Doc: Ardhy, 2016).

rhythm and found no difficulty in singing the song canonically. There was also an increase in the harmony sensitivity of the 'Bagimu Negeri' arrangement by SATB, which could be seen in the accuracy in singing the rhythms and chords produced. However, the choir members were not fluent in singing the 'Magnificat' song canonically because there were some members who had not memorised the melody of the song and that led to a disturbance of the harmony produced when the song was sung simultaneously. 2) At the second meeting, the choir members were careful in producing tones and rhythym in the 'Burung Hantu' song canonically. However, there was an increase in singing the 'Magnificat' and 'Bagimu Negeri' songs canonically that could be seen from the balance of the sound produced by each member of the choir. The choir members began to pay attention to aspects of intonation, rhythm, chord and balance when singing. 3) At the third meeting, the choir members could sing the 'Magnificat', 'Are you sleeping?', 'Burung Hantu' and 'Bagimu Negeri' songs well. Before the canon training activities of Cycle II were held at the third meeting, the choir members took their time to rehearse together independently. At this stage, the harmony sensitivity of the choir members sounded better. 4) At the fourth meeting, the choir members could take the Cycle II test well. The observation results of the Cycle II test were that each member of the choir had good harmony sensitivity. The choir members were able to sing the songs and paid more attention to intonation, rhythm, chord and balancing well and properly.

The results of harmony sensitivity increasing in the Cycle II can be seen in Figure 2.

Based on Figure 2, it can be seen that the results of the Cycle II action obtained an average value of 79.71 and showed an increase of 13% from Cycle I to Cycle II. In addition, all members of Senior High School 1 Rembang's choir met the criteria standards in the good category (70-89). Thus, Cycle II action was the final action conducted in this research.

4.4 *Discussion*

Based on the results of the research, the implementation of canon training did increase the harmony sensitivity of Senior High School 1 Rembang's choir members. This can be seen from the results of the pre-cycle and training activities in Cycles I and II. The average value of the pre-cycle is 61.17 and of Cycle I is 70.45. Hence, the increase from the pre-cycle to Cycle I is 9.28 or an increase of 15%. The average value for Cycle I is 70.45 and for Cycle II is 79.71 and thus an increase of 9.26 or by 13% from Cycle I to Cycle II. Furthermore, from the average value of the pre-cycle to Cycle II, there was an increase of 18.54 or an increase of 30%.

The canon training began by delivering general knowledge about the canon and how to sing with four voices simultaneously and alternately. The action aimed to make the choir members have an idea of the canon concept so that they could sing the canon song well and properly. The definition of canon was explained in general and an example of how to sing the canon songs was given. There was an explanation not only of canon in general but of the relation of harmony and the importance of canon training in increasing harmony'. If the choir members started producing tones well and could maintain the melody without being affected by other

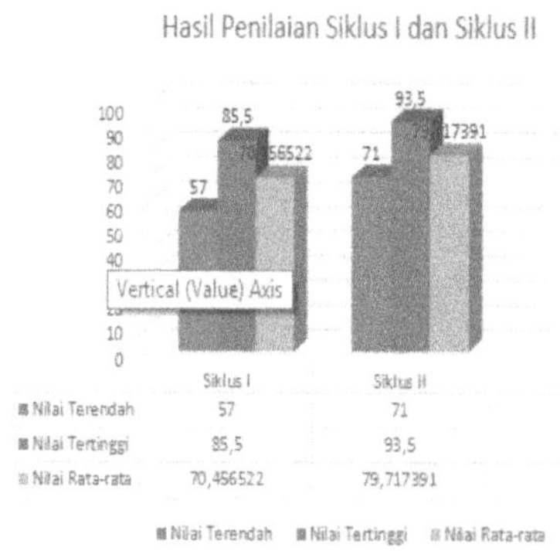

Figure 2. Cycle I and Cycle II Result graph (Doc: Ardhy, 2016).

choir members' melodies, the harmony produced from the canon song sung simultaneously would sound harmonious. Thus, the harmony of the choir members in singing the song arranged by SATB could increase.

Furthermore, materials of three canon songs entitled 'Magnificat', 'Are you sleeping?' and 'Burung Hantu' were given, as well as other song material arranged by SATB entitled 'Bagimu Negeri'. These four songs were chosen considering the results of the members' ability of Senior High School 1 Rembang's choir in the pre-cycle and taking into account expert advice. The song materials were delivered as a training activity in stages based on the development of the harmony sensitivity level of the choir members. During the training, the material of canon songs and songs arranged by SATB was given, as well as an explanation of the definition of intonation, rhythm, chords, and balance in general. This was then implemented in the song materials aiming to make choir members have a reference in increasing harmony sensitivity.

Based on the results of the research, there was an increase of harmony sensitivity of Senior High School 1 Rembang's choir members that appeared in Cycles I and II or after the actions were undertaken in the form of canon training. Therefore, the implementation of the canon did increase the harmony sensitivity of Senior High School 1 Rembang's choir members as expected.

5 CONCLUSION

There were two cycles given to the class research actions. In Cycle I, only 11 out of 23 members met the successful standard of the research that reached the value of ≥70. In Cycle II, all 23 choir members met the successful standard of the research and obtained an above 70 score.

Based on the results of the research, it can be seen that the average results of choir groups' harmony sensitivity in the tests held before given action or the pre-cycle was 61.17 then after an action was given or Cycle I, there was an increase in the average value of 70.45. Based on those average results, from the pre-cycle to Cycle I, there was an increase in harmony sensitivity of 15% ending in the Cycle II action with an average value of 79.71. From the values obtained from Cycles I and II, it can be concluded that there was an increase in the harmony sensitivity of the choir members by 13% or 30% as the whole increase from the pre-cycle to Cycle II.

The implementation of the actions conducted for two cycles was successful because the seven indicators of the criteria for the actions designed had been thoroughly met. Thus, it can be concluded that the implementation of the canon training did increase the harmony sensitivity of Senior High School 1 Rembang's choir members.

REFERENCES

Banoe, P. (2003). *Pengantar Pengetahuan Harmoni* [Introduction to harmony]. Kanisius.

Daugherty, J.F. (1999). Spacing, formation, and choral sound: Preferences and perceptions of auditors and choristers. *Journal of Research in Music Education*, *47*(3), 224–238.

Falck, R. (1972). "Rondellus", canon, and related types before 1300. *Journal of the American Musicological Society*, *25*(1), 38–57.

Kamien, R. & Kamien, A. (1988). *Music: An appreciation*. New York, NY: McGraw-Hill.

Kärjä, A.V. (2006). A prescribed alternative mainstream: Popular music and canon formation. *Popular Music*, *25*(1), 3–19.

Liturgi, T.P.M. (2013). *Menjadi Dirigen II* [Become Conductor II]. Yogyakarta: PML.

Prier, S.J. & Edmund, K. (2003). *Menjadi Dirigen III-Membina Paduan Suara* [Become Dirigen III - How to manage a choir]. Yogyakarta, Indonesia: Pusat Musik Liturgi.

Suharsimi, A. (2011). *Penelitian Tindakan Kelas* [Classroom action research]. Yogyakarta, Indonesia: Aditya Media.

Simanungkalit, N. (2013). *Teknik Vokal Paduan Suara* [Vocal techniques for choir]. Jakarta, Indonesia: Gramedia Pustaka Utama.

21st Century Innovation in Music Education – Astuti et al. (eds)

Advantages and challenges of the open-ended approach in nurturing creativity

D. Wulandari
Universitas Negeri Yogyakarta, Yogyakarta, Daerah Istimewa Yogyakarta, Indonesia

ABSTRACT: Deeper study of the ability of the open-ended approach to enhance creativity in various fields is necessary in order to encourage its implementation a variety of subjects. Through a systematic literature review, this paper aims to describe the nature of the open-ended approach and discuss the advantages and challenges that may be encountered in its implementation. The open-ended learning discussed is related to all educational fields of study and may allow its use in the context of a more comprehensive integrative learning process. The advantages of an open-ended approach are related to student activity, including opportunities for the students to gain information, and the celebration of students' uniqueness and empowerment. Meanwhile the challenges are related to learning designs, teacher's control and the assessment of students. Suggestions for future research are also discussed.

1 INTRODUCTION

A new era in which technology is growing rapidly and unpredictably demands creative problem-solving abilities. Humans are faced with more varied challenges with competitors that include not only fellow human beings, but also machines and other artificial intelligence. Eventually, humans will have to be able to create other possibilities to be successful in this era of industrial revolution 4.0. In other words, creativity will become one of the main weapons to survive and thrive in all fields. This is in line with Pehkonen's opinion that ultimately creativity is no longer a typical characteristic of artists or scientists, but a part of everybody's everyday life ensuring that they can survive as an independent and productive individual (Pehnoken, 2018).

One educational approach that can nurture student creativity is the open-ended approach. Such a learning approach can train students to think flexibly and produce diverse solutions to new problems. However, this approach is rarely implemented even in arts education, which is considered one of the areas most likely to foster creativity. When arts education merely focuses on cultural inheritance, for example, the assignments ultimately emphasize the practice of making or presenting existing artworks so that students are only able to imitate or reproduce existing products and are less able to innovate (Wulandari, 2017). Such conditions also occur in other subjects or areas despite the different nature of the challenges. Therefore, deeper study of the applicability of the open-ended approach in various fields to enhance creativity is necessary in order to encourage its implementation in any field or subject area.

Through a systematic literature review, this paper aims to describe the nature of the open-ended approach, and the advantages and challenges that may be encountered in its implementation. The open-ended learning discussed below is not just limited to arts, science or mathematics, but any educational field of study, and may allow its use in the context of a more comprehensive integrative learning process. The remainder of the paper is organized as follows. In Section 2, the nature of the open-ended approach as students' learning journey is described. Section 3 discusses the advantages of using the approach. Finally, the discussion of the challenges in implementing the open-ended approach is discussed in Section 4.

2 OPEN-ENDED APPROACH AS STUDENTS'S LEARNING JOURNEY

The Open-ended approach is a learning approach where students are faced with flexibly designed, open-ended problems or assignments, which require students to design their own solutions and methodology for completing the task at hand, prioritizing their own knowledge, ideas and skills. Among others, the arts might be the field that is most familiar with the open-ended approach. In most of the works in this field, artists or learners often work with open-ended challenges where they to be consistently creating different works by developing methods and approaches that suit the conditions and needs of the task. Later studies began to examine the concept of the open-ended approach and its use in mathematics education as an effort to foster creativity and problem-solving abilities. By recognizing its characteristics, the open-ended approach has the potential to be used in all fields and in education in general.

Some researchers and practitioners have tried to conceptualize the open-ended approach. Takahashi (2008, p.2) states that, *"the open-ended approach is an instructional approach using an open-ended problem, which has multiple solutions or multiple approaches to a solution. The lesson proceeds by using several student's responses to the given problem to provide experience in finding something new in the process"*. Meanwhile, Hancock (2009, p.228) states that open-ended problems *"are often considered as tasks in which there are more than one correct solution. Such problems allow students to use many approaches to solve them by placing few restrictions on students' solution methods"*. It can be concluded that open-ended learning emphasizes that every given problem may have many correct solutions, so that students can achieve the solution in various ways.

The Open-ended approach is not oriented towards a result, but rather on the learning process experienced by students, or the so-called *students' learning journey,* during the attempt to solve the problem. In practice, the implementation of open-ended learning activities is more flexible to the learning design which has been determined. This is *"not so much about following a road map"* but rather about enjoying the journey together and *"seeing where the group arrived"* (Thomson et. al, 2012, p.44). The learning journey does not really include the outcomes which have been chosen and asserted in advance by the teacher. Students are given the opportunity to use their own methods and achieve their own results. Therefore, in the open-ended learning classroom, teachers need to understand how to interact with student progress, facilitate open discussion and debriefing, and pose and respond to open-ended questions (Becker et. al, 1997)

In the context of mathematics education, according to Sullivan et al. (2000), the types of openness intended in the open-ended approach are *"openness in students' activity, openness in the content, and openness of interaction between students and the content."* *Openness in students' activity* means the questions that arise during the learning process come from the students themselves, so that students are motivated to overcome the problem. *Openness in the content* means that learning materials discussed during the learning process are not only fixed as they are in a textbook but are open to the possibility of generalization and diversification in order to solve the proposed problem. *Openness of interaction between students and the content* means that students have the freedom to receive, process, and formulate existing information to solve the given problems.

The open-ended approach is basically aimed at simultaneously fostering students' creativity and thinking in problem solving (Nohda, 2008). In open-ended learning, creative thinking processes occur when learners provide multiple interpretations of a problem and can then provide multiple different answers (Bernadi, 2017).

The open-ended approach is different from a closed-ended approach or lessons with prespecified learning outcomes. In a closed-ended approach, learning or the assignment tends to refer to one answer or completion strategy that is usually considered as a target achievement of the material to be understood and mastered by the students. In the learning process where the teacher's decision on the subject matter is so strict, the students are less encouraged to explore their abilities and ideas, making it less effective in building students' understanding, reasoning and communication skills (Mahmudi, 2008).

In closed-ended learning, the success of learners tends to be more determined by their ability in memorizing materials provided. In an analysis of the math tests of fourth grade learners of an elementary school who received close-ended conventional learning, Bernadi (2017) found that students tend to work on mathematics in one way, either as taught by the teacher or from examples of problems as described in textbooks. This may be because students are accustomed to using only one method or to only finding one correct answer. Procedural problem-solving and the emphasis on the ability to memorize tends to hinder the process of reasoning and critical thinking among learners. When faced with other forms of problems, these types of learners end up having difficulties trying to solve them. This lack of contextual learning is one of the causes of the low quality of education where students are unable to relate what they learn in the classroom to the diverse problems that exist in the real world (Handika & Wangid, 2013).

One of the aims of open-ended learning is to tackle those problems. In open-ended learning, learners are trained to face an open situation that allows a variety of responses to a variety of problems. Learners are accustomed to thinking about various possible answers to the problems. Along with this, Munandar (1985, p.83) asserts that "*the question that stimulates creative thinking is a divergent or open question*". Through open questions, learners gain the freedom to express themselves and communicate their own thoughts without being burdened or limited by having to conform to one method or trying to find a single answer a problem.

3 ADVANTAGES OF THE OPEN-ENDED APPROACH

The open-ended approach has several advantages over conventional learning. Firstly, open-ended learning is student-centered meaning that students are encouraged to be actively involved in learning activities. Students can express themselves verbally, in writing and in other forms in their work. Such student-centered learning has the potential to develop students' knowledge and skills. When students are given the opportunity to actively explore their own ideas, this enables the emergence of authentic ideas from learners. Students are encouraged to develop their own methods to find the most appropriate answers to the given problems.

The open-ended approach learning process usually encourages two-way discussion both between students and other students, and between students and teachers. This gives students the opportunity to convey and explain their opinions, and to confirm and respond to the questions and opinions of others (Becker et al, 1997, Bernadi, 2017). This kind of discussion also involves sharing ideas and experiences that will enrich students' insights and understanding of the subject matter.

The active learning process not only has a positive impact on the students themselves, but also gives the teacher the opportunity to more accurately assess the extent of the students' knowledge and their understanding of the subject matter. By using this teaching method, the teacher will be able to know the students' initial knowledge, the knowledge gained during the learning process, and what student should do next (Nohda, 2008). In other words, teachers can evaluate students' higher order thinking skills. Open-ended learning familiarizes students with active thinking to improve their ability to construct new knowledge. It also gives educators an understanding of the nature of students' thinking and learning of content (Pound, 2008). Teachers can understand how children learn about a particular field in their natural situation.

The second advantage is that students can explore, process and apply the necessary information from various sources to a given situation. When faced with an open issue and given the opportunity to solve it using their own methods, students are encouraged to utilize various sources of knowledge such as teachers, fellow students, books, the Internet, and other means of information. This information is also not only focused on the subject matter at the time, but rather a variety of information that students feel may be important in formulating solutions to the problem. This information can be the previous subject matters, other subject matter materials, information from the internet, out-of-school discussions, and other information they get from their daily life. This will also encourage students to make sense of the knowledge they learn anywhere, which can be useful and usable in a variety of situations.

When learners attempt to find a problem-solving strategy by using their ability to extract information, learners tend to be more competent in understanding the ideas associated with the problem (Takahashi, 2008). In solving problems, students are likely to seek information they have not yet learned and enable them to develop higher material, perhaps even beyond the targets of competence achievement at the time (Bernadi, 2017). Nohda (2008) claims that open-ended learning can enable the development of richer thinking so as to encourage more creative activities.

The third advantage to the open-ended approach is that it celebrates the unique characteristics of the students. They get opportunities to use their potential skills more comprehensively. No student is considered to have a low-ability, rather each student has a different approach according to their respective potential to solve problems both individually and when working in a team. Students are encouraged to build self and community identities by appreciating their own abilities, as well as the abilities of others.

Finally, all the advantages of an open-ended approach boil down to the empowerment students. This can lead to authentic learning experiences and meaningful learning. Through learning activities that provide full opportunities for students to determine their own problem-solving skills, the open-ended approach builds intrinsic motivation of students. Students are likely to be motivated to prove and test their ideas and abilities in solving a problem (Becker et al, 1997, Sullivan et al, 1995). The goal of open-ended learning is not just to get a solution or an answer, but rather, more important is that students understand how to arrive at that solution or answer (Bernadi, 2017). There is no right or wrong answer, but rather an appropriate or inappropriate answer tailored to the context of the problem or the solution. To achieve the most suitable results, students go through an experimental process where they continually try and improve the methods used in solving the problem. This encourages students' intrinsic ability in evaluating different methodologies. Rather than being declared unsuccessful because of possible errors, students will be encouraged to evaluate their own work. Open-ended learning celebrates the challenges and risks met through hard work (Thomson et al, 2012). Different from lessons with pre-specific learning outcomes in which failure is something to be avoided, in open-ended learning failure is not regarded as a scourge and it is appreciated as a part of the student's journey in discovering the meaning of their learning process.

Positive conditions built into the open-ended approach explicitly and straightforwardly argued by Adams and Hamm (Adam et al, 2010, p.5) show that "*openness captures serendipity and give learners the freedom to create. Part of openness involves a type of curiosity that stimulates the ability to critically evaluate data, accept input, and be ready to adapt to change*". Ultimately, an open-ended approach is expected to develop students' ability to be more flexible when facing new challenges and situations, and to create their own possibilities when facing unpredictable futures.

4 THE CHALLENGES

Apart from the benefits stated above, implementing open-ended learning may provide educational practitioners with some challenges. In this section, some challenges in applying the open-ended approach are discussed. They are related to learning designs, teacher's control and the assessment of students' knowledge. Some solutions to these challenges are also presented. However, it is noted that the discussion below cannot be considered an exhaustive list of challenges and solutions. Further research is still needed to develop the analysis.

One of the fundamental for implementing an open-ended approach is to provide students with opportunities to solve problems and assignments freely and using their own initiative. This is intended to give space for students to develop authentic ideas and skills. However, this may become the most fundamental challenge in implementing an open-ended approach. If teachers try to encourage flexibility of learning by not much following a specific road map too closely, the first challenge faced may be that the students' learning journey is out of reach (Thomson et al, 2012). When learning activities are open, students are likely to elicit unpredictable responses. The burden on teachers to fully understand all materials and skills in detail becomes more severe. Teachers should be able to facilitate student outbursts and answer both expected and unexpected questions in detail.

In response to this, Hashimoto (1997, p.13) suggests that "*the teacher must carefully allocate and manage time because students are likely to generate many responses, both expected and unexpected, and all should be discussed and summarized*". Time management becomes an important point in anticipating classroom conditions that are beyond the reach of teachers. In addition, Joseph (cited in Wulandari, 2017, p. 229) suggests one should "*strike the delicate balance between providing too much information, which makes the problem restrictive and close, and too little, makes the problem ambiguous*". Although allowed to use information from multiple sources, it is necessary to avoid overly broad or overly narrow material in order for learning to remain focused on the given problems. This requires the teacher to flexibly control the expected material coverage.

In any case, it should be realized that the learning process should not be limited to meeting the targets of the provided material. Teachers must maintain that open-ended learning is not a process which is fully controlled by the teacher alone. Learning is no longer just a transfer of knowledge, where students only receive, know and memorize material delivered by the teacher (Wulandari, 2017). In this digital age, information about the material can be obtained from many resources. Therefore, education should focus more on learning that provides students with the means to be able to manage the information they receive from any number of sources and apply them in solving problems they will encounter in real life.

Related to the point above, the second challenge is how to position teachers to neither too controlling nor too liberating. If learning is too free, the role of the teacher will become invisible and learning becomes ambiguous so that students will eventually have difficulty drawing conclusions about and summarizing what they have learned from the lesson. If a teacher is too controlling, students' opportunities to explore their own ideas and methods will be blocked. Therefore, teachers need to have the strategies and skills to fall somewhere in between the two. Teachers should be sensitive to when it is appropriate to let the students discover information, materials, and sources for themselves, when to provide clues and information, and when to intervene. In intervening though, teachers should be able to provide appropriate interventions such as open responses or questions and should avoid the straight answers and judgmental responses so that students will still have to think independently.

Consequently, the third challenge is the ability of teachers to design a suitable learning process. Teachers are expected to be able to set goals, learning steps, and create an authentic evaluation while providing enough flexibility for unexpected processes and results. The role of teachers is very important in designing the learning process. Teachers' readiness to manage unexpected learning steps can mitigate confusion for the learners (Bernadi, 2017). However, it needs to be adjusted for each field of educations. In sciences, for instance, the extent to which conventional procedures must still be followed one for the sake of job security or other significant reasons should be taken into consideration.

The difficulties encountered in designing the lesson plan are usually difficulties in preparing a meaningful problem situation for the students to overcome (Becker et al, 1997). Assignments or problems presented in an open-ended approach are usually more significant and complex. Quality issues are expected to 'have potential to elicit a range of response[s], from incorrect to simplistic to generalized' (Joseph, 2009, p. 182). As expected, 'richer and more open problems (such as those in domain-based assessment) yield more information about individual students understanding of skills and concepts than do traditional assessments' ((Joseph, 2009, p. 182)). From this type of task, students are expected to be able to connect the various forms of information that exist from various sources as well as their imagination with real conditions in everyday life. Accordingly, in raising the problem the teacher also needs to explain its pedagogical context. Teachers are expected to be able to explain the purpose of problem-solving, both in terms of its relevance to classroom learning activities and the benefits it may have in real life (Becker et al, 1997). Giving students a deeper understanding of the context may stimulate them make sense of what they are doing during the learning process.

Finally, another challenge that may be faced in implementing an open-ended approach is related to measuring the students' achievement, in other words how the teacher assesses students' progression and results authentically and validly. In the open-ended approach, there is no right or wrong answer. Thomson et al. (2012, p.45) explain that open-ended learning puts an 'emphasis on whether the work looked and/or felt right to the student in the context of what else was happening in the class'. Thus, although the teacher does make a standard for student to achieve, it should not erect borders around students' ability to explore their potential, as students are expected to develop their own ideas and skills. This dual nature may result in a bias within the measurement. When using a qualitative measurement, teachers will usually have difficulty in interpreting the response of learners (Bernadi, 2015).

In addition, some experts assume that an open-ended approach will produce lower quality results if the content of the achievement material is determined or limited. Such open target goals require complex intelligence capabilities as well as a relatively longer time to explore and experiment than conventional learning. Therefore, the open-ended approach suitable for emphasizing students' performance.

In any case, educators need to understand the nature of creativity within an educational context. In open-ended learning, the concept of the students' creativity that is expected to flourish is not the creativity that directly changes the world, usually referred to as Big-C (Begheto, 2007). Rather, nurturing students' creativity within classroom context means enabling children to experience meaningful personal creativity so that their work is new to them, then guiding them to be able to express this creativity in order to have meaning for others. Therefore, the standard of success cannot be as high as that expected of professionals and should be adjusted to meet the conditions and level of the students. Moreover, the teacher should be able to explain to students that being brave in facing challenges and failures will help them avoid anxiety in responding to or providing solutions to problems or a sense of dissatisfaction with the results obtained. Students need to understand that the most important thing is their journey to achieve understanding, not merely the result itself.

5 CONCLUSION

It can be concluded that the open-ended approach emphasizes the learning journey in which students have a chance to actively find and create something new within the learning activities. Thus, the experiences, experimentations, or methods students use to solve a given problem may be more important than achieving the outcomes. Students are taught to make sense of what they do during the lessons and eventually understand how the things they learn in classroom may connect to the real world. In nurturing creativity, the core idea of using open-ended approaches is giving space to the emergence of unexpected methods or results which are initiated by students. Consequently, the way educator designs, controls, and assesses students' learning and achievement may face real challenges. It may be difficult to create learning processes that allow enough flexibility for students to explore their own potential, and that is free enough for them to build their identities whilst also accounting for authenticity, validity and reliability. However, finding solutions for these challenges is not impossible. These challenges are likely to be solved through the hard work of educational practitioners.

From this analysis of the concept and advantages and challenges in implementing the open-ended approach, one of the main gaps that remains is the limitation of empirical data on the use of the open-ended approach used in fields other than arts and mathematics. Although the open-ended approach has the potential to be used in any subject, a broader picture of how this approach can be adapted to other subjects with their own characteristics is needed. Therefore, future empirical research about how to implement the open-ended approach in various subjects is needed. Hopefully, it can provide a significant contribution for creating future creative education, preparing students through open learning to deal with real conditions of the world that require individuals who are adept at creating their own solutions to unique problems.

REFERENCES

Adams, D. and Hamm, M. (2010). *"Demystify Math, Science, and Technology, Creativity, Innovation, and Problem Solving,"* Rowman & Little Field Publisher, Inc.

Becker, J.P. and Shimada, S. (1997). "*The Open-Ended Approach: A New Proposal for Teaching Mathematics*," NCTM, Inc.

Bernadi, R.M.A., (2015) "*Increased Creativity of Grade IV Elementary School Students in Prapaglor*, Pituruh, *Purworejo Through Integrative Thematic Learning with an Open-Ended Approach*," Master dissertation, Postgraduate Departement, Yogyakarta State University, Yogyakarta.

Bernadi, R.M.A. (2017). "Increased Creativity of Elementary School Students through Integrative Thematic Learning with Open-Ended Approach," *Jurnal Prima Edukasia* 5(1),91-101.

Handika, I., and Wangid, M. (2013). "The Influence of Problem-Based Learning on Class V Mastery of Concepts and Skills of Science". *Jurnal Prima Edukasia* 1(1),85-93.

Hashimoto. (1997). "An example of lesson development," In Becker, J.P. and Shimada, S. (Eds) *The Open Ended Approach: New Proposal for Teaching Mathematics*, The National Council of Teacher of Mathematics Inc, 10-22.

Joseph, Y.K.K. (2009). "Integrating open ended problems in the lower secondary mathematics lessons," In Kaur, B., *et al* (Eds) *Mathematical Problem Solving: yearbook 2009, Association of Mathematics Educator*, Word Scientific Publishing Co. Pte. Ltd, 226-240.

Mahmudi, A. (2008). "Develop open-ended problems in mathematics learning", *In the Proceedings of* Seminar Nasional Matematika dan Pendidikan Matematika, Universitas Negeri Yogyakarta.

Munandar, U. (1985). *"Develop Students' Talent and Creativity,"* Grasindo.

Nohda, N. (2008). "*A study of "open-approach" method in school mathematics teaching – focusing on mathematical problem solving activities"*. Available from: http://www.nku.edu/~sheffield/nohda.html [Accessed: July 18th, 2018].

Pehnoken, E. (1997). "The State-of-Art in Mathematical Creativity. Zentralblatt fur Didaktik der Mathematik (ZDM)," *The International Journal on Mathematics Education*. Available from: http://www.emis.de/journals/ZDM/zdm973a1.pdf [Accessed: July 17th, 2018].

Pound, L. (2008). "*Thinking and Learning about Mathematics in the Early Years*," Roudledge Taylor and Francis Group.

Sullivan, P., Bourke, D., and Scott, A. (1995). *"Open Ended Tasks as Stimuly for Learning Mathematics"*.

Sullivan, P., Warren, E., and White, P. (2000). "Students' responses to content specific open-ended mathematical tasks," *Mathematics Education Research Journal* 12(1),2-17.

Takahashi, A. (2008). *"Communication as process for students to learn mathematical,"*. Available from: http://www.criced.tsukuba.ac.jp/math/apec/apec2008/papers/PDF/14.Akihiko_Takahashi_USA.pdf. [Accessed: July 18th, 2018].

Thomson, P., Hall, C., Jones, K. and Green, J.S. (2012). *"The Signature Pedagogies Project: Final Report,"* Creativity, Culture and Education.

Wulandari, D. (2017) "The Role of Arts Education In Fostering Creative Problem Solving Skills and Its Implementation In Curriculum 2013," *In the Proceedings of The International Seminar Quovadis of Traditional Arts* XIII, "*Diversity in Culture*", SPs Indonesia University of Education.

21st Century Innovation in Music Education – Astuti et al. (eds)
 ISBN 978-0-367-11122-9

'Aesthetic' electoral politics: How songs are utilized in Indonesian elections

Halili
Yogyakarta State University, Yogyakarta, Indonesia

ABSTRACT: This article aims: 1) to describe the utilisation of songs within Indonesian electoral politics; 2) to analyse the advantages of online or new media as a new context for electoral campaigning by using songs; and 3) to analyse the transformation of campaign strategies in electoral politics. The study is based on research using content analysis method with a qualitative-comparative approach. This study provides results: First, the use of songs in the electoral politics of Indonesia is generally oriented into two political objectives—partisan and non-partisan. Secondly, online media advances have been utilized by political campaign teams to upscale campaign programs through various accounts, timelines and channels of social and other online media. Third, song utilisation as campaign materials confirms the transformation of electoral campaigns, from vulnerability to security, from the real world to the virtual space, from crowds of people to the convenience of computer screens, gadgets and headsets.

1 INTRODUCTION

General elections are part of a larger government system and ideally function to develop democracy in a country. The general election is considered a symbol as well as a benchmark of democracy itself (Sarbaini, 2014). General elections have significant functions in the political system, e.g. political socialisation and recruitment, articulation of interests, interests' aggregation and political communication (Liberty, 1982). General elections are interesting to understand as the measurement of a democratic political system. Political articulations in the general election process can be seen clearly so it is important to conduct research in the process of political communication.

The process of political communication ahead of the 2019 political year in Indonesia has recently started. Freedom of speech which is guaranteed in the Indonesian Constitution will be utilized by all participants from the political elite to the grassroots community to share their aspirations (Indriani, 2008). The campaign is a moment that will decorate the upcoming 2019 political year and is already being carried out by the political elites at even though the political year is still one year away. The political elites have started to conduct their campaigns in many ways. As a result of technological developments in recent decades, the model of the current campaign has undergone several changes (Hasyim, 2013).

These changes look very significant as the campaigns have begun to fuse with electronic culture (Cipto, 1996). Electronic culture also has a variety of types. For example, mainstream media outlets are currently used by political elites to support their campaigns such as making content go viral through online or new media. It is hoped that these contents or strategies will have a major influence in society as the content goes viral. These viral contents are also different from one another. In this paper, the author focuses on the transformation of campaign strategies in electoral politics, the advancement of online or new media as a new context of electoral campaigns, and song utilisation in Indonesian electoral politics.

2 RESEARCH METHODOLOGY

This research employs content analysis method with a qualitative-comparative approach (Moleong, 1998). Content analysis research is a method that seeks to uncover the symbolic meaning of several messages. Data procurement is held by unitisation and recording, while unitisation is activated by determining reference units, syntax units and thematic units. The instruments used in this study are the researcher himself with the help of check lists and recording notes. The check list and recording note are used to track and record data. Testing the validity of the data employs semantic validity and predictive other approaches. Within the research, data resources are in the form of papers selected using a purposive approach, while the secondary data are obtained from several YouTube channels.

3 RESULT AND DISCUSSION

In this section, the writer reviews several sub topics; a selection of popular songs related to elections in Indonesia, song utilization strategies as an instrument for electoral campaigns, the advancement of digital and online media technology as a new context, and the transformation of electoral campaigns in Indonesia related to the use of songs.

3.1 *Some popular songs related to elections in Indonesia*

3.1.1 *Wahai presiden (Dear President), Iwan Fals*

Lyrics:
Wahai presiden kami yang baru
Kamu harus dengar suara ini
Suara yang keluar dari dalam goa
Goa yang penuh lumut kebosanan
Walau hidup adalah hiburan
Tetapi kita tak mau dipermainkan
Dan kami juga bukan hiburan
Turunkan harga secepatnya
Berikan kami pekerjaan
Pasti ku angkat engkau menjadi Manusia setengah dewa
Masalah moral masalah akhlak
Biar kami cari sendiri
Urus saja moralmu urus saja akhlakmu
Peraturan yang sehat yang kami mau
Tegakkan hukum setegak tegaknya
Adil dan tegas tak pandang bulu
Pasti ku angkat engkau menjadi manusia setengah dewa
Source: Kapanlagi.com

This song criticises the government especially the Executive Office or the President. Criticism is conveyed directly and implicitly in each of the song lyrics. In this song, criticism is conveyed through sarcasm and synonymy as in the lyrics: "*Urus saja moralmu urus saja akhlakmu*" ("Mind your morality, mind your attitude"). The message conveyed in this song is that the government should permit opportunities for the people to express their opinions. The mandate in this song is more focused on the president in order to improve Indonesian policy in terms of politics, the economy and law enforcement.

In this song, there are no "black" campaign or negative campaign elements. Instead, the political message carried out in the song is considered high politics not affiliated with any partisan representative.

3.1.2 *2019 Ganti presiden (2019 Change the President), Sang Alang*

Lyrics:
Dulu kami hidup tak susah
Mencari kerja sangat mudah
Tetapi kini, pengangguran
Semakin banyak nggak karuan
Sepuluh juta lapangan kerja
Tetapi bukan untuk kita
Kerja, kerja, kerja, buruh asing yang kerja
Anak, anak, bangsa tetap nganggur aja
Di sana sini orang menjerit
harga-harga selangit hidupnya yang sulit
Sembako naik, listrik naik
Di malam buta bbm ikut naik
(buset)...
Pajak mencekik usaha sulit
Tapi korupsi subur penguasanya makmur
Rumah rakyat kau gusur, nasib rakyat yang kabur
Awas awas kursimu nanti tergusur
Beban hidup kami sudah nggak sanggup
Pengennya cepat-cepat tahun depan
2019 ganti presiden
Kuingin presiden yang cinta pada rakyatnya
2019 ganti presiden
Kuingin presiden yang tak pandai berbohong
2019 ganti presiden
Kuingin presiden yang cerdas gagah perkasa
2019 ganti presiden
Bukan presiden yang suka memenjarakan ulama
Dan rakyatnya ye ye
Beban hidup kami udah nggak kuat
Maunya cepat-cepat tahun depan.
Source: news.detik.com

This song offers some criticism of the current official government, in particular the president, the incumbent candidate in the upcoming 2019 presidential election. In this song, the composer focuses criticisms on the weaknesses of the President. For example, the lyrics, "*Sepuluh juta lapangan kerja, Tetapi bukan untuk kita, Kerja, kerja, kerja, buruh asing yang kerja ini,* is a negative campaign message that reveals the shortcomings of the current president. This negative campaign message is expressed hyperbolically which means that the political element in this song is fully exposed.

The lyrics, "*Di sana sini orang menjerit, harga-harga selangit hidupnya yang sulit, Sembako naik, listrik naik, Di malam buta bbm ikut naik*", are also hyperbolic and tend to be misunderstood. These lyrics can possibly lead to a black campaign because the lyrics are not based on actual data and possibly only based on the opinions of the songwriter.

In other words, this song contains some elements of negative campaigning and soft black campaigning. In short, the aforementioned lyrics are an example of hyperbolic negative campaigning as if the data is made up. Thus, it is very obvious that this song is affiliated with partisan representatives. The composer is possibly a sympathiser or unofficial campaign staff member of another presidential candidate, Prabowo-Sandi.

3.1.3 *Jokowi wae (Jokowi Only)*

Lyrics:
Hati ini sudah nyatu
Kita semua milih nomor satu
Pak Jokowiii..
Presiden Idolaku..
Jangan ragu dan jangan bimbang
Pak jokowi sudah pasti menang
Indonesia....
Berkah dan sejahtera
Presidennya jujur dan amanah
Rakyatnya sudah pasti sejahtera
Tak peduli siapa itu lawannya
Pak jokowi.. Tetap jadi presiden kita
Jokowi wae mas jokowi wae
Ojo liyane ojo liyane Jokowi wae
Source: Youtube.com

This song tells of the excellence of the number one presidential candidate number, the incumbent president. This example seems to be a positive campaign song in which the contents of the song only include information about one candidate without referring to or revealing alleged deficiencies of the other candidate. In this song, there are no elements of negative campaigning or even black campaigning. This song purely provides information about the candidate and his dominance and is obviously composed for the purpose of showing affiliation with partisan representatives to influence the public to choose Jokowi-Ma'ruf, as the first candidates. As a result, the narrative contained in the song leads to the glorification of Indonesian President Joko Widodo, also known as Jokowi, and other figures associated with him.

3.1.4 *Goyang Dua Jari (dancing using two Fingers), Prabowo dan sandi's version*

Lyric:
Ayo kita
Pilih prabowo sandi
Untuk NKRI
Pilihan dari hati
Yang lain jangan iri
Salam 2 jari
Pilih prabowo sandi
Ksatria sejati
Ayo kita
Goyang 2 jari
Salam 2 jari
Pilih prabowo sandi
Ayo kita goyang 2 jari
Biar kita happy
Pilih prabowo sandi
Ayo kita
Pilih prabowo sandi
Untuk NKRI

Yang lain jangan panik
Jangan sampai iri
Jangan iri hati
Mari.. mari.. sini..
Pilih prabowo sandi
Prabowo sandi
Prabowo sandi
Prabowo sandi
Source: Youtube.com

This song tells about the bargaining power of one of the second presidential candidates. This song is another example of positive campaigning where content in the song is information about one candidate without dropping or revealing alleged deficiencies of the opposing candidates. In this song, there are no elements of negative campaigning or black campaigning. This song purely provides the candidate's information and dominance.

This song, thus, is truly affiliated with a partisan representative and aims to persuade voters to choose the number two presidential candidates.

3.1.5 *Cari pemimpin (Seeking a Leader) – Iksan Skuter*

Lyrics:
Di pinggir jalan mulai bertebaran
Poster-poster besar yang gambarnya sok pahlawan
Mengobral janji melindungi dan semoga itu terbukti
Aku cari pemimpin yang naiknya bis kota
Kami cari pemimpin yang hidup sederhana
Bukan cari pemimpin yang lupakan janjinya
Apakah ada pemimpin yang kami idam-idamkan?
Yang telah lalu janji terbukti palsu
Kami telah bosan dengan janji-janji kalian
Jika berbohong kami akan memaksa tuan untuk turun jabatan
Aku cari pemimpin yang naiknya bis kota
Kami cari pemimpin yang hidup sederhana
Bukan cari pemimpin yang lupakan janjinya
Apakah ada pemimpin yang kami idam-idamkan?
Source: Liriklagu.info

The above song tells a story of torment and the anger of a young man who encounters candidates' posters that show heroic images of the candidates but who often sell promises that are ultimately not fulfilled when they have been elected. In the lyrics, this song also provides criteria for leaders who are admired by the community, such as leaders who ride the city bus, live modestly and do not forget their promises. In this song, there is a threat made to the kind of leader who often forgets his promises is then forced to step down.

The song's social criticism is aimed at candidates who enter electoral politics often using heroic figures on their posters while selling promises to the community, but at the end forget about their promises after being elected.

This song has no element of electoral campaigning whether positive, negative or black campaigning. This song is more about the common good of the country and the people.

3.1.6 *Jokowi presidenku 2019 (Jokowi is My 2019 President) – XMINOR*

Lyrics:
Dibawah sumpah bukan hanya sandiwara
tak banyak cakap tapi semua terlaksana
berani ikrarkan janji untukmu garuda
jujur berbakti bagi nusantara
ibu pertiwi kembali tersenyum riang
lihat wibawamu bagaikan yang pertama
merintis kebebasan bagian yang tertekan jajahan meski nyawa taruhannya
lorong yang gelap perlahan mulai terang
yang kumuh perlahan diperindahnya
lembah yang kelam perlahan terlihat jalan
licin seperti di metropolitan
lahir bukan dari keluarga yang kaya
dibesarkan oleh hidup yang sederhana
jadi kebanggaan bagi rakyat jelata
bapak presidenku luar biasa
Source: Youtube.com

This song tells about the special qualities of the first candidates for the Office of the President who are the incumbent candidates. In this song's lyrics, there is no negative campaigning or black campaigning. The creator provides information about the successes of the incumbent candidates during his and his Vice-President's time in the office. The lyrics of this song also present the life of the candidates who started pioneering careers from the bottom. This song is another example of positive campaigning in which the content includes information about the first candidate without dropping or revealing the alleged deficiencies of the opposing candidates. In this song, there are no elements of negative campaigning or black campaigning. This song purely provides the public with the candidates' information and supremacy.

Judging from the content it is clear that this song is affiliated with partisan representatives. The initial intention to make this song is to persuade the public to choose the current President, Jokowi.

3.2 *Song Utilization strategy*

This strategy involves two techniques for electoral politics. The first technique is to create new songs such as, "*Wahai Presiden*" ("Dear President") by Iwan Fals, "*2019 Ganti Presiden*" ("2019 Change the President") by Sang Alang, "*Cari Pemimpin*" ("Seeking a Leader") by Iksan Skuter, "*Jokowi Presidenku 2019*" ("Jokowi is My 2019 President") by Xminor. The second technique is to modify the lyrics of the song such as Julia Perez's song, "*Aku Rapopo*" ("I am OK"), that is modified into "*Jokowi Wae*" ("Jokowi Only") campaign song and Sandrina's "*Goyang Dua Jari*" which is converted to "*Salam Dua Jari Prabowo Sandi*". Based on the results of the song utilisation analysis in electoral politics, each song has the elements of positive campaigning, negative campaigning and even black campaigning.

Positive campaigning is the process of spreading information about candidates using their visions and missions, their work programs and their superiority. This kind of campaign can be in form of slogans, TV commercials, billboards, debates and interviews (lilian, 2015) . As time progresses into a modern political campaign, these campaigns are often carried out using social media and the Internet because of low costs and as efficient methods for mobilizing voters. However, politicians rarely use positive campaigning as their main strategy in gaining votes in electoral politics. Instead, their campaigns emphasize messages that attack other candidates through negative campaigning and black campaigning.

Negative campaigning is quite an offensive strategy conducted by a candidate's campaign attacking an opponent's personality, record or opinions. Negative campaigning spreads alleged negative aspects about the opposing party which is then followed by asserting the positive abilities of the first candidate to address the topic of a negative campaign message. Negative campaigning often takes the form of opinions and are not yet based on clear data, or sometimes based on existing data that is selectively used by raising negative aspects. This kind of negative campaigning, as long as it does not violate the Election Supervisory Agency Regulation *(Peraturan Badan Pengawas Pemilu/Bawaslu)* No. 28 year of 2018 overseen by the Supervision of Campaigns for the General Election, cannot be prosecuted. In fact, negative campaigning is still considered fair and in accordance with the recent regulation.

Black campaigning is the process of making false claims linked to the opposite party without the support of clear facts or evidence. Black campaigning is commonly associated with covert psychological operations and all of the initiators of this kind of campaigning usually do not reveal their identities nor identify with a political group. The issues raised are irrational and difficult to be assessed such that voters are not aware that an unknown provocateur is influencing them toward accepting the black campaign's message without processing the information in advance (Oktaviani, 2015). This type of campaign likely violates some portions of the regulation such that this kind of campaign can be prosecuted. However, because the source of this campaign cannot be easily identified, a person who reports the black campaign will have difficulty in naming who should be prosecuted.

Positive campaigning, negative campaigning and black campaigning in electoral politics will likely affect voters' decisions in choosing and supporting who they feel are capable candidates. Positive campaigning coaxes voters to choose that campaign's candidate based on the candidate's programs, vision and mission and superiority compared to the other candidates. During the process of this positive campaigning, if it is considered that there is a lack of enough support for the candidate, a negative campaign could further influence the views and thoughts of the voters by directing them to pay attention to the weakness of the candidate's electoral opponents. What this means is that both types of campaigning provide alternative strategies that could strengthen the image of political candidates in electoral politics. On the other hand, black campaigning that is carried out by irresponsible parties can be a problem because this kind of campaigning tends to be spread massively among people with low political awareness. Black campaigning can be said to be akin to using "dirty tricks", indeed an effective way in gaining votes by violating the image of political opponents without giving a clear source of information.

3.3 *Online or new media as context*

The discussion of song utilisation in electoral politics is closely related to the new media, mostly online based. It is undeniable that new media has been an effective method to publish the songs that have then gone viral. The power of going viral through the new media is one of the effective ways to draw out public opinion from the community. With this viral perception, the public will be more likely accept the meaning of what is presented in a viral campaign. The phenomenon of viral ad campaigns already occurred during Indonesia's 2014 Presidential campaign. For example, The Jokowi-Jk's two-finger concert had successfully increased Jokowi-Jk's electability during the 2014 Presidential Election (Haryanto, 2016).

The nature of the current new media largely involves manipulating voter behaviours. Oswald (1993) stated that the importance of the media's power and influence is fully appreciated through a complete understanding of who or what the media are. New media in this context is intended to circulate the songs rapidly for the sake of electoral politics. This media comes with a very strategic base to share information that can be easily accessed and used by anyone.

3.4 *Transformation of the electoral campaign*

In the current electoral era in Indonesia, there are changes in campaign strategy that are used in every electoral politics period, from a shift in the role of conventional campaign style approaches such as rallying, mass meeting and processions that are considered ineffective in reaching voters nowadays to new or social media. This change is due to the fact that the current electoral political system is no longer hierarchal, from up to down, but lies directly in the hands of the people. If the candidate is eager to get a lot of support, media that easily reaches the wider community should be used, for example, the new media particularly social media (Hasyim, 2013). At the same time, the shift of the campaign from mass meetings in the open field to the more creative model of the new media did not drastically alter conventional campaign strategy. Some politicians still use conventional campaigning because of public opinion that insists that conventional campaigning should still exist in electoral politics for the sake of democracy.

The campaign modernisation carried out by politicians was driven by the increasing number of younger voters often referred to as millennial voters. These millennial voters can easily be reached through social media rather than conventional campaigning. In addition, politicians have to be smart in creating 'artistic' campaign content, using effective visual, musical and other art elements. This creative work in making YouTube, Instagram and Facebook content can be vital for the success of politicians in reaching millennial voters. If the content can be effectively conveyed to millennial voters, it is possible that they will choose politicians whose campaigns know how to smartly use new media for the general election.

The modern, technological campaigns in this period of Internet culture results in the declining role of political parties in electoral politics. In conventional schema, political parties have a large share of interest during the year of general election year by plunging into society. In the conventional campaigning model, the role of political parties in influencing the masses is largely counted. Now however, the importance of new media in various communities has shifted the traditional role of political parties to content creators. Many content creators are very creative in imaging the character of candidates that he or she supports while in fact they are not part of any political party.

The transformation of the traditional campaign model in electoral politics is hopefully not absolute. Some candidates still maintain a conventional campaigning model that is considered as a "ritual" during the general election year. Shifts only occur in the media they use and the amount of new media or old media used in the campaign. Some elite politicians still consider conventional campaign methods as important such as processions and parades to show the extent of support by the people. Nonetheless, the new or online media would help reach voters through a wider scope in the form of songs, videos, visual images and other forms of representation in new (social) media.

4 CONCLUSION

Electoral politics is a benchmark for the level of success of democracy in a country and is a standardized aspect of a democratic political system. The process of political communication ahead of the upcoming 2019 political year has been recently observed as most of the elite politicians have already started their campaigning period. The campaigning changes are so significant as the melding of online culture into political campaigns cannot be nullified. The use of mainstream new media in campaigns by content creators to send out a viral ad campaign is a common practice these days.

Most of these campaigns are circulated rapidly and widely in the new or online media such as YouTube and Instagram. Songs such as "2019 Ganti Presiden" ("2019, Change the President"), "Jokowi Presidentku 2019" ("2019, Jokowi is my President") and other songs that are purely social criticisms are just echoes of campaign strategies. Aspects of viral ad campaigning include several negative campaigns, gray campaigns and even the most negative, black campaigning. The use of lyrics that are sometimes not based on clear evidence can be considered black campaigning, but if the lyrics of this kind of song is modified with a softer tone, the song could enter instead into the area of negative campaigning. Additionally, there are some songs that are filled only with positive campaign messages.

Finally, making shifts in a traditional campaign strategy does not change the entire strategy as some political elites are still keen to using conventional ways. What shifts is that the frequency of traditional campaign methods is slightly reduced and replaced with more modern methods such as online and new media campaigns that use creative content.

REFERENCES

Ahmad, T.A. (2010). *Iwan Fals Vs Oom Pasikom: Alternative Political Education Media.* Yogyakarta: Penerbit Ombak.

Basri, S. (2011). *"Introduction to political science"*,Yogyakarta: Indiebookcorner.

Cipto, Bambang. (1996) *"Prospects and Challenges of Political Parties". Yogyakarta: Pustaka Pelajar.*

Carritt, E.F. (1949). "An introduction to aesthetics." (No city and publisher).

Haryanto. (1982). *Political System: an Introduction.* Liberty.

Haryanto, Alexander. (2016). "Simbiosis Asyik Musik dan Politik", https://tirto.id/simbiosis-asyik-musik-dan-politik-boaN? usqp=mq331AQCCAE%3D&amp_js_v=0.1 (Retrieved on October 15th, 2018 at 03.40 P.M.).

Hasyim, N.M. (2013). *"From the Street to the Glass Screen"*. Profetik. Sunan Kalijaga State Islamic University, 6 (1).

Indriani, F. (2016). "Juridical Review of Defamation Through Social Media Based on Article 27 Paragraph (3) Law Number 11 Year 2008 concerning Information and Electronic Transactions Associated with Freedom of Praise". *The Students Online Journal of Faculty of law, Riau University.* Riau University, 3 (1), p. 1–15.

Krippendorff, Klaus. (1980). *Content Analysis: An Introduction to Its Methodology.* Beverly Hills-London: SAGE Publication.

Maarif, A.S. (2017). *Seni dan Politik (1) I Republika Online, Republika.* Cited from: https://www.republika.co.id/berita/kolom/resonansi/17/10/03/ox7hmx440-seni-dan-politik-1 (Accessed on Oktober 17th, 2018).

Moertopo, A. (1974). *National political strategy. The proclamation.* Centre for Strategic and International Studies.

Moleong, Lexy J. (1998). *Social Research Methodology.* Yogyakarta: Gadjah Mada University Press.

Oswald, K.A. (1993). *"Mass media and the transformation of American politics"*, Marq. L. Rev. Hein Online, 77, p. 385.

Otakviani, L. (2015). The Influence of Black Campaign through Mass Media Against Beginner Voter Elections in the 2014 Presidential Election (Study in Purworejo and Srirejo Hamlets in Branti Raya of Natar sub-district). *Faculty of Social and Political Science of Lampung University*, p. 10–26.

Sarbaini, S. (2014). "Democratization and Freedom to Choose Citizens in General Elections". *Inovatif: Journal of Law.* Jambi University, 7 (3).

Street, John. (2003). "Fight the Power": Politics of Music and Music of Politics". *Government and Opposition.* Oxford and Maiden: Blackwell Publishing.

Street, John. (2012). *Music and Politic.* Cambridge and Maiden: Polity Press.

21st Century Innovation in Music Education – Astuti et al. (eds)
© 2020 Taylor & Francis Group, London, ISBN 978-0-367-11122-9

Literature review: The effect of music therapy on labour pains

K. Resmaniasih
Poltekkes Palangka Raya, Palangka Raya, Indonesia

ABSTRACT: Labour pains during the delivery of a baby are a natural occurrence. Every labour will cause pain due to uterine contractions and normal labour activities. Pain and anxiety during labour can be overcome by the use of analgesics and anaesthesia, but these cause side effects. As an alternative, music therapy can reduce the incidence of pain and anxiety in labour without side effects. The purpose of this literature review is to review the effectiveness of music therapy in reducing labour pains and anxiety during delivery. This study uses a literature review method based on issues, methodologies, equations and advanced research proposals. Included criteria include the application of music therapy to labour, specifically labour pains. The results show that all studies agree that music therapy is effective in reducing labour pains and anxiety during a normal delivery.

1 INTRODUCTION

Labour is part of a physiological and happy event. For some women labour often causes pain (Prawirohardjo, 2008). Pain is normal and is something that is not separated from labour as 80% to 95% of women who give birth, feel great pain due to the cervical dilation process and a decrease in the lower part of the foetus (Tournaire & Theau Yonneau, 2007). Labour pains and stress are interrelated sequences. Stress in labour leads to excessive release of the hormones catecholamine and cortisol, which ultimately results in smooth muscle tension and vascular vasoconstriction. Consequently, the release of hormones can further inhibit the effectiveness of contractions and thus prolong labour (Romano & Lothian, 2008). The research of Hasan, et al (2016) in the Babil Iraqi provision examined the awareness and attitudes of pregnant women for painless labour, from 730 (89.02%) multi paras as respondents, 59.9% answered that labour could not take place without pain while 64.4% requested that future labour be helped by pain-free service. Many complementary therapies that can be used to reduce pain during labour including music therapy. Audio analgesia is the use of sound or music to eliminate or calm the intensity of pain. This is used to relax and calm the mind and body during stressful situations and therefore can be used to relieve pain during labour and childbirth.

Handling labour pain by using pharmacological drugs causes side effects on the well-being of the mother and the foetus. Another approach that can be used to reduce labour pain is music therapy. The use of music is a therapeutic intervention as a complementary therapy that is expected to increase comfort and decrease pain and anxiety in the mother.

Based on the discussion above, there has been a lot of research on the use of music therapy in reducing labour pain in primigravidian mothers. This study was carried out to conduct an analysis of music therapy through the literature review.

2 MATERIAL & METHODS

This chapter discusses strategies that were used to search for journals referred to in the literature review. Search questions were used to find and review journals that are tailored to PICOT and journal search terms through MESH, and to note limitations on using journals and other sources. The journals that are used in this paper's literature review were obtained through a database of providers of international journals EBSCO and Pubmed. The author opened the website www. poltekkespalangkaraya.ac.id then opened EBSCO. The researcher typed in keywords according to MESH (Medical Subject Heading), such as "music", "normal labour" and "pain" and selected full text. There were 18 findings with the next findings sorted from the latest findings. Language selection was not used because all the journals that were found used the English language. Each question was compiled based on PICOT where each question contained P = problem/patient/ population, I/E = implementation/intervention/exposure, C = control/comparative intervention, O = result and T = Time. Another relevant action that this paper's author used to obtain journal sources about the effectiveness of music therapy for pain in normal labour, was to look for sources about the effectiveness of music therapy for maternal anxiety. The intervention used in the study focused on music therapy and other therapeutic combinations. The research was carried out to identify the effectiveness of music therapy for labour pain and anxiety only in studies with experimental designs, both actual and pseudo experiments.

3 THE RESULT

Emphasis on the problem of labour pain was raised in the Xavier Research Research, Viswanath (2016), in Karkal E, Kharde S, Dhumale H. (2017), in the work of Ramesh et al (2018) and in Nayak, Rastogi and Kathuria (2014). For the most part, these studies looked at the same problem base. These works came out of studies about childbirth experiences and provide an overview of labour pain and maternal anxiety that can create physiological and biochemical responses or stress symptoms to the body increasing risks associated with labor. Hosseini, Bagher and Honarparvaran (2013), emphasise the problem that labour itself is an important factor in the tolerance of one's pain threshold. Ramesh, et al (2018), emphasise a comprehensive and multidimensional approach to pain management whereby health personnel must prepare prospective mothers to undergo labour in a relaxed state, so as to be able to support the mother during the process of childbirth to be comfortable and safe. This study also focuses on therapeutic music interventions in traditional songs combined with deep breathing techniques. Karalal, Kharde and Dhumale (2017), focus on primigravida mothers who are more anxious about things that are not possible, so that it causes anxiety. Where anxiety and labour pain are interrelated, John and Angeline (2017), emphasise that labour has psychological, social and emotional significance for the mother and the family and that most women experience physical stress and pain during labour. Pain and anxiety can affect each other and have an impact on labour. Nayak, Rastogi and Kathuria (2014) emphasised that environmental factors can influence perceptions of pain. Xavier and Viswanath (2016), describe the process, time and location of pain in labour. They also describe the incidence of back pain that is commonly experienced during labour.

Research method steps on the effectiveness of music therapy for pain and anxiety in labour was carried out with the same design via experiments and with partial selection of research samples using random samples and in some cases using purposive samples. Research that used random sampling include Hosseini, Bagher and Honarparvaran (2013), Ramesh (2018) and Karkal, Kharde and Dhumale (2017). On the other hand, research by Nayak, Rastogi and Kathuria (2014), Xavier and Viswanath (2016) and John and Angeline (2017) used non-random sampling. The number of samples produced by random sampling is representative, where the number of samples is at least 30 and a total of 120 participants. Based on the sample used and the study design of six studies, five of the six studies divided the sample into intervention groups and control groups. One study, Nayak, Rastogi and Kathuria (2014) did not differentiate their samples into groups.

The music intervention procedure used by researchers was different. There were those chosen by participants or chosen by the researchers themselves or a combination of participant and research choices. In John and Angeline (2017) and Xavier and Viswanath (2016), the music was determined by researchers, while in the research of Ramesh (2018) music was chosen by participants. A combination approach in music choices by both researchers and participants informed two studies, Hosseini, Bagher and Honarparvaran (2013) and Karkal, Kharde and Dhumale (2017). The treatment of duration, frequency and duration of administering music therapy in each study had different criteria. Of the six studies, two studies gave a duration of 30 minutes, three studies used a duration of one hour with a duration of 30 minutes between stages in Hosseini, Bagher and Honarparvara (2013) and in Xavier and Viswanath. (2016). The longest duration occurred in the research by Nayak, Rastogi and Kathuria (2014), John and Angeline (2017), Ramesh, et al (2018) each of which was one hour. The study by Karkal, Kharde and Dhumale (2017) did not specify the duration of time.

4 DISCUSSION

1) Analysis of the similarities and differences between each study

The research results by Ramesh (2018), Hosseini,Bagher and Honarparvaran (2013) and Xavier and Viswanath (2016), showed a significant decrease in pain and anxiety scores in the experimental group compared to the control group. In another study, Nayak, Rastogi and Kathuria (2014) showed lower mean pain and anxiety scores after listening to music than before. In terms of the results of six studies, in general, the study assessed the effectiveness of music therapy for labour pain and the level of anxiety in primiparous maternity women. The studies were more likely to analyse differences in results from the intervention group and the control group.

All of the studies have the same background which is the presence of pain in labour, especially primiparas and maternal anxiety, and the use of analgesic drugs by mothers to deal with pain and maternal anxiety levels that can affect labour and its outcomes. Thus, it is necessary to use other therapies to reduce pain and anxiety in the mother such as music therapy. The references that were used in all of the aforementioned studies are appropriate and complete because they consist of concepts and the kinds of journals needed for this particular kind of research. Overall, this present study has the same background problem: the pain experienced during labour is one of the most severe forms of pain. Anxiety and pain are closely related to each other and the interaction between anxiety and pain can be a spiraling process. Pain and anxiety can together become severe and in turn, cause panic that leads to maternal and foetal complications. Providing comfort to women in labour is still a challenge for midwife nurses and giving medicines has side effects for the mother and foetus. Thus, it is necessary to provide an alternative intervention that can provide comfort during the mother's process of birthing. That alternative intervention is music therapy.

All existing studies outline and assess the effectiveness of music therapy by comparing the intervention groups that were given music therapy and the control group that was treated as usual without music therapy. The results of all existing studies show that music therapy can reduce pain scores and maternal anxiety.

Furthermore, there are two studies that provide input for the next study. Nayak, Rastogi and Kathuria (2014) provide input for further research via the use of larger samples in different settings so that the findings can be generalised to a larger population and that starts from the antenatal period and the start of labour to help women respond to their contractions with better relaxation and work. Ramesh, et al (2018) suggest more studies at various stages of labour. However, most of the studies did not provide suggestions for further research. The following table provides categorical summaries for each study.

No	Author	Title	Design	Respondent	Treatment	Control	Assessments Procedure	Result
1	**DebajaniNayak, Sharada Rastogi, OmKumariKathuria** (2014)	*Effectiveness of music therapy on anxiety level and pain perceptionin primipara mothers during first stage of labour in selected hospitals of Odisha*	*A quasi experimental research approach with one group pre-test post-test design*	30 primiparous mothers who were in the first stage of labour at the hospital and selected met The sample criteria were deliberately chosen for research (purposive)	The experimental group received the preferred music (soft music, devotional music, and instrumental music) provided by the researcher for one hour in the active phase of labour.		Anxiety was tested using the state scale Speiberger's State Trait Anxiety Inventory which had a total of 20 questions. Anxiety is recorded twice, which is in the active phase of labour before starting and after the completion of music therapy. To compare the differences between how to pre-test and post-test anxiety score of the experimental group, t-test was calculated.	A significant decrease in the average score of anxiety scores post-test (50.06) and pre-test (59.23). Similar trends were followed in terms of pain perception scores (posttest score 6.27; pretest score 7.8). Furthermore, a significant correlation (r = 0.8192) was obtained between anxiety and pain perception scores. The research findings revealed that music therapy would be very helpful in reducing the level of anxiety and pain perception in primiparous mothers during the first stage of labour.
2	***S.E.Hosseini,M. Bagher,N.Honarpar-varan. (2013)***	*Investigating the effect of music on labour pain and progress in the active stage of first labour*	*Experim ental pretest-posttest design*	The 30 patterned primiparas were randomly selected and based on their tendency, 15 of which were included	The experimental group listens to relaxing music during	The control group did not get music	In this study, the Numerical Pain Rating Scale (NPRS), Visual Analogue Scale (VAS), and the	t independent for pain sensation in the experimental group and controls before the intervention had (p = 0.875) 0.601

(Continued)

(*Continued*)

				in the experimental group and 15 in the control group.	30 minutes every hour for a two-hour period	during labor	Verbal Pain Pain Scale has been applied to assess the severity of pain. Data obtained by the independent t test were used to compare the control group and the experimental group and compare the post-test and pre-test groups using the dependent t statistics and the average score.	numerical and visual pain and ($p <0.01$) 2.92 verbal pain, and one hour after intervention, had ($p<0.0001$) 8,527 in visual and numerical pain and ($p <0,0001$) 11,824 in verbal pain. Also, the same values of t independent for duration of delivery in the control group before and after the intervention indicate that the drug has no effect on serotonin levels.
3	***Bhuvaneswari Ramesh, Sumathy Sundar, Jayapreetha R, Sunitha Samal, Seetesh Ghose (2018)***	*Effects of Culture-Based Chants on Labour Pain During the Latent Stage of Labour in Primigravidae Mothers: A Randomized Controlled Trial*	*Experim ental pretest-posttest design*	A total of 120 primiparous women in the latent stage of labour with regular contractions and dilatations of less than 4 cm were included in the study and randomly divided into two groups.	The experimental group (n = 60) received MT in the form of vocal music sung for one hour. Listening to music while singing along and doing deep breathing exercises between each and every contraction	the control group (n= 60) only received standard treatment	Pain measures were recorded by researchers using the analog visual pain scale. (VAS) and Behaviour Pain Scale (BPRS) consisting of dimensions - facial expression, muscle tone, anxiety, vocalisation and comfort before and after exposure to MT in	MT intervention resulted in a statistically significant reduction in pain and a total BPRS score between music and the control group with ($p =0.001$). Also, BPRS domain scores in facial expressions, restlessness, comfort and vocalisation showed

							both the experimental group and the control group. Normality test is carried out to ascertain whether the data is normally distributed. Because data is normally distributed, data is analysed by an independent Student t test. A P value of less than 0.05 was considered significant.	a significant decrease due to MT interventions ($p = 0.001$). SRB domain - muscle tone does not make a significant impact on music.
4	**TintuXavier,Lekha-Viswanath.(2016)**	*Effect of Music therapy on LabourPain among Women in Active Labour Admittedin TertiaryCare HospitalKochi*	*A Quasi-experimental pretest posttestcontrolgroup design*	The sample consisted of 40 women in the active stage of labour, with 4-7cm cervical dilation. The sample is selected based on criteria (purposive).	The experimental group was then given music therapy (music and vocal based on the chosen taste) for 30 minutes using a headset. Posttest was conducted at 30 and 60 minutes after the pretest.	The control group was not exposed to music during the period	The first 20 women were allocated to control the group and the next 20 went to the experimental group to avoid contamination. Background information was collected using semi-structured interviews and review notes. The analog V scale is used to measure pain and anxiety. After measuring the pretest pain and anxiety the	The mean pain post test score of the experimental group (7.9) was significantly lower than the posttest score of the control group at 30ke minutes (5.75 +/- 0.44 vs. 8.55 +/- 0.51, $p < 0.001$) and 60ke minutes (7.9 +/- 0.45 vs. 8.7 +/- 0.47, $p < 0.001$). No differences were observed in uterine contractions between groups. The average anxiety

(*Continued*)

(*Continued*)

							women in the experimental group were then given 30 minutes of music therapy using a headset. Posttest was conducted at 30 and 60 minutes after the pretest.	score from the experimental group was lower than the control group (p <0.001)
5	**Karkal E, Kharde S, Dhumale H. (2017)**	*Effectiveness of Music Therapy in Reducing Pain and Anxiety among Primigravid Women during Active Phase of First Stage of Labour*	*Experimental*	This study was conducted in a sample of 60 (30 experiments and 30 controls) primigravida women during the active phase of the first stage of labour admitted to the delivery hospital room of Dr. Prabhakar Kore Belgaum	The experimental group listens to relaxing music	The control group is not exposed to music	Prabu Penguin Data was collected using the standard visual analog pain (VAS) scale and Zung's self-assessment anxiety scale. The data obtained were analysed using descriptive and inferential statistics.	There were significant differences in the experimental group, with the difference in mean pain scores pretest (5.29) lower than posttest as well as the anxiety score in the intervention group (33.32) points lower between the pretest and posttest. Unpaired T test results Average post-pain test score between the experimental and control groups was 46.375 (P <0.001). And the average post-anxiety

							test score 40.01 (P <0.001).
6	**Nisha John,Annal Angeline (2017)**	*Effectiveness of Music Therapy on Anxiety and Pain among Mothers During First Stage of Labour in Selected Hospitals at Kollam*	*Quasi experimental – pretest posttestwith controlgroup design*	The sample consisted of 60 mothers in the latent phase of the first stage of labour (purposive sampling)	Experimental group listened to musical instruments (violins) for 1 hour in the latent phase of labour.	The control group was not exposed to music during the period.	The tool used is Spielberger's State Trait Anxiety Inventory and Numerical Pain Intensity Scale. Music therapy is given to the experimental group for a period of one hour. Pre and post tests for anxiety and pain assessment tests were obtained in both groups. Data were analysed using descriptive and inferential statistics	There was a decrease in the mean pain and anxiety scores in the intervention group after getting a musical intervention. The research findings show that there is a significant relationship between anxiety and social support, there is no significant relationship between pain and demographic variables, such as age, education, area of residence, religion, family type, parity and social support.

5 CONCLUSION

The six studies looked at the development of midwifery with music therapy interventions for labour pain. On the effectiveness of music therapy for labour pain, four of the six studies did not clearly explain research implications for the development of midwifery and health services. However, two of the studies discussed implications for health services where music therapy can be used to complement conventional pain management strategies. The basic implication is that music therapy is a safe, non-invasive and effective alternative therapy to improve maternal well-being during labour. Thus, the results of the literature review encourage the recommendation for music therapy as the treatment of choice for dealing with labour pain and anxiety along with midwifery care at health facilities.

REFERENCES

Beigi, N.M.A., Broumandfar, K., Bahadoran, P., & Abedi, H.A. (2010). Women's experience of pain during childbirth. *Iranian journal of nursing and midwifery research, 15*(2), 77.

Hasan, M.S., Alsaadi, Z.A., Abbas, M.A., & Algoraby, J.M. (2016). Awareness and attitude of pregnant women towards labor analgesia in Babil Province. *Medical Journal of Babylon, 13*(1), 95-104.

Hosseini, S.E., Bagheri, M., & Honarparvaran, N. (2013). Investigating the effect of music on labor pain and progress in the active stage of first labor. *Eur Rev Med Pharmacol Sci, 17*(11), 1479-1487.

John, N., & Angeline, A. (2017). Effectiveness of Music Therapy on Anxiety and Pain among Mothers During First Stage of Labour in Selected Hospitals at Kollam. *International Journal of Nursing Education, 9*(2), 24-29.

Karkal, E., Kharde, S., & Dhumale, H. (2017). Effectiveness of music therapy in reducing pain and anxiety among primigravid women during active phase of first stage of labor. *International Journal of Nursing Education*, 9(2), 57-60.

Nayak, D., Rastogi, S., & Kathuria, O.K. (2014). Effectiveness of music therapy on anxiety level, and pain perception in primipara mothers during first stage of labor in selected hospitals of Odisha. *IOSR Journal of Nursing and Health Science (IOSR-JNHS) e-ISSN*, 2320-1959.

Prawirohardjo, S. (2008). Ilmu kebidanan[midwifery]. *Jakarta: Yayasan bina pustaka sarwono prawirohardjo*.

Ramesh, B., Sundar, S., Jayapreetha, R., Samal, S., & Ghose, S. (2018). Effects of culture-based chants on labour pain during the latent stage of labour in first-time mothers: A randomized controlled trial. *SBV Journal of Basic Clinical and Applied Health Science*, 2(1), 16-19.

Romano, A.M., & Lothian, J.A. (2008). Promoting, protecting, and supporting normal birth: A look at the evidence. *Journal of Obstetric, Gynecologic & Neonatal Nursing, 37*(1), 94-105.

Tournaire, M., & Theau-Yonneau, A. (2007). Complementary and alternative approaches to pain relief during labor. *Evidence-based complementary and alternative medicine*, 4(4), 409-417.

Xavier, T., & Viswanath, L. (2016). Effect of Music therapy on Labor Pain among Women in Active Labor Admitted in Tertiary Care Hospital Kochi. *Int J Intg Med Sci, 3*(11), 444-448.

21st Century Innovation in Music Education – Astuti et al. (eds)

Effects of music therapy on decreasing pain in post-partum mothers with Caesarean sections (meta-analysis study)

S. Mawaddah
Poltekkes Palangka Raya, Palangka Raya, Indonesia

ABSTRACT: Labour with a Caesarean section often gives pain for several weeks due to incisions in the abdominal area. Pain management therapy is pharmacological and non-pharmacological. Music therapy is recognised as one form of complementary therapy that has the aesthetic aspects of music. The aim of the present study is to summarise ten journal articles about the effect of music therapy for post-Caesarean mothers. The study uses a meta-analysis method involving journal articles that discuss the effects of music therapy for post-Caesarean mothers. The results show that the ten journal articles in the meta-analysis summarise the effects of music therapy on the reduction in pain levels in the post-Caesarean section. It can be stated that music therapy can help reduce the level of pain in the post-Caesarean surgery because pain impulses sent to the brain are hampered by the impulses of the classical music that are also sent simultaneously to the brain.

1 INTRODUCTION

Labour is the process of expelling the results of conception that can live outside the uterus through the vagina to the outside of the body. The Caesarean section is a form of childbirth by performing a surgical incision that pierces a mother's abdomen and uterus to remove one or more babies, and this is done when normal vaginal delivery could lead to complications.

The number of section Caesarean section operations in the world has increased sharply in the last 20 years and the World Health Organization (WHO) estimates the number of deliveries by surgery is around 10–15% (World Health Organization, 2004). This treatment is carried out because of medical indications and non-medical indications. Medical indications consist of two factors, to wit *maternal factors* that are the age of the mother, the pelvic bone, previous labour history with Caesarean section, factors of the muscular canal, uterine contraction abnormalities, premature rupture of membranes and fetal factors consisting of fetal macrosomia, infant location abnormalities, placenta factors, umbilical cord abnormalities and twins.

For reasons of non-medical indications or because of requests from the patients themselves, condition such as fears of vital distress or labour more than six hours are not restrained. Previous experience of poor vaginal delivery and concerns about labour vaginal delivery will damage sexual relations. Labour with a Caesarean section often causes pain for several weeks due to an incision in the abdominal area. Pain management to deal with the pain is pharmacological and non-pharmacological therapy. Pharmacological treatments are done by giving drugs such as analgesics, non-narcotic analgesics and non-steroidal anti-inflammatory drugs. Meanwhile, non-pharmacological treatments can be done in other ways such as with music therapy and massage. Music therapy is recognised as one form of complementary therapy which has the aesthetic aspects of music therapy. It also has therapeutic aspects that are used to help cure, soothe and improve physiological conditions. Music will directly affect the hypothalamus and stimulate the production of endorphins, which is the body's natural opium. It will also provide a relaxing effect so that it can reduce pain.

Music and pain have an important similarity that can be classified as input and output sensors. Input sensors mean that when music is heard, signals are sent to the brain when pain is felt. If the vibrations of music can be brought into response close to the vibration of pain, then the psychological perception of pain will be changed and eliminated. Music produces changes in consciousness through sound, silence, space and time. The effectiveness of classical music therapy in reducing pain is caused by music providing a calm atmosphere, physically and mentally, through regular rhythms and gentle strains that are in harmony with the pulse. This causes the effects of distraction on the mind about pain, reduces anxiety, stimulates a more regular breathing rhythm, and decreases body tension and increases relaxation.

2 METHOD

The study uses a meta-analysis research method involving ten journal articles that have been reviewed using descriptive methods and *ex-post facto* methods. The ten journal articles are reduced to five by way of theme selection. The selected five journal articles have the theme of music therapy to reduce pain management in post-Caesarean section *post-partum*. Each of the journal articles has a different variable. The data are secondary data and are collected through library research, document studies and journal reviews. The data are analysed by descriptive analyses.

3 RESULTS

The journal article entitled *Effectiveness of Classical Music Therapy on Decreasing Intensity in Primipara's Mother Post Section Caesarea Operation in the Flamboyan Room of Regional Public Hospital Prof DR. W. Z. Johannes Kupang* was reviewed by Here et al. (2015). The study used the quasi-experiment research method involving a sample of 50 respondents (25 respondents in the control group and 25 respondents in the intervention group). The sample was taken by an accidental sampling technique. The results of the statistical test show that value of p (2-tailed) = 0.000 (α 0.05) where the mean of pain reduction in the intervention group (2.04) was greater than that in the control group (0.60). The conclusion is that classical music therapy can reduce pain in the mothers' Caesarean section post-operatively.

The journal article entitled *The Effects of Music Therapy on Pain Management in Patients in Section Caesarea Post Operations in the Delima Room of Pasar Rebo Hospital in 2013* was reviewed by Yustina (2013). The study used a quasi-experiment research method of pretest/post-test control groups. This involved a sample of 42 respondents (21 respondents in the control group and 21 respondents in the intervention group). The statistical tests used paired samples t-test and Mann-Whitney U-test with p-value = 0.002. It was concluded that there is an effect of music therapy on pain management in post-Caesarean sections.

The Effect of Music Therapy on Pain Levels of Patients in Post-Caesarean Operations at Surabaya Islamic Hospital was studied by Handayani (2015). The study in this journal used a true experimental research method of the pretest/post-test control group design by random allocation involving 24 respondents (12 respondents in the control group and 12 respondents in the intervention group). Results show that the independent t-test with a p-value = 0.024 was significant. It was concluded that there was an effect of music therapy on maternal pain post-Caesarean section.

The journal article entitled *The Effect of Classical Music Therapy on Pain Reduction in Post-Caesarean Patients in Kenanga Ward Karanganyar Hospital* was examined by Sari. (2016). This study used a pre-experimental method of the one-group pretest/post-test design, with a sample of 40 respondents taken by an incidental sampling technique. The results show that the two-related sample test Wilcoxon statistical test was significant with a p-value = 0.000. It was concluded that there was a significant influence of classical music therapy on the reduction of pain levels in post-Caesarean sections.

The journal article entitled *Guided Imagery and Music (GIM) Reduce Pain Intensity of Post Section Caesarea Patients Based on Roy Adaptation* was examined by Suarilah et al. (2014). The study used the quasi-experiment research method of non-random control group pretest/

post-test with a total of 30 respondents (15 control group respondents and 15 intervention respondents) as the sample. Respondents were taken by a purposive sampling technique. The results show that the independent *t*-test value is 0.0027 with a *p*-value ($p < 0.05$). The conclusion was that GIM was able to reduce intensity of post-Caesarean section pain.

Effectiveness of Islamic Music Therapy Against Decreased Pain Intensity in Post Section Caesarea in Puri Husada Tembilahan Hospital was examined by Sari (2018). The study used the quasi-experiment design method of the one-group pretest/post-test type with a sample of 35 respondents, taken with a total sampling technique. With a non-parametric statistical test using the Wilcoxon test with a confidence level of 95% ($\alpha < 0.05$), the obtained *p*-value was 0.000. It was concluded that Islamic music therapy was effective in reducing maternal pain intensity post-Caesarean section.

The journal article entitled *The Effect of Classical Music Therapy on the Decreasing Scale of Post Section Caesarea Patients* was reviewed by Astuti and Merdekawati (2016). The study used the quasi-experiment research method by the one control group design involving 36 respondents. The sampling technique was purposive sampling. The statistical tests results, using Wilcoxon with *p*-value = 0.002, gave a significant *t* value. It was concluded that there was an effect of music therapy on the reduction of the scale of pain in the post-Caesarean section.

The journal article entitled *Pain Management Using Music Therapy in Post Section Caesarea Patients in Pasar Rebo Hospital in 2013* was examined by Irmawaty and Ratilasari (2013). The study used the quasi-experiment research method of the pretest/post-test control group design. The research subjects were 42 respondents (21 control group respondents and 21 intervention group respondents) sampled using the purposive sampling technique. The statistical tests, using paired samples *t*-test and Mann-Whitney U-test, showed a *p*-value of 0.000. It was concluded that pain management using music therapy was effective in post-Caesarean mothers.

The journal article entitled *The Influence of Mozart's Classical Music Therapy on Pain Intense in Patients Post-Caesarean Section Surgery in RSKD or Dharmais Cancer Hospital Siti Fatimah Makassar's Mother and Child* was reviewed by Ramadhani (2014). The study used the pre-experimental research method of the pretest/post-test design. It involved a sample of 18 respondents, taken by consecutive sampling. The results showed that the statistical test using Wilcoxon obtained a *p*-value of 0.000. It was concluded that there was influence of Mozart's classical music therapy on the intensity of maternal pain after Caesarean section surgeries.

The journal article entitled *The Effect of Music Therapy on Patient Pain Levels after Caesarean Section Operations at Surabaya Islamic Hospital* was investigated by A'in and Fanty (2015). The study used the true experimental research method of the pretest/post-test control group design. A total of 24 respondents participated in the study, taken using the consecutive sampling technique. The statistical *t*-test showed a significant value at $p = 0.002$. The conclusion was that there was an effect of music therapy on the pain level of patients after Caesarean section surgery.

4 CONCLUSION

One effort to overcome pain is to provide non-pharmacological treatments. Pharmacological techniques for reducing pain consist of massage, relaxation and distraction. Distraction is directed to focus the mother's attention on something or make a shift of attention to things other than the pain. Music produces changes in consciousness through sound, silence, space and time. The effectiveness of classical music therapy in reducing pain is caused by the fact that music provides a calm atmosphere, physically and mentally, through regular rhythms and gentle strains that are in harmony with the nerve pulses, causing the effects of distraction on the mind about pain, reducing anxiety, stimulating a more regular breathing rhythm, decreasing body tension, and relaxation. Classical music also reduces alertness to pain because pain impulses sent to the brain are blocked by the impulses of classical music that are also sent simultaneously to the brain. From the ten journals reviewed in this study, on average, the results show that music therapy can reduce the level of pain in the post-Caesarean section because the music impulses sent to the brain will be perceived first before the pain impulses. The results of these studies also show that the post-Caesarean section mothers who are given music therapy have less pain than the post-Caesarean section mothers who are not given music therapy. The advantage of this music therapy is that it is done independently, it does not depend on others, and the success of classical music therapy is entirely determined by self-participation.

REFERENCES

A'in & Fanty, N.M. (2015). *The Effect of Music Therapy on Patient Pain Levels after Caesarean Section Operations at Surabaya Islamic Hospital.* Surabaya: Airlangga University.

Astuti, A. & Merdekawati, D. (2016). *The Effect of Classical Music Therapy on the Decreasing Scale of Post Section Caesarea Patients.* Yogyakarta: UGM Press.

Handayani, N. (2015). *The Effect of Music Therapy on Pain Levels of Patients in Post-Caesarean Operations at Surabaya Islamic Hospital.* Surabaya: Airlangga University.

Here, K.A.P., Batubara, S.O. & Gatum, A.M. (2015). *Effectiveness of Classical Music Therapy on Decreasing Intensity in Primipara's Mother Post Section Caesarea Operation in the Flamboyan Room of Regional Public Hospital Prof DR. W. Z. Johannes Kupang.* Kupang: Kupang University.

Irmawaty, L. & Ratilasari, M. (2013). *ain Management Using Music Therapy in Post Section Caesarea Patients in Pasar Rebo Hospital in 2013.* Jakarta: University of Indonesia.

Ramadhani, R. (2014). *The Influence of Mozart's Classical Music Therapy on Pain Intense in Patients Post-Caesarean Section Surgery in RSKD or Dharmais Cancer Hospital Siti Fatimah Makassar's Mother and Child* . Makassar, Indonesia: UIN Alauddin.

Sari, P. (2016). *The Effect of Classical Music Therapy on Pain Reduction in Post-Caesarean Patients in Kenanga Ward Karanganyar Hospital.* Yogyakarta: UGM Press.

Sari.P. (2018). *Effectiveness of Islamic Music Therapy Against Decreased Pain Intensity in Post Section Caesarea in Puri Husada Tembilahan Hospital.* Jakarta: University of Indonesia.

Suarilah, I., Wahyuni, E.D. & Falupi, R.R. (2014). *Guided Imagery and Music (GIM) Reduce Pain Intensity of Post Section Caesarea Patients Based on Roy Adaptation.* Yogyakarta: UGM Press.

Yustina, E. (2013). *The Effects of Music Therapy on Pain Management in Patients in Section Caesarea Post Operations in the Delima Room of Pasar Rebo Hospital in 2013.* Jakarta: University of Indonesia.

World Health Organization. (2004). Low birthweight: country, regional and global estimates. Geneva: WHO.

21st Century Innovation in Music Education – Astuti et al. (eds)
© 2020 Taylor & Francis Group, London, ISBN 978-0-367-11122-9

Influence of playing motion and songs on the social skills of children in early childhood

E. Yetti
Universitas Negeri Jakarta, Indonesia

V. Anggraini
STIT Diniyyah Putri R.E.Y., Padang Panjang, Indonesia

ABSTRACT: This study was carried out to find out the influence of motion and songs on the social skills of children during early childhood years. The research method used, was a quasi-experimental design with paired data t-test analysis. This study was conducted on 16 respondents aged 5-6 years old. The results showed that t value > t table, sig <0.01 and that there was a significant difference between before and after the children participated in playing motion and song activities that were implemented. This study shows that playing motion and song can significantly influence the social skills of children in early childhood. Further research can be carried out by involving other variables that affect social skills along with other motion and song activities.

1 INTRODUCTION

Early childhood, often called the Golden Age, is usually characterised by rapid changes in physical, cognitive, social and emotional development (Sigelman & Rider, 2014) (Smith, 2018) (Hurd & Anderson, 2017). In order for this period to be passed properly by every child, it is necessary to seek the right education for children, starting in the early childhood years. Various studies state that during early childhood, children experience extraordinary development in all aspects such as intellectual intelligence, emotional intelligence and spiritual intelligence Based on Bloom's longitudinal study, by the age of 4 years the capacity of intelligence has reached 50%, 80% by the age of 8 years and 92% by the age of 13 years. Early childhood is the period of maturity of physical and psychological functions that make it possible for a child to respond to stimulus from the environment. This period is the time to lay the groundwork for developing the physical (motor), intellectual, emotional and social skills, as well as language, art, spiritual and moral aspects.

Early childhood is unique and full of surprises and dynamic. Children are typically full of curiosity, always exploring, playing with words, learning, always evolving along as part of their development and growth. The world of children is so colourful that ups and downs indeed occur in children's behaviour. Along with the child's development there is a process of change in which the child learns to master higher levels of movement, thinking, feeling and interacting, either with others or with objects in their environment such as games, children's songs and more (Freud, 2018)

One aspect mentioned above is social development. Social development in early childhood is related to children's ability to socially interact with peers, adults and the wider environment. However, abnormal social development of children or the lack of children's social skills will cause fatal problems in the future (Littlefield, Cavanagh, Knapp, & O'Grady, 2017) (Coplan, Ooi, Xiao, & Rose-Krasnor, 2018)

Children's problems such as poor social skills can make children spoiled, shy and behave aggressively, negatively and destructively (Dollar, Perry, Calkins, Keane, & Shanahan, 2018). To prevent such problems, teachers of Early Childhood Education should always strive to improve the quality of learning so that learners can enjoy enjoyable learning activities.

Furthermore, in order to improve the social skills of children, learning ought to be interesting and fun so that children are more excited and motivated to follow the learning activities. One way that can improve children's social skills through good and right methods is by playing motion and children's songs (Lavrysen et al., 2017) (Smith, 2018).

According to Stirrup, Evans and Davies (2017), motion and song play activities are very tightly attached and cannot be separated, especially in providing learning to young children. The results of research conducted by Moore (2017) show that incorporating motion and song play in learning activities can develop the personality and potential improvement of children's social skills to the fullest. Learning that involves motion and song is a playing-while-learning and learning-while-playing activity. Activities performed through playing motion and song are expected to please the child and touch upon the development of language, sensitivity to the rhythm of music, motor development, confidence, as well as train children to improve social skills (Trevarthen & Malloch, 2018).

This paper describes and explains how to improve social skills of children during the early childhood of 5-6 years old through playing motion and song. The research results are expected to make a very valuable contribution to the development of Educational Science especially in the field of Early Childhood Education, particularly for teachers of Early Childhood Education in assisting teachers in improving children's social skills through playing motion and songs during the learning process.

1.1 *Social skills*

The development of children's social skills is strongly influenced by the condition of the children and their social environment of parents, peers and the surrounding community. Combs and Slaby (1977) stated that, "social skill is the ability to interact with other in a given social context in specific ways that are socially acceptable or valued and at the same time personality beneficial, mutually beneficial, or beneficial primarily to other" (Cartledge, G., & Milburn, 1980). However, for young children to develop good social skills, or what is sometimes known as prosocial behaviour, they need to feel good about themselves and to feel a certain amount of autonomy and independence. This explains that children learn to develop social skills well when they feel comfortable and free to explore their feelings. Through freedom in associating with others, children will easily interact, yet the freedom should be a responsible one.

Lynch and Simpson (2010) explained that social skills are behaviours that promote positive interactions between others and the environment. Some of these skills include showing empathy, participating in group activities, generosity, helpfulness, communicating with others, negotiating and problem solving. This means that social skills can be introduced and developed during early childhood: interacting with friends, sharing and helping, working together, communicating, solving problems, as well as understanding rules and expressing their feelings. Nancy J. Patrick 2008:24; Cummings, 2014) stated that, "Social skills are the capabilities that we are expected to use to interact with others in our society. They are based on the social norms of our society and they tell us what attitudes and behaviours are normal, acceptable and expected in a particular social situation". Thus, social skills are the ability to positively interact with others. In early childhood it is expected that after learning social skills, children will be able to accept the prevailing social norms in society and be able to behave in ways that are considered normal and acceptable in certain social situations.

Brewer (2007, 23) in Littlefield et al. (2017) explained that what is meant by social skills is the, "ability to assess what is happening in a social situation; skill to perceive and correctly interpret the actions and needs of the children on the group at play; ability to imagine possible courses of action and select the most appropriate one". This statement identifies social skills as the ability to judge what happens in social situations, as well as emphasises the skills to understand what is right and what is to be done to others. Children learn social skills when teachers give these skills the same attention and focus that they give to academic subjects. Children can learn about social skills when a teacher provides examples through learning, behaviour and daily attitudes. "Social skills are the mechanisms through which individuals make contact, initiate, engage, and follow through in reciprocal social interchange. They are actual tools or skills needed for the socialisation process to move forward and for meaningful relationships to be attempted and to succeed. For human beings, social competence provides the capacity to learn socialization skills, that is, the "legs" on which individuals can move forward, while social skills are the tool and techniques, the actual process of "learning to walk".

Referring to all of the above-mentioned studies, researchers conclude that social skills are skills or abilities in fostering and building relationships with others in communication and interaction terms, and in creating good relationships through attitudes, speech and behaviour. Aspects of social skills that need to be developed in children since early childhood include cooperation, communication, empathy, understanding of rules and responsibility. Eventually, by training and developing children's social skills since early childhood, it is expected that the children will be able to survive any condition in their environment.

1.2 *Playing motion and song*

Playing motion and song is a fun tool for children, since children can move while listening to music. According to Howard and Justine (2011), "in play, children have the opportunity to learn about themselves and others. They become". Moreover Gardner (1983) asserted that motion or kinesthetics is an ability that involves a feeling of giving awareness against position of motion by the control from brain. Kinesthetic intelligence is related to the body movements produced by the brain in the form of knowledge about the regulation of gestures. Fatur (2010) further stated that providing stimulus through music and motions for children in their early childhood should consider some principles, such as maturity, sequence, motivation, experience and practice. Motion is a medium to express power, sadness, anger, enjoyment, etc. Motion is also the expression of freedom. Motion will be more expressive when combined with music which is personally interpreted by the children.

Song is a composition of tone or voice in sequence, combination and temporal relationship (usually accompanied by a musical instrument) to produce a musical composition that has unity and continuity (containing rhythm). A variety of rhythmic tones or sounds are also called songs. Songs can be sung in solo, duet (two people), trio (three people) or choir (many people). The notion of music art, is a work of sound art in the form of a song or musical composition expressing the thoughts and feelings of its creator through the elements of music, the elements of which are rhythm, melody, harmony, song structure and expressions. Liao and Campbell (2014) in their research titled, *An analysis of song-leading by kindergarten teachers in Taiwan and the USA* revealed that, "singing plays an important role for the development of children's music and education during their kindergarten experience at the age of 5-6 years. Although kindergarten teachers have limited ability in teaching music, singing is among the most dominant musical activities used in kindergarten. Songs and singing are highly recommended as materials and learning process in kindergarten class. Singing is an activity that enhances developmental skills".

Laird (2015) mentions that music allows humans to express feelings and connect with the feelings of others. This aesthetic component of music provides an emotionally rich environment in which those who listen, perform or write music can relate and communicate with others. According to Bugos and Demarie (2017), the influence of music in early childhood was shown in their study conducted on 36 children aged 3-5 years, consisting of 20 girls, 16 boys, who were recruited from a variety of preschools in large cities in the Southeastern United States. Their study found that aspects of social ability (verbal-based inhibition [Day/Night Stroop]) increased in both research groups through a comprehensive short-term music training program or control condition. According to the researchers, music education programs in early childhood should include activities that are integrated in music, as this can provide an opportunity to learn music that can contribute to cognitive and social development in areas such as social inhibition skills. Involvement in music activities can help maintain children's concentration which is an essential skill as their provision for learning in all domains.

The experience of playing music that is important for early childhood social development has also been reported in a study conducted by Charissi and Rinta (2014). Their research investigated the ways in which young children develop their social and musical skills with music playing activities supported by digital technology in the context of multicultural education. Initial results showed that children develop their musical knowledge as part of their active interaction with others. Children were observed to develop negotiation skills, empathy and inclination to express their thoughts while exchanging ideas through their music games. Initial findings in Charissi and Rinta's research provide evidence that using appropriate music technology in the classroom can be beneficial for children to develop their social and musical skills.

Another opinion comes from the research of Zachariou and Whitebread (2016), stating that music games for young children allow the emergence of self-regulation behaviour. Their research presents observations of ten 6-year-old children as they engage in music play sessions. The results show that music games allow self-regulating behaviour to arise in all possible self-regulation areas. This research can inform a theoretical foundation for the relationship between playing music and self-regulation, as it provides early evidence that music games provide opportunities for the development of children's self-regulating abilities.

The research conducted by Zachariou and Whitebread was supported by Williams and Lewin's (2014) work that showed differences in children's self-regulating skills contribute to the success of early learning and school transitions, as well as children's welfare. Self-regulation refers to the management of emotional, cognitive and behavioural processes that are conducive to positive adjustment and social relationships. The researchers identified that the various domains of learning and development can be enhanced by music training and that the neurological processes responsible for these effects increase. Williams and Lewin argued that coordinated rhythmic motion activities in preschools are effective approaches to supporting the neurological foundations of self-regulation. Related evidence and theories of synchronisation, the cognitive benefits of formal music training and music therapy for the clinical population are discussed in the article suggesting that musical activities can be better used in early childhood education. Williams and Lewin's research concluded that preschool activities designed to stimulate knock synchronisation and motor coordination skills and embedded in group activities can improve children's motor function, hearing and self-regulation.

Motion and song have also been mentioned in other articles as media that can improve some aspects of children's development. According to Hallam (2010) the power of music influences the intellectual, social and personal development of children in early childhood. Hallam's research examined empirical evidence related to the influence of active engagement with music on the intellectual, social and personal development of children and young people. This research used the most sophisticated technology to study the brain, in addition to the study of psychology and quantitative and qualitative education. There is some evidence in Hallam's research that music skills have a positive impact on language development, literacy, numeracy, intelligence, general achievement, creativity, fine motor coordination, concentration, self-confidence, emotional sensitivity, social skills, teamwork, self-discipline and relaxation. The research further found that the positive effects of engaging with music on personal and social development in children only occur when music play is a fun and rewarding experience.

Trainor and Cirelli (2015) examined the synchronous movements of adults as parents of 14-month-old infants and found that parents who move simultaneously (playing motion and song) may improve their 14-month-old children's prosocial behaviour to a higher level than parents who move asynchronously (in motion and song). In general, the findings suggest that interaction between motion and music is an important media in social development during early childhood.

Classroom action research conducted by Cummings (2014) aimed to determine how children use music to help with self-regulation and conflict resolution with peers and presented a curriculum that focuses on delaying gratification, emotional alteration and disclosure through music, that can be used by other teachers in the classroom. The research took place in a Northern California preschool classroom and focused on individual responses of three focus participants selected for their demonstration of low self-regulation skills. The intervention class consisted of warm-up activities, imagination games, rhythm drills, group discussions, puppet shows and songs. The findings of this research were that by the end of the study, the children who were involved in the music intervention class were able to respond to conflict in a more socially appropriate way even if the children selected for the case study did not use scripts or sing the songs they made.

As revealed by Ilari (2015), there are two assumptions that underlie much of the research in early childhood music education: 1) that music is a social endeavour and 2) the participation of children in playing music is beneficial to the social development of the children as a whole. Ilari's research tested the involvement of early childhood music from a social perspective using methods that integrate research from various fields and theoretical orientations.

Based on the opinions of various experts and their research results discussed above, playing music in early childhood is important because it can improve many aspects of children development such as inhibition of inappropriate behaviour, help with self-regulation, encourage empathy and increase prosocial abilities of children through playing motion and song. Therefore, in the following sections of this article, the researchers present classroom action research conducted with children in their early childhood years with the aim to improve their social skills through playing motion and song.

2 METHOD

This study used a quasi-experimental method with data collection techniques via questionnaires on 16 respondents at a kindergarten school in Jakarta, Indonesia. The respondents were children aged 5-6 years old. The samples were obtained using multistage cluster random sampling technique. The data was analysed by using statistical analysis of paired t-tests calculated by the SPSS program.

2.1 *Result*

The results of data analysis on 16 respondents before and after the experiment of playing motion and song activities can be seen in Tables 1, 2 and 3 below:

Table 1. Paired samples statistics.

		Mean	N	Std. Deviation	Std. Error Mean
Pair 1	Pretest	69,3250	16	9,91890	2,47973
	Posttest	85,9187	16	5,33032	1,33258

The data in the table above shows the normal standard error such that the data error rate is low and can be used for paired t test.

Table 2. Paired samples correlations.

		N	Correlation	Sig.
Pair 1	Pretest & Posttest	16	,611	,012

Table 3. Paired samples test.

		Paired Differences							
					95% Confidence Interval of the Difference				
		Mean	Std. Deviation	Std. Error Mean	Lower	Upper	t	df	Sig. (2-tailed)
Pair 1	Pretest - Posttest	-16,59375	7,88412	1,97103	-20,79490	-12,39260	-8,419	15	,000

If t value = 8,419, t table = 2.660, therefore 8,419 > 2.660 → t value > t table. H_0 is rejected means H_1 is accepted. Therefore, there are significant differences between before and after doing dance creativity activities. If using Sig (p-value) Sig. (2-tailed) = 00,1 then, 0,00 < 0,01.

In the results, the researchers obtained 85.95%, meaning that the stimulation of motion and new songs was called a success and was not to be continued to the next cycle. From these activities the researchers obtained various findings of both process and results.

The findings on the process of children's participation in playing motion and song are: a) The children's engagement was evidenced by the children seeming to be happy to follow the teacher's direction and were able to complete certain tasks in the game process. Here, it can be seen that the children's responsibility confidence began to appear; b) The children's response was demonstrated in such a way that the children were very happy playing motion and song and that although at times there were children who did not initially respond, during the next meeting they were very enthusiastic to play together; and c) The children's attention was influenced so that when the teacher explained and provided guidance or steps in playing, the children enthusiastically listened to and followed what the teacher said.

The next finding is on the teacher's action process. While participating in playing motion and songs that can improve the children's social skills, the teachers became: a) facilitators by providing materials and media needed during the playing motion and song process. Teachers also provided stimuli that influenced the children's attention and interest to participate in playing and learning activities; b) motivators, encouraging the children while they performed learning tasks and provided the children with reinforcement when they did not succeed in completing the task and gave appreciation when the children successfully did the task; and c) evaluators, by evaluating the entire process of teaching and learning activities and providing assessment of the results of children's learning and playing activities.

Based on the research findings, playing motion and songs can improve the social skills of children the early childhood period of 5-6 years of age. Learning by playing motion and songs can cultivate children's social skills as the process of playing motion and songs provides opportunities for children to communicate actively and build good relationships with others. Playing motion and songs can also provide a new experience in socializing with the environment (A, 2016)(Kenny, 2016)(Countryman, J., Gabriel, M., & Thompson, 2016).

3 DISCUSSION

At kindergarten age, children begin to adjust (egocentric) with a cooperative or sociocentric attitude (paying attention to the interests of others). Through the development of social skills, children can adjust to groups and peers as well as their surrounding environment. During the learning process in school, the development and maturity social skills can be helped by providing group tasks, both those that require physical work and those that require thinking. This process makes it possible for children to learn about attitudes and habits in working together and to build mutual respect and responsibility.

The development of children's social skills is strongly influenced by the conditions and social environment of children, by both parents and peers and by the surrounding community. This is in line with Sharon A. Lynch and Cynthia G. Simpson (2010), whose research demonstrated that social skills are behavioural and promote positive interactions with others and the environment. Some of these skills include showing empathy, participation in group activities, generosity, helpfulness, communicating with others, negotiating and problem solving. This means that social skills can be introduced and developed beginning in early childhood in various ways such as associating with friends, sharing and helping, cooperating, communicating, learning to solve problems and also understanding rules and expressing their feelings. Playing activities using singing and motion in children aged 5-6 years can provide experiences that spur sensations that can be accepted by the body during the activities and can generate pleasure. Feeling happy while playing activities using songs and movements can develop various aspects of skills development. Fatur (2010) suggests that the provision of stimulation through music and motion in early childhood must pay attention to the principles of maturity, sequence, motivation, experience and practice. Motion is a means of expression and diverting strength, sadness, energy, pleasure and so on. Motion also expresses liberating activities and is a very creative process when combined with the music that children interpret according to their own way.

Based on the present research at the kindergarten school in Jakarta, the activity of playing motion and songs for children aged 5-6 years can provide experiences that spur positive sensations and can be accepted by the body by generating pleasure. Feelings of joy and delight in the activity of playing motion and songs can develop various aspects of development and life skills in children. In this study motion and songs can affect children's social skills. In line with the research conducted by Cotugno (2009), Cotugno opined that, "Social skills through individuals make contact, initiate, engage, and follow up on reciprocal social exchange. They are the actual tools or skills needed for the process socialization to move forward and to try and to succeed for humans, social competence provides the capacity to learn socialization skills, that is, 'feet' in individuals can move forward, while social skills are tools and techniques, the real process of "learning to walk".

In this research on playing motion and songs there are several findings related to the attitude of the child and to children's participation. The findings related to children's attitudes are: a) involvement by children when playing and that even though at times there were children who did not want to be initially involved, these children finally enjoyed the game with the teacher's direction; b) curiosity demonstrated when children asked about the steps of the game and eagerly waited for their turn to play; and c) the interest of the children while playing using singing and motion was so expressive and they looked happy.

The findings related to the children's participation in playing activities using motion and songs included: a) children's participation such that they seemed happy to follow the teacher's direction and were able to complete certain tasks in the process of playing the game. What was observed is the responsibility of the children and their confidence; b) the response of the children, as they were very happy while playing motion and songs. Although at times there were children who did not show their response, at the next meeting these children were very enthusiastic to play together; and c) the children's attention was noticeable. When the teacher explained and gave guidance or steps in playing, the children were enthusiastic to listen and follow what the teacher said.

The final set of findings is about the teacher's action process of participation in the motion and song activities that improved the children's social skills as: a) facilitators who provided the materials and media needed for the playing activities and songs. The teacher also provided stimulus that affected the children's attention and interest in participating in play and learning activities; b) motivators who gave encouragement to the children while they worked on the tasks and who also provided reinforcement to the children when they did not succeed in completing the task and gave rewards when the children successfully carried out the task; and c) evaluators who were tasked with evaluating the entire process of learning activities and to safeguard and provide assessment of the results of children's learning and playing activities. Thus, what was found is that motion and song activities provided opportunities for children to communicate actively and to build good relations with others. Playing while singing and moving, of course, can also provide new experiences for socialising with the environment.

4 CONCLUSION

Playing motion and songs can stimulate various aspects of children's development especially social skills development. Through playing motion and songs children can learn to socialise with each other, learn about cohesiveness, learn to control themselves or control emotions, learn to obey rules and learn to appreciate others. Thus, the stimulation of social skills through playing motion and songs can enable children to develop emotional intelligence.

Based on the findings and discussion of the results of the study, it can be concluded that motion and song activities can affect social skills of children during early childhood. In the future, research can be conducted that involve other variables that affect social skills during playing motion and song activities.

REFERENCES

A, B. (2016). *Game bermain anak-anak di era media baru. Dalam Permainan Anak-Anak di Era Media Baru* (Children play games in the new media era. In the Games of Children in the New Media Era). Routledge.

Bugos, J.A., & Demarie, D. (2017). The effects of a short-term music program on preschool children's executive functions. *Psychology of Music*, *45*(6), 855–867. https://doi.org/10.1177/0305735617692666.

Cartledge, G., & Milburn, J.F. (1980). *Teaching social skills to children*. Pergamon Press.

Charissi, V., & Rinta, T. (2014). Children's musical and social behaviours in the context of music-making activities supported by digital tools: examples from a pilot study in the UK. *Journal of Music, Technology and Education*, *7*(1), XXXXX. https://doi.org/10.1386/jmte.7.1.39_1.

Combs, A., & Slaby, L. (1977). *Habilidades Sociales*. New York: Grades.

Coplan, R.J., Ooi, L.L., Xiao, B., & Rose-Krasnor, L. (2018). Assessment and implications of social withdrawal in early childhood: A first look at social avoidance. *Social Development*, *27*(1), 125–139. https://doi.org/10.1111/sode.12258.

Cotugno, A. J. (2009). Social competence and social skills training and intervention for children with Autism Spectrum Disorders. *Journal of Autism and Developmental Disorders*, *39*(9), 1268–1277. https://doi.org/doi.org/10.1007/s10803-009-0741-4.

Countryman, J., Gabriel, M., & Thompson, K. (2016). Children's spontaneous vocalisations during play: aesthetic dimensions. In *Music Education Research* (pp. 1–9). Retrieved from https://books.google.co.id/books?id=UbwtDwAAQBAJ&pg=PT315&lpg=PT315&dq=Countryman,+J.,+Gabriel,+M.,+%26+Thompson,+K.+(2016).+Children's+spontaneous+vocalisations+during+play:+aesthetic+dimensions.+Music+Education+Research,+18(1),+1-19.&source=bl&ots=pFuof.

Cummings, D. (2014). *Using Music as a Tool for Promoting Self-Regulation and Conflict Resolution Skills in Preschool Children*. Sonoma State University.

Dollar, J.M., Perry, N.B., Calkins, S.D., Keane, S.P., & Shanahan, L. (2018). Temperamental Anger and Positive Reactivity and the Development of Social Skills: Implications for Academic Competence During Preadolescence. *Early Education and Development*, *29*(5), 747–761. https://doi.org/10.1080/10409289.2017.1409606.

Fatur, R. (2010). *Cerdaskan Anakmu dengan Musik (Smart Your Child with Music*). Yogyakarta: Diva Press.

Freud, A. (2018). *Normality and pathology in childhood: Assessments of development*. Routledge.

Gardner, H. (1983). *Frames of mind*. New York: Harper Collins.

Hallam, S. (2010). The power of music: Its impact on the intellectual, social and personal development of children and young people. *International Journal of Music Education*, *28*(3), 269–289. https://doi.org/10.1177/0255761410370658.

Howard, D., & Justine, A. (2011). *Play In Early Childhood,From Birth to Six Years*. London: Routledge.

Hurd, A., & Anderson, D.M. (2017). *Kraus' Recreation & Leisure in Modern Society*. Jones & Bartlett Learning.

Ilari, B. (2015). Music in the early years: Pathways into the social world. *Research Studies in Music Education*, *38*(1), 23–39. https://doi.org/10.1177/1321103X16642631.

Kenny, C. (2016). The field of play. In *The Oxford Handbook of Music Therapy*,.

Laird, L. (2015). Empathy in the Classroom. *Music Educators Journal*, *101*(4), 56–61. https://doi.org/10.1177/0027432115572230.

Lavrysen, A., Bertrands, E., Leyssen, L., Smets, L., Vanderspikken, A., & De Graef, P. (2017). Risky-play at school. Facilitating risk perception and competence in young children. *European Early Childhood Education Research Journal*, *25*(1), 89–105. https://doi.org/10.1080/1350293X.2015.1102412.

Liao, Mei-Ying; Campbell, P.S. (2014). An Analysis of Song-Leading by Kindergarten Teachers in Taiwan and the USA. *Music Education Research*, *16*(n2), 144–161. Retrieved from https://eric.ed.gov/?id=EJ1033120.

Littlefield, L., Cavanagh, S., Knapp, R., & O'Grady, L. (2017). KidsMatter: Building the Capacity of Australian Primary Schools and Early Childhood Services to Foster Children's Social and Emotional Skills and Promote Children's Mental Health Lyn. *Social and Emotional Learning in Australia and the Asia-Pacific: Perspectives, Programs and Approaches*, 1–471. https://doi.org/10.1007/978-981-10-3394-0.

Moore, R.C. (2017). *Childhood's domain: Play and place in child development*. Routledge.

Sharon A. Lynch, & Cynthia G. Simpson. (2010). Social Skills: Laying the Foundation for Success. *Dimensions of Early Childhood*, *38*(2), 1–121. Retrieved from https://southernearlychildhood.org/upload/pdf/Social_Skills_Laying_the_Foundation_for_Success_Sharon_A_Lynch_and_Cynthia_G_Simpson_Volume_38_Issue_2_1.pdf.

Sigelman, C.K., & Rider, E.A. (2014). *Life-span human development*. Cengage Learning.

Smith, P.K. (2018). *The relevance of fantasy play for development in young children. In The Development of Children's Imaginative Writing*. Routledge.

Stirrup, J., Evans, J., & Davies, B. (2017). Early years learning, play pedagogy and social class. *British Journal of Sociology of Education*, *38*(6), 872–886. https://doi.org/10.1080/01425692.2016.1182010.

Trainor, L.J., & Cirelli, L. (2015). Rhythm and interpersonal synchrony in early social development. *Annals of the New York Academy of Sciences*, *1337*(1), 45–52. https://doi.org/10.1111/nyas.12649.

Trevarthen, C., & Malloch, S. (2018). No TitleMusicality and musical culture: Sharing narratives of sound from early childhood. Music Learning and Teaching in Infancy, Childhood, and Adolescence. In *An Oxford Handbook of Music Education*.

Williams, Kate E. & Lewin, S. (2014). Self-regulation support through music, 1–6.

Zachariou, A., & Whitebread, D. (2016). Musical play and self-regulation : does musical play allow for the emergence of self-regulatory behaviours ?, *4937*(February). https://doi.org/10.1080/21594937.2015.1060572.

21st Century Innovation in Music Education – Astuti et al. (eds)
© 2020 Taylor & Francis Group, London, ISBN 978-0-367-11122-9

Targeting music business opportunities in the town

R.P Hanangun & A.G.B. Triprasetyo
Indonesia Institute of Arts Yogyakarta, Yogyakarta Daerah Istimewa Yogyakarta, Indonesia

ABSTRACT: Some Indonesians have the perspective that music occupations cannot support musicians. Based on this perspective, this research focused on the struggles of a music department alumnus in earning income from music and built his own music school in his childhood town of Pacitan. This research is based on qualitative research method approach. The data was collected by using in-depth interviewing strategies with the school's founder, Fajar Brian Lasawon, and some of his active pupils. The Royal Potential Music (RPM) School has been operating for three years. Lasawon uses pop music as the teaching material and offers his music education services by using posters, banners and social media such as Facebook, WhatsApp and Instagram to promote his school activities, instead of having a semi-annual student concert. Thus, Lasawon creates opportunities for his institution to develop, furthermore Lasawon can improve his earnings and survive as a musician in Pacitan Town.

1 INTRODUCTION

Today, the number of graduates from schools does not equal the number of those employed. The author Julianto (2017) in one online posting from Kompas.com wrote that there was a surge in the number of unemployed people in Indonesia with 10,000 more unemployed from the previous year. This has led to a variety of problems for graduates especially for individuals with a bachelor's degree in music.

Some parents tend to forbid their children from studying music. The simple reason is that it is feared that a degree in music will not lead to success in old age. Therefore, a degree in music is seen by the general public as not desirable.

What underlies the above paradigm is the idea that the majority of musicians today experience financial problems, which is partly caused by piracy that is rampant and never gets significant attention from the government. In 2006, it was recorded in the American magazine *Cosmogirl,* that 30% of the perpetrators of the free song download phenomenon are Americans (Dwiyoga, 2015). This problem has persisted for the past 12 years which ultimately affects the royalties obtained by musicians. The above-mentioned conditions are in line with what was said by Widy Soediro Nichlany, a professional singer from the Indonesian band Vierratale, in an interview that, "Because here the system is still messy, music, downloads, piracy are still a mess, so not all of us get the sales proceeds". Nichlany added that Indonesian musicians need to have other businesses outside of music, including Nichlany herself, who was developing her fashion business.

Those in the music profession can feel cornered. However, seen more broadly, music has a variety of sub-fields that one can be involved in that does not necessarily involve one working as a musician. Despite the fact that most people see the field of music from one perspective, there are various kinds of music professions. Therefore, it is necessary to know some sub-fields in the field of music. By knowing the sub-fields of music, there will be an increase in the quantity of employment opportunities offered and where career opportunities in the music world are still wide open.

It can be concluded simply that a career in music is not only oriented towards being a musician. Other sub-fields of music can be found, for example, being a businessman, a teacher or other professions that involve music. From the various music professions, the most commonly encountered is becoming a businessman or teacher. A degree in music should provide opportunities for alumni to survive globalisation and find financial security as they age.

Therefore, a case study was conducted with a music bachelor degree student named Fajar Brian Lasawon who shaped his own fate and established a music school called The Royal Potential Music (RPM) in the small town of Pacitan in Indonesia. Pacitan is not a big town nor is it industrial. This research took a case study approach that is intended to provide information about how Fajar Brian Lasawon has been able to adapt to life with the music school he started. The research discusses how Fajar Brian Lasawon, with a bachelor's degree in music, was able to succeed. This study used direct observation of the object and subject and interviews. The problems are how can music majors survive in small towns in the midst of the current pressures of globalisation and what strategies can be used to be successful amidst intense market competition?

2 LITERATURE REVIEW

Heryanto (2012) describes how popular culture continues to grow in Indonesia. There are many kinds of theories and opinions from experts regarding changes that have occurred because of the influences from popular culture. This book provides references for the situation of music in Pacitan. Djohan (2007) discusses the effects of Mozart's music on intelligence and how the general public views music. Therefore, Djohan's book is a reference for how music is developed in the town of Pacitan.

Warindrani (2006) provides an understanding of the laws of economics, market share and business management. The concept of entrepreneurship was used in this research to look at RPM, giving an idea of how one survives market competition and other opportunities.

Lasawon shares the main strategies of RPM and concrete information for carrying out its development called Human Resource. Rupiyono (2009) discusses the management system of a music education institution and its response to consumers. This journal is a main reference in making observations and conclusions about how RPM reciprocates to its consumers.

3 RESEARCH METHODS

The subject of this research involved the case study of Fajar Brian Lasawon and the establishment of his school RPM. Lasawon is a bachelor's degree recipient from the ISI Yogyakarta University. The population of this study covers bachelor's degree music students in Pacitan, a total of six people including the Lasawon. This research sample is Fajar Brian Lasawon with his business venture that is his means to survive in the town of Pacitan, Indonesia.

The research used a qualitative method approach. According to Sugiyono (2012), the qualitative method is an artistic method because the research process is art (less patterned) and the research data is more tangible for the interpretation of the data found.

The process of collecting data was conducted as follows:

1. The research began with an interview of Fajar Brian Lasawon and some of his consumers. According to Rathiman et.al. (2014), the interview is a meeting of two people to exchange information through questions and answers so that meaning can be constructed in a particular topic This is done to historically get the data and to fully and directly describe information from the relevant subject in accordance with the problems in this study.
2. Thorough observations were carried out accompanied by documentation in the form of photos and videos on teaching and learning activities, conditions of the research location and how management strategies in the RPM were applied. According to Marshal & D'Amo (2011), through observation researchers learn about the behaviour, and the meaning of the behaviour . The function of this observation is to match data from the results of interviews and also collect data outside of interviews.

The research flow is outlined in the form of a chart as follows:

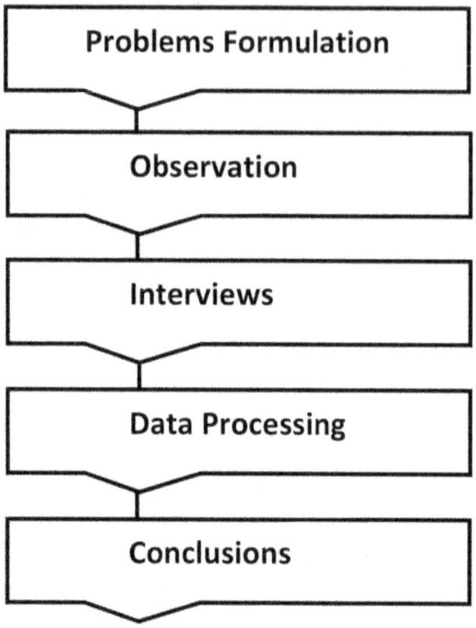

Figure 1. Research flow.

4 RESULTS AND DISCUSSION

4.1 *Result*

The first results show that in the past four years since the RPM was established, there has been an increase in the number of students. See Table 1.

From the results of interviews in 2018, the number of RPM students reached 25 people with fees of Rp. 200,000 per student for every four meetings. If the nominal fee is multiplied by 25 students, Fajar Brian Lasawon earns Rp. 5,000,000. Additionally, because of another accommodation, he receives a net of around Rp. 2,500,000 up to 3,000,000 per month. With the nominal value, Fajar Brian Lasawon felt that it was enough if it was used to meet the needs for a full month in Pacitan.

The second result of this study shows that RPM uses several strategies to increase the number of students. These strategies include:

a. Utilisation of NPSN (National School Number) Certificate;
b. Background of music education;
c. Pop music as a learning curriculum;
d. The use of mass media as a promotional facility;
e. Mid-year scheduled concert;
f. Recruitment of teaching staff who have a formal background in music education.

Table: 1 . Data on the number of RPM students within 4 years.

Years	Number of students
2015 (Before officially becoming RPM)	2-5
2016	7-15
2017	15-21
2018	21-25

4.2 *Explanation*

4.2.1 *Utilisation of NPSN certificate*

NPSN (National School Number) is used to increase the weight of certificates and letters issued by the RPM. Documents issued by the RPM are more valuable because the institution has been recognised by the Education Office. Evidence of such recognition is issued in the form of an NPSN certificate.

The requirements to obtain the NPSN are as follows:

1. Make an application for an operational permit provided that you have at least five students and that the application is addressed to the Education Office;
2. Obtain the NPWP (taxpayer Registration Number/ tax office);
3. Minimum S1 education background that is linear with established educational institutions.

Management documents that are required with the submission of an application for the NPSN:

1. Registration form;
2. First Establishment/Operational Permit;
3. Last Operational Permit that is still valid;
4. Photos of the Chairman;
5. Photo of signpost board institution;
6. Institution's AD/ART;
7. Photos of institution, front view.

Figure 2. NPSN certificate (National School Number) RPM music school (RPM document archive, 2018).

4.2.2 *Music education background*

Since he graduated as a music bachelor's degree from a state college, Fajar Brian Lasawon scored additional points. Consumers will trust products from companies that have competent staff members according to their fields. This happened at RPM so that consumers in Pacitan prefer RPM compared to other organizations where courses are taught by people who do not have a formal education background in music.

4.2.3 *Pop music as a learning curriculum*

Pop music is the music preference of the people in Pacitan. RPM uses this preference as a distinctive feature of learning at its institution. Therefore, Lasawon has an advantage in targeting market segments. This is evidenced by the number of students that always increases each year as shown in the table above.

In addition, Fajar Brian Lasawon uses a backing track approach, play and study and outdoor classes. The backing track is used to accompany students both in class and in performances. Playing and learning emphasises more social interaction and playing in public places along with outdoor classes every two weeks to maintain the enthusiasm of RPM students. Lasawon named the outdoor sessions, *Laras class* (Outside Class Training), where the training ground is a place such as a public square, night market, a beach and so on.

4.2.4 *Utilisation of mass media as a promotional facility*

Student recruitment conducted by the RPM uses billboards erected on the roadside. Brochures and other promotional material are distributed through social media such as:

- Facebook: RPM Music School
- Instagram: @ schoolrpm
- Website: www.rpm.sch.id
- Email: info@rpm.sch.id

4.2.5 *Mid-Year scheduled concert*

To maintain the interest of their wives and channel their talents, RPM holds recital concerts that are held every six months. These recitals are also intended to promote and show learning outcomes to the public. Through recital concerts, it is expected that the level of boredom of students will be reduced, they will receive audience appreciation from stage performances and also attract the interest of prospective students to join the RPM.

Figure 3. Outdoor activities in *Taman Pantai Pancer Door, Barehan, Pacitan* (RPM document archive, 2017).

Figure 4. The first promotion that Lasawon did through his Facebook account (Riva Pugar Hanangun, 2018).

Figure 5. Recital 2 # concert held at Pacitan YABI building (RPM document archive, 2017).

4.2.6 *Recruitment of teaching staff who have a formal backgrounds in music education*
To minimise competitor risks and improve the quality of the institution, Fajar Brian Lasawon recruited scholars in the field of music who were confused about how to get work. Thus, several mutually beneficial agreements were made between RPM and employees in wasy that encouraged prospective teachers to agree with several provisions.

4.3 *Discussion of RPM management constraints and solutions*

As a non-formal music education institution, RPM is faced with several obstacles in carrying out its work program. These constraints are of course due to several technical factors that might clash with conditions in the field. RPM faces several obstacles that are apparent. The concerns and constraints currently at the RPM are:

a. Worries related to if the number of students decreases;
b. Concerns about not achieving the RPM's vision and mission;

Figure 6. Some staff members who assist at RPM (RPM document archive, 2018).

c. Limitations of competent teaching staff according to the linear path of education;
d. Facilities are not standard according to SOP;
e. Often there is a lack of funds for holding recitals;
f. It is difficult to carry out competency tests because there is no current competency test in Pacitan;
g. Advertising and promotions have not spread widely to remote villages and are only limited to the Pacitan District which is the center of Pacitan Town.

The above concerns and constraints do not mean there is no solution. Currently RPM is preparing several kinds of solutions that will be used to overcome existing and potential problems. These solutions have been planned and prepared with full calculation even though overall, they have not been optimally realised. There are several solutions that have been prepared by the manager to deal with the obstacles that currently exist in the RPM:

a. Improving the quality of teachers and quality of learning;
b. Temporarily recruiting assistant teachers (such as Fendi Dwi Cahyono, S.Kom.);
c. Accumulate funds from institutions for additional facilities;
d. Consider raising the monthly tuition fee according to the advice of some parents of students;
e. Provide input through the regional government. If this is not successful, the last step is to take the competency tests independently outside Pacitan, the costs of which would be borne by the RPM;
f. Enhancing the marketing strategy for the future with new methods.

Thus, the constraints at the RPM must be addressed and the solutions must be implemented. With the existence of these problems, it is hoped that RPM can further develop its quality and provide the best learning experience for its students. Even with the abovementioned problems, the RPM will maintain its existence and its reputation in the business market realm of music education institutions.

5 CONCLUSION

Base on the disccusion, it can be conclude that bachelor of music majors who live in small towns such as Pacitan, can work in the music education sector by serving as a teacher in a formal institution and by also establishing a music education institution such as what has been accomplished by Fajar Brian Lasawon. By adhering to this research data and the RPM institution as a sample, it can be concluded that the music business in the form of an educational institution can be used as a foundation for life in small towns especially in Pacitan.

There are strategies that can be taken to survive amidst market competition such as:

a) Obtaining an official license from the Education Office so that the institution is recognised with NPSN as evidence;
b) Hire staff members with educational backgrounds that are linear with business fields. This often convinces consumers to choose these institutions;
c) Develop and use learning materials that are appropriate and in accordance with the market preferences. In this case, RPM chose to use pop music because it was judged to be in accordance with the music preferences of the majority of the Pacitan people;
d) Utilise social media. The effective use of social media and mass media can be good promotional tools;
e) Schedule recital concerts that can then be used as a promotion. Recitals can also be aimed at maintaining the interest and reducing boredom of students while channeling their talents and showing the results of their training.
f) Recruit music professionals who are experts their fields. This can improve teaching quality and reduce competition from other organizations.

REFERENCES

Dwiyoga, M. (2015). *Preferensi Musik Hip Hop Pada Remaja Studi Kasus: Siswa Smp Negeri 13 Yogyakarta Dan Smp Negeri 1 Gedangsari Gunungkidul* [Hip-Hop Music Preferences in Youth; a Case Study of SMP 13 Yogyakarta and SMP Gedangsari 1 Gunung Kidul Students]. (Doctoral dissertation, Institut Seni Indonesia Yogyakarta).

Heryanto, A. (2012). Kewarganegaraan dan Etnis Cina dalam Dua Film Indonesia Pasca-1998. *dalam Budaya Populer di Indonesia, ed. Ariel Heryanto*[Nationality and Chinesse ethnic in Indonesia's films pasca 1998 in Popular Culture in Indonesia - Degrading Identity on Post-New Orde]. *Yogyakarta: Jalasutra*, 110-113.

Marshal, J., & D'Adamo, K. (2011). Art practice as research in the classroom: A new paradigm in art education. *Art Education, 64*(5), 12-18.

Julianto, P. A. (2017). Agustus 2017, Jumlah Pengangguran Naik Menjadi 7, 04 Juta Orang[the number of unemplotment reach 7,04 Million]. Accessed on 19 august 2017.

Rathinam, S. R., Babu, M., Thundikandy, R., Kanakath, A., Nardone, N., Esterberg, E., & Weinrib, R. (2014). A randomized clinical trial comparing methotrexate and mycophenolate mofetil for noninfectious uveitis. *Ophthalmology, 121*(10), 1863-1870.

Rupiyono, L. A. (2009). Kursus musik bagi anak: Antara idealisme dan komersialisme [Music Course for Children: Between Idealism and Commercialism], Journal of Music Art, Vol. 1, No. 2. Satya Wacana Christian University: Salatiga.

Salim, D., & Gusmian, I. (2007). *Matinya Efek Mozart*[The death of the Mozart Effect]. Yogyakarta. Galang Press.

Sugiyono. (2012). *Metode Penelitian Pendidikan Pendekatan Kuantitatif, Kualitatif, dan R&D*[Quantitative, Qualitative Research Methods and R & D]. Bandung: ALFABETA.

Warindrani, A. K. (2006). Akuntansi Manajemen. *Yogyakarta: Graha Ilmu[*Management Accounting.] Yogyakarta: Graha Ilmu.

21st Century Innovation in Music Education – Astuti et al. (eds)
© 2020 Taylor & Francis Group, London, ISBN 978-0-367-11122-9

Integrating music and visual art in batik Tembe Renda

Fathurrahman & I.K. Sunarya
Yogyakarta State University, Yogyakarta, Indonesia

ABSTRACT: *Batik* is a cultural product that is loved by the Indonesian people. However, they do not necessarily understand the meaning of batik.This misunderstanding is extended to other products, suchs as *tembe renda*, a woven product found in Bima, NTB(Nusa Tenggara Barat,Indonesia. *Tembe renda* is also considered as *batik* although *batik* and weaving are two different techniques. Therefore, this study aims to provide an understanding so that people can become more familiar with *batik* and weaving. Creating a *batik* motif with *tembe renda* weaving is called *batik tembe renda*. This work is made with the traditional Bima musical concept. Thus, in addition to discussing *batik* and weaving, this work also aims to foster a sense of community through the love for traditional music as local wisdom.

1 INTRODUCTION

Indonesia is a country that is rich in culture. E.B. Tylor describes that culture is a complex whole, including knowledge, beliefs, arts, laws, morals, traditions and various abilities and other habits that are acquired by humans as members of society (Altarriba, 1993). One area where culture is conveyed is in art. It is the responsibility of every community to safeguard and maintain art in order to maintain the existence of culture as identity and national identity (Kim & Ko, 2016). There are numerous cultures in Indonesia, so it is important for every ethnic group to honor and respect each other's culture and distinguishing one tribe from another. The country is the one owner of a great culture that is formed from the plurality of cultural patterns of many different ethnic groups (Suyatno, 2014).

One of Indonesia's striking and well-known arts is *batik*. *Batik* is a national culture art form (Ismail, Wiyantoro, Meutia & Muchlish, 2012) that originates from Java, *batik* being a Javanese term (Nurhaida, Noviyanto, Manurung & Arymurthy, 2015). *Batik* has developed over a long period of time (Haake 1989). On 2 October 2009, *batik* in Indonesia was designated as one of the masterpieces of UNESCO's Oral and Intangible Heritage of Humanity. National Batik Dayis commemorated every year on 2 October. *Batik* is recognised by UNESCO as involving the culture of the whole technique, technology and development of motifs (Musman & Arini, 2011). From this description, it is clear that *batik* is not only about motifs, but the technique and technology of making motifs. The *batik* technique involves the use of wax that is heated and melted, then inscribed on fabric with the aim to cover the whole fabric (Nurhaida, et al., 2015). After the cloth is covered with wax, the cloth is then coloured with a liquid dye. The parts of the cloth that are covered with wax is not be exposed to colour dye, which is important in the actual creation of *batik*.

Depending on the techniques used, *batik* is divided into two types: *batik tulis* and *batik cap* (Anugraha, Sutan & Mufidah, 2015). *Batik tulis* is said to be high-quality batik because of its more complex manufacturing process (Musman & Arini, 2011) that involves making motifs manually by hand using a tool called *canting tulis*, a technique similar to writing. The making of *batik cap* is simpler and uses a *canting cap* whose motif is already made and is stamped onto the fabric (Sutari, Yekti, Astuti, and Sari, 2015).

2 LITERATURE REVIEW

2.1 *Batik*

Batik is classified into two types: actual *batik* and *batik* patterned textiles. Actual batik is batik that is made through a technique using wax, colouring, until the *pelorodan* stage (Wulandari, 2011). In contrast, *batik* patterned textiles are made with printing techniques using modern technology without the original process of using wax with motifs that resemble *batik* (Astuti, 2012). This type of fabric is called *batik* printing. *Batik* printing production is done to meet market needs. However, *batik* printing production has a negative side, because the reality and meaning of batik as a culture (Smiers, 2009) is not known or understood. Modern industries also bring negative social impacts (Dissanayake, Perera & Wanniarachchi, 2017) because the presence of printing *batik* diminishes traditional *batik* craftsmen's income, and therefore makes it difficult for the new generation of *batik* craftsmen.

The idea of *batik* is known to most people, especially outside Java, as limited to *batik*-patterned cloth or *batik* printing. This is caused by the lack of information and references made available to most people. Unfortunately, this is seen in the lack of public interest in preserving *batik* as national culture which is limited by a true understanding of the meaning of *batik*.

2.2 Tembe renda *bima NTB indonesia*

Tembe renda is one of the woven products from Bima NTB that has distinctive motif and color characteristics. The *tembe renda* has as many as 24 different motifs. The colour of the *tembe renda* motif consists of bright colours called *tambaha*—the bright colors of the clothes in Bima symbolise health and prosperity (Hitchcock, 1991). Colour in Bima belief has its own philosophical meaning; *monca* (yellow) means a symbol of glory and greatness, *kala* (red) means bravery in defending the truth, *lanta* (white) means a sincere heart, *jao* (green) symbolizes prosperity, *owa* (blue) depicts a calm life and peace, *keta* (purple) represents a person who is patient and steadfast in carrying out duties and *me'e* (black) symbolises a life that highly respects the earth and the environment as a source of life and prosperity (Ismail & Malingi, 2010). Here are some examples with various motifs.

Figure 1. *Tembe renda* with *anti fare* (rice) pattern and *pariah* (pare fruit).

Figure 2. Tembe Renda motif *fu'u haju* (tree) and *kakando* (bamboo shoots).

3 MATERIAL & METHODOLOGY

3.1 *Data*

The data in this study was obtained through observation, interviews and documentation. Based on the data collected, it was found that the misunderstanding of *batik* occurred in almost all regions of Indonesia, especially outside Java, because the practice of making *batik* is hardly found whereas in Java, *batik* is one of the subjects in school (Fathurrahman, 2017). In Bima NTB, people know *batik* only as patterned fabrics, so that woven fabrics with motifs such as *tembe renda* are also considered to be *batik*. There are several examples that show the lack of understanding of Bima people in understanding the differences between batik and weaving, including an article in a journal written by Masita where *batik mbojo* (Bima) is described as consisting of various motifs and functions called *tembe nggoli mbojo* (Masita, 2012). From the text, Masita clearly considers *batik* and weaving as the same where *tembe nggoli mbojo* is referred to in the article as a type of weaving from Bima, but is also likened to *batik* because it is composed of various motifs. Another example of the misunderstanding between *batik* and weaving can be seen in the following photo (Figure 3).

What is seen in Figure 3 is an appreciation for *batik* as national culture, which exclaims "Selamat Hari Batik Nasional" (Happy National Batik Day). Unfortunately, *batik* day is depicted using a piece of *tembe renda* which is a work of weaving and not *batik*.

There are several problems caused by the misunderstanding of batik as weaving. The first problem assumes that the uniform is genuine *batik* cloth. The second problem assumes that the weaving motif on the uniform is a *batik* motif. The third problem is that the uniform used is neither woven nor is it batik. At a glance, the uniform does look like it is made from woven *tembe renda*, but if you look more closely, the uniform is not woven. The *tembe renda* woven pattern was produced with modern technology called weaving printing, where a weaving pattern is printed on fabric in order to simulate the appearance of woven *tembe renda*. One of the producers of *tembe renda* weaving patterned fabrics also label their products *batik* (Figure 5), even though the motifs produced are woven motifs. Indirectly, the labeling has an impact on the community mindset that is increasingly far from the meaning of *batik*.

The examples described above are some of the common mistakes in understanding what *batik* is. These examples are sufficient to prove that public opinion is far from knowing the meaning of *batik*. For that reason, an important focus of this work is to clarify what *batik* is and to not equate it with weaving.

Figure 3. The greeting of the batik day using a weaving *tembe renda* background [Instagram users].

Figure 4. Labeling with the name "Batik".

3.2 *Method*

In order to respond to the lack of public understanding of the culture around *batik*, it is necessary to give real examples of how form and characteristics in *batik* and weaving are different from each other. This work is a product of research and development that aims to increase people's appreciation and understanding of cultural diversity (Miller, Wice & Goyal, 2018). This work focuses on *batik*, because original *batik* is very difficult to find in Bima NTB. This work hopes to provide good education about batik and weaving so that cultural preservation includes the right meaning (Hong, 2018).

3.2.1 *The concept of creation of works*

Batik tembe renda consists of the characteristics of *batik* and *tembe renda* weaving in one work at a time. *Tembe renda* motifs generally tend to be symmetrical and neatly arranged in a straight line so that they seem monotonous (see pictures 1 and 2), while the characteristics of *batik* motifs tend to be asymmetrical and can be made more expressive, flexible and not rigid. For thpse reasons, symmetrical *tembe renda* motifs applied to this *batik* work are designed through a stylisation process or a change in shape that is asymmetrical with a more flexible contour without removing the original motif character.

Tembe renda has about 24 different motifs; the motifs seen in Figures 1 and 2 are only a part of other motifs. All *tembe renda* motifs are applied to this *batik* work, the aim being that the characteristics of *batik* are increasingly prominent because in making *batik* there are no limits in the application of motifs. In contrast to the application of motifs on tembe renda, weavings there are only a few motifs due to certain calculations of the patterns. In addition to applying all motifs, *tembe renda* visual elements in the form of colour are also applied to all of the works so that *tembe renda* characteristics remain visible (Barng, 2017). The application of all *tembe renda* elements is done in an effort to maintain original cultural values (Hye-Joo & Choe, 2018).

The making of *batik tembe renda* was also conceptualised as having a musical theme with the display of traditional Bima musical instrument motifs. The aim was to show that Bima has other local knowledge besides weaving, specifically music (Santos, 2016). In addition, through this music theme, the characteristics of the *tembe renda batik* is displayed through musical note-shaped visualisation along with the paranada lines (Hood, 2016). *Batik tembe renda* work is expected to increase people's love and appreciation of the culture of Bima through the visualisation of designs that represent Bima values (Hyun-Kyung, 2017).Introducing *batik* can broaden the understanding of the art along with applying all motifs, colours and various batik techniques, including *canting tulis*, *canting cap* and brush, and colouring techniques with dyeing and using swab. Besides that, *batik tembe remba* is also made with various *batik* materials, including brown *batik* wax and white paraffin. Each technique of sticking, colouring, and the use of various materials has its own characteristics. For example, *batik* wax is used on the motif part to look neat, while paraffin is used to produce a cracking effect. The use of various materials, tools and various techniques is intended to display the characteristics of thick *batik*.

3.2.2 *Musical theme of* batik tembe renda

Bima has local knowledge relating to the art of music. For this reason, in addition to the main purpose of creating *batik tembe renda* and in order to introduce *batik* and *tembe renda*, the theme of music was used to foster a sense of love for traditional music. The music theme in this work is displayed through the visualised motifs of Bima traditional musical instruments (Bauza, 2018), that consist of hitting, blowing, rubbing and plucking musical instruments.

In Figure 14, there is a musical instrument called *genda* (drum) where the sound coming from the vibration of the skin membrane is caused by being struck (Nugroho, 2014). In Figure 15 there is another musical instrument called *no* (gong) made of metal and played by being struck with a bat. In Figure 16 there is also the Bima traditional musical instrument named *katongga*. This instrument resembles a *bonang/kenong* in the gamelan and is played by being struck with a bat (Prasetyo, 2012). For Javanese and Balinese people, the *genda, no* and *katongga* musical instruments are part of gamelan musical instruments (Afendras et al., 1995). *Sarone* (Figure 17) and *silu* (Figure 18), in Indonesian language are called *serunai*, and are wind instruments that are played by being blown (Nugroho, 2014). The difference between *sarone* and *silu* lies in the material used; the *sarone* uses bamboo and palm leaves, while the *silu* uses wood and silver (Ramadhan, 2015). The next instrument is the violin (Figure 19) (Zopf, 2017). The last instrument in Figure 20 is *gambo* or *gambus* (Anwar, Viridi, Rusdiana, and Kaniawati, 2016), which is played by plucking the instrument's strings.All these Bima traditional musical instruments are applied in the form of motifs and combined with *tembe renda* motifs. In order for the musical theme to be more visible, the pattern of composing the motifs are made to resemble musical notes with a *paranada* line (Wulandari, 2008).

3.2.3 *Creation of the batik* tembe renda *musical theme*

Tembe renda batik is with *tembe renda* weaving motifs and motifs of Bima traditional musical instruments that are presented with musical concepts in the form of musical notes along with *paranada* lines. These motifs are designed through a stylisation process by changing the shape into a motif that has a contour that appears to be flexible and rhythmic. As explained earlier, the purpose of the creation of *tembe renda batik* is to provide education to the public, especially the people of Bima, to understand batik and not equate it with weaving. For this reason, this research will highlight the characteristics of *batik* through various tools, materials and techniques used. The tools used are *canting* writing and stamp *canting*. Stoves and small pans were used to heat the wax,. Attaching cloth, gloves and buckets for colouring, and stoves and cages for *pelorodan* or shedding wax. The fabric used was a 115 x 200 cm white cloth where a pencil was used for drawing the pattern on cloth; *batik* wax and paraffin were used to cover the fabric, *naptol* colouring material (dark blue), *indigosol* (pink, orange, yellow, green, blue and purple), and *rapide* (red) were the colours used. Water glass was used for the *pelorodan* process or the final stage of removing wax from the fabric.

The first step was to make a motif on the fabric with a pencil as a reference. The second step was to cover the motif using *canting tulis* and *canting cap* using *batik* wax so the motif does not crack and look neat. The third step closes the background and makes a *paranada* line using paraffin; this paraffin is used in order to give the impression of cracking.

The fourth step was to do the colouring. The colouring process was carried out in three stages: the first stage was the colouring of the motif with a dab technique using a brush with *indigosol* (pink, orange, yellow, green, blue and purple) and *rapide* (red)—these colouring dyes were chosen because they have a brightness that is the same as the *tembe renda* colours.

The fifth step involved closing or blocking the motif with chocolate waxes so that the previously coloured motif is not exposed to the color used for colouring in the next stages. The sixth step was colouring the second stage, staining the *paranada* line and the background. Before the background is coloured, paraffin was used so that the dye can enter the fabric's pores through

Figure 5. Results of motifs (batik waxes) and background blocks (paraffin).

Figure 6. The first coloring on the motif with a brush.

the cracks in the *paraffin*. The second stage of colouring still uses the same material as the first stage, but the intensity of the colour is faded from the colour of the motif. In addition, this colouring will also display colour gradations, both on the *paranada* line and on the background. The cracking effects and colour gradation is intended to add to the characteristics of *batik*, because the impression of cracks and subtle gradations cannot be displayed in weaving works. To produce gradations, the staining stage is done by using a brush and finger swab.

The seventh step involves blocking or closing the pores of the fabric on the *paranada* line so as not to be exposed to the next stage of colouring using brown wax. The eighth step (Figure 9) is the third staining or the last colouring using a dye technique with dark blue *naptol* material (Setyawati, Darmokoesoemo, Rochman & Permana, 2017).

The eight step is *pelorodan*, the last step, which aims to melt all the wax attached to the fabric. This stage is done using hot water mixed with the water glass, then cleaned and rinsed with water.

Figure 7. The second stage of colouring on the line and background.

Figure 8. Illustration of coloring with *naptol* dyeing technique.

4 THE RESULTS

From the results of the creation of the *tembe renda* batik with a music theme (Figure 10), it can be seen that the motifs and colours of *tembe renda* still dominate but are displayed through the characteristics of *batik* that are thick with a music theme, including motifs of Bima traditional musical instruments. The *batik* characteristics are seen in the stylised motifs in such a way so that it is not as rigid as the original weaving motifs. The motifs are stylised with a contour that seems flexible, as well as a straight line stylised into a wave line to make it seem rhythmic and not rigid. The following is a photo of the *tembe renda batik* with some examples of *tembe renda* motifs and motifs of Bima traditional musical instruments that have been stylised into *batik* motifs.

Figure 9. Tembe Renda batik with a musical theme.

Figure 10. Stylisation of motifs (zigzag).

Figure 11. Styilisation of *fu'u haju* (tree) motif.

Figure 12. Styilisation of *fu'u haju* (tree) motif.

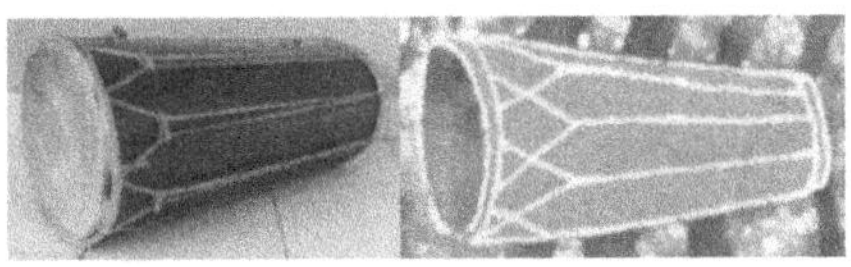

Figure 13. Styilisation of a musical instrument motif (drum).

Figure 14. Styilisation of musical instrument motif *no* (gong).

Figure 15. Stylisation of a *katongga* musical instrument motif.

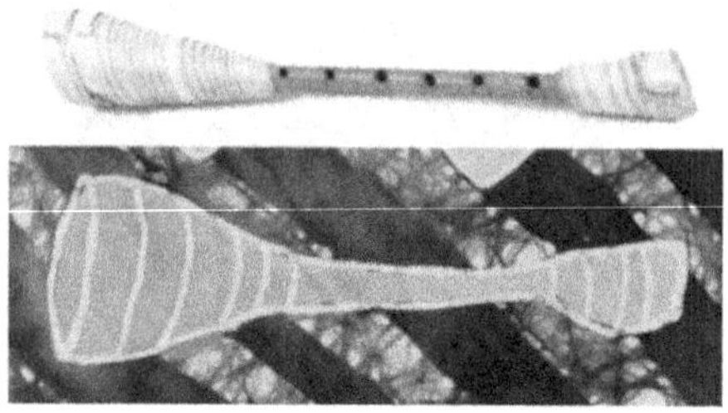

Figure 16. Stylisation of a *sarone* (*serunai*) instrument motif.

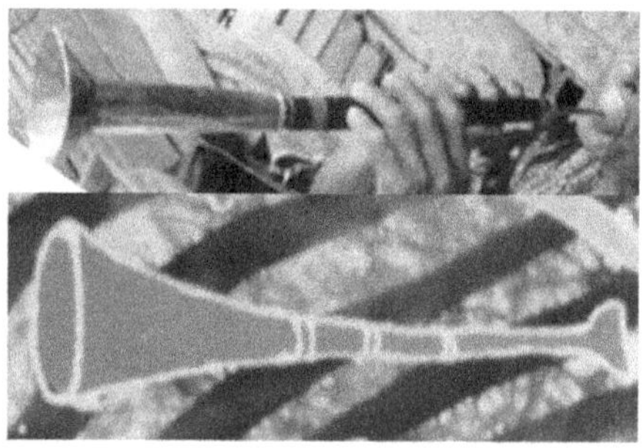

Figure 17. Stylisation of the *silu* (serunai) musical instrument motif.

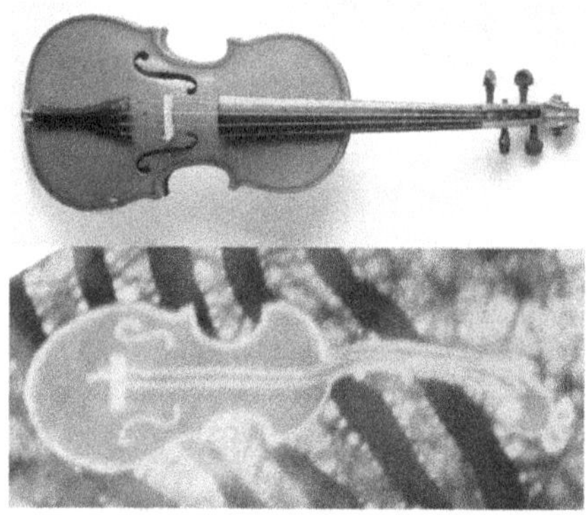

Figure 18. Stylisation of the violin musical instrument motif.

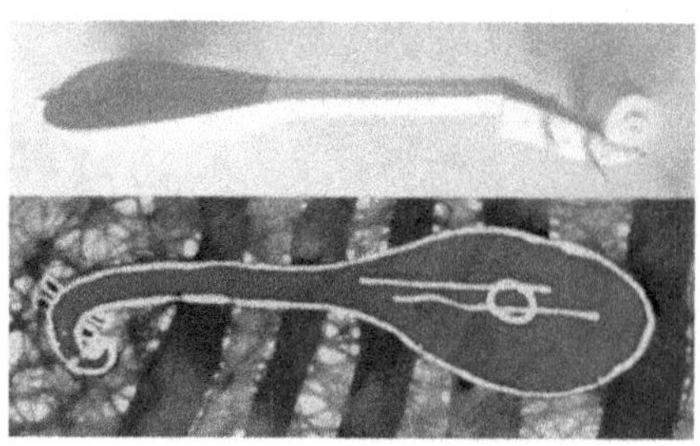

Figure 19. Stylisation of the *gambo* musical instrument motif (*gambus*).

5 CONCLUSION

Tembe renda batik is a representation of Bima NTB's national culture and local wisdom. The purpose of the creation of this work is to provide education to the public, especially the people of Bima to be more familiar with *batik*. Bima society generally believes *batik* is only limited to motifs, even though the essence of *batik* is the unique manufacturing technique of using wax. The lack of understanding of the community can be seen from the assumption that weaving is also *batik* because both have motifs. However, *batik* and weaving involve two different techniques. The weaving called *temba renda* is believed to be *batik* because this type of weaving has many motifs. In response to this incorrect belief, the idea to create a work that represented *batik* and *tembe renda* weaving was conceptualised so that people did not mistake weaving for *batik*. The *batik* work was created using the *tembe renda* motifs and was later called *tembe renda batik*. Through this work, the community will be educated through tangible works that show how the *tembe renda* weaving motifs can be applied in making *batik*. This is so that people can call the textile *batik* from Bima.

In addition to showing *tembe renda* motifs as one of several examples of Bima's local knowledge, the textile was also conceptualised to have a music theme. Adding motifs of traditional Bima musical instruments, the *genda* (drum), *no* (gong), *katongga*, *sarone* (serunai), *silu* (serunai), violin and *gambo* (gambus) represents local knowledge. To further elaborate on the musical theme, the batik was made with the concept of musical notes along with the *paranada* lines.

The creation of the textile (Figure 11) shows that the *tembe renda batik* is still dominated by *tembe renda* colours and motifs but with prominent *batik* characteristics. *Tembe renda* motifs applied to this work are designed through a process of stylisation or change in shape without removing the original form public (Chung & Ryoo, 2018). The motif of a musical instrument is made with a flexible contour and a line of *paranadas* that are initially straight and then stylised into a wave line in order to show the music theme. Motif stylisation is intended to provide a comparison between the characteristics of weaving versus *batik* motifs. The weaving motif in general and *tembe renda* weaving motifs in particular, tend to be stiff because they are symmetrical and different from the *batik* motifs which tend to be flexible. The characteristics of other *batik* can be seen from the application of subtle colour gradations on the *paranada* and background lines. Other than the more prominent *batik* characteristics that cannot be done on woven fabrics, the addition of the impression of cracks in the background also cannot be applied to woven fabrics. All the characteristics of *batik* are displayed through various materials, tools and techniques, each of which has its own character.

The creation of *tembe renda batik* is written in this article with complete steps so that this art form can be practiced in the community as well as used as a reference for schools (Nam, Lee & Kim, 2018). This article hopes to stimulate public enthusiasm for Bima culture and to recognise and appreciate culture (Astuti, Pamadhi, and Rini, 2010). The introduction of culture is more effective if it is done through direct experience (Carla, Emilinda, Devralin, Emilia & Jennifer, 2013).

REFERENCES

Afendras et al. (1995). On 'new/non-native' englishes: a gamelan. Journal of Pragmatics, 24 (3), 295–321.

Altarriba, J. (1993). The influence of culture on cognitive. *Advances in Psychology*, *103*, 379–384.

Anugraha, R. A., Sutan, W., and Mufidah, I. (2015). The design of batik stamp tool scraping working table using ergonomics principles.*Procedia Manufacturing*, *4*, 543–551.

Anwar, K.., Viridi, S., Rusdiana, D., dan Kaniawati, I. (2016). Telaah alat musik tradisional etnik mbojo (gambo) sebagai media pembelajaran fisika. *In the proceding of Simposium Nasional Inovasi dan Pembelajaran Sains (SNIPS*), 481–490.

Astuti, K. S., Pamadhi, H., and Rini, Y. S. (2010). Pengembangan model evaluasi pembelajaran seni budaya SMP. Jurnal Kependidikan, 40 (1), 87–98.

Astuti, S. P. (2012). Pemaknaan batik dalam bahasa budaya, bahasa seni rupa dan bahasa industri dalam perkembangan batik masa kini. Jurnal Dimensi, 9(2), 197–210.

Barng, K. J. (2017). A study on aesthetic treatment through a coloring method and a color symbol in go-ryeo period. *Asia-Pacific Journal of Advanced Research in Music, Art, Culture and Literature*, *2 (2)*, 15–20.

Bauza, Wahiman, J. G. (2018). Assessing poiesis as an effective approach to integrating music and visual art. *Malaysian Journal of Music*, *7*, 20–37.

Carla, J., Emilinda, M. M., Devralin, T., Emilia, V., and Jennifer, A. (2013). Determining the social acceptability of a large scale quarrying project: The case of Barangay San Rafael, Rodriguez, Rizal Province, Philippines. Asia Life Sciences,22 (2), 515–537.

Chung, S. E. & Ryoo, H. Y. (2018). Developing user interface design guideline for flexible display and case studies of its application. *International Journal of Advanced Science and Technology*, *116*, 23–34.

Dissanayake, D. G. K., Perera, S., and Wanniarachchi, T. (2017). Sustainable and ethical manufacturing: a case study from handloom industry. Textiles and Clothing Sustainability, 3 (2), 1–10.

Fathurrahman. (2017). Batik dan Indonesia timur: upaya memecah krisis kebudayaan adiluhung bangsa. *In the proceedings of Seminar Nasional Seni dan Desain 2017*. 468–475.

Haake, A. (1989). The role of symmetry in javanese batik patterns. *In International Series in Modern Applied Mathematics and Computer Science*, 815–826.

Hitchcock, M. (1991). *Indonesian Textiles*. BurySt Edmunds: Periplus.

Hong, Y. S. (2018).A Study on the Sustainability of Historical Urban Cultural Heritage by Interface Concept. *Asia-pacific Journal of Multimedia Services Convergent with Art, Humanities, and Sociology*, *8 (1)*, 1–6.

Hood, M. M. (2016). Notating heritage musics: preservation and practice in Thailand, Indonesia and Malaysia. Malaysian Music Journal, 5 (1), 53–73.

Hye-Joo, L. & Choe, J.H. (2018). A study on the usability reduction of mobile graphic user interface: in comparison with de-functionalization process in art history. *International Journal of Advanced Science and Technology*,*111*, 43–54.

Hyun-Kyung, L. (2017). Recent strategy & characteristic of worldwide design museums. International Journal of Art and Culture Technology, 1 (1), 15–20.

Instagram, https://www.instagram.com/p/BZvFgZghZUz/Retrieved 03 September, 2018.

Ismail, M. H. & Malingi, A. (2010). *Ragam motif tenun Bima Dompu*. Mataram: Mahani Persada.

Ismail, T., Wiyantoro, L. S., Meutia, and Muchlish, M. (2012). Strategy, interactive control system and national culture: a case study of batik industry in Indonesia. *Procedia Social and Behavioral Sciences*,*65*, 33–38.

Kim, H., & Ko, J. W. (2016). Determining the most important dimension of the development of social responsibility. *International Journal of IT-based Social Welfare Promotion and Management*, *3*, 35–40.

Masita. (2012). Pendidikan karakter berbasis budaya lokal pada masyarakat muslim.Jurnal Studi Masyarakat Islam, 15 (2), 302–320.

Miller, J. G., Wice, M., and Goyal, N. (2018). Contributions and challenges of cultural research on the development of social cognition.*Developmental Review*.

Musman, A.,& Arini, A.B. (2011).*Batik-warisan adilihung nusantara*. Yogyakarta:

G-MediaNam, S.H., Lee, S.H., & Kim, J. (2018). A study on the architecture of virtual reality system for art education. *International Journal of Advanced Science and Technology*, *114*, 161–172.

Nugroho, A. (2014).*Pengembangan multimedia interaktif musik tradisional nusantara berbasis adobe flash*, thesis, Department of Music Education, Yogyakarta State University, Yogyakarta, Indonesia.

Nurhaida, I., Noviyanto, A., Manurung, R., and Arymurthy, A. M. (2015). Automatic Indonesian's batik pattern recognition using SIFT approach. *Procedia Computer Science*, *59*, 567–576.

Prasetyo. (2012). *Seni gamelan jawa sebagai representasi tradisi kehidupan manusia jawa: suatu telaah dari pemikiran collingwood*, thesis, Department of Humanity, University of Indonesia, Jakarta, Indonesia.

Ramadhan, M. (2015). Faktor-faktor penyebab beralihnya minat masyarakat desa Tambe, kabupaten Bima, Nusa Tenggara Barat dari tari wura bongi monca ke hiburan keyboard tunggal dalam acara nika, thesis, Department of Music Education, Yogyakarta State University, Yogyakarta, Indonesia.

Santos, R. P. (2016). Alternative directions in art education and dissemination of traditional arts in a globalised society. Malaysian Music Journal, 5 (2), 1–16.

Setyawati, H., Darmokoesoemo, H., Rochman, F., and Permana, A. J. (2017). Affordable dye sensitizer by waste. Mater Renew Sustain Energy, 6 (17), 1–6.

Smiers, J. (2009). *Art under pressure*. Yogyakarta: Insistpress.

Sutari, W., Yekti, N. D. Y., Astuti, M.D., and Sari, Y. M. (2015). Analysis of working posture on muscular skeleton disorders of operator in stamp scraping in 'batik cap' industry. *Procedia Manufacturing*, *4*, 133–138.

Suyatno. (2014). Identitas Keindonesiaan dalam Novel Karya Anak Indonesia. Jurnal Litera, 13 (2), 293–301.

Wulandari, A. (2011).*Batik nusantara - makna filosofis, cara pembuatan, dan industri batik*.Yogyakarta: Andi.

Wulandari, R. (2008). Karakteristik Lagu yang Sesuai untuk Anak Ditinjau dari Segi Ambitus. *Jurnal Majalah Ilmiah Pembelajaran, (2)*, 1–10.

Zopf, S. R. (2017). Critical Study of the Use of a Length Unit in the Design of 16th to 18th Century Italian Violins. *Journal of Cultural Heritage*, *27*, S26–S33.

21st Century Innovation in Music Education – Astuti et al. (eds)

Talempong *Anam Salido* as an Islamic representation of Minangkabau people

M. Irfan & A.S. Suminto
Universitas Negeri Yogyakarta, Yogyakarta, Daerah Istimewa Yogyakarta, Indonesia

ABSTRACT: This article aims to analyse the values embedded in *Talempong anam salido* Art of Pesisir Selatan District. This ethnographic research analysed data based on a philosophical perspective. The material object of this research was *Talempong anam salido*, an art from the Pesisir Selatan District, West Sumatra. Meanwhile, the formal object was in the form of axiological study of *Talempong anam salido* Art. The results show that *Talempong anam salido* art refers to the identity of the Minangkabau people whose majority embraced Islam. This art also contains educational values such as religiousness, cooperation, tolerance, and responsibility. This article concludes that the Minang is an Islamic society. The values embedded in this art can be applied to both formal and non-formal education.

1 INTRODUCTION

The forms of culture consist of ideas, activities, and artefacts. Music is one of human works which is part of the culture itself. These three forms contain the things needed to live a life such as values, norms and all the rules of living in a society.

Almost every culture in Indonesia has arts such as music, dance, visual arts, and theatrical arts. In fact, the Indonesian people are also very fond of these arts, especially music. Indonesian people's passion for music is evident from the large number of musicians, ranging from young to old, in Indonesia. Television programs in Indonesia also offer music competitions and these are the most popular programs for Indonesian people from the lower, middle, and upper classes. However, these programs are seen as mere entertainment while in essence Indonesian ethnic music does not only function as entertainment for the community, but also has meaning and values that are beneficial to the lives of Indonesian people. Music contains information or messages that must be delivered from the previous generation to the current generation.

This meaning is now forgotten by Indonesia's young generation. Therefore, many young people do not like Indonesian ethnic arts as is the case with the younger generation in Nagari Salido, Pesisir Selatan District, West Sumatra. The results of observations in the field show that many young people in West Sumatra, especially in the South Pesisir District, expressed their disinterest and lack of understanding of their own ethnic music called *Talempong anam salido*. This article aims to shed light on the value and meaning of *Talempong anam salido* art as a form of enlightenment for the younger generation in this area and increase their awareness that *Talempong anam salido* art is not an outdated musical entertainment but is a form of education carried out through an art inherited by their ancestors. This article is also expected to be an additional reading reference for the next generation of art colleges so that they can also improve their cultural insights through art.

Until now, there has been a lack of literature on the meaning and values contained in *Talempong anam salido*. Therefore, the authors assume that writing about the meaning and values in this art is very important as an effort to add references to the younger generation in West Sumatra, especially for the younger generation in Pesisir Selatan District.

2 RELATED WORKS/LITERATURE REVIEW

There are several previous studies that are relevant to this study: (1) Andara Indra Sastra's dissertation (2015) entitled "The Concept of *Batalun* in the Presentation of *Talempong Renjeang Anam Salbuhan* in Luhak Nan Tigo, Minangkabau" shows that *Batalun* is a concept of appreciation in the performance of *Talempong Renjeang* art which contains a high aesthetic value. (2) A national seminar paper written by Susilo Pradoko (2002) entitled "Application of the Levi-Strauss Structuralism Paradigm to Analyze the Phenomenon of the Performing Arts" found that *gamelan* players had a rank structure in the palace (*keraton*). It is this relationship among contexts, network, and interconnectedness which can finally be found in the internal structure model. (3) Nadya Fulzi's (2011) article in the Journal of *Ekspresi Seni* entitled "Nature and Customs as a Local Aesthetic Source of Talempong Art, Dendang Song" found that *Talempong*'s voice that sounds like the voice of people debating each other was an analogy to the dialectic concept, a skill to argue a correct logic. Dialectics is part of the Minangkabau culture. Therefore, the concept of dialectics is the basic musical concept and the philosophy of *Talempong pacik*.

3 RESEARCH METHODOLOGY

3.1 *Research method*

This study used qualitative descriptive method to obtain data and facts about *Talempong anam salido* art from Pesisir Selatan District. Qualitative research is a method for exploring and understanding the meaning from a number of individuals or groups of people who are ascribed to social or humanitarian problems. Therefore, this type of research must be qualitative (Creswell, 2016: 4).

After the researchers described this study by using an ethnographic method, the researchers then analyzed the data using a philosophical approach in epistemology, ontology, and axiology. Epistemology is the study of the nature of an object of research itself. Therefore, this research first described the essence of *Talempong anam salido* art. Then, we described the reality of *Talempong anam salido*'s art in the midst of the people who own this art. After that, the researchers concluded the values contained in Talempong *anam salido* art as a representation of moral education and character.

3.2 *Data collection techniques*

The instruments for collecting data were used as a guideline to obtain data including participant observation and interviews with artists and the local community. To determine the validity of data for a qualitative study, we followed four criteria by Moleong in Djamal (2015: 127): credibility, transferability, dependability, and conform ability.

Interviews were conducted with several participants regarding *Talempong anam salido*. The scope of this research was limited only to interviews conducted with several competent resource persons regarding the history of *Talempong anam salido* art. Once completed, the data were then described using an ethnographic method and analyzed from the philosophical perspective.

4 RESULTS AND DISCUSSION

4.1 *Results*

Based on the results of *Talempong anam salido* art analysis from a philosophical point of view, the values of this art can be used as one of the weapons in realizing moral and manner education for the next generation, especially the younger generation in Salido. This value can also be used in formal and non-formal education. In formal education, it can be used as teaching materials used before carrying out art practice. Non-formal education can make it as a material for socialization to public. The meaning revealed in this art can also help strengthen the relationship in the community because *Talempong anam salido* art teaches an attitude of mutual assistance. This art also symbolizes the significance of the Minangkabau people, especially the Nagari Salido community, which has been Islamic society since a long time ago.

Based on the concept, the presentation of *Talempong anam salido* also symbolizes the Minangkabau customary government system, known as "tigo tungku sajarangan." It means that in any decision making for customary interests, a discussion should be held among *niniek mamak* such as *ulama* (muslim elders), *cerdik pandai* (intellectuals), and *penghululI* (headman). Therefore, *Talempong* teaches tolerance.

4.2 *Discussion*

4.2.1 *The history of* talempong anam salido *art*

Talempong Pacik Art is one of the oldest arts in Minangkabau. At first, this art came from the idea of the Minangkabau people who made an entertainment when they held a party of gratitude for their agricultural crops. Because the Minangkabau people originally came from the slopes of Mount Merapi, they relied on agricultural crops to meet their clothing, food and shelter needs. It is customary for the Minangkabau people to realize their gratitude for the abundance of their harvests to the almighty God by holding a people's party after the agricultural harvest. This harvest is done by working together and taking turns working in each group. During breaks after farming, while eating food provided by women, they hold a musical performance.

They use *dulang* or *talam* (trays) as drums and are accompanied by the sound of plates with regular rhythms and in harmony with the blast of *pupuik* (a traditional musical instrument from rice stems). Meanwhile, others took plates and danced to accompany the rhythm produced by this simple musical instrument. Along with the development of technology and human thought, simple musical instruments consisting of several plates, *pupuik*, and *dulang* or *talam* are perfected with a mixture of copper and brass which is now known as *Talempong* while *dulang* is replaced with musical instruments made of wood and goat skin called *Adok*.

In Pesisir Selatan District, precisely in Nagari Salido, *Talempong* Art was previously played with *sayak* (coconut skin), *pupuik katopong* and *adok*. This historical difference is not a significant problem for the Minangkabau community. This provision has been affirmed by one of the philosophies of Minagkabau "Adaik salingka nagari", which means that all procedures and decision-making are determined by the results of deliberations and consensus in each region. this is based on the main philosophy of the Minangkabau custom "Alam Takambang Jadi Guru", which means that the nature is a great teacher for Minangkabau people because all activities, actions and ways of thinking are learned from nature.

4.2.2 *The development of* talempong anam salido *art*

Talempong anam salido is an art originating from Nagari Salido. It is called *Talempong anam salido* because the Salido people refer to *Talempong Pacik* art as *Talempong* only. Therefore, the name used in this article is *Talempong anam salido*. Based on information obtained from the *Talempong anam salido* artist, since they started working in the art world starting from 1980 until now, the artist for this art has no successor. It is very unfortunate because this art is one of the identities of the region. This art is also very important in every traditional event such as marriage, and the inauguration of a new leader.

4.2.3 *The Concept and Meaning of the* talempong anam salido *structure*

The study was conducted on February 7, 2018. On that day, researchers asked for a research permit letter from the Directorate General of Politics and Public Administration in West Sumatra. The first interview was conducted with Mr. Irmun Krisman, familiarly called Pak Mon, an artist especially in the field of *karawitan*. He was the first informant to be interviewed in this study. In addition to an artist, he also became a teacher at SMK Negeri 7 Padang. In Minangkabau, there are two types of *Talempong Pacik* art games: some play five *Talempong* and some play six *Talempong*. According to Mr. Mon, *Talempong Pacik* art is not just a musical instrument for *Talempong*. This art also has a meaning that is closely related to the Minangkabau traditional philosophy 'Adat Basandi Syarah, syarah basandi kitabullah' (Adat is according to sharia, sharia is according to revelations). This means that Minangkabau custom is in accordance with the teaching of Islam. According to Shihab (1995: 01), expressed through Jobrohim and Belian, in Islamic teachings, the Qur'an, which is the main source of Islamic teachings, also appreciates art. One verse that states this is Ashshafaat (37): verses 6-7 and Fushshilat (41): verse 12, which means that heaven is not only created as a *hifzah* but also as a beautiful decoration. The sea was also created, among others, so that it could be obtained from it, not just "fresh meat" but also an ornament that beautified a person's appearance, (An Nahl (16): verse 04).

In Islamic teachings there are pillars of Islam and harmony. The five pillars of Islam are faith, prayer, charity, fasting, and pilgrimage to Mecca. Meanwhile, the six pillars of faith are belief in Allah, the only God; belief in the angels; belief in holy books; belief in the prophets, belief in the day of judgement, and belief in God's predestination. Basic *Talempong* has 2 *Talempongs*. Basic *Talempong* is called basic because the rhythm of the game in *Talempong* basic is only one and may not be changed or improvised until *Talempong* is finished. According to Mr. Mon, the nature of the *Talempong* base is like Shahada: it has its own pattern that is absolute and cannot be influenced. This principle is contained in Basic *Talempong*. If the basis has played the song *cak din din*, then until the music is complete, the pattern of the basic *Talempong* will remain *cak din din*.

This principle is in accordance with the first and second pillars of Islam: saying two sentences of shahada and performing prayer. The main requirement that must be fulfilled by someone if they want to embrace Islam is to say these two sentences of shahada and to perform prayer. This requirement is the basis of the teachings of Islam and has never been influenced from the days of the prophets to the present. However, before fulfilling the pillars of Islam, one must first fulfill the pillars of faith. Because without having faith it will be in vain for someone to embrace Islam. In the first pillar of faith there is a requirement to believe in Allah, the only God, while the second pillar is belief in angels. Someone who wants to embrace Islam must have faith in Allah first. After that, he deserves to say the two sentences of shahada. Furthermore, if someone has embraced Islam, it is an obligation for him to perform prayer because prayer is the main pillar of Islam. In order for the prayer to be perfect, a Muslim must have a strong faith in angels. Because angels have the duty to monitor all human activities in the world and no human can deceive them.

Based on this explanation, from the relations between the first and second pillars of faith as well as the first and second pillars of Islam, it can be concluded that in the basic concept of *Talempong,* there are elements of Islamic teachings. This is in accordance with Minagkabau's cultural philosophy "Adat Basandi Syarah, Syarah basandi Kitabullah". In the *paningkah* section, there are other two different *Talempongs*. In terms of the pattern of his playing, this is a *Pangincuang* (troublemaker) but the pattern of the rhythm does not collide with the basic *Talempong*. According to Pak Mon, this enhancement is a clash of life in the world, if the basis or foundation of one's self is shaky then until the end of his life it will never be smooth or in other words, chaotic. Viewed from the third and fourth pillars of Islam, they are fasting in the holy month of Ramadan and paying zakat.

After establishing the foundation of Islamic religion, namely the two sentences of shahada and performing prayer, there is a necessity or obligation outside the first and second pillars of Islam which will strengthen the basis of religion itself, namely paying *zakat* (charity) and fasting in the holy month of Ramadhan. This third and fourth pillars of Islam are a form of overcoming the collisions of life in the world in order to establish a good relationship between human beings in the world. In other words, it is the formation of harmony among each other. To be able to fulfill the third and fourth pillars of Islam, of course, an Islamic community must believe in the books and prophets/messengers. It is because all procedures, the basic principles of fasting and paying zakat, have been written in the holy books of Allah SWT.

The last part is called *Talempong Anak*. *Anak* is the source, mobilizer, and regulator of *Basic Talempong* and *Talempong Paningkah*. In other words, the presence of *Talempong Anak* is an improvement in *Talempong Pacik's* art. The character of the *Talempong Anak* is that it is in accordance with the fifth pillar of Islam, namely, the pilgrimage to Mecca. The pilgrimage is the last worship that will complete one's Islamic values.

4.2.4 *The values implied in* talempong anam salido *art*

Talempong anam salido is one of the traditional arts of the Minangkabau people. Because Talempong is a traditional art, it certainly has rules, teachings, and values contained in it. Traditions are customs, rites, social teachings, views, values, behavioural rules, and so on, which are passed down from generation to generation. It is an element of socio-cultural heritage that is preserved in society or in the social groups of society over a long period of time. This article focuses on the values contained in the art of the Minangkabau tradition, *Talempong anam salido*.

From a philosophical point of view, value is one of the themes in the study of philosophy. This study of values is called axiology in philosophy. Husserl through Frondizi (2011: 09) states that the value belongs to all objects that are not independent. In other words, art is also an object of value study. With the disclosure of the values contained in the arts, those values can be used as materials in education. This is what is called as value education through art. Value education helps many people to distinguish which is not good (= ethics), which ones should be prioritized, and which are allowed and not allowed (= moral). Therefore, it is very necessary to study the meaning and value in a work of art in order to realize the goal of art education.

In assessing the value, of course, it needs to be agreed that the value cannot be determined only through the sensing process because in a work of art the value is hidden. To be able to express the value in art, especially in *Talempong anam salido* art and we first need to know that the value has characteristics. Bertens in his book states that the characteristics of value are: value is related to the subject, value appears in a practical context when the subject wants to make something, and value concerns the characteristics that the subject adds to the properties possessed by the object.

These characteristics refer to the creation of works of art, where in the work of a work of art, an artist as a subject tries to expel and translate his imagination and emotions through artwork as his object. Divides values into three: material, vital, and spiritual. Material value is anything that is useful for the human body. Vital value is anything that is useful for humans to be able to hold activities. Spiritual Value is everything that is useful to humans.

Spiritual values are divided into four types: the value of truth (reality) which is the value that comes from the element of human reason (rationality, character, and creativity), the value of beauty, and the value that comes from human feelings (aesthetics). Moral values (goodness) are the value that comes from the element of will (intention and ethics). Religious values are the highest divine values that are absolute and eternal. Related to this, based on the concept and meaning of *Talempong anam salido* art, it can be seen that *Talempong's* art contains several educational values that can be used in the education of values to educate morals and characters. The values in *Talempong anam salido* will be traced according to Notonagoro's distribution of values as follows:

1) Vital Value:
   Talempong anam salido Art is a vital part of every traditional ceremony in Salido. Therefore, *Talempong* is an icon for the Minangkabau community.
2) *Talempong* Spiritual value:
   In every traditional ceremony, *Talempong* becomes a kind of notification media in the form of sounds for the Minangkabau people to inform them that there is a celebration held by one of the local residents. Therefore, by playing *Talempong*, people in the area will come to attend the celebration. This value can be subdivided into:
3) The value of truth (reality):
   Talempong anam salido in the beginning was an art that was purely derived from nature. Previously, *Talempong anam salido* consisted of several *sayak* (coconut skin) which were played with certain patterns and when *nagari* (youth) gathered and argued with each other. This was where the dialectic process takes place. A rhythmic pattern was created and called *anak*, *paningkah* and *dasar* (basic). From this event, it can be observed that there is a value

of mutual cooperation in *Talempong anam salido* art. This can train mutual cooperation in daily life. This tradition also occurs in the determination of a customary rules and norms. Therefore, they have to conduct a deliberation between non-Muslim groups in a customary institution called KAN (Village Adat Council).

4) Beauty value (aesthetics):
   This value arises because players must understand each other when playing *Talempong*. Otherwise, it will be difficult for *Talempong* players to play it. In other words, it needs a good chemistry among *Talempong* players even though the *Talempong* game pattern has been set. Furthermore, this value can also be seen from the interlocking techniques as evident from the playing of *Talempong* which are interlocking with each other. Suhanda (2013: 71) states that a series of tones produced by *Talempong Pacik* music is in the form of a melody that can reveal a good and amazing interlocking technique. Interlocking is an expression of the messages to be conveyed to the listeners. This is reflected in the Minangkabau's traditional philosophy "barek samo in pikua, cahaya samo in jinjiang" (bear the same problem, enjoy the same happiness). This has become its own beauty in *Talempong* in line with Nyoman Kutha Ratna's view that aesthetics or beauty means goodness and virtue (2007: 2).
5) Moral value (goodness):
   This value is reflected in different points at each *Talempong*. Every player in one pair of *Talempong* plays a certain pattern. The clash between different rhythmic patterns resembles tolerance. If someone has a deweakness, others will cover it up.
6) Religious value:
   This value is very strong in *Talempong anam salido* art. This can be seen in the relations of the three *Talempong anam salido* pairs which symbolize the pillars of faith and the pillars of Islam in Islamic teachings. This is relevant to Islam, the religion adopted by the Minangkabau people, as reflected in the Minangkabau's philosophy "*Adat basandi syara, syara basandi kitabullah*".

5 CONCLUSIONS

Based on the discussions above, it can be concluded that:

a. *Talempong Salido* Art has meanings and values that can increase awareness of young people in Minangkabau towards frequently-occurring deviant behaviors in this globalization era.
b. *Talempong Salido* is a musical art that reflects the contents of the Minangkabau culture, both in terms of values, customary norms, and even the social aspects of the Minangkabau community.
c. In terms of the concept and meaning of *Talempong* art, it can be concluded that one hundred percent of the Minangkabau people are Islamic. Therefore, it can be said that the Minangkabau people are Muslims, if they are not Muslims, they are not Minangkabau people.

REFERENCES

Creswell, J. W., & Creswell, J. D. (2017). *Research design: Qualitative, quantitative, and mixed methods approaches*. Sage publications.

Djamal, M. (2015). *Paradigma Penelitian Kualitatif* [Qualitative Research Paradigm]. *Yogyakarta: Pustaka Pelajar*.

Frondizi, R., & Wijaya, C. A. (2001). *Pengantar filsafat nilai* [Introduction to Value Philosophy]. Pustaka Pelajar.

Shihab, M. Q. (1995). *Islam dan Kesenian* [Islam and Arts]. *dalam Jabrohim dan Saudi Berlian (ed.)*.

Ratna, N. K. (2007). *Estetika sastra dan Budaya* [*Literary and Cultural Aesthetics*]. Pustaka Pelajar.

Suhanda, S., Marzam, M., & Wimbrayardi, W. (2013). *Fungsi Talempong Pacik Dalam Upacara Perkawinan Masyarakat Nagari Koto Anau* [*Talempong Pacik Function in the Community Marriage Ceremony of Nagari Koto Anau*]. *Jurnal Sendratasik*, *1*(2), 64–72

21st Century Innovation in Music Education – Astuti et al. (eds)
 ISBN 978-0-367-11122-9

Sasandu musical instruments as *Nusa Tenggara Timur* cultural arts promotion media

W.O. Wiko, S. Pradoko, G.K.W. Anggraheni & A. Husna
Universitas Negeri Yogyakarta, Yogyakarta, Indonesia

ABSTRACT: This study aims to explore the efforts of the government of the Rote Ndao *Nusa Tenggara Timur* (NTT) District to revitalise the promotion of cultural arts through the medium of *Sasandu* musical instruments. This research is a descriptive qualitative study. The results of the study show that *Sasandu* will be included in the cultural arts curriculum in every educational institution (SD, SMP, and SMA, or elementary, middle, and high schools) by requiring there to be professional cultural arts educators, cultural arts books as *Sasandu* learning guides, and complete facilities and infrastructure for their students. The community response of Rote Ndao (overall speaker) was that *Sasandu* deserved to be used as a medium for promoting arts and culture due to its uniqueness (using the lontar leaf as a leaf or tree of life, a tree that gave a lot of meaning to the people of Rote Ndao).

1 INTRODUCTION

Regional cultural development is an integral part of national cultural development. The direction and objectives of the regional cultural development strategy of the *Nusa Tenggara Timur* (NTT) Province are guided by the direction and strategy of the development of national culture, as formulated in the GBHN (Indonesian: Broad guidelines of state policy) (Abdullah, 1992, p. 334). The purpose of the type of cultural development referred to above is to maintain, preserve, develop and appreciate the cultural values of the region as a reflection of the noble values of the region itself in order to foster the noble values of national culture, which further strengthens the realisation of insight archipelago, environmental insight and national resilience (Abdullah, 1992, p. 335). NTT's regional culture is the result of a mix of cultures from various regions in Indonesia, but it still possesses native traditionalism, a characteristic that is not shared by other regions (Santoso, 1992, p. 90). One such characteristic is the *Sasandu* musical instrument. The charm of *Sasandu*'s music is not only known among the local community, but also abroad. Its benefits are greatly appreciated by foreign humanists. Jeremias O. Pah (one of the *Sasandu* experts) was invited to hold a music performance at Sakura Japan. This is the pride of the Indonesian people, especially the people of NTT.

The priority when advancing NTT's art and culture is to introduce the peculiarities it has, one of them being *Sasandu* musical instruments. and *Sasandu* musical instruments in one of its media. The progress of local culture can attract foreign humanists, which has an impact on increasing the country's foreign exchange and improving the image of Indonesian culture in the eyes of the world. In addition to government revitalisation efforts, encouragement is also needed from the Rote Ndao NTT community.When carrying out a *Sasandu* promotion, a strategy is needed that will facilitate communication with consumers. *Sasandu*'s superiority gives NTT an image of being an alternative city to travel to, both on a national scale and on an international scale. *Sasandu* was chosen as a promotional medium for NTT in order to encourage other branches of art. Based on the description of the problem, two interesting issues were found that needed to be explored: 1) What effort is needed by the government of NTT's Rote Ndao District to revitalise the promotion of cultural arts through the medium of *Sasandu* musical instruments? 2) What is the extent of people's response (cultural *Sasandu* artists) to the use of *Sasandu* musical instruments as a medium for promoting arts and culture in NTT? Reviewing these problems has practical and theoretical benefits, which, of course, are beneficial for the continuation of *Sasandu,* especially with regard to the progress of NTT's cultural arts.

2 METHODOLOGY

The type of research used in this study is qualitative research with descriptive analysis. Bogdan and Taylor define this type of qualitative research as a procedure that produces descriptive data in the form of written or oral words from people and in the form of observable behaviour (Moleong, 1989 p. 3). This study uses a descriptive approach because the data collected is in the form of words and images, not numbers. In this case, the object of the research is using *Sasandu* musical instruments as a medium for promoting the culture of NTT. The researcher focused on the efforts of the NTT Rote Ndao government to use *Sasandu* as a medium to promote NTT's cultural arts and the response of the community, especially *Sasandu*'s humanists and artists, to the use of *Sasandu* musical instruments in NTT Rote Ndao.

3 RESULTS AND DISCUSSION

3.1 *Results*

3.1.1 *General description of research locations*

The area of Rote Ndao Regency is surrounded by 103 small islands and there are six inhabited islands, which are Pulau Rote, Pulau Ndao, Pulau Nuse, Pulau Landu, Pulau Nusa Manuk and Pulau Usu. The Rote Ndao Regencyis made up of land area of 45,250 Ha, hills 32,625 Ha and other land 50,135 Ha. Rote Ndao, which has a semi-tropical climate, has palm leaves or *tuak* as its unique plant. The existence of palm trees benefits the community as they grow naturally and are not cultivated by humans. Leaves from palm trees provide many benefits. Besides being the main ingredient used in making *Sasandu* musical instruments, they are also used for the manufacture of household supplies and necessities for *Ti'ilangga* (Rote fashion). In traditional poems, palm trees are also referred to as the tree of lifeIn relation to the utilisation of nature, through interviews with Herman Adolf Ledoh (*Sasandu* maker as well as head of the Manukoa studio located in the Rote Barat Laut District, Busalangga sub-district) he explained that 'Nature of Rote Ndao, especially palm trees, is likened to priceless gold. So, it must be maintained and kept protected' (personal communication, 7 April 2016).

In addition, one of the traditional leaders (Arkimes Mole) explained in an interview on how Rote Ndao's nature, especially *Sasandu,* is expressed in regional languages:'*Meko Liti do Sasandu Lilo. Liti* (brass that resembles gold) and *Lilo* according to people Rote Ndao is interpreted as something that has extraordinary artistic value. *Sasandu* is considered as gold that has extraordinary artistic value' (personal communication, 7 April 2016)The palm tree is a unique plant of Rote Ndao NTT that is priceless. Likewise, *Sasandu,* which is made with palm oil, holds the highest position above the other branches of culture in Rote Ndao NTT.

3.1.2 *Efforts of the Government of NTT's Rote Ndao District to revitalise the promotion of cultural arts through the medium of Sasandu musical instruments*

The government of the Rote Ndao District of NTT, especially the Tourism Office, plays an important role in promoting cultural arts through the medium of *Sasandu* musical instruments. *Dinas Pariwisata,* or the government Tourism Office, is a government agency that is responsible for preserving cultural arts, including being responsible for the efforts made in promoting cultural arts.

In an interview with the speaker (Mery Rosalin Bako, S. Pt., Msi., Head of the Culture and History Art Division at the Tourism Office of Rote Ndao District, NTT), she explained that '*Sasandu* is worthy of being one of the mediums to promote Rote Ndao culture. Its uniqueness is known everywhere, even abroad. Rote's unique musical instrument is actually not Sasando but *Sasandu*, only since the establishment of one of the hotels in Kupang, *Sasandu*'s name has changed with the name of the hotel. It is my hope that the researcher will change Sasando's name to *Sasandu* in the name of the title of his scientific work. We strive to succeed in some of our efforts as a tourism service government' (personal communication, 6 April 2016).

Based on the results of the interview, there were several revitalisation efforts carried out using *Sasandu* musical instruments as a medium for promoting arts and culture, including:

a. *Sasandu* in the Local Content Curriculum (Mulok) in Rote Ndao NTT
   Mery Rosalin Bako, in the interview that took place at the Rote Ndao District Tourism Office, explained that:

> 'There are several revitalisation efforts that we are trying to promote in the arts and culture; the first is through the application of the curriculum in every school, such as elementary, middle and high school. Rote Ndao lacks educators who specialize in teaching cultural arts, for but we will overcome it with the addition of educators from *Sasandu* art actors specifically in the extracurricular section. Procurement of cultural arts books and *Sasandu* musical instruments in each educational institution will be applied.' (personal communication, 6 April 2016).

The selection of educators is the most important thing in incorporating *Sasandu* as part of the mulok curriculum. The Tourism Office will involve *Sasandu* arts actors as special educators in the extracurricular section. Aside from teaching staff (teachers) in cultural arts subjects, there is also a book that is used as a *Sasandu* learning guide. Books will not only be in the school library or only owned by the teacher, but they must also be owned by each student. Apart from *Sasandu* textbooks, general knowledge books will also be used. *Sasandu* must be owned by schools as one of the main mediums in learning art and culture.

b. Hold a Cultural Arts Festival
   In the interview, Mery Rosalin Bako explained:

> 'One of the biggest festivals ever held is the *Sasandu* music festival, which is competing for the Presidential Cup. Three hundred participants from the cities of Kupang, Kupang Regency and Rote Ndao District attended. Participants were grouped into several categories, which are *Sasandu gong tunggal*, *Sasandu* gong group, *Sasandu* single violin and *Sasandu* group violin. In addition to the competition, the festival also features *Sasandu* music and choir performances designed by Dwiki Darmawan with the theme 'Ku Yakin Sampai di Sana'. I hope this festival becomes an annual programme' (personal communication, 6 April 2016).

In addition to what was conveyed by the speakers, the researchers also received responses via written documents from several government figures from the Rote Ndao District Tourism Office. Suparman Tjetjep, as director general of culture, art and film, said that the *Sasandu* music festival was an inaccurate manifestation of the concern of all levels of society towards the art and culture of the nation. Music that was originally created by *dua penggembala kambing* (or two goat shepherds) in Rote Ndao was already known by the people of Kupang and even throughout the archipelago. This music has added to the diversity of cultures that we have (Haning, 2009, p. 32).

NTT governor Frans Lebu Raya said that, as citizens of the world, of course, we the Indonesian people are proud that we are a highly cultured and creative nation, so it is fitting for all our national cultural wealth to be preserved, developed and passed on to younger generations. In addition to festivals conducted within the country, Mery explained that *Sasandu* had also been deployed abroad:

'*Sasandu*'s NTT Music Show Efforts to Wonderful Indonesia Promotion in New Zealand were held on 29 November–1 December 2015 by the Assistant Deputy for Asia Pacific Market Development at the Ministry of Tourism (Kemenpar) Vincent Jemadu (personal communication, 6 April 2016).

Cultural festivals need to be carried out when seeking to use *Sasandu* musical instruments as a promotional medium. Among the cultural festival activities carried out, the biggest festival ever carried out, which has been made into an annual government programme, is the *Sasandu* festival, the Presidential Cup of the Republic of Indonesia. The idea for the *Sasandu* music festival was first sparked during a visit by President Susilo Bambang Yudhoyono to Kupang on 14 June 2009. On that occasion, the President of the Republic of Indonesia gave a briefing on the need for the protection, development and utilisation of traditional arts. This was also intended to compensate for the tendency of the younger generation to prefer Western music and art.

c. Dialogue on *Sasandu* Culture
Cultural dialogue is one form of activity that supports advancing art and culture. This activity is one of the important agendas pursued by the government of the Tourism Office of the Rote Ndao District of NTT. Mery Rosalin Bako explained:

'The cultural dialogue was carried out to all groups and became one of the annual agendas of the Rote Ndao Tourism Office' (personal communication, 8 April 2016).

Jersi W. Mesakh, SE (former head of the Rote Ndao District Tourism Office and one of the descendants of the king) explained:

'We really appreciate the *Sasandu* cultural dialogue and hope that it will continue to run smoothly every year' (personal communication, 7 April 2016).

This means that this programme is one of the government's annual routine programmes. It offers more variety in its activities and it is expected that not only certain people will attend and participate, but that all of the Rote Ndao communities will do so, especially the young.

d. *Sasandu* Seminar
The seminar was part of the government's idea to use *Sasandu* as a medium for promoting arts and culture in the Rote Ndao District. Mery Rosalin Bako said:

'The *Sasandu* seminar will be held every year as a routine agenda in the development of Rote Ndao' (personal communication, 07 April 2016).

Herman Adolf Ledoh (a *Sasandu* expert) said:

'*Sasandu* should not stop at the age we are today, but it must arrive at young people now and then. So the seminar or presentation of material related to *Sasandu* is needed' (personal communication, 6 April 2016).

e. Collaborating with *Sasandu* Arts Actors
Synergetic support from *Sasandu* artists is greatly encouraged by the Rote Ndao District Tourism Office in order to advance the cultural arts of Rote Ndao. Mery Rosalin Bako explained:

'The number or absence of art actors is not a problem, the most important thing is how the spirit of *Sasandu* artists in working to preserve *Sasandu*' (personal communication, 8 April 2016).

Providing support through individual *Sasandu* artists and groups will influence *Sasandu*'s progress.

f. Co-operating with Art Studios
Working together in the form of providing support (both moral and material) to art studios is carried out so that the training participants' desire to learn can grow. An adequate

number of musical instruments greatly influences the development of the studio itself. Mery Rosalin Bako explained:

> 'In addition to regular training, what is also considered is the equipment of the studio facilities. Facilities and infrastructure greatly affect the learning spirit of the studio participants' (personal communication, 6 April 2016).

g. Installation of *Sasandu* Icons in Various Tourist Visitor Spots
The installation of *Sasandu* icons in various places that are often encountered by tourists is one of the efforts being made by the government of NTT's Rote Ndao District to introduce *Sasandu* to the tourists or foreign humanists who visit Rote Ndao. The places in question are airports, ports, offices and tourist attractions in Rote Ndao Regency.

3.1.3 *Community response to Sasandu musical instruments as a medium for promoting arts and culture in Rote Ndao District, East Nusa Tenggara*

There were several responses obtained from various people within the community in the process of conducting this research, including related art studios (Sanggar Manukoa and Sanggar Nusatuameni, Sanggar Edon Sasando-Kupang), *Sasandu* craftsmen (Andreas Tonakh), traditional figures (Arkimes Mole), a community leader (Ibrahim Panie), one of the descendants of the king, a former head of the Tourism Office of Rote Ndao Regency (Jersi Wmesakh, SE), several teachers from one of the local elementary and high schools (Jeremias Manafe, S.Pd. and Na Omi Eb Buik), and students from the local area.

The response delivered by Fransina Pah-Sui (Head of the Nusatuameni Studio) said:

> '*Sasandu* musical instruments have their own uniqueness, so they are worthy of being used as promotional media. Its uniqueness is supported by the basic ingredients of making it, palm leaves' (personal communication, 10 April 2016).

Sasandu's success as an icon of cultural art from Rote Ndao can be seen from its uniqueness. Herman Adolf Ledoh (Head of Manukoa Studio), in an interview that took place at Sanggar Manukoa, said:

> '*Sasandu* sounds, especially *Sasandu* Gong, are the same as Gong Rote Ndao, so in indirectly playing *Sasandu* we also learn the sounds of Gong Rote Ndao. We play the sound of *Sasandu* Gong as well as we play the sounds of Gong Rote Ndao to the wider community. Election of *Sasandu* as a media for promotion of cultural arts is very feasible' (personal communication, 10 April 2016).

Marline Edon (Head of Edon *Sasandu* Studio), in the interview that took place at Sanggar Edon *Sasandu* Kupang, said:

> '*Sasandu* is very appropriate if it is said to be a media for promotion of Rote Ndao culture as well as other cultural arts found in NTT. This can be accounted for by looking at the many achievements obtained when attending a music festival. *Sasandu* players (members of Edon *Sasandu* Studio) have been invited to several countries to hold *Sasandu* performances. Only *Sasandu* we have, only *Sasandu* is able to bring the name of NTT to the international level, especially in the arts. *Sasandu* won an award from cultural conservation from the UN agency for UNESCO education, science and culture' (personal communication, 10 April 2016).

Jeremias Manafe, head of SMAN (public senior high school) 3 Rote Ndao, said:

> '*Sasandu* is seen as a media for promoting cultural arts in Rote Ndao and is one of the things that must be taken seriously by the relevant government. Which should be considered by the government and also our hope is that *Sasandu* will become a part of the local content curriculum in Rote Ndao' (personal communication, 10 April 2016).

In addition to educators, researchers also interviewed one of the local Berto students (2nd grade student at Mokdale 1 Public High School Rote Ndao), in which they said:

'For the time being, we only learn *Sasandu* through art studios. There are no specific lessons to learn *Sasandu* at school' (personal communication, 9 April 2016).

Based on several of the responses described above, it can be concluded that *Sasandu* is very reasonable and is a must if it is to be a media for promoting the culture of Rote Ndao. *Sasandu* can represent all the wealth of the other cultural arts found in NTT, but additional government efforts need to be applied in order to support this. In order to make *Sasandu* a medium for the promotion of cultural arts in Rote Ndao NTT, there must be intergovernmental co-operation related to the community, which will maximise the efforts to promote *Sasandu* as an icon of NTT's Rote Ndao.

3.2 *Discussion*

3.2.1 *Efforts of the Rote Ndao District Government of East Nusa Tenggara to revitalise the promotion of cultural arts through the medium of Sasandu musical instruments*

The Tourism Office of Rote Ndao District of NTT is necessary for the promotion of cultural arts as part of the government's revitalisation efforts. These efforts were realised through the assistance of various parties from central government, local government, educational institutions (formal education and non-formal education), local communities and local cultural observers, as well as foreigners and foreign humanists.

In carrying out these revitalisation efforts, the government (*Dinas Pariwisata*) of Rote Ndao District, NTT divided the discussion into four sections, including:

a. Reorganisation

Educational institutions, both formal (elementary, middle and high schools) and non-formal (art studios), are the first targets in this stage of developing cultural arts. The government will co-operate with schools and related art studios, both individually and in groups. This co-operation will be carried out by providing support in the form of facilities and infrastructure, both moral and material. *Sasandu* art actors will be involved in every activity related to the progress of art and culture.

b. Function

Musical instruments must be learned in accordance with the curriculum. The Large Indonesian Dictionary described the curriculum as a set of subjects taught in educational institutions: a set of subjects regarding special fields of expertise. Local content is a curriculum that contains subjects tailored to regional interests. So the *Sasandu* learning device must be adapted to the conditions/interests of the Rote Ndao area.

The term curriculum is simply the name for the organisational pattern of school education programmes. A complete picture of the curriculum includes at least three components, which are:

1. What is learned, namely concerning the problem of the content or subjects to be taught?
2. How to learn it, which concerns the problem of teaching methods?
3. If the various lessons are taught or presented, which concerns the matter of order or teaching settings (Phenix, 1961).

The curriculum is an educational programme that is designed to complement certain educational goals by using certain educational tools in order to perfect those goals. The curriculum consists of a broad environment, in which interactive teaching takes place, covering all of the content and approaches used (Sahabuddin, 2007, p. 147).

The effort made to implement the teaching of *Sasandu* in the *mulok* curriculum in each educational institution is the most effective step needed to strive for *Sasandu* to be learned by students. Aside from being a form of preservation or effort for protection, it is also the initial stage of promoting cultural arts in NTT's Rote Ndao District.

There are three stages in seeking the implementation of *Sasandu* in the *mulok* curriculum. In each educational institution its implementation will be sought by the Rote Ndao District government, including: professional arts and culture teachers, procurement of special textbooks for *Sasandu* and the procurement of *Sasandu* musical instruments in every educational institution.

c. Actualisation
Activities that are closely related to the development of cultural arts are very helpful in their successful promotion, one of which is the cultural festival. Festivals are described in the Large Indonesian Dictionary as happy days or weeks in commemoration of important and historic events; people's parties are also referred to as celebrations, along with activities, crowds, exhibitions, parades, fairs, celebrations and performances.

It can be concluded that the festival is an activity that moves within the area of art performances. Cultural festivals must be carried out for *Sasandu* musical instruments in the hope of attracting a passion for creativity and enthusiasm for learning from every artist in Rote Ndao District, especially *Sasandu* musical instrument players. The targets of this cultural festival are students (elementary, middle and high school) and the general public.

d. Interpretation
The efforts made at this stage of interpretation are to hold a dialogue or cultural seminar. Cultural dialogue is a form of cultural art conversation that is carried out both horizontally and vertically, such as among cultural arts teachers, between fellow students in school or between *Sasandu* composers and *Sasandu* players.

Effective dialogue can be a guaranteed success if carried out routinely and periodically. Dialogue on art and culture that is carried out routinely will have an impact on the progress of cultural arts itself. Dialogue on art is also known as a cultural seminar. The seminar is generally a form of academic teaching, either at a university or given by a commercial or professional organisation. The word seminar originates from the Latin *seminarium*, which means 'the land where you plant seeds' (Haning, 2009, p. 30).

3.2.2 *Community response to Sasandu musical instruments as a medium for promoting arts and culture in Rote Ndao District, East Nusa Tenggara*

Sasandu is worthy of being used as a medium for promoting arts and culture in the Rote Ndao District of NTT. The uniqueness of this musical instrument is well known at both a national and international level, and the Rote Ndao musical instrument was even awarded cultural conservation status by the UN Organization for Education, Science and Culture (UNESCO).

Sasandu's uniqueness could cause it to become one of Indonesia's globally recognised traditional musical instruments. The existence of this musical instrument shows the process of human growth and the state of society in Rote Ndao. The traditional materials used for making *Sasandu* means that a *Sasandu* can compete with modern musical instruments when viewed from the perspective of the sound quality produced. For that reason, *Sasandu,* which is produced with simple materials, is in high demand for its good quality and can elicit interest from foreign humanists or from anyone who wants to have a *Sasandu.*

The quality of *Sasandu* musical instruments is very important, and this can be achieved through the training (*Sasandu* workshop) of all cultural experts or artists, and even students, by the government. Other than in terms of quality, the market for *Sasandu* craftsman has an impact on the quality and effectiveness of buying and selling transactions. *Sasandu* are no longer transacted freely, but through good procedures overseen by the government. This is expected to be able to contribute to the income of the region.

The awards obtained by art studios strongly support the ability of *Sasandu* to be used as a medium for the promotion of the cultural arts of Rote Ndao NTT. Appreciation from various parties, including the government, is needed to improve this achievement.

The three points presented show that synergic co-operation from various parties is required, such as from the government, local communities, *Sasandu* arts actors and educational institutions, both formal (elementary, middle and high school) and non-formal education (art studio in Rote Ndao District NTT).

4 CONCLUSION

The *Sasandu* musical instrument has its own special features for the Rote Ndao community. The uniqueness of this *lontar* stringed instrument is widely known among art lovers, including cultural experts from every foreign country. To use *Sasandu* as a promotional medium, a government revitalisation effort is needed (Rote Ndao District Tourism Office) and this requires support from various elements of society. Government revitalisation efforts are divided into four parts, which are reorganisation, function, actualisation, and interpretation.

In addition to the revitalisation efforts carried out by the government, the response and influence of the community was very helpful for using *Sasandu* as a medium for promoting cultural arts. The positive and negative responses conveyed by the Rote Ndao community are the main strengths or capital that should be noted by the government when making *Sasandu* a medium for promoting arts and culture in NTT's Rote Ndao District.

REFERENCES

Abdullah, T. (1992). *Profile of the Republic of Indonesia-East Nusa Tenggara Province*. Jakarta, Indonesia: Yayasan Bhakti Wawasan Nusantara.

Haning, P. (2009). *Sasandu - Traditional musical instrument society of Rote Ndao*. Kupang, Indonesia: CV. Kairo S.

Moleong, L.J. (1989). *Qualitative research methodology*. Bandung, Indonesia: PT. Remaja Rosdakarya.

Phenix, P.H. (1961). Education and the common good: A moral philosophy of the curriculum. Harper.

Sahabuddin. (2007). *Teaching and learning*. Makassar, Indonesia: Badan Penerbit UNM.

Santoso, B. (1992). *Profil Propinsi Repoblik Indonesia-Nusa Tenggara Timur (The Profile of Republic of Nusa Tenggara Timur)*. Jakarta, Indonesia: Yayasan Bhakti Wawasan Nusantara.

Susilana, R. & Riyana, C. (2009). *Teaching media*. Bandung, Indonesia: CV. Wacana Prima.

21st Century Innovation in Music Education – Astuti et al. (eds)

Correlation between musical imagery and guitar improvisability in Soroako's music community

A.C. Pasande & K.S. Astuti
Yogyakarta State University, Yogyakarta, Indonesia

ABSTRACT: This study aims to examine musical imagery, the ability to improvise music and the correlation between sensitivity to musical imagery and the ability of musicians in Soroako to improvise on a guitar. The approach of this research is quantitative and uses the correlation analyse method. The results of the study showed a correlation between musical imagery and: 1) 'guitar improvisability' of musicians in Soroako's music community, 2) intonation, 3) dynamics, 4) tempo, 5) creativity. Furthermore, it can be concluded that musical imagery is the variable that most contributes to guitar improvisability in Soroako's music community.

1 INTRODUCTION

Music plays an important role in the everyday lives of people. Many people listen to music from morning to late at night, especially since the discovery of radio and recording, for entertainment, disturbances or mood enhancements (Jäncke, 2008, p. 1). Music, according to Rini (in Kristiawan, 2016, p. 37), is the result of sound art in the form of songs or musical compositions that express the thoughts and feelings of the creator, through the musical elements of rhythm, melody, harmony, form or structure of songs and expressions.

The guitar is one of the most popular musical instruments in the world and also in Indonesia, where the two string instruments, vihuela and lute, were previously developed. According to Miftahunnajah (2013), the standard guitar instrument has six strings, each of which produces a different tone. In many guitar games, variations that can be played include improvisation. The guitar is also a musical instrument that is much favoured by young people. Apart from being relatively inexpensive compared with other musical instruments, the guitar is also a harmonious instrument so that it can be used to accompany a song or be played solo (Hidayat, 2012, p. 1). Learning music, generally aims to instil aesthetic values in students and to develop the cognitive domain, through which students gain knowledge about the world around them, the affective domain that helps students acquire knowledge to develop attitudes, and the psychomotor domain, through which students gain knowledge about skill development. Music learning is very important to stimulate the development of one's creativity (Kristiawan, 2016, pp. 36–38).

The problems faced about music education in schools in Indonesia (e.g. in Soroako, Nuha District, East Luwu Regency, South Sulawesi) include the lack of teaching staff or music teachers in schools and the lack of interest in music. Some effects of these problems are seen directly: for example, a band performs music, and children play bands with skill and conventional knowledge. As a result, it can be said that their music play is not maximal. This happened because of a lack of knowledge and experience of teachers or art instructors in learning music at school and in the village environment. In school, music theory and music practice, and important elements in music, such as aesthetics (beauty), are forgotten.

According to the experience of researchers in the field, some people who have good musical imagery skills are not good at improvising, and vice versa. Musical imagery, according to Bailes (2002, p. 1), is the production of hearing in the mind, which reflects imaginative processing. According to Seashore (in Bailes, 2002, p. 2), the most prominent imagery and creation of musical signs is hearing which is the ability to hear music from memory, to apply it in creative works, and to complement the actual physical sound in music.

The music community in Soroako refers to a group of people who join forces to form music groups. In a music community, to be able to create a strong identity, first each individual must have the same views and motivations in terms of music (Agoes, 2011). Music and singing stimulate creativity and help establish good attitudes. Hearing is one of the most important senses for musicians because it provides important information during practice and performance where they have to play with matching frequencies so they can play properly (Gunawan, 2015, p. 4).

This study tested several guitarists in the music community in Soroako, South Sulawesi. The researchers spontaneously tested the ability of the study subjects to hear tones, chords and rhythm and the ability to improvise on the guitar to find out the level of correlation between musical imagery and 'guitar improvisability'.

2 RESEARCH METHODS

2.1 *Research design*

This research is quantitative and uses the correlation method. Correlation research is used to determine the relationship and its extent between two or more variables without any attempt to influence these variables so that there are no variable manipulations (Faenkel & Wallen, 2008, p. 328). The place of research in this study is Soroako, Nuha District, East Luwu Regency, South Sulawesi Province, Indonesia, and the study subjects selected are members of the music community who play the guitar. The population in this study, consisting of musicians in the music community in Soroako, includes 30 people. This research is population research because all populations are sampled. The variables in this study are musical imagery as an independent variable (X) and improvisability as a dependent variable (Y).

2.2 *Data collection and analysis technique*

Data collection techniques used by researchers in this study are practical tests. The practice of musical instruments is about how one applies the knowledge obtained of ways/techniques to play musical instruments. In instrument practice, the most important thing to learn is how to improve motor skills to produce better quality music (Agustianto, 2016, p. 4). Test assessment is based on the following predetermined criteria:

1) Musical imagery
   a) Tone (guessing the tone correctly)
   b) Rhythm (hearing and playing rhythmically)
   c) Chords (guessing chords correctly)
2) Guitar improvisability
   a) Intonation
   b) Technique of improvisability with dynamics
   c) Tempo (timing, i.e. on time when playing tones)
   d) Creativity.

Testing of convergent validity instruments in this study is done using confirmatory factor analysis (Azwar, 2012). The analysis process used IBM SPSS 17 statistical software. The results of testing showed a Kaiser–Meyer–Olkin (KMO) value of 0.709; a KMO value greater than 0.5 means that sampling is sufficient and included in the appropriate category to be used for factor analysis. Also, Bartlett's test of sphericity fulfils the validity requirements because of significance below 0.05 (5%), which is equal to 0.000. In a component matrix, it can be seen that instrument items have a factor value above 0.50, and the results of factor analysis show that variable items are extracted perfectly because they have spread to all factors. Based on this, construct validity is declared valid. Based on the results of the instrument reliability test improvability is equal to 0.765, indicating that the instrument reliability is high. Thus, the research variables are reliable, so they can be used for research.

Data analysis in this study is descriptive and statistical analysis is inferential. In this study, the assessment of ability was carried out by three assessors, so the descriptive statistics here used the mean of musical imagery and guitar improvisability, to be further analysed. The analysis technique in this study is Pearson's correlation test.

3 RESEARCH RESULTS AND DISCUSSION

3.1 *Research results*

The study was conducted on 30 participants in the music community in Soroako, Nuha District, East Luwu Regency, South Sulawesi Province, Indonesia. Musical imagery was assessed on the average score of knowledge about tone, rhythm and chords based on the ability to hear the participants. The highest musical imagery score was 100 and the lowest 40.

The results of the guitar improvisability test conducted on 30 participants were obtained from the average score given by the three assessors regarding knowledge about intonation, dynamics, tempo and creativity. Scores of the four aspects assessed on the ability to improvise on a guitar were then added up, resulting in a score of guitar improvisability. The highest score for guitar improvisability was 90 and the lowest 25.7.

Analysis of the correlation between musical imagery and guitar improvisability can be seen in Tables 1–5 that show the correlation of musical imagery with each indicator of guitar improvisability, including intonation, dynamics, tempo and creativity.

Table 1. Correlation between musical imagery and intonation.

		Musical imagery	Intonation
Musical imagery	Pearson correlation	1	0.419*
	Significance (two-tailed)		0.021
	N	30	30
Intonation	Pearson correlation	0.419*	1
	Significance (two-tailed)	0.021	
	N	30	30

* Correlation is significant at the 0.05 level (two-tailed).

Table 2. Correlation between musical imagery and dynamics.

		Musical imagery	Dynamics
Musical imagery	Pearson correlation	1	0.233
	Significance (two-tailed)		0.214
	N	30	30
Dynamics	Pearson correlation	0.233	1
	Significance (two-tailed)	0.214	
	N	30	30

Table 3. Correlation between musical imagery and tempo.

		Musical imagery	Tempo
Musical imagery	Pearson correlation	1	0.332
	Significance. (two-tailed)		0.073
	N	30	30
Tempo	Pearson correlation	0.332	1
	Significance (two-tailed)	0.073	
	N	30	30

Table 4. Correlation between musical imagery and creativity.

		Musical imagery	Creativity
Musical imagery	Pearson correlation	1	0.355
	Significance (two-tailed)		0.054
	N	30	30
Creativity	Pearson correlation	0.355	1
	Significance (two-tailed)	0.054	
	N	30	30

Table 5. Correlation between musical imagery and guitar improvisability.

		Musical imagery	Guitar improvisability
Musical imagery	Pearson correlation	1	0.359
	Significance (two-tailed)		0.052
	N	30	30
Guitar improvisability	Pearson correlation	0.359	1
	Significance (two-tailed)	0.052	
	N	30	30

3.2 *Discussion*

In this study, to find out about the correlation between musical imagery and guitar improvisability, a series of tests were conducted on musicians in Soroako's music community. The musical imagery test was done specifically on the hearing ability of these musicians, while the guitar improvisability test involved instrumental songs in C major.

Of 30 participants, fourteen had a high category score in the musical imagery test, nine has a moderate category score and seven a low category score. The score for musical imagery in this study is the sum of the scores for tone, rhythm and chords among musicians in Soroako's music community. The test results showed that the highest musical imagery score was 100 and the lowest was 40. The following is a description of each indicator of guitar improvisability and its correlation with musical imagery for musicians in Soroako' music community.

Guitar improvisability, in this study, was measured on aspects such as intonation, dynamics, tempo and creativity. Of 30 participants, 12 had a high category score, nine had a moderate category score and nine had a low category score. The score for measuring guitar improvisability in this study is the sum of the average scores given by the three assessors for intonation, dynamics, tempo and creativity of the study participants. The results of the guitar improvisability test showed that the highest score was 90 and the lowest score was 25.7. The following is a description of each indicator evaluating the ability of musicians in Soroako's music community to improvise on the guitar.

Results of each of the categories mentioned show that the largest score, for both musical imagery and guitar improvisability, is in the high category, the second largest score is in the medium category and the smallest score is in the low category. This means that musicians in Soroako's music community have high musical imagery and guitar improvisability skills.

The results of data analysis show that the correlation between musical imagery and guitar improvisability is shown by analysing the correlation between musical imagery (especially in this study, the ability to hear) and the aspects guitar improvisability measured. This refers to the correlation between musical imagery and intonation, dynamics, tempo and creativity, and the correlation between musical imagery and guitar improvisability. This research was conducted on 30 musicians in Soroako's music community, with a significance level of 5%.

The results of data analysis for the correlation of musical imagery and intonation indicate that the *r* table value is equal to 0.361 and the value of *r* count is equal to 0.021, indicating a correlation between musical imagery and intonation. The existence of musical imagery, especially the ability of musicians to hear music, can support musicians in recognising the high and low sounds produced by guitars. The accuracy of the tone produced by playing a guitar shows the ability to play music on a musical instrument.

The results of data analysis for the correlation of musical imagery and dynamics show that the *r* table value is equal to 0.361 and the value of *r* count is equal to 0.214, indicating a correlation between musical imagery and dynamics. The ability of musicians to know the dynamics of playing a guitar can be supported by the existence of good listening skills which is part of musical imagery. The weaknesses of and sound strength produced by a guitar need to be known by musicians to produce music. Weakness and sound strength can be interpreted by loud and gentle sounds produced by guitars played by musicians in Soroako's music community.

The results of data analysis for the correlation of musical imagery and tempo show that the *r* table value is equal to 0.361 and the value of *r* count is equal to 0.073; this means that there is a correlation between musical imagery and tempo. Guitar improvisability skills of musicians in Soroako's music community include the ability to understand the speed or slowness of the song being played. Knowledge of various musical tempo variations can be a supporting factor in improving guitar improvisability of musicians. Thus, understanding the accuracy of the tempo in playing a guitar is very important for musicians

The results of data analysis for the correlation of musical imagery and creativity show that the *r* table value is equal to 0.361 and the calculated *r* value is equal to 0.054; this means that there is a correlation between musical imagery and creativity. Creativity can be shown by producing innovations in playing music. Creativity is achieved by honing the ability to hear. Good listening skills can support musicians in increasing their creativity in playing the guitar. So, musicians can produce various techniques or works from playing the guitar.

The results of data analysis for the correlation of musical imagery and guitar improvisability show that the *r* table value is 0.361 and the *r* count is 0.052. This suggests that there is a correlation between musical imagery and the ability of musicians to improvise on a guitar. Musical imagery, especially the ability to hear a musician, is related to guitar improvisability. Listening ability is developed to help produce good notes. Knowledge of tones is developed through the ability to hear, which can be a reference for musicians in Soroako's music community to improvise guitar games.

The improvisability of guitar playing can create innovations that make a song sound beautiful or successfully performed by musicians. Listening to songs repeatedly can hone musicians' musical imagery so that they can develop skills in guitar improvisability.

4 CONCLUSION AND SUGGESTIONS

The study was conducted on 30 music community musicians in Soroako Village, Nuha District, East Luwu Regency, South Sulawesi Province, Indonesia by giving a series of musical imagery and guitar improvisability tests. The results of the study showed that the largest scores, for both musical imagery and guitar improvisability, were in the high category. The second largest score was in the medium category and the smallest score was in the low category. The results of data analysis showed that there is a correlation between musical imagery and guitar improvisability of music community musicians in Soroako, with a significance level of 5%. As evidenced by the *r* count, there is also a correlation between musical imagery and: 1) intonation (r = 0.021); 2) dynamics (r = 0.214); 3) tempo (r = 0.073); 4) creativity (r = 0.054); 5) guitar improvisability of music community musicians in Soroako (r = 0.052).

On the basis of the results and discussion, it can be concluded that musical imagery is the variable that most contributes to musical ability. Suggestions from the present research are as follows: 1) to develop guitar improvisability, musicians in Soroako's music community should listen to varied music on a regular basis so that the ability and sensitivity to play music increases; 2) to improve students' ability to play music, training should be imparted to hone students' hearing or sensitivity in listening.

REFERENCES

Agoes, B.N. (2011). *Modul Seminar 'Peran Komunitas Musik Etnik dalam Kebangkitan Budaya Bangsa'* [The function of ethnic music in resurrection of culture of nation]. Yogyakarta, Indonesia: Jurnal Geografi Universitas Indonesia.

Agustianto. (2016). *Kontribusi Prinsip-Prinsip Pengajaran dalam Praktik Instrumen Musik Gesek* [The contribution of instructional teaching in strings instrument learning]. *Imaji, 13*(2), 98-114. Retrieved from https://journal.uny.ac.id/index.php/imaji/article/view/7888.

Azwar, S. (2012). *Reliabilitas dan validitas edisi 4* [Validity and reliability, fourth edition]. Yogyakarta, Indonesia: Pustaka Pelajar.

Bailes, F.A. (2002). *Musical imagery: Hearing and imagining music*. Doctoral dissertation. Sheffield: University of Sheffield.

Fraenkel, J.R., & Wallen, N.E. (2007). *How to design and evaluate research in education*. New York, NY: McGraw-Hill Education.

Gunawan, M. (2015). *Noise-induced hearing loss pada musisi gereja Satu Jam Saja Surabaya* [Noise-induced hearing loss at Surabaya Satu Jam Saja Church musician]. Doctoral Dissertation. Surabaya: Widya Mandala Catholic University.

Hidayat, T. (2012). Graph application in easy finger positioning for playing melodies with guitar [Graph application in positioning to play guitar melodies easily]. *Makalah IF2091 Struktur Diskrit* – Sem. I Tahun 2012/13. Retrieved from http://informatika.stei.itb.ac.id/~rinaldi.munir/Matdis/2012-2013/Makalah2012/Makalah-IF2091-2012-063.pdf

Jäncke, L. (2008). Music, memory and emotion. *Journal of Biology, 7*(6), 21-25.

Kristiawan, Y. (2016). *Pengembangan kreativitas musik dalam pembelajaran seni budaya (musik) di SMA Negeri 1 Pati* [Music creativity development in art and culture teaching and learning at the public school of Senior High School I Pati]. *Jurnal Seni Musik, 5*(1), 73-78.

Miftahunnajah, F. (2013). Studi Organologi Pembuatan Gitar Klasik Produksi PT. Genta Trikarya Bandung [Organology study: Creating classical guitar at PT. Genta Trikarya]. *Swara, 1*(3). 35-41.

Author Index

For Product Safety Concerns and Information please contact our EU representative GPSR@taylorandfrancis.com
Taylor & Francis Verlag GmbH, Kaufingerstraße 24, 80331 München, Germany

www.ingramcontent.com/pod-product-compliance
Lightning Source LLC
LaVergne TN
LVHW080114011026
849160LV00013B/1095

9781032238357